# GLOSSARY OF BIOTECHNOLOGY TERMS

## THIRD EDITION

## KIMBALL NILL

# CRC PRESS

Boca Raton London New York Washington, D.C.

## Library of Congress Cataloging-in-Publication Data

Nill, Kimball R.
  Glossary of biotechnology terms / Kimball R. Nill.-- 3rd ed.
    p.  cm.
  ISBN 1-58716-122-2
  1. Biotechnology--Dictionaries. I. Title.

TP248.16 .F54 2002
660.6′03--dc21

2002017441

### Visit the CRC Press Web site at www.crcpress.com

*To my wife, Janet J. Nill.*

# Preface

I began writing this book as a hobby, more than a decade ago, when it became obvious to me that the various specialists working in the then-emerging field of biotechnology (e.g., geneticists, chemists, molecular biologists, intellectual property attorneys, marketers, etc.) were occasionally having difficulty simply understanding the terms utilized by colleagues in their respective fields.

Subsequently, a number of organizations with various motivations have raised some concerns around the world about biotechnology. In my experience, the level of concern inevitably diminishes when people understand the terms used to discuss a given topic. To this end, I have attempted to write definitions in this book employing words that would enable the reader to conceptualize the idea embodied in the term, without the necessity of holding advanced degrees in biochemistry or molecular biology. In order to accomplish this, however, I had to make certain compromises between scientific rigor and definitions based on analogy, with the inherent possibility of oversimplification. Nonetheless, throughout the text, emphasis has been placed on explanation by analogy whenever possible; I have found this method to be most effective for defining words, phrases, and terms to diverse publics.

I offer this work in good faith, and in the hope that it will assist those individuals who seek to gain some understanding of the terminology as it is currently used. However, the reader should be aware that the field of biotechnology is rapidly expanding and evolving; new terms are entering the nomenclature at a rapid pace. In fact, the meaning(s) of some of the newest terms will undoubtedly be expanded or narrowed as the technology further develops. Although I have endeavored to be as accurate as possible, this work is meant to provide a general introduction rather than to be absolute and legally definitive.

**Kimball R. Nill**
Technical Issues Director
American Soybean Association
St. Louis, Missouri

# Author

**Kimball Nill**, technical issues director at American Soybean Association (ASA), is responsible for early detection of emerging technology-related issues that could impact U.S. soybean exports, and for dealing proactively with those threats and/or opportunities.

The author grew up on a farm in North Dakota. He holds a bachelor of science degree in chemistry from North Dakota State University, Fargo, and a masters degree in business administration from the Wharton Business School in Philadelphia, PA. He has authored numerous papers and articles on various aspects of marketing agricultural biotechnology products for U.S. and European journals, and other publications.

Prior to joining the ASA in 1996, Nill was international marketing manager for Moorman's Inc., an Illinois manufacturer of specialty livestock nutrition products. Before that, he spent 5 years in positions supporting in-house venture capital and biotechnology research activities in a major biotechnology company.

Professional affiliations include membership in the American Chemical Society, the Licensing Executives Society, and the American Association for the Advancement of Science.

# Glossary of Biotechnology Terms

A-DNA A particular right-handed helical form of DNA (possessing 11 base pairs per turn), in which DNA molecules exist when they are partially dehydrated. A-form DNA is found in fibers at 75% relative humidity and requires the presence of sodium, potassium, or cesium as the counterion. Instead of lying flat, the bases are tilted with regard to the helical axis and there are more base pairs per turn. The A-form is biologically interesting because it is probably very close to the conformation adopted by DNA-RNA hybrids or by RNA-RNA double-stranded regions. The reason is that the presence of the 2'2 hydroxyl group prevents RNA from lying in the B-form. See also B-DNA, DNA-RNA HYBRID, DEOXYRIBONUCLEIC ACID (DNA), BASE PAIR (bp).

AβPP See AMYLOID β PROTEIN PRECURSOR (AβPP).

A_w See WATER ACTIVITY (A_w).

aAI-1 See ALPHA AMYLASE INHIBITOR-1.

ABC See ASSOCIATION OF BIOTECHNOLOGY COMPANIES (ABC).

ABC Transporters A class of membrane transporter proteins which "transfer" across cell membranes: sugar molecules (i.e., used by cells as "fuel"); inorganic ions (needed to catalyze certain cellular processes); polypeptides (i.e., protein molecules); certain anticancer drugs (thereby making it harder to halt certain cancer tumors via use of pharmaceuticals); certain antibiotics (thereby conferring antibiotic resistance to some pathogenic bacteria). ABC transporter molecules are embedded in the plasma membrane (i.e., surface

"skin") of cells. See also CELL, PLASMA MEMBRANE, PROTEIN, CATALYST, ION, POLYPEPTIDE (protein), CANCER, CHEMOTHERAPY, ANTIBIOTIC RESISTANCE.

Abiogenesis Spontaneous generation. See also BIOGENESIS.

Abiotic Absence of living organisms. See also ABIOTIC STRESSES.

Abiotic Stresses The stress caused (e.g., to crop plants) by nonliving, environmental factors such as cold, drought, flooding, salinity, ozone, toxic-to-that-organism metals (e.g., aluminum, for plants), and ultraviolet-B light. See also CITRATE SYNTHASE (CSB) GENE.

Abrin A toxin derived from the seed of the rosary pea. See also RICIN, PHYTOCHEMICALS, TOXIN.

Abscisic Acid A phytohormone (plant hormone) utilized to control: the size of stomatal pores — i.e., the openings in leaves through which plants exchange oxygen and carbon dioxide (and water inadvertently) with the atmosphere; abscision (e.g., shedding of flowers, fruit, etc.); dormancy. See also PLANT HORMONE, GPA1.

Absolute Configuration The configuration of four different substituent groups around an asymmetric carbon atom, in relation to D- and L-glyceraldehyde. See also DEXTROROTARY (D) ISOMER, LEVOROTARY (L) ISOMER.

Absorbance (A) A measure of the amount of light absorbed by a substance suspended in a matrix. The matrix may be gaseous, liquid, or solid in nature. Most biologically active

1

A

compounds (e.g., proteins) absorb light in the ultraviolet (UV) or visible light portion of the spectrum. Absorbance is used to quantitate (measure) the concentration of the substance in question (e.g., substance dissolved in a liquid). See also OPTICAL DENSITY (OD), SPECTROPHOTOMETER.

**Absorption** From the Latin *ab*, away, and *sorbere*, to suck into. The taking-up of nutrients, water, etc. by assimilation (e.g., transport of the products of digestion from the intestinal tract across the cell membranes that comprise the gut, and into the blood). See also "ADME" TESTS, DIGESTION (WITHIN ORGANISMS).

**Abzymes** Catalytic antibodies that are synthetic constructs. They either stabilize the transition state of a chemical reaction or bind to a specific substrate, thereby increasing the reaction rate of that chemical reaction. See also CATALYTIC ANTIBODY, TRANSITION STATE, SUBSTRATE (CHEMICAL).

**Ac-CoA** Abbreviation for Acetyl-coenzyme A. Ac-CoA is a chemical synthesized in cell mitochondria by combining the thiol (molecular group) of coenzyme A with an acetyl group (i.e., from breakdown/digestion of fats, carbohydrates, or proteins). See also COENZYME, COENZYME A, FATS, ACETYLCHOLINE, GLUCONEOGENESIS, ACETYL-CoA CARBOXYLASE, CHOLINESTERASE, CELL, MITOCHONDRIA, FATS, PROTEIN.

**Ac-P** Acetylphosphate.

**ACC** Abbreviation/acronym for the compound 1-aminocyclopropane-1-carboxylic acid, which is produced from S-adenosylmethionine (SAM) in the fruit of certain plants. When the "sam-k" gene is inserted into the genome of those plants, the level of SAM is greatly reduced in their fruit, which inhibits (slows) ripening/softening of that fruit via a reduction/slowdown in production of ethylene (hormone that causes fruit to ripen/soften). See also ACC SYNTHASE, ETHYLENE, SAM-K GENE, GENETIC ENGINEERING, GENOME, PLANT HORMONE.

**ACC Synthase** Aminocyclopropane carboxylic acid synthase/deaminase; it is one of the most critical enzymes in the metabolic pathway that creates the hormone ethylene inside fruit. Because ethylene causes certain fruit (e.g., tomatoes) to ripen (soften), it is possible

to significantly delay the softening (i.e., spoilage) process by controlling creation of ACC synthase via manipulation of the ACC synthase gene. See also ACC, METABOLISM, ENZYME, METABOLITE, INTERMEDIARY METABOLISM, PATHWAY, PLANT HORMONE, POLYGALACTURONASE (PG), ETHYLENE, SAM-K GENE.

**ACCase** See ACETYL-CoA CARBOXYLASE.

**Acceptor Control** The regulation of the rate of respiration by the availability of ADP as phosphate acceptor. See also RESPIRATION, ADENOSINE DIPHOSPHATE (ADP).

**Acceptor Junction Site** The junction between the right 3′ end of an intron and the left 5′ end of an exon. See also INTRON, EXON.

**Accession** The addition of germ-plasm deposits to existing germ-plasm storage bands. See also AMERICAN TYPE CULTURE COLLECTION (ATCC).

**Acclimatization** The biological process whereby an organism adapts to a new environment. For example, the body of a mountain climber who has spent significant time high on Mount Everest (e.g., 20,000 feet above sea level) produces twice as many red blood cells (to transport oxygen) as it does at sea level. Often, this adaptation actually occurs on a molecular level. One example is when natural microorganisms adapt so that they feed on, and degrade, toxic chemical wastes; or change from using one sugar as a fuel source to another. See also SUGAR MOLECULES, CATABOLISM, RED BLOOD CELLS, COLD HARDENING, PHARMACOENVIROGENETICS.

**ACE** Angiotensin-converting enzyme. A crucial enzyme (within the human vascular system) for catalyzing the formation of angiotensin, a hormone that causes narrowing/restriction of blood vessels, thus increasing the body's blood pressure as the blood is squeezed through those narrowed blood vessels. The action of ACE can be inhibited by the pharmaceuticals known as ACE inhibitors. Research indicates that consumption of whey protein can also result in inhibition of ACE. See also ENZYME, HORMONE, ACE INHIBITORS.

**ACE Inhibitors** A family of chemically-similar pharmaceuticals utilized to lower blood pressure in humans, by blocking formation of a hormone (angiotensin) that narrows/restricts blood vessels. See also ACE.

**Acetolactate Synthase** See ALS.

**Acetyl Carnitine** One of the metabolites of mitochondria, it is a substrate (i.e., substance that is acted upon) for acylcarnitine transferase (which converts the acetyl carnitine to carnitine). Research indicates that consumption of acetyl carnitine helps increase the levels of acetylcholine and nerve growth factor (NGF) in the brain. See also METABOLITE, MITOCHONDRIA, ACYLCARNITINE TRANSFERASE, SUBSTRATE (CHEMICAL), CARNITINE, ACETYL-CHOLINE, NERVE GROWTH FACTOR (NGF).

**Acetyl Coenzyme A** See Ac-CoA.

**Acetyl-CoA** Acetyl-coenzyme A. See also Ac-CoA.

**Acetyl-CoA Carboxylase** An enzyme that catalyzes the chemical reaction (i.e., conversion of Ac-CoA to malonyl CoA via carboxylation) which is the first step in the series of chemical reactions through which some plants manufacture oils (e.g., soybean oil, canola oil, etc.). See also ENZYME, FATS, SOYBEAN OIL, CANOLA.

**Acetylcholine** A neurotransmitter (i.e., one of several relatively small, diffusible molecules utilized by the human body to "transmit" nerve impulses) that is synthesized (manufactured) near the ends of axons (i.e., one type of neuron). That synthesis is accomplished by the "transfer" of an acetyl group (portion of molecule) from Ac-CoA to a choline molecule (available in the body via consumption of soybean lecithin or certain other foods), in a chemical reaction catalyzed by cholinesterase. Increased amounts of acetylcholine in the (human) brain has been shown to reduce the symptoms of Alzheimer's disease. See also NEUROTRANSMITTER, NEURON, CHOLINE, Ac-CoA, LECITHIN, ALZHEIMER'S DISEASE, THYMUS, ENZYME, CHOLINESTERASE.

**Acetylcholinesterase** An enzyme that hydrolyzes (cuts into smaller pieces) molecules of the neurotransmitter acetylcholine, after the acetylcholine molecules have accomplished "transmission" of a nerve impulse. That hydrolysis (cutting into pieces) of acetylcholine molecules thus serves to prepare the neurons (cells of the body's nervous system) to be able to transmit other, later nerve impulses. See also ENZYME, HYDROLYSIS, NEUROTRANSMITTER, ACETYLCHOLINE, NEURON.

**Acid** A substance that contains hydrogen atom(s) in its molecular structure, with a pH in the range from 0–6, which will react with a base to form a salt. Acids normally taste sour and feel slippery. For example, food product manufacturers often add citric acid, malic acid, fumaric acid, and itaconic acid in order to impart a sharp taste to food products. See also BASE, CITRIC ACID, FUMARIC ACID $(C_4H_4O_4)$.

**Acidic Fibroblast Growth Factor (AFGF)** See FIBROBLAST GROWTH FACTOR (FGF).

**Acidosis** A metabolic condition in which the capacity of the body to buffer changes in pH is diminished. Hence, acidosis is accompanied by decreased blood pH (i.e., the blood becomes more acidic than is normal).

**ACP (acyl carrier protein)** A protein that binds acyl intermediates during the formation of long-chain fatty acids. ACP is important in that it is involved in every step of fatty acid synthesis. See also FATTY ACID, ACYL-CoA, FATS.

**Acquired Immune Deficiency Syndrome (AIDS)** A disease in which a specific virus attacks and kills macrophages and helper T cells (thus causing collapse of the entire immune system). Once the immune system has been inactivated, other diseases, which under normal circumstances can be fought off, become fatal. See also HUMAN IMMUNO-DEFICIENCY VIRUS TYPE 1 (HIV-1), HUMAN IMMU-NODEFICIENCY VIRUS TYPE 2 (HIV-2), HELPER T CELLS (T4 CELLS), MACROPHAGE, TUMOR NECROSIS FACTOR (TNF).

**Acrylamide Gel** See POLYACRYLAMIDE GELS.

**ACTH [adrenocorticotropic hormone (corticotropin)]** A polypeptide secreted by the anterior lobe of the pituitary gland. This is an example of a protein hormone. See also POLYPEPTIDE (PROTEIN), ENDOCRINE GLANDS, ENDOCRINE HORMONES.

**Activation Energy** The amount of energy (calories) required to bring all the molecules in one mole of a reacting substance to the transition state. More simply, it may also be viewed as the energy required to bring reacting molecules to a certain energy state from which point the reaction proceeds spontaneously.

See also TRANSITION STATE (IN A CHEMICAL REACTION), MOLE, FREE ENERGY.

**Activator** A small molecule that stimulates (increases) an enzyme's catalytic activity when it binds to an allosteric site. See also ENZYME, EFFECTOR, ALLOSTERIC SITE.

**Active Site** The region of an enzyme surface that binds the substrate molecule and transforms the substrate molecule into the new (chemical) product (entity). This site is usually located not on a protruding portion of the enzyme, but rather in a cleft or depression. This establishes a controlled environment in which the chemical reaction may occur. See also CATALYTIC SITE, AGONISTS, PHARMACOPHORE, SUBSTRATE (CHEMICAL), ENZYME, ANTAGONISTS.

**Active Transport** Cell-mediated, energy-requiring translocation of a molecule across a membrane in the direction of increasing concentration (i.e., opposite of natural tendency). See also OSMOTIC PRESSURE.

**α-Helix** See ALPHA HELIX.

**A. flavus** See ASPERGILLUS FLAVUS.

**Activity Coefficient** The factor by which the concentration of a solute must be multiplied to give its true thermodynamic activity.

**Acuron™ Gene** A gene, trademarked by Syngenta AG, that can be inserted into plants via genetic engineering techniques. Inserted into the genome (DNA) of a plant, the gene confers tolerance to herbicide(s) whose active ingredient is protoporphyrinogen oxidase inhibitor (thus, such herbicides are known as PPO inhibitors). See also HERBICIDE-TOLERANT CROP, GENE, GENETIC ENGINEERING, GENOME, DEOXYRIBONUCLEIC ACID (DNA).

**Acute Transfection** Short-term infection of cells with DNA.

**Acyl-CoA** Acyl derivatives of coenzyme A (acyl-S-CoA). See also CARNITINE, COENZYME A, TRYPSIN INHIBITORS.

**Acylcarnitine Transferase** An enzyme that converts the mitochondrial metabolite acetyl carnitine into carnitine. See also ENZYME, ACETYL CARNITINE, CARNITINE.

**AD** An acronym referring to the group of diseases known collectively as Autoimmune Disorders. These include diseases such as multiple sclerosis, lupus, rheumatoid arthritis, etc. See also AUTOIMMUNE DISEASE, MULTIPLE SCLEROSIS, LUPUS.

**Adaptation** Refers to the adjustment of a population of organisms to a changed environment. For example, during the 19th century, the Industrial Revolution caused large black soot deposits on the white bark of certain trees in England. The change in environment resulted in adaptation (e.g., via selective breeding) of a particular indigenous moth population, consisting of a mixture of all-white and all-black members. Because the soot blackened the formerly white bark of the trees on which the moths rested, predatory birds were able to easily catch and eat the all-white members of the population. Thus, there were fewer all-white moths present in the breeding population, and a preponderance of all-black members. During the 20th century, antipollution efforts in England resulted in a cessation of the airborne soot and the return of tree bark to its original white color. Because the predatory birds were now able to catch and eat the all-black members of that moth population more easily, the proportion of all-black and all-white moths in the breeding population once again changed. See also ORGANISM.

**Adaptive Enzymes** See INDUCIBLE ENZYMES.

**ADBF** See AZUROPHIL-DERIVED BACTERICIDAL FACTOR (ADBF).

**Additive Genes** Genes that interact but do not show dominance (in the case of alleles) or epistasis (if they are not alleles). See also GENE, ALLELE, DOMINANT ALLELE, EPISTASIS.

**Adenylate Cyclase** The enzyme (within cells) that catalyzes the synthesis (manufacture) of cyclic AMP. See also CYCLIC AMP.

**Adenine** A purine base, 6-aminopurine, occurring in ribonucleic acid (RNA) as well as in deoxyribonucleic acid (DNA) and a component of adenosine diphosphate (ADP) and adenosine triphosphate (ATP). Adenine pairs with thymine in DNA and uracil in RNA. See also BASE (NUCLEOTIDE), BASE PAIR (bp), RIBONUCLEIC ACID (RNA), DEOXYRIBONUCLEIC ACID (DNA).

**Adenosine Diphosphate (ADP)** A ribonucleoside 5′-diphosphate serving as phosphate-group acceptor in the cell energy cycle. See also CATABOLISM, ADENOSINE TRIPHOSPHATE (ATP), ADENOSINE MONOPHOSPHATE (AMP).

**Adenosine Monophosphate (AMP)** A ribonucleoside 5′-monophosphate that is formed by hydrolysis of ATP or ADP. See also HYDROLYSIS, ADENOSINE DIPHOSPHATE (ADP), ADENOSINE TRIPHOSPHATE (ATP).

**Adenosine Triphosphate (ATP)** The major carrier of chemical energy in the cells of all living things on this planet. A ribonucleoside 5′-triphosphate functioning as a phosphate-group donor in the energy cycle of the cell, ATP contains three phosphate/oxygen molecules linked together. When a phosphate-phosphate bond in ATP is broken (hydrolyzed), the energy produced can be used by the cell to carry out its functions. Thus, ATP serves as the universal medium of biological energy storage and exchange in living cells. See also ATPase, ATP SYNTHETASE, HYDROLYSIS, CYCLIC PHOTOPHOSPHORYLATION, BIOLUMINESCENCE, ATP SYNTHASE, ADENOSINE MONOPHOSPHATE (AMP).

**Adenovirus** A type of virus that can infect humans. As with all viruses, it can reproduce only inside living cells (of other host, organisms). Adenovirus causes manufacture of a protein (metabolite) that disables the p53 gene. Because the p53 gene then cannot perform its usual function (i.e., prevention of uncontrolled cell growth caused by virus/DNA damage), the adenovirus takes over and causes the cell to make numerous copies of the virus until the cell dies, thus releasing the virus copies into the body of the host organism to cause further infection. See also VIRUS, RETROVIRUSES, GENE DELIVERY, GENE THERAPY, CELL, PROTEIN, p53 GENE, DEOXYRIBONUCLEIC ACID (DNA).

**Adhesion Molecule** From the Latin *adhaerere*, to stick to, the term adhesion molecule refers to a glycoprotein (oligosaccharide) molecular chain that protrudes from the surface membrane of certain cells, causing cells possessing matching adhesion molecules to adhere to each other. For example, in 1952 Aaron Moscona observed that (harvesting enzyme-separated) chicken embryo cells did not remain separated, but instead coalesced again into an (embryo) aggregate. In 1955,

Philip Townes and Johannes Holtfreter showed that like amphibian (e.g., frog) neuron cells will rejoin after being physically separated (e.g., with a knife blade); but unlike cells remain segregated (apart).

Adhesion molecules also play a crucial role in guiding monocytes to sources of infection (e.g., pathogens) because adhesion molecules in the walls of blood vessels (after activation caused by pathogen invasion of adjacent tissue) adhere to like adhesion molecules in the membranes of monocytes in the blood. The monocytes pass through the blood vessel walls, become macrophages, and fight the pathogen infection (e.g., triggering tissue inflammation, etc.). See also OLIGOSACCHARIDES, MONOCYTES, MACROPHAGE, POLYPEPTIDE (PROTEIN), CELL, PATHOGEN, CD4 PROTEIN, CD44 PROTEIN, GP120 PROTEIN, VAGINOSIS, HARVESTING ENZYMES, HARVESTING, SIGNAL TRANSDUCTION, SELECTINS, LECTINS, GLYCOPROTEINS, SUGAR MOLECULES, LEUKOCYTES, LYMPHOCYTES, NEUTROPHILS, ENDOTHELIUM, ENDOTHELIAL CELLS, P-SELECTIN, ELAM-1, INTEGRINS, CYTOKINES.

**Adhesion Protein** See ADHESION MOLECULE, ENDOTHELIAL CELLS.

**Adipocytes** Specialized cells within an organism's lymphatic system that store the triacylglycerols (also sometimes called triglycerides) after digestion of those fats, later releasing fatty acids and glycerol into the bloodstream when needed by the organism. See also CELL, TRIGLYCERIDES, FATTY ACID, DIGESTION (within organism), FATS.

**Adipose** Refers to energy storage tissues consisting of fat molecules within some animals. Adipose tissue tends to increase if an animal consumes more energy-containing food than needed for its level of energy expenditure (e.g., via exercise). In humans older than 40, an increase in the body's adipose tissue is correlated with an increased risk of premature death (e.g., from coronary heart disease). See also FATS, CORONARY HEART DISEASE (CHD), LEPTIN.

**Adjuvant (to a herbicide)** Any compound that enhances the effectiveness (i.e., weed-killing ability) of a given herbicide. For example, adjuvants such as surfactants can

A

be mixed (prior to application to weeds) with herbicide (in water), in order to hasten transport of the herbicide's active ingredient into the weed plant. That is because the herbicide must move from an aqueous (water) environment into one (i.e., the weed plant's cuticle or "skin") comprised of lipids/lipophilic molecules, before it can accomplish its task. See also SURFACTANT, LIPIDS, LIPOPHILIC.

**Adjuvant (to a pharmaceutical)** Any compound that enhances the desired response by the body to that pharmaceutical. For example, adjuvants such as certain polysaccharides or surface-modified diamond nanoparticles, can be injected along with (vaccine) antigen in order to increase the immune response (e.g., production of antibodies) to a given antigen. Another example is that consumption of grapefruit juice by humans will increase the impact of certain pharmaceuticals. Those pharmaceuticals include some sedatives, antihypertensives, the antihistamine terfenadine, and the immunosuppressant cyclosporine. The adjuvant effect of grapefruit juice is thought to be caused via inhibition of the enzyme cytochrome P4503A4, which catalyzes reactions involved in the metabolism (breakdown) of those pharmaceuticals.

Another example is that consumption of the pharmaceutical known as clopidogrel (commercial name Plavix) by people immediately following a mild heart attack (severe chest pain) — along with aspirin — greatly reduces the risk of death, strokes, and (new, additional) heart attacks, versus taking aspirin alone after a mild heart attack. See also CELLULAR IMMUNE RESPONSE, HUMORAL IMMUNITY, POLYSACCHARIDES, NANOTECHNOLOGY, ANTIGEN, ANTIBODY, ENZYME, METABOLISM, HISTAMINE, CYCLOSPORINE, CYTOCHROME P4503A4.

**ADME** Acronym for Absorption, Distribution (within the body), Metabolism, and Elimination of pharmaceuticals. See also ADME TESTS, IN SILICO SCREENING.

**ADME Tests** Refers to Absorption, Distribution (within the body), Metabolism, and Elimination tests required by the U.S. Food and Drug Administration (FDA) for approval of new pharmaceuticals or new food ingredients. See also FOOD AND DRUG ADMINISTRATION (FDA), ABSORPTION, METABOLISM, INTERMEDIARY

METABOLISM, PHARMACOKINETICS, PHARMACOGENOMICS, CODEX ALIMENTARIUS COMMISSION, ADME, ADMET, IN SILICO SCREENING.

**ADMET** Acronym for Absorption, Distribution (within the body), Metabolism, Elimination, Toxicity of pharmaceuticals. See also ADME TESTING, IN SILICO TESTING.

**Adoptive Cellular Therapy** The increase in immune response that is achieved by selectively removing certain immune system cells from a (patient's) body, multiplying them in vitro outside the body to increase their number greatly, then reinserting those (more numerous) immune system cells into the same body. See also CELLULAR IMMUNE RESPONSE, CELL CULTURE, IN VITRO, GENE DELIVERY, GENE THERAPY, EX VIVO (THERAPY).

**Adoptive Immunization** The transfer of an immune state from one animal to another by means of lymphocyte transfusions. See also LYMPHOCYTE.

**ADP** See ADENOSINE DIPHOSPHATE (ADP).

**Adventitious** From the Latin adventitius, not properly belonging to. The term can be utilized to refer to: plant shoots emanating from sites other than typical ones (e.g., from a plant's leaves); a small amount of transgenic grain accidentally mixed into other grain. See also TRANSGENIC.

**Aerobe** An organism that requires oxygen to live (respire).

**Aerobic** Exposed to air or oxygen. An oxygenated environment.

**Affinity Chromatography** A method of separating a mixture of proteins or nucleic acids (molecules) by specific interactions of those molecules with a component known as a ligand, which is immobilized on a support. If a solution of, say, a mixture of proteins is passed over (through) the column, one of the proteins binds to the ligand on the basis of specificity and high affinity (they fit together like a lock and key). The other proteins in the solution wash through the column because they were not able to bind to the ligand. Once the column is devoid of the other proteins, an appropriate wash solution is passed through the column, which causes the protein/ligand complex to dissociate. The protein is subsequently collected in a highly purified form. See also CHROMATOGRAPHY, PROTEIN, NUCLEIC

ACIDS, ANTIBODY AFFINITY CHROMATOGRAPHY, LIGAND (IN CHROMATOGRAPHY).

**Aflatoxin** The term that is used to refer to a group of related mycotoxins (i.e., metabolites produced by fungi that are toxic to animals and humans) produced by some strains *Aspergillus flavus* and *Aspergillus parasiticus*, common fungi that typically live on decaying vegetation. Corn earworm (*Helicoverpa zea*) and European corn borer (*Ostrinia nubilalis*) are vectors (carriers) of *Aspergillus flavus*. Aflatoxin B1 is the most commonly occurring aflatoxin and one of the most potent carcinogens known to man. When dairy cattle eat aflatoxin-contaminated feed, their metabolism process converts the aflatoxin (e.g., Aflatoxin B1) into the mycotoxins known as Aflatoxin M1 and Aflatoxin M2, which soon appear in the milk produced. Consumption of aflatoxins by humans can also result in acute liver damage. See also CARCINOGEN, TOXIN, FUNGUS, MYCOTOXINS, STRESS PROTEINS, LIPOXYGENASE (LOX), PEROXIDASE, *HELICOVERPA ZEA* (*H.* zea), BETA CAROTENE, OH43, BRIGHT GREENISH-YELLOW FLUORESCENCE (BGYF), CORN, EUROPEAN CORN BORER (ECB).

**AFLP** Acronym for Amplified Fragment Length Polymorphism. See also AMPLIFIED FRAGMENT LENGTH POLYMORPHISM.

**Agar** A complex mixture of polysaccharides obtained from marine red algae. It is also called agar-agar. Agar is used as an emulsion stabilizer in foods, as a sizing agent in fabrics, and as a solid substrate for the laboratory culture of microorganisms. Agar melts at 100°C (212°F), and when cooled below 44°C (123°F) forms a stiff and transparent gel. Microorganisms are seeded onto and grown (in the laboratory) on the surface of the gel. See also POLYSACCHARIDES, CULTURE MEDIUM.

**Agarose** A highly purified form of agar used as a stationary phase (substrate) in some chromatography and electrophoretic methods. See also CHROMATOGRAPHY, ELECTROPHORESIS, AGAR.

**Aging** The process, affecting organisms and most cells, whereby each cell division (mitosis) brings that cell (or organism composed of such cells) closer to its final cell division

(i.e., death). Notable exceptions to this aging process include cancerous cells (e.g., myelomas) and the single-celled organism; both of which are "immortal." See also TELOMERES, MITOSIS, HYBRIDOMA, MYELOMA, CANCER.

**Aglycon** A nonsugar component of a glycoside. See also GLYCOSIDE.

**Aglycone** The biologically active (molecular) form of molecules of isoflavones. See also ISOFLAVONES, BIOLOGICAL ACTIVITY.

**Agonists** Small protein or organic molecules that bind to certain cell proteins (i.e., receptors) at a site that is adjacent to the cell's "docking" site of protein hormones, neurotransmitters, etc. (i.e., receptor) to induce a conformational change in that cell protein, thereby enhancing its activity (i.e., effect upon the cell). See also RECEPTORS, ACTIVE SITE, CONFORMATION, CELL, HORMONE, ANTAGONISTS, NEUROTRANSMITTER.

**Agraceutical** See NUTRACEUTICAL, PHYTOCHEMICAL.

***Agrobacterium tumefaciens*** A naturally occurring bacterium that is capable of inserting its DNA (genetic information) into plants, resulting in a type of injury to the plant known as crown gall. In 1980, Marc van Montagu showed that *Agrobacterium tumefaciens* could alter the DNA of its host plant(s) by inserting its own ("foreign") DNA into the genome of the host plants (thereby opening the way for scientists to insert virtually any foreign genes into plants via use of *A. tumefaciens*). In 1983, Luis Herrera-Estrella created the first man-made transgenic plant by inserting an antibiotic-resistant gene into a tobacco plant. During 2000, Weija Zhou and Richard Vierling proved that *A. tumefaciens* is at least 10 times more effective (i.e., at "infecting" plants to insert DNA) in space (i.e., weightlessness/microgravity) than it is when on the surface of the Earth. Among others, Monsanto Company has developed a way to stop *A. tumefaciens* from causing crown gall, while maintaining its ability to insert DNA into plant cells, and now uses *A. tumefaciens* as a vehicle to insert desired genes into crop plants (e.g., the gene causing high production of CP4 EPSP synthase, thus conferring resistance to glyphosate-containing herbicide).

A

**A**

See also BACTERIA, DEOXYRIBONUCLEIC ACID (DNA), INFORMATIONAL MOLECULES, GENOME, TRANSGENIC (ORGANISM), PROTOPLAST, EPSP SYNTHASE, CP4 EPSPS, "SHOTGUN" METHOD, BIOLISTIC® GENE GUN, WHISKERS™, GENETIC ENGINEERING, GENE, BIOSEEDS, GLYPHOSATE, GLYPHOSATE-TRIMESIUM, GLYPHOSATE ISOPROPYLAMINE SALT, NOS TERMINATOR.

**AHG** Antihemophilic Globulin. Also known as FACTOR VIII or Antihemophilic Factor VIII. See also FACTOR VIII, GAMMA GLOBULIN.

**AIDS** See ACQUIRED IMMUNE DEFICIENCY SYNDROME (AIDS).

**Alanine (ala)** A nonessential amino acid of the pyruvic acid family. In its dry, bulk form it appears as a white crystalline solid. See also ESSENTIAL AMINO ACIDS.

**Albumin** A protein that the liver synthesizes (manufactures). Most minerals and hormones utilized by the human body are first "attached" to a molecule of albumin before they are transported in the bloodstream to where they are needed in the body. See also PROTEIN, HORMONE, SUPERCRITICAL CARBON DIOXIDE.

**ALCAR** Acronym for Acetyl-L-Carnitine. See also ACETYL CARNITINE.

**Aldose** A simple sugar in which the carbonyl carbon atom is at one end of the carbon chain. A class of monosaccharide sugars; the molecule contains an aldehyde group. See also MONOSACCHARIDES.

**Aleurone** The layer ("skin") that covers the endosperm portion of a plant seed. See also ENDOSPERM.

**AlfAFP** Acronym for Alfalfa Antifungal Peptide. See also DEFENSINS.

**Algae** A heterogeneous (widely varying) group of photosynthetic plants, ranging from microscopic single-cell forms to multicellular, very large forms such as seaweed. All of them contain chlorophyll and hence most are green, but some may be different colors due to the presence of other, overshadowing pigments.

**Alicin** A compound that is produced naturally by the garlic plant when the cells within garlic bulbs are broken open (e.g., during food preparation or consumption). Enzymes present within those garlic cells convert (precursor compound) to alicin. Research indicates that human consumption of alicin confers some

specific health benefits (anti-thrombotic, reduce blood cholesterol levels, reduce/avoid coronary heart disease, enhance the immune system, etc.). See also CELL, PHYTOCHEMICALS, ENZYME, THROMBOSIS, CORONARY HEART DISEASE (CHD), CHOLESTEROL.

**Alkaline Hydrolysis** A chemical method of liberating DNA from a DNA-RNA hybrid. See also HYDROLYSIS, RIBONUCLEIC ACID (RNA), DNA-RNA HYBRID, DEOXYRIBONUCLEIC ACID (DNA).

**Alkaloids** A class of toxic compounds that are naturally produced by some organisms (e.g., ants, certain plants such as lupines, and certain fungi such as ergot). For example, certain species of ants naturally produce alkaloids, as a self-defense mechanism. Poison-dart frogs (*Dendrobates azureus*) and two species of New Guinea songbirds (*Pitohui dichrous* and *Ifrita kowaldi*) can tolerate those ant-produced alkaloids, so they also acquire that self-defense (toxin) by eating those particular ants. Another example is the moth *Utetheisa ornatrix*, whose larvae (caterpillars) feed on certain plants that contain pyrrolizidine alkaloids. Because those alkaloids are extremely bitter tasting and toxic, spiders that normally prey on them refuse to eat those *Utetheisa ornatrix*; even after they later become adult moths. If those moths (who consumed those pyrrolizidine alkaloids as larvae) get caught in the spider's web, the spider will cut it out of the web and release that particular (toxic) moth. Vinca alkaloids, isolated from the specific plants that produce them, have been utilized as cancer-treating (antitumor) drugs. See also TOXIN, FUNGUS, TREMORGENIC INDOLE ALKALOIDS, ERGOTAMINE.

**Allele** From the Greek *allelon*, mutually each other, the term refers to one of several alternate forms of a gene occupying a given locus on the chromosome, which controls expression (of product) in different ways. See also EXPRESS, GENE, CHROMOSOMES, LOCUS.

**Allelic Exclusion** The expression in any particular manner of only one of the alleles in an antibody gene within a B lymphocyte (cell), coding for the expressed antibody. See also ALLELE, CODING SEQUENCE, GENE, B LYMPHOCYTES, ANTIBODY, IMMUNOGLOBULIN.

**Allelopathy** Refers to the secretion of certain chemicals (e.g., terpenoid compounds) by a plant, in order to hinder the growth or reproduction of other plants growing near it.

**Allergies (airborne)** See MAST CELLS.

**Allergies (foodborne)** An IgE-mediated (aggressive) immune system response to antigen(s) present on protein molecules in the particular food to which (a given) person is allergic. The antibodies (IgE) bind to those antigens and trigger a humoral immune response that can cause vomiting, diarrhea, skin reactions, wheezing, and respiratory distress. In severe cases, the immune response can cause death. In some rare instances, the allergic reaction is mediated by sensitized T cells. In some rare instances, the onset of a food allergy incident is induced by exercise (before or after eating that particular food).

The U.S. Food and Drug Administration (FDA) requires testing in advance to determine if a genetically engineered foodstuff has the potential to cause allergic reactions in humans, before that genetically engineered foodstuff (e.g., a modified crop plant) is approved by the FDA. In general, known food allergens (e.g., peanuts, Brazil nuts, wheat, etc.) are protein molecules that are resistant to rapid digestion (because those protein molecules are too tightly "folded together" for digestive enzymes to access their chemical bonds to break down). One potential way to genetically engineer currently allergenic crops (e.g., wheat) to make them less allergenic, is to insert gene(s) for extra production of thioredoxin. Found in all living organisms, thioredoxin is a protein that targets and breaks down the chemical bonds holding together a tightly folded-together protein molecule (thereby making those protein molecules easier to digest). Future crops engineered to contain more thioredoxin than the traditional average level may be nonallergenic. See also PROTEIN, PROTEIN FOLDING, ANTIBODY, ANTIGEN, FOOD AND DRUG ADMINISTRATION (FDA), GENETIC ENGINEERING, IMMUNOGLOBULIN, HUMORAL IMMUNITY, MAST CELLS, LEUKOTRIENES, DIGESTION (WITHIN ORGANISMS), ORGANISM.

**Allicin** See ALICIN.

**Allogeneic** With a different set of genes (but same species). For example, an organ transplant from one nonrelated human to another is allogeneic. An organ transplant from a baboon to a human would be xenogeneic. See also GENE, SPECIES, XENOGENEIC ORGANS.

**Allosteric Enzymes** Regulatory enzymes whose catalytic activity is modulated by the noncovalent binding of a specific metabolite (effector) at a site (regulatory site) other than the catalytic site (on the enzyme). Effector binding causes a three-dimensional conformation change in the enzyme and is the root of the modulation. The term allosteric is used to differentiate this form of regulation from the type that may result from the competition between substrate and inhibitors at the catalytic site. See also ENZYME, STERIC HINDRANCE, EFFECTOR, CONFORMATION, ACTIVE SITE.

**Allosteric Site** The site on an (allosteric) enzyme molecule where, via noncovalent binding to the site, a given effector can increase or decrease that enzyme's catalytic activity. Such an effector is called an allosteric effector because it binds at a site on the enzyme molecule that is other (allo) than the enzyme's catalytic site. See also ALLOSTERIC ENZYMES, ACTIVATOR, CATALYTIC SITE, EFFECTOR, CONFORMATION, ENZYME, METABOLITE, CATALYST.

**Allotypic Monoclonal Antibodies** Monoclonal antibodies that are isoantigenic. See also MONOCLONAL ANTIBODIES (MAb), ANTIGEN.

**Allozyme** See ALLOSTERIC ENZYMES.

*Aloe vera L.* A plant whose sap (juice) contains certain carbohydrates that naturally assist healing of human skin (wounds). Those carbohydrates "activate" macrophages, which cause those macrophages to produce cytokines (that regulate human immune system and inflammatory responses which promote healing). See also PHYTOCHEMICALS, CARBOHYDRATES (SACCHARIDES), MACROPHAGE, CYTOKINES.

**Alpha Amylase Inhibitor-1** A protein naturally produced in the seeds of the plant known as the common bean *Phaseolus vulgaris* that inhibits the amylase enzyme in the gut of the pest insect known as the pea weevil. Because the amylase enzyme (in its gut) is inhibited (prevented from helping digestion)

**A**

by the Alpha Amylase Inhibitor-1, the seeds of the *P. vulgaris* plant are protected from depradation by the pea weevil. See also PRO-TEIN, ENZYME, AMYLASE, WEEVILS.

**Alpha Galactosides** Term referring to a family of polysaccharides (produced in plant seeds) composed (at the molecular level) of one sucrose unit linked by a 1,6 molecular bond to several galactose units. Alpha galactosides include raffinose, stachyose, and verbascose. See also POLYSACCHARIDES, GALACTOSE (GAL), STACHYOSE.

**Alpha Helix (α-helix)** A highly regular (i.e., repeating) structural feature that occurs in certain large molecules. First discovered in protein molecules by Linus Pauling in the late 1940s. See also A-DNA, PROTEIN, PROTEIN FOLDING, PROTEIN STRUCTURE.

**Alpha Interferon** Also written as α-interferon, it has been shown to prolong life and reduce tumor size in patients suffering from Kaposi's sarcoma (a cancer that affects approximately 10% of people with acquired immune deficiency syndrome). It is also effective against hairy-cell leukemia and may work against other cancers. It has recently been approved by the U.S. FDA for use against certain types of sarcoma. Recent research indicates that injections of alpha interferon can limit the liver damage typically caused by hepatitis C, a viral disease. See also INTERFERONS.

**ALS** A plant enzyme (also present in some microoganisms) known as acetolactate synthase or acetohydroxy acid synthase. ALS catalyzes (enables to occur) one of the early chemical reaction steps in the synthesis (manufacturing) of branched-chain amino acids (isoleucine, leucine, valine) required by plants to sustain life (i.e., to make needed proteins). Herbicides that deactivate/destroy ALS are effective at killing plants (e.g., weeds). See also ENZYME, GENE, ALS GENE, MICROORGANISMS, CATALYST, AMINO ACID, ISO-LEUCINE (ile), LEUCINE (leu), VALINE (val).

**ALS Gene** Gene that codes for (i.e., causes to be produced in microorganisms or plants' chloroplasts) the critical-to-plants enzyme acetolactate synthase (ALS). See also GENE, HTC, MICROORGANISMS, CHLOROPLASTS, ENZYME, CATALYST, AMINO ACID, ISOLEUCINE (ile), LEUCINE

(leu), VALINE, STS SULFONYLUREA (HERBICIDE)-TOLERANT SOYBEANS.

**Alternative mRNA Splicing** See TRANSCRIP-TOME, CENTRAL DOGMA (NEW).

**Alternative Splicing** See TRANSCRIPTOME, CENTRAL DOGMA (NEW).

**Alu Family** A set of dispersed and related genetic sequences, each about 300 base pairs long, in the human genome. At both ends of these 300 bp segments there is an A-G-C-T sequence. Alu 1 is a restriction enzyme that recognizes this sequence and cleaves (cuts) it between the G (guanine) and the C (cytosine). See also GENOME, RESTRICTION ENDONUCLEASES.

**Aluminum Resistance** See CITRATE SYNTHASE (CSb) GENE, GENE, CITRIC ACID.

**Aluminum Tolerance** See CITRATE SYNTHASE (CSb) GENE, GENE, CITRIC ACID.

**Aluminum Toxicity** See CITRATE SYNTHASE (CSb) GENE, GENE, CITRIC ACID.

**Alzheimer's Disease** Named after Alois Alzheimer who, in 1906, first described the Amyloid β Protein (AβP) plaques in the human brain that are caused by this disease. Alzheimer's disease causes progressive memory loss and dementia in its victims as it kills brain cells (neurons). Some drugs (e.g., tacrine, donepezil, etc.) appear to slow the progression of Alzheimer's disease (by increasing the availability of acetylcholine in the brain), but there is currently no way to stop the disease. See also AMYLOID β PRO-TEIN (AβP), AMYLOID β PROTEIN PRECURSOR (AβPP), NEURON, NEUROTRANSMITTER, ACETYL-CHOLINE, OXIDATIVE STRESS.

**AMD** Acronym for Age-related Macular Degeneration. See also LUTEIN.

**American Society for Biotechnology (ASB)** A society founded for the purpose of "providing a multi- and interdisciplinary forum for those persons from academia, industry, and government who are interested in any and all aspects of biotechnology, and will achieve its aims by cooperation with existing organizations active in the field." To join, write to ASB, P.O. Box 2820, Sausalito, California, 94966-2820. See also BIOTECHNOLOGY, INTERNATIONAL SOCIETY FOR THE ADVANCEMENT OF BIOTECHNOLOGY (ISAB), BIOTECHNOLOGY INDUSTRY ORGANIZATION (BIO).

**American Type Culture Collection (ATCC)** An independent, nonprofit organization established in 1925 for the preservation and distribution of reference cultures. See also CELL CULTURE, CULTURE, CULTURE MEDIUM, TYPE SPECIMEN, CONSULTATIVE GROUP ON INTERNATIONAL AGRICULTURAL RESEARCH (CGIAR).

**Ames Test** A simple bacterial-based carcinogens test that was developed by Bruce Ames in 1961. Although this test evaluates mutagenesis (causation of mutations) in the DNA of bacteria, its results have been utilized to approve or not approve certain compounds for consumption by humans. See also BIOASSAY, BACTERIA, ASSAY, MUTUAL RECOGNITION AGREEMENTS (MRAs), GENOTOXIC CARCINOGENS, CARCINOGEN, PARP.

**Amino Acid** There are 20 common amino acids, each specified by a different arrangement of three adjacent DNA nucleotides. These are the building blocks of proteins. Joined together in a strictly ordered chain, the sequence of amino acids determines the character of each protein (chain) molecule. The 20 common amino acids are: alanine, arginine, aspartic acid, glutamic acid, glutamine, glycine, histidine, isoleucine, leucine, phenylalanine, proline, serine, threonine, tryptophan, tyrosine, valine, cysteine, methionine, lysine, and asparagine. Note that virtually all of these amino acids (except glycine) possess an asymmetric carbon atom, and thus are potentially chiral in nature. See also PROTEIN, POLYPEPTIDE (protein), STEREOISOMERS, CHIRAL COMPOUND, MESSENGER RNA (mRNA), ESSENTIAL AMINO ACIDS, DEOXYRIBONUCLEIC ACID (DNA), ABSOLUTE CONFIGURATION.

**Amino Acid Profile** Also known as "protein quality," this refers to a quantitative delineation of how much of each amino acid is contained in a given source of (livestock feed or food) protein. For example, the amino acid profile of soybean meal is matched closest to the profile of amino acids needed for human nutrition, of all protein meals. See also "IDEAL PROTEIN" CONCEPT, PROTEIN, AMINO ACID, SOYBEAN MEAL, PDCAAS.

**Aminocyclopropane Carboxylic Acid Synthase/deaminase** See ACC SYNTHASE, ACC.

**AMP** See ADENOSINE MONOPHOSPHATE (AMP).

**Amphibolic Pathway** A metabolic pathway used in both catabolism and anabolism. See also ANABOLISM, CATABOLISM.

**Amphipathic Molecules** Molecules bearing both polar and nonpolar domains (within the same molecule). Some examples of amphipathic molecules are wetting agents (SDS), and membrane lipids such as lecithin. See also MICELLE, REVERSE MICELLE (RM), POLARITY (CHEMICAL).

**Amphiphilic Molecules** Also known collectively as amphiphiles, these molecules possess distinct regions of hydrophobic ("water hating") and hydrophilic ("water loving") character within the same molecule. When dissolved in water above a certain concentration (known as the CMC), they are capable of forming high molecular weight aggregates, or micelles. See also CRITICAL MICELLE CONCENTRATION, HYDROPHOBIC, HYDROPHILIC, MICELLE, REVERSE MICELLE (RM).

**Amphoteric Compound** A compound capable of both donating and accepting protons and thus able to act chemically as either an acid or a base.

**Amplicon** A specific sequence of DNA produced by a DNA-amplification technology such as the Polymerase Chain Reaction (PCR) technique. See also DEOXYRIBONUCLEIC ACID (DNA), SEQUENCE (OF A DNA MOLECULE), POLYMERASE CHAIN REACTION (PCR) TECHNIQUE, NESTED PCR.

**Amplification** The production of additional copies of a chromosomal sequence, found as either intrachromosomal or extrachromosomal DNA. See also IN VITRO SELECTION.

**Amplified Fragment Length Polymorphism** Also known by its acronym, AFLP is a "DNA marker" utilized in a "genetic mapping" technique which employs the specific sequence of bases (nucleotides) in a piece of DNA (from an organism). Since the specific sequence of bases in their DNA molecules is different for each species, strain, variety, or individual (due to DNA polymorphism), AFLP can be used to "map" those DNA molecules (e.g., to assist and speed up plant breeding programs). See also GENETIC MAP, SEQUENCE (OF A DNA MOLECULE), DEOXYRIBONUCLEIC ACID (DNA), GENOME, PHYSICAL MAP (OF GENOME), MARKER (DNA SEQUENCE),

MARKER (GENETIC MARKER), POLYMORPHISM (CHEMICAL), NUCLEIC ACIDS, NUCLEOTIDE, GENETIC CODE.

**Amplimer** See AMPLICON.

**Amylase** A term that is used to refer to a category of enzymes that catalyzes the chemical reaction in which amylose (starch) molecules are hydrolytically cleaved (broken) to molecular pieces (e.g., the polysaccharides maltose, maltotriose, a-dextrin, etc.). For example, α-amylase is used to break apart corn starch molecules in the first step of manufacturing fructose (sweetener for soft drinks). Since 1857, amylase has been utilized to remove (amylose) starch from woven fabrics in the textile industries. Modern uses of some amylases include enabling the substitution of barley grain for malt in the beer brewing process. See also ENZYME, STARCH, AMYLOSE, BARLEY, HYDROLYTIC CLEAVAGE, POLYSACCHARIDES, ALPHA AMYLASE INHIBITOR-1.

**Amyloid β Protein Precursor (AβPP)** A (collective) set of protein molecules, from which are derived Amyloid β Protein (AβP). See also PROTEIN, AMYLOID β PROTEIN PRECURSOR (AβPP).

**Amyloid β Protein (AβP)** A small protein that forms plaque in the brains and in the brain blood vessels of victims of Alzheimer's disease. AβP forms cation-selective ion channels in lipid bilayers (e.g., membranes surrounding cells). This ion channel formation disrupts calcium homeostasis, allowing (destructive) high concentrations of calcium ions in brain cells. See also PROTEIN, AMYLOID β PROTEIN PRECURSOR (AβPP), ALZHEIMER'S DISEASE.

**Amyloid Placques** See AMYLOID β PROTEIN (AβP).

**Amylopectin** The form of starch (molecule) that consists of multi-branched polymers, containing approximately 100,000 glucose units per molecule (polysaccharide). See also STARCH, POLYMER, GLUCOSE (GLc), POLYSACCHARIDES, WAXY CORN.

**Amylose** The form of starch that consists of unbranched polymers, containing approximately 4000 glucose units per molecule (polysaccharide). It is present in potatoes at 23–29% content (variation is thought to be caused by different growing conditions). See also POLYMER, GLUCOSE (GLc), AMYLASE, POLYSACCHARIDES.

**Anabolism** The phase of intermediary metabolism concerned with the energy-requiring biosynthesis of cell components from smaller precursor molecules. See also CATABOLISM, ASSIMILATION, METABOLISM, CELL, PLASMA MEMBRANE.

**Anaerobe** An organism that lives in the absence of oxygen and generally cannot grow in the presence of oxygen. The catabolic metabolism of anaerobic microorganisms reduces a variety of organic and inorganic compounds in order to survive (e.g., carbon dioxide, sulfate, nitrate, fumarate, iron, manganese); anaerobes produce a large number of end products of metabolism (e.g., acetic acid, propionic acid, lactic acid, ethanol, methane, etc.). See also CATABOLISM, METABOLISM, METABOLITE, REDUCTION (IN A CHEMICAL REACTION), ANAEROBIC.

**Anaerobic** An environment without air or oxygen. See also ANAEROBE.

**Analogue** (Analog) A compound (or molecule) that is a (chemical) structural derivative of a parent compound. The word is also used to describe a molecule that may be structurally similar (but not identical) to another, and which exhibits many or some of the same biological functions of the other. For example, the large class of antibiotics known as the sulfa drugs are all analogues of the original synthetic chemical drug (known as Prontosil, which cures streptococcal infections) discovered by the German biologist Gerhart Domagk. His discovery and others made possible a program of further chemical syntheses based upon the original (sulfanilamide) molecular structure and resulted in the large number of sulfonamide (also called "sulfa") drugs available today. All of the analogue (also analog) sulfa drugs that were patterned after the original sulfanilamide molecular structure may be called sulfanilamide analogues.

Today, analogues are known by man for various vitamins, amino acids, purines, sugars, growth factors, and many other chemical compounds. Research chemists produce analogues of various molecules in order to ascertain the biological role of, or importance of, certain structures (within the molecule) to the molecule's function within a living organism. See also BIOMIMETIC MATERIALS, RATIONAL DRUG DESIGN, HETEROLOGY, GIBBERELLINS, QUANTITATIVE STRUCTURE-ACTIVITY RELATIONSHIP (QSAR).

**ANDA** (to FDA) Abbreviated New Drug Application (to the U.S. Food and Drug Administration). See also NDA, "TREATMENT" IND REGULATIONS, FOOD AND DRUG ADMINISTRATION (FDA).

**Angiogenesis** Formation/development of new blood vessels in the body. Discovered to be triggered and stimulated by angiogenic growth factors, in the early 1980s. Angiogenesis is required for malignant tumors to metastasize (spread throughout the body), because it provides the (newly-created) blood supply that tumors require. Angiogenesis is also crucial to the development of glaucoma and macular degeneration (major cause of blindness). The drug Thalidomide is a potent inhibitor of angiogenesis, as are the proteins angiostatin and endostatin. See also ANGIOGENIC GROWTH FACTORS, TUMOR, CANCER, METASTASIS, ANTIANGIOGENESIS, CHIRAL COMPOUND, ANGIOSTATIN, ENDOSTATIN.

**Angiogenesis Factors** See ANGIOGENIC GROWTH FACTORS.

**Angiogenic Growth Factors** Proteins that stimulate formation of blood vessels (e.g., in tissue being formed by the body to repair wounds). See also FILLER EPITHELIAL CELLS, FIBROBLAST GROWTH FACTOR (FGF), MITOGEN, ANGIOGENIN, ENDOTHELIAL CELLS, TRANSFORMING GROWTH FACTOR-ALPHA (TGF-ALPHA), TRANSFORMING GROWTH FACTOR-BETA (TGF-BETA), PLATELET-DERIVED GROWTH FACTOR (PDGF), ANGIOGENESIS.

**Angiogenin** One of the human angiogenic growth factors, it possesses potent angiogenic (formation of blood vessels) activity. In addition to stimulating (normal) blood vessel formation, angiogenin levels are correlated with placenta formation and tumor growth (tumors require new blood vessels). See also ANGIOGENIC GROWTH FACTORS, ANGIOGENESIS, TUMOR, GROWTH FACTOR.

**Angiostatin** An antiangiogenesis (anti-blood-vessel-formation) human protein discovered by Judah Folkman. In combination with endostatin, it has been shown to cause certain cancer tumors in mice to shrink by cutting off the creation of new blood vessels required to "feed" a growing tumor. Angiostatin acts to halt the creation of new blood vessels by binding to ATP synthase (an enzyme needed to initiate new blood vessels). See also PROTEIN, ANTIANGIOGENESIS, ENDOSTATIN, CANCER, ATP SYNTHASE, TUMOR.

**Angstrom (Å)** $10^{-8}$ cm ($3.937 \times 10^{-9}$ inch).

**Anion** See ION.

**Anneal** The process by which the complementary base pairs in the strands of DNA combine. See also BASE PAIR (bp), DEOXYRIBONUCLEIC ACID (DNA).

**Anonymous DNA Marker** Refers to a DNA marker with a clearly identifiable sequence variation (i.e., it is detectable by the specific variation in its DNA sequence, whether or not it occurs in or near a coding sequence). See also DEOXYRIBONUCLEIC ACID (DNA), SEQUENCE (OF A DNA MOLECULE), MARKER (DNA SEQUENCE), MICROSATELLITE DNA.

**Antagonists** Molecules that bind to certain proteins (e.g., receptors, enzymes) at a specific (active) site on that protein. The binding suppresses or inhibits the activity (function) of that protein. See also RECEPTORS, ACTIVE SITE, CONFORMATION, AGONISTS, ENZYME, ALLOSTERIC ENZYMES.

**Anterior Pituitary Gland** See PITUITARY GLAND.

**Anthocyanidins** Natural pigments (flavonoids) produced in blueberries (genus *Vaccinium*), blackberries (*Rubus fruticosus*), cranberries (*Vaccinium macrocarpon*), cherries (genus *Prunus*), black or purple carrots (*Daucus carota*), and some types of grapes. Consumption of anthocyanidins by humans has been shown to be beneficial to eyesight by aiding the health of the retina. Within the human body, anthocyanidins act as antioxidants (i.e., "quenchers" of free radicals), so consumption apparently reduces the risk of some cancers, coronary heart disease, eyesight loss, and cataracts. See also PHYTOCHEMICALS, NUTRACEUTICALS, CAROTENOIDS, ANTIOXIDANTS, OXIDATIVE STRESS, CANCER, CORONARY HEART DISEASE (CHD), INSULIN, PROANTHOCYANIDINS, FOSHU.

**Anthocyanins** See ANTHOCYANIDINS.

**A**

**Anti-Idiotype Antibodies** See ANTI-IDIOTYPES.

**Anti-Idiotypes** Antibodies to antibodies. In other words, if a human antibody is injected into rabbits, the rabbit immune systems will recognize the human antibodies as foreign (regardless of the fact that they are antibodies) and produce antibodies against them. To the rabbit, the foreign antibodies represent just another invader or nonself to be targeted and destroyed. Anti-idiotypes mimic antigens in that they are shaped to fit into the antibody's binding site (in lock-and-key fashion). As such, anti-idiotypes can be used to create vaccines that stimulate production of antibodies to the antigen (that the anti-idiotype mimics). This confers disease resistance (to the pathogen associated with that antigen) without the risk that a vaccine using attenuated pathogens entails (i.e., that the pathogen "revives" to cause the disease). See also ANTIBODY, MONOCLONAL ANTIBODIES (MAb), ANTIGEN, IDIOTYPE, PATHOGEN, ATTENUATED (PATHOGENS).

**Anti-Interferon** An antibody to interferon. Used for the purification of interferons. See also ANTIBODY, INTERFERONS, AFFINITY CHROMATOGRAPHY.

**Anti-Oncogenes** See ONCOGENES, ANTISENSE (DNA SEQUENCE).

**Antiangiogenesis** Refers to impact of any compound that prevents angiogenesis (i.e., formation/development of new blood vessels). Because angiogenesis is required for malignant tumors to grow and/or metastasize (spread), antiangiogenesis was proposed by Judah Folkman in 1970 as a means to combat cancer. Because angiogenesis is required for embryonic development, antiangiogenic drugs inhibit proper development/growth of infants in the womb. Fumagillin, ovalicin, and Thalidomide have been found to possess antiangiogenic properties. Also, the human proteins angiostatin and endostatin. See also ANGIOGENESIS, ANGIOGENIC GROWTH FACTORS, TUMOR, CANCER, ANGIOSTATIN, ENDOSTATIN, GENISTEIN.

**Antibiosis** Refers to the processes by which one organism produces a substance that is toxic or repellent to another organism (e.g., a parasite) that is attacking the first organism. For example, certain varieties of corn/maize (*Zea mays* L.) naturally produce chemical substances in their roots that are toxic to the corn rootworm. See also ANTIBIOTIC, *BACILLUS THURINGIENSIS* (*B.t.*), CORN, CORN ROOTWORM.

**Antibiotic** Coined by Selman Waksman during the 1940s, this term refers to organic compounds that are naturally formed and secreted by various species of microorganisms and/or plants. It has a defensive function and is often toxic to other species (e.g., penicillin, originally produced by bread mold, is toxic to numerous human pathogens). Antibiotics generally act by inhibiting protein synthesis, DNA replication, synthesis of cell wall (cytoplasmic membrane) constituents, inhibition of required cell (e.g., bacteria) metabolic processes, and nucleic acid (DNA and RNA) biosynthesis, hence killing the (targeted bacteria) cells involved. Inorganic (e.g., certain metals) molecules may also have antibiotic properties. See also PATHOGEN, MICROORGANISM, PROTEIN, NUCLEIC ACIDS, PENICILLIN G (benzylpenicillin), SYMBIOTIC, GRAM STAIN, GRAM-NEGATIVE, ALLELOPATHY, BACTERIA, GRAM-POSITIVE, CELL, ANTIBIOSIS, AUREOFACIN, PHOTORHABDUS LUMINESCENS, BETA-LACTAM ANTIBIOTICS, METABOLISM, DEOXYRIBONUCLEIC ACID (DNA), PLASMA MEMBRANE, RIBONUCLEIC ACID (RNA).

**Antibiotic Resistance** A property of a cell (e.g.. pathogenic bacteria) that enables it to avoid the effect of an antibiotic that had formerly killed or inhibited that cell. Ways this can occur include: changing the structure of the cell wall (plasma membrane); synthesis (manufacture) of enzymes to inactivate the antibiotic (e.g., penicillinases, which inactivate penicillin); synthesis of enzymes to prevent antibiotic entering cell; and active removal of the antibiotic from the cell. For example, the membrane transporter protein molecules known as ABC transporters are sometimes able to help pathogenic bacteria resist certain antibiotics by transporting out the antibiotic before it can kill the bacteria. The ABC transporter is a V-shaped molecule embedded in the (bacteria) cell's plasma membrane, with the open end of the "V" pointed toward the interior of the cell. When molecules of certain antibiotics (inside the cell) contact the ABC transporter molecule, the two "arms" of the ABC transporter close around the antibiotic molecule, the ABC transporter flips over, and thereby sends the antibiotic molecule out through the exterior of the cell's plasma membrane, replacing some critical cell metabolic processes, with (new) metabolic processes that bypass the antibiotic's (former) effect. See also CELL, PATHOGEN, PATHOGENIC, BACTERIA, ANTIBIOTIC, PLASMA MEMBRANE, ENZYME, PENICILLINASES, METABOLISM, ABC TRANSPORTERS, MYCOBACTE-RIUM TUBERCULOSIS.

**Antibody** Also called immunoglobulin, Ig. A large defense protein that consists of two classes of polypeptide chains, light (L) chains and heavy (H) chains. A single antibody molecule consists of two identical copies of the L chain and two of the H chain. They are synthesized (made) by the immune system (B lymphocytes) of the organism. The antibody is composed of four proteins linked together to form a Y-shaped bundle of proteins (looks somewhat like a slingshot or two hockey sticks taped together at the handles). The amino acid sequence that makes up the stem (heavy chains) of the Y (i.e., the handles of the taped together hockey sticks) is similar for all antibodies. The stem is known as the Fc region of the antibody, and it does not bind to antigens, but does have other regulatory functions.

The two arms of the Y are each made up of two side-by-side proteins called light chains and heavy chains (proteins are chains of amino acids), with identical antigen-binding (ab) sites on the tips of each "arm." The antibody is thus bivalent in that it has two binding sites for antigen. Taken together, the two arms of the Y are known as the Fab portions of the antibody molecule. The Fab portions can be cleaved from the antibody molecule with papain (an enzyme that is also used as a meat tenderizer) or the Fab portions can be produced by genetically engineered *Escherichia coli* (*E. coli*) bacteria. When a foreign molecule (e.g., a bacterium, virus, etc.) enters the body, B lymphocytes are stimulated into becoming rapidly dividing blast cells, which mature into antibody-producing plasma cells. The plasma cells are triggered by the foreign molecule's epitope(s) [i.e., group or groups of

specific atoms (also known as a hapten), that are recognized to be foreign by the body's immune system] into producing antibody molecules possessing antigen-binding (ab) sites (also called combining sites or determinants).

These fit into the foreign molecule's epitope. Thus, via the tips of its arms, the antibody molecule binds specifically to the foreign entity (antigen) that has entered the body. By this process it inactivates that foreign molecule or marks it for eventual destruction by other immune system cells.

System marking of the foreign molecule (e.g., pathogen or toxin) for destruction is accomplished by the fact that the stem of the Y (i.e., the Fc) fragment hangs free from the combined antibody-antigen clump, thereby providing a receptor for phagocytes, which roam throughout the body ingesting and subsequently destroying such "marked" foreign molecules. Research published during 2001 indicates that antibodies may also kill some pathogens themselves by catalyzing the formation of hydrogen peroxide from oxygen free radicals (singlet oxygen) and water. Hydrogen peroxide is highly reactive, and could potentially kill pathogens when generated by an (attached) antibody. There are five classes of immunoglobulin: IgG, IgM, IgD, IgA, and IgE. See also HUMORAL IMMU-NITY, IMMUNOGLOBULIN, PROTEIN, POLYPEPTIDE (PROTEIN), AMINO ACID, B LYMPHOCYTES, BLAST CELL, ANTIGEN, HAPTEN, EPITOPE, COMBINING SITE, DOMAIN (OF A PROTEIN), SEQUENCE (OF A PROTEIN MOLECULE), ESCHERICHIA COLIFORM (E. COLI), PATHOGEN, TOXIN, PHAGOCYTE, MICROPHAGE, MONOCYTES, T CELLS, POLYMOR-PHONUCLEAR LEUKOCYTES (PMN), CELLULAR IMMUNE RESPONSE, POLYMORPHONUCLEAR GRAN-ULOCYTES, GENETIC ENGINEERING, "MAGIC BUL-LET", ENGINEERED ANTIBODIES, RECEPTORS, OXYGEN FREE RADICALS.

**Antibody Affinity Chromatography** A type of chromatography in which antibodies are immobilized onto the column material. The antibodies bind to their target molecules while the other components in the solution are not retained. In this way a separation (purification) is achieved. See also ANTIBODY, CHROMA-TOGRAPHY, AFFINITY CHROMATOGRAPHY.

**A**

**Antibody-Mediated Immune Response** See HUMORAL IMMUNE RESPONSE.

**Anticoding Strand** Refers to the single strand of DNA (double helix) that is transcribed. Sometimes called the antisense strand or the template strand. See also DEOXYRIBONUCLEIC ACID (DNA), TRANSCRIPTION, ANTISENSE (DNA SEQUENCE).

**Anticodon** A specific sequence of three nucleotides in a transfer RNA (tRNA), complementary to a codon (also three nucleotides) for an amino acid in a messenger RNA. See also CODON, TRANSFER RNA (tRNA), AMINO ACID, MESSENGER RNA (mRNA), NUCLEOTIDE.

**Antigen** Also called an immunogen. Any large molecule or small organism whose entry into the body provokes synthesis of an antibody or immunoglobulin (i.e., an immune system response). See also HAPTEN, ANTIBODY, EPITOPE, CELLULAR IMMUNE RESPONSE, HUMORAL IMMUNITY.

**Antigenic Determinant** See HAPTEN, EPITOPE, SUPERANTIGENS.

**Antihemophilic Factor VIII** Also known as Factor VIII or Antihemophilic Globulin (AHG). See also FACTOR VIII.

**Antihemophilic Globulin** Also known as Factor VIII or Antihemophilic Factor VIII. See also FACTOR VIII.

**Antioxidants** Compounds (e.g., phytochemicals) that act to prevent lipids from oxidizing (to plaque) or breaking down (e.g., to carcinogenic compounds), or that act to capture and halt singlet oxygen (O-) free radicals; which can damage DNA in cells (causing mutations). Since oxidation of lipids in the blood is the intitial step in atherosclerosis, consumption of large amounts of certain antioxidants (e.g., flavonoids) may prevent atherosclerosis. Because oxidation reactions within the body often lead to formation of tissue-damaging free radicals (molecules containing an "extra" electron), consumption of antioxidants can help to prevent such tissue damage. Evidence indicates that tissue damage from free radicals may play a role in causing some arthritis, coronary heart disease, diabetes, and cancers. Synthetic analogues have also been manufactured (e.g., synthetic vitamins, etc.) which perform a similar antioxidant function to naturally occurring antioxidant phytochemicals. See also OXIDATIVE STRESS, PHYTOCHEMICALS, LIPIDS, CARCINOGEN, CANCER, ANALOGUES, OXIDATION, CORONARY HEART DISEASE, INSULIN, LYCOPENE, MUTAGEN, MUTATION, FLAVONOIDS, ISOFLAVONES, ATHEROSCLEROSIS, ASTAXANTHIN, HUMAN SUPEROXIDE DISMUTASE (hSOD), PEG-SOD (POLYETHYLENE GLYCOL SUPEROXIDE DISMUTASE), PLAQUE, PHYTATE, POLYPHENOLS, BETA CAROTENE, VITAMIN E, POLYUNSATURATED FATTY ACIDS (PUFA), CONJUGATED LINOLEIC ACID (CLA).

**Antiparallel** Describes molecules that are parallel but point in opposite directions. The strands of the DNA double helix are antiparallel. See also DOUBLE HELIX.

**Antisense (DNA sequence)** A strand of DNA that produces a messenger RNA (mRNA) molecule which (when reversed end-for-end) has the same sequence as (is complementary to) the unwanted ("bad") messenger RNA. The SENSE (forward) and ANTISENSE (backward) mRNA strands hybridize (tightly bond to each other), which prevents the bonded pair from leaving the cell's nucleus, so that bonded pair is rapidly degraded (destroyed) by nuclei within the cell nucleus. In genetic targeting (to block "bad" genes), antisense molecules are used to bind to a "bad" gene's (an oncogene) messenger RNA (mRNA), thus canceling the (cancer-causing) message of the gene and preventing cells from following its (tumor growth) instructions. Another example would be the use of antisense DNA to block the gene that codes for production of polygalacturonase (an enzyme that causes ripe fruit to (soften). Physically, antisense is accomplished by removing a given gene from an organism's genome, reversing it (end-for-end), and reinserting it back into the organism's genome. See also DEOXYRIBONUCLEIC ACID (DNA), CODING SEQUENCE, GENE, GENOME, COMPLEMENTARY DNA (c-DNA), MESSENGER RNA (mRNA), GENETIC TARGETING, CANCER, POLYGALACTURONASE (PG), ONCOGENES, SENSE, COSUPPRESSION, GENE SILENCING, HYBRIDIZATION (MOLECULAR GENETICS), NUCLEASE, ANTICODING STRAND.

**Antisense RNA** See ANTISENSE (DNA SEQUENCE).

**Antithrombogenous Polymers** Synthetic polymers (i.e., plastics) used to make medical devices that will be in contact with a patient's blood (e.g., catheters), but will not initiate the coagulation process as synthetic polymers usually do. The natural anticoagulant heparin is incorporated into the polymer and is gradually released into the bloodstream by the polymer, thus preventing blood coagulation on the surface of the polymer. See also POLYMER, THROMBOSIS.

**Antitoxin** See POLYCLONAL ANTIBODIES, DIPHTHERIA ANTITOXIN.

**AP** Atrial peptide. See also ATRIAL PEPTIDES.

**APHIS** The Animal and Plant Health Inspection Service is the agency of the U.S. Department of Agriculture responsible for regulating the field (outdoor) testing of genetically engineered plants and certain microorganisms. See also COORDINATED FRAMEWORK FOR REGULATION OF BIOTECHNOLOGY, MICROORGANISM, GENETIC ENGINEERING.

**Aplastic Anemia** An autoimmune disease of the bone marrow. See also AUTOIMMUNE DISEASE.

**APO B-100** See LOW-DENSITY LIPOPROTEINS (LDLP), APOLIPOPROTEINS, VERY LOW-DENSITY LIPOPROTEINS (VLDL).

**APO-1/Fas** See CD95 PROTEIN.

**Apoenzyme** The protein portion of a holoenzyme. Many (but not all) enzymes are composed of functional "pieces" (i.e., a protein piece (chain) and another piece that is an organic and/or inorganic molecule). The other piece is known as a cofactor, and it may be removed from the enzyme under certain conditions, after which the resulting inactive enzyme is known as an apoenzyme. The inactive apoenzyme becomes functionally active again if it is allowed to recombine with its cofactor. See also COFACTOR, ENZYME, HOLOENZYME.

**Apolipoprotein B** See LOW-DENSITY LIPOPROTEINS (LDLP), APOLIPOPROTEINS, VERY LOW-DENSITY LIPOPROTEINS (VLDL).

**Apolipoproteins** The protein portion of lipoproteins (i.e., after the lipid portion is removed from those molecules). See also LOW-DENSITY LIPOPROTEINS (LDLP), PROTEIN, LIPIDS, VERY LOW-DENSITY LIPOPROTEINS (VLDL).

**Apomixis** A method of reproduction used by scientists to propagate (hybrid) plants without having to utilize sexual fertilization. By combining apomixis with tissue culture technology, Cai Detian, Ma Piugfu, and Yao

Jialin were able to propagate rice varieties in 1994. In 1998, Dimitri Petrov, Phillip Sims, and Chester Deald were able to cause apomixis in corn (maize). By "fixing" hybrid dominance, the need for (sexual) breeding is eliminated and the hybrid vigor is passed down via the seed from generation to generation. See also ASEXUAL, GERM CELL, HYBRID VIGOR, TISSUE CULTURE, HYBRIDIZATION (PLANT GENETICS), CORN, F1 HYBRIDS.

**Apoptosis** Also called "programmed cell death," it is a series of programmed steps that cause a cell to die by "self digestion" without rupturing and releasing intracellular contents (e.g., nucleus, chromosomes, refractile bodies, etc.) into the local (surrounding tissue) environment. Manifestations of cell apoptosis include shrinking of the cell's cytoplasm and chromatin condensation. If the normal cell apoptosis is prevented (e.g., by an enzyme that is present due to disease) in the body, cells can grow uncontrollably (i.e., causing cancer). For example, people with chronic myelogenous leukemia (CML, also known as chronic myeloid leukemia) typically have 10–25 times as many white blood cells as normal. See also CELL, CD95 PROTEIN, SIGNAL TRANSDUCTION, SIGNALING, REFRACTILE BODIES (RB), NUCLEUS, CHROMOSOMES, CHROMATIN, CYTOPLASM, FUSARIUM, p53 GENE, TUBULIN, CANCER, SELECTIVE APOPTOTIC ANTI-NEOPLASTIC DRUG (SAAND), HYPERSENSITIVE RESPONSE, SIGNAL TRANSDUCTION, SIGNAL TRANSDUCERS AND ACTIVATORS OF TRANSCRIPTION (STATs), GENE EXPRESSION CASCADE, ENZYME, WHITE BLOOD CELLS, PHILADELPHIA CHROMOSOME, GLEEVEC™.

**Approvable Letter** (from the FDA) One of the final steps in the U.S. Food and Drug Administration's (FDA) review process for new pharmaceuticals. The letter precedes final FDA clearance for marketing of the new compound. See also FOOD AND DRUG ADMINISTRATION (FDA), IND, IND EXEMPTION.

**Aptamers** Oligonucleotide molecules that bind (stick to) other, specific molecules (e.g., proteins). Aptamer is from the Latin *aptus*, to fit. In 1992, Louis Bock and John Toole isolated aptamers that bind and inhibit the blood-coagulation enzyme thrombin. Since thrombin is crucial to the formation of blood

**A**

clots (coagulation), such aptamers may someday be useful for anticoagulant therapy (e.g., to prevent blood clots following surgery or heart attacks). See also ENZYME, OLIGONUCLEOTIDE, PROTEIN, INHIBITION, THROMBIN, THROMBUS, THROMBOSIS.

***Arabidopsis thaliana*** A small weed plant (Cruciferae) possessing 70,000 kilobase pairs in its genome, with very little repetitive DNA. This makes it an ideal model for studying plant genetics. At least two genetic maps have been created for *Arabidopsis thaliana* (one using yeast artificial chromosomes). Because of this, a large base of knowledge about it has been accumulated by the scientific community.

    *A. thaliana* was first genetically engineered in 1986. In 1994, researchers succeeded in transferring genes for polyhydroxylbutylate ("biodegradable plastic") production into *A. thaliana*. Because production of polyhydroxylbutylate (PHB) requires simultaneous expression of three genes (the PHB production process is "polygenic") — yet researchers have only been able to insert a maximum of two genes — they have to insert two genes into one plant and one gene into a second plant, then finally get the (total) three genes into (offspring) plants via traditional breeding. During 2001, Eduardo Blumwald and Hong-Xia Zhang inserted a salt-tolerance gene from *A. thaliana* into a tomato (*Lycopersicon esculentum*), and thereby made that tomato plant resistant to salt in concentrations up to 200 mM (far higher than it could previously survive). See also BRASSICA, GENE, EXPRESS, BASE PAIR (bp), KILOBASE PAIRS (Kbp), GENOME, GENETIC CODE, GENETIC MAP, GENETICS, TRAIT, POLYGENIC, DEOXYRIBONUCLEIC ACID (DNA), POLYHYDROXYLBUTYLATE (PHB), YEAST ARTIFICIAL CHROMOSOMES (YAC), MODEL ORGANISM, TOMATO, SALT TOLERANCE.

**Arachidonic Acid (AA)** One of the omega-6 (n-6) highly unsaturated fatty acids (HUFA), AA is synthesized (manufactured) by the human body from linoleic acid (e.g., obtained by consuming soybean oil). AA is present in human breast milk, and research indicates that it plays an important role in the mental development of infants. Arachidonic acid is a crucial precursor for prostaglandins and other eicosanoids. The COX-1 enzyme converts arachidonic acid to constitutive prostaglandins and the COX-2 enzyme converts arachidonic acid to inducible prostaglandins. See also CYCLOOXYGENASE, POLYUNSATURATED FATTY ACIDS (PUFA), N-6 FATTY ACIDS, FATTY ACIDS, UNSATURATED FATTY ACIDS, LINOLEIC ACID, SOYBEAN OIL, CONSTITUTIVE ENZYMES, INDUCIBLE ENZYMES, LEUKOTRIENES, ESSENTIAL FATTY ACIDS, EICOSANOIDS.

***Archaea*** Single-celled life forms that can live at extreme ocean depths (high pressure) and in the absence of oxygen. Enzymes robust (sturdy) enough for industrial process utilization have been isolated by scientists from some strains of *Archaea*. Other *Archaea* strains are sometimes present in the rumen ("first stomach") of cattle and sheep. Those *Archaea* produce methane gas by breaking down some of the feed consumed by the cattle and sheep. See also ENZYME, EXTREMOZYMES, CELL, ANAEROBE, ANAEROBIC, STRAIN.

**Arginine (arg)** An amino acid, commonly abbreviated arg. In dry, bulk form arginine is colorless, crystalline, and water soluble. It is an essential amino acid of the α-ketoglutaric acid family. See also AMINO ACID, ESSENTIAL AMINO ACIDS, NITRIC OXIDE SYNTHASE.

**ARM** Acronym for antibiotic resistance marker. See also MARKER (GENETIC MARKER).

**ARMD** Acronym for Age-Related Macular Degeneration. See also LUTEIN.

**ARMG** Acronym for Antibiotic Resistance Marker Gene. See also ANTIBIOTIC, ANTIBIOTIC RESISTANCE, GENE, MARKER (GENETIC MARKER), RECOMBINASE.

**Armyworm** Caterpillars (pupae) of the Lepidopteran insect *Pseudaletia unipuncta* family; most of which are harmful to crops (e.g., wheat, corn/maize, etc.) grown by humans. Armyworms are susceptible to some of the "cry" proteins (e.g., they are killed if they eat plants genetically engineered to contain Cry1A(b), Cry9C, or Cry1F proteins). Armyworms are preyed upon by some species of ground beetles, sphecid wasps, toads, birds, etc. See also PROTEIN, VOLICITIN, CRY PROTEINS, CRY1A(b) PROTEIN, CRY1F PROTEIN, CRY9C PROTEIN, CORN, WHEAT.

**AroA** Refers to the transgene (cassette) which was initially isolated/extracted from the genome of the Agrobacterium bacteria species (strain CP4) and inserted via genetic engineering techniques into a crop plant (e.g., soybean, *Glycine max* L.) in order to make that (soybean) plant tolerant to glyphosate-based herbicides (and also sulfosate-based herbicides). See also GENE, TRANSGENE, CASSETTE, GENOME, AGROBACTERIUM TUMEFACIENS, EPSP SYNTHASE, mEPSPS, CP4 EPSPS, SOYBEAN, HERBICIDE-TOLERANT CROP, GENETIC ENGINEERING, SOYBEAN PLANT, GLYPHOSATE, SULFOSATE.

**ARS** See ARS ELEMENT.

**ARS Element** A sequence of DNA that will support autonomous replication (sequence, ARS). See also DEOXYRIBONUCLEIC ACID (DNA), SEQUENCE (OF A DNA MOLECULE).

**Arteriosclerosis** A group of diseases (including atherosclerosis) which is characterized by a decrease in elasticity (stretchiness) and a thickening of the walls of the body's arteries. See also ATHEROSCLEROSIS, CORONARY HEART DISEASE (CHD), PLAQUE.

**Arthritis** See OSTEOARTHRITIS, AUTOIMMUNE DISEASE.

**Ascites** Liquid accumulations in the peritoneal cavity. Used as an input in one of the methods for producing monoclonal antibodies. See also MONOCLONAL ANTIBODIES (MAb), PERITONEAL CAVITY/MEMBRANE, ANTIBODY.

**Ascorbic Acid** A water-soluble vitamin and antioxidant. See also VITAMIN, ANTIOXIDANTS.

**-ase** The three-letter suffix that is added to a (root) word to denote an enzyme. For example, the stomachs of reindeer contain lichenase, an enzyme that enables reindeer to digest lichen that the reindeer consume as a source of winter food. See also ENZYME, PROTEASE, OXYGENASE, HUMAN PROTEIN KINASE C, HUMAN SUPEROXIDE DISMUTASE (hSOD), POLYMERASE, ATPase, ATP SYNTHASE, REGULATORY ENZYME.

**Asexual** Denotes fertilization and/or reproduction by *in vitro* means. Without sex. See also *IN VITRO*, APOMIXIS, GERM CELL.

**Asian Corn Borer** Also known by its Latin name, *Ostrinia furnacalis* is an insect (originally from Asia) whose larvae (caterpillars) eat and bore into the corn/maize (*Zea Mays* L.) plant. In doing so, they can act as vectors (carriers) of the fungi known as *Aspergillus flavus* (a source of aflatoxin), *Fusarium moniliforme* (a source of fumonisin), or *Aspergillus parasiticus* (a source of aflatoxin). See also EUROPEAN CORN BORER (ECB), CORN, FUNGUS, AFLATOXIN, FUSARIUM, FUSARIUM MONILIFORME.

**Asparagine (asp)** An amino acid, commonly abbreviated asp. In dry, bulk form asparagine appears as a white, crystalline solid. It is found in high amounts in many plants. See also AMINO ACID.

**Aspartic Acid** A dicarboxylic amino acid found in plants and animals, especially in molasses from young sugarcane and sugar beets. See also AMINO ACID.

***Aspergillus flavus*** See AFLATOXIN, PEROXIDASE, BETA CAROTENE.

**Assay** A test (specific technique) that measures a response to a test substance or the efficacy (effectiveness) of the test substance. See also IMMUNOASSAY, BIOASSAY, LUMINESCENT ASSAY, HYBRIDIZATION SURFACES.

**Assimilation** The formation of self cellular material from small molecules derived from food. See also INSULIN-LIKE GROWTH FACTOR-1 (IGF-1), RIBOSOMES, MESSENGER RNA (mRNA).

**Association of Biotechnology Companies (ABC)** An American trade association of companies involved in biotechnology and services to biotechnology companies (e.g., accounting, law, etc.). Formed in 1984, the ABC tended to consist of the smaller firms involved in biotechnology (and service firms that worked for all biotechnology companies). In 1993, the ABC was merged with the Industrial Biotechnology Association (IBA) to form the Biotechnology Industry Organization (BIO). See also INDUSTRIAL BIOTECHNOLOGY ASSOCIATION (IBA), BIOTECHNOLOGY INDUSTRY ORGANIZATION (BIO), BIOTECHNOLOGY.

**Astaxanthin** A carotenoid pigment responsible for the characteristic pink coloring of salmon, trout, and shrimp. It is produced by the microorganisms in the natural (wild) diets of those aquatic animals. Research has shown that astaxanthin (an antioxidant) helps boost the immune systems of humans that consume it. Research has also shown that astaxanthin helps to reduce oral cancer

A

in rats and inhibit breast cancer in mice. See also CAROTENOIDS, ANTIOXIDANTS, OXIDATIVE STRESS.

**AT-III** A human blood factor that promotes clotting. A deficiency of AT-III can be inherited or can result from certain surgical procedures, certain illnesses, and sometimes use of certain oral contraceptives. See also FACTOR VIII.

**ATCC** See AMERICAN TYPE CULTURE COLLECTION (ATCC), TYPE SPECIMEN, ACCESSION.

**Atherosclerosis** A form of arteriosclerosis characterized by deposition and buildup of fatty deposits (plaque) on the internal walls of the body's arteries, in addition to the decreased elasticity of artery walls that characterizes all forms of arteriosclerosis. When a piece of plaque breaks off, a blood clot generally forms, and that clot often blocks blood flow through the artery, causing a heart attack or stroke. See also ARTERIOSCLEROSIS, CORONARY HEART DISEASE (CHD), CHOLESTEROL, THROMBOSIS, THROMBUS, FLAVONOIDS, OXIDATIVE STRESS, ANTIOXIDANTS, PLAQUE.

**Atomic Weight** The total mass of an atom equal to the sum of the isotope's number of protons and neutrons (in the atom's nucleus). The atomic weights of the earth's elements are based on the assignment of exactly 12.000 as the atomic weight of the carbon-12 isotope (variation of atom). The atomic (weight) theory was established as a framework in 1869 by Meyer and Mendeléev, but standard precise values were not adopted internationally until an international commission on atomic weights was formed in 1899 in response to an initiative by the German Chemical Society. An element's atomic weight does not come out to a whole number (with the exception of carbon), because of the existence of isotopes which differ slightly with respect to the number of neutrons each contains. See also MOLECULAR WEIGHT, ISOTOPE.

**ATP** See ADENOSINE TRIPHOSPHATE (ATP).

**ATP Synthase** An enzyme complex that forms ATP from ADP and phosphate during oxidative phosphorylation in the inner mitochondrial membrane (in animals), in chloroplasts (in plants), and in cell membranes (in bacteria). This is an energy-producing reaction in that ATP is a high-energy compound

used by cells to maintain their living condition. ATP synthase is also present on the surface of endothelial cells (lining of blood vessels) where it helps to build new blood vessels (e.g., to replace tissue damaged by injury or disease). Under certain circumstances, this also creates new blood vessels that provide blood supply to tumors. When separated from the cell's membrane, ATP synthase hydrolyzes (breaks down) ATP via a chemical process in which one subunit (designated g) of ATP synthase rotates within the other (hollow) part of ATP synthase. See also ENZYME, CHLOROPLASTS, ADENOSINE TRIPHOSPHATE (ATP), HYDROLYSIS, ADENOSINE DIPHOSPHATE (ADP), MITOCHONDRIA, TUMOR, ENDOTHELIAL CELLS, ANGIOSTATIN.

**ATP Synthetase** See ATP SYNTHASE.

**ATPase** Adenosine triphosphatase, an enzyme that hydrolyzes (clips the bond between two phosphates in) ATP to yield ADP, phosphate, and energy. The reaction is usually coupled to an energy-requiring process. ATP is hydrolyzed in the act of shivering and the energy produced is converted into heat to increase body temperature. This type of heat production involves what is known as a futile cycle because the energy is converted to (and wasted as) heat rather than used in motion, etc. See also ATP SYNTHASE, ENZYME, ADENOSINE TRIPHOSPHATE (ATP), ADENOSINE DIPHOSPHATE (ADP), FUTILE CYCLE, HYDROLYSIS, HYDROLYZE.

**Atrial Natriuretic Factor** An atrial peptide hormone that may regulate blood pressure and electrolyte balance within the body. An example is a peptide hormone. See also HORMONE, ATRIAL PEPTIDES, PEPTIDE.

**Atrial Peptides** Endocrine components (proteins) that act to regulate blood pressure, as well as water and electrolyte homeostasis within the body. Atrial peptides are made by the heart in response to elevated blood pressure levels, and they stimulate the kidneys to excrete water and sodium into the urine, thus lowering blood pressure. They also slow the heartbeat. An example is a peptide hormone. See also ENDOCRINE HORMONES, HOMEOSTASIS, ELECTROLYTE.

**Attenuated (pathogens)** Inactivated, rendered harmless (e.g., killed viruses used to make a vaccine). Some of the ways in which viruses and other pathogens may be attenuated are by heat, chemical, or radiation treatment. See also PATHOGEN.

**Attenuation (of RNA)** Premature termination of an elongating RNA chain. See also RIBONUCLEIC ACID (RNA).

**Aureofacin** An antifungal antibiotic produced by a strain of *Streptomyces aureofaciens*. At least one company has incorporated the gene for this antibiotic (which acts against wheat take-all disease) into a *Pseudomonas fluorescens* used to confer resistance to wheat take-all disease by allowing the bacteria to colonize the wheat's roots. In this way the plant obtains the benefits of the antibiotic because the bacteria become part of the plant. See also *PSEUDOMONAS FLUORESCENS*, ENDOPHYTE, ANTIBIOTIC, *BACILLUS THURINGIENSIS* (*B.t.*).

**Autogenous Control** The action of a gene product (a molecule) that either inhibits (negative autogenous control) or activates (positive autogenous control) expression of the gene that codes for it (Greek *auto*, self). The presence of the product either causes or stops its own production. See also GENE, EXPRESS.

**Autoimmune Disease** A disease in which the body produces an immunogenic (immune system) response to some constituent of its own tissue. In other words, the immune system loses its ability to recognize some tissue or system within the body as "self" and targets and attacks it as if it were foreign. Autoimmune diseases can be classified into those in which one organ is predominantly affected (e.g., hemolytic anemia and chronic thyroiditis), and those in which the autoimmune disease process is diffused through many tissues (e.g., multiple sclerosis, systemic lupus erythematosus, and rheumatoid arthritis).

For example, multiple sclerosis is thought to be caused by T cells attacking acetylcholine receptors in the sheaths (myelin) that surround the nerve fibers of the brain and spinal cord. This eventually results in loss of coordination, weakness, and blurred vision. Arthritis is caused by immune system cells attacking joint tissues. Certain bacterial infections (e.g., Lyme disease, *Salmonella*, etc.) are followed by arthritis in approximately 10% of cases. The antigen (on surface of those bacteria) targeted by the human immune system is similar (in its molecular shape) to a protein located on the surface of cells in human joint tissue(s). See also THYMUS, SUPERANTIGENS, T CELLS, TUMOR NECROSIS FACTOR (TNF), MULTIPLE SCLEROSIS, MYOELECTRIC SIGNALS, ACETYLCHOLINE, LUPUS, INSULIN-DEPENDENT DIABETES MELLITIS (IDDM), DIABETES, ANTIGEN, BACTERIA, *SALMONELLA TYPHIMURIUM*, PROTEIN, CELL.

**Autonomous Replicating Segment** See ARS ELEMENT.

**Autonomous Replicating Sequence** See ARS ELEMENT.

**Autoradiography** A technique to detect radioactively labeled molecules by creating an image on photographic film. The slab of gel or other material in which the molecules are held (suspended) is placed on top of a piece of photographic film. The two are then securely fastened together such that movement is eliminated and the film is exposed for a period of time. The exposed (to the radiation) film is subsequently developed and the radioactive area is seen as a dark (black) area. Among other uses, autoradiography has been used to track the spread of (radioactively labeled) viruses in a living plant. After treatment (the radioactive labeling process), the whole plant (in a slab) is placed on top of a piece of photographic film. When the film is subsequently developed, the picture seen is of a plant, with darker areas indicating regions of greater virus concentration. See also LABEL (RADIOACTIVE), VIRUS.

**Autosomes** All chromosomes except the sex chromosomes. A diploid cell has two copies of each autosome.

**Autotroph** An organism that can live on very simple carbon and nitrogen sources, such as carbon dioxide and ammonia. See also HETEROTROPH.

**Auxins** From the Greek *auxein*, to increase, this term refers to a family of chemical compounds that regulate plant growth (e.g., stimulate cell enlargement, cell division, initate roots/growth, flowering, etc.). See also CELL.

**Auxotroph** Auxotrophic mutant. A mutant defective in the synthesis of a given biomolecule. The biomolecule must be supplied to the organism if normal growth is to be achieved. See also MUTATION, GENE, GENE DELIVERY (GENE THERAPY), ESSENTIAL FATTY ACIDS.

**Avidin** A protein naturally present in egg white, oilseed protein (e.g., soybean meal), and grain (e.g., corn/maize), it is 70 kilodaltons in mass (weight) and has a high affinity for biotin (i.e., it "sticks" tightly to the biotin molecule). Since grain-eating insects require biotin (a B-complex vitamin) to live, adding extra avidin to grain (e.g., by inserting a gene to cause overproduction of avidin in the grain kernels) may be a way to protect grain from insects (e.g., weevils in stored corn/maize). See also PROTEIN, SOY PROTEIN, CORN, KILODALTON (KD), BIOTIN, WEEVILS, VITAMIN.

**Avidity** (of an antibody) The "tightness of fit" between a given antibody's combining site and the antigenic determinant with which it combines. The firmness of the combination of antigen with antibody. See also ANTIGENIC DETERMINANT, ANTIBODY, ANTIGEN, COMBINING SITE, POLYCLONAL RESPONSE, CATALYTIC ANTIBODY.

**Azadirachtin** The pharmacophore (active ingredient) in secretions of the tropical neem tree, which resists insect depradations. See also PHARMACOPHORE, NEEM TREE.

**Azurophil-Derived Bactericidal Factor (ADBF)** Potent antimicrobial protein produced by neutrophils (a type of white blood cell). See also LEUKOCYTES.

# B

β **Sitostanol** See BETA SITOSTANOL (β SITOSTANOL).

β**-conglycinin** See BETA-CONGLYCININ.

**B Cells** B lymphocytes. See also LYMPHOCYTE, B LYMPHOCYTES, BLAST CELL.

**B Lymphocytes** A class of white blood cells originating in the bone marrow and found in blood, spleen, and lymph nodes, they are the precursors of (blood) plasma cells (B cells) that secrete antibodies (IgG) directed against invading antigens (e.g., of pathogenic bacteria). Via a complex "gene splicing" process, the B cells of the human body are able to produce more than one billion different IgG antibodies (i.e., able to bind onto and neutralize a billion different antigens). See also ANTIGEN, ANTIBODY, BLAST CELL, LYMPHOCYTE, PATHOGEN, BACTERIA, GENE SPLICING, IMMUNO-GLOBULIN, ALLELIC EXCLUSION.

**B-DNA** A helical form of DNA. B-DNA can be formed by adding back water to (dehydrated) A-DNA. B-DNA is the form of DNA of which James Watson and Francis Crick first constructed their model in 1953. It is found in fibers of very high (92%) relative humidity and in solutions of low ionic strength. This corresponds to the form of DNA that is prevalent in the living cell. See also DEOXYRIBONUCLEIC ACID (DNA), A-DNA, ION, CELL.

**BAC** Acronym for Bacterial Artificial Chromosomes. See also BACTERIAL ARTIFICIAL CHROMOSOMES (BAC).

*Bacillus* Rod-shaped bacteria.

*Bacillus subtilis (B. subtilis)* A (rod-shaped) aerobic bacterium commonly used as a host in recombinant DNA experiments. During the 1990s, research showed that corn (maize) plant tissues infected with the endophyte *Bacillus subtilis* were less likely to become infected with *Fusarium moniliforme* fungus.

Other research has indicated the potential for prior infection of corn (maize) plant tissues to hinder any subsequent aflatoxin production in that plant by *Aspergillus flavus* fungus. See also BACTERIA, HOST VECTOR (HV) SYSTEM, DEOXYRIBONUCLEIC ACID (DNA), CORN, ENDOPHYTE, FUNGUS, *FUSARIUM MONILIFORME*, AFLATOXIN.

*Bacillus thuringiensis (B.t.)* Discovered by bacteriologist Ishiwata Shigetane on a diseased silkworm in 1901. Later discovered on a dead Mediterranean flour moth, and first named *Bacillus thuringiensis*, by Ernst Berliner in 1915. Today, *B. thuringiensis* refers to a group of rod-shaped soil bacteria found all over the earth, that produce "cry" proteins which are indigestible by — yet still "bind" to — specific insects' gut (stomach) lining (epithelium cell) receptors, so those "cry" proteins are thereby toxic to certain classes of insects (corn borers, corn rootworms, mosquitoes, black flies, some types of beetles, etc.), but are harmless to all mammals. At least 20,000 strains of *B. thuringiensis* are known. Genes that code for the production of these cry proteins that are toxic to insects have been inserted by scientists since 1989 into vectors (i.e., viruses, other bacteria, and other microorganisms) in order to confer insect resistance to certain agricultural plants (e.g., via expression of those *B.t.* proteins by one or more tissues of the transgenic plant). For example, the *B.t.* strain known as *B.t. kurstaki*, which is fatal when ingested by the European corn borer was first (genetically) inserted into a corn plant (via vector) in 1991. *B.t. kurstaki* kills borers via perforation of that insect's gut by cry ("crystal-like") proteins that are coded for by the *B.t. kurstaki* gene. The vectors as listed above are entities that can take up and carry the DNA into plant or other cells. Vectors are DNA-carrying vehicles. See also ENDOPHYTE, CORN, GENE, PSEUDOMONAS FLUORESCENS, AGROBACTERIUM TUMEFACIENS, AUREOFACIN, EUROPEAN CORN BORER (ECB), COWPEA TRYPSIN INHIBITOR (CpTI), PROTEIN, "SHOTGUN" METHOD, CODING SEQUENCE, *FUSARIUM*, VECTOR, EXPRESS, GENETIC ENGINEERING, "EXPLOSION" METHOD, BIOLISTIC® GENE GUN, CRY PROTEINS, CRY1A (b) PROTEIN, CRY1A (c) PROTEIN, CRY9C PROTEIN, *B.t. KURSTAKI*, *B.t. TENEBRIONIS*, *B.t. ISRAELENSIS*, *B.t. TOLWORTHI*, ION CHANNELS.

**Back Mutation** Reverse the effect of a mutation that had inactivated a gene, thus restoring wild phenotype. See also PHENOTYPE, MUTATION.

**Bacteria** From the Greek *bakterion*, stick, since the first bacteria viewed by man (via crude microscopes) appeared to be stick-shaped. Any of a large group of microscopic organisms having round, rod-like, spiral, or filamentous unicellular or noncellular bodies that are often aggregated into colonies, are enclosed by a cell wall or membrane (procaryotes), and lack fully differentiated nuclei. Bacteria may exist as free-living organisms in soil, water, and organic matter, or as parasites in the live bodies of plants and animals. See also BACTERIOLOGY.

**Bacterial Artificial Chromosomes (BAC)** Pieces of DNA (e.g., plant DNA) that have been cloned (made) inside living bacteria (e.g., by plant researchers who need to "manufacture" some pieces of plant DNA). They can be utilized as vectors (for genetic engineering), to carry (inserted) genes into certain organisms. Some potential uses of BACs include: the "manufacture" of probes (i.e., sequences of DNA utilized to "find" complementary sequences within large pieces of DNA) via hybridization; the "manufacture" of "DNA sequence markers" for use in marker assisted selection (e.g., to guide choices made by commercial crop breeders, so they can more quickly select plants bearing gene(s) for a particular trait) to develop future improved crop varieties faster than was previously possible. See also BACTERIA, CLONE (A MOLECULE), SYNTHESIZING (OF DNA MOLECULES), CHROMOSOMES, YEAST ARTIFICIAL CHROMOSOMES (YAC), HUMAN ARTIFICIAL CHROMOSOMES (HAC), PROBE, MARKER ASSISTED SELECTION, COMPLEMENTARY DNA (c-DNA), HYBRIDIZATION (MOLECULAR GENETICS), DEOXYRIBONUCLEIC ACID (DNA), SEQUENCE (OF A DNA MOLECULE), MARKER (DNA SEQUENCE), GENE, TRAIT, GENETIC ENGINEERING, VECTOR.

**Bacterial Expressed Sequence Tags** These are ESTs (expressed sequence tags) based on sequenced/mapped bacterial genes instead of the genes of ("traditional" EST) *C. elegans* nematode. They are utilized to "label" a given gene (i.e., in terms of that gene's function/protein). See also BEST, EXPRESSED SEQUENCE TAGS (EST), BACTERIA, SEQUENCING (OF DNA MOLECULES), SEQUENCE (OF A DNA MOLECULE), MAPPING, *CAENORHABDITIS ELEGANS* (*C. ELEGANS*).

**Bactericide** See MICROBICIDE, BIOCIDE, ANTIBIOTIC.

**Bacteriocide** See BACTERICIDE.

**Bacteriocins** Proteins produced by many types of bacteria that are toxic (primarily) to other closely related strains of the particular bacteria that produce those proteins. Bacteriocins hold promise (e.g., after genetic engineering of the DNA responsible for their production) for future possible use as food preservatives (i.e., acting against bacteria species that cause food spoilage). For example: the bacteriocin known as curvaticin 13, which is produced by *Lactobacillus curvatus* bacteria, inhibits the food-poisoning bacteria *Listeria monocytogenes*; the bacteriocin known as sakacin K, which is produced by *Lactobacillus sakei* bacteria, inhibits the food-poisoning bacteria *Listeria monocytogenes*. However, the effectiveness of both curvaticin 13 and sakacin K are lessened by the presence of salt (e.g., in processed meat products), so salt resistance would be a desired property that may some day be engineered into those bacteriocins. See also PROTEIN, BACTERIA, BACTERIOLOGY, *BIFIDUS*, STRAIN, TOXIN, GENETIC ENGINEERING, DEOXYRIBONUCLEIC ACID (DNA), CODING SEQUENCE, COLICINS, *LISTERIA MONOCYTOGENES*, EXTREMOPHILIC BACTERIA.

**Bacteriology** The science and study of bacteria, a specialized branch of microbiology. The bacteria constitute a useful and essential group in the biological community. Although some bacteria prey on higher forms of life, relatively few are pathogens (disease-causing organisms). Life on earth depends on the activity of bacteria to mineralize organic compounds and to capture the free nitrogen molecules in the air for use by plants. Also, bacteria are important industrially for the conversion of raw materials into products such as organic chemicals, antibiotics, cheeses, etc. Genetically engineered bacteria are starting to be used to produce high value-added pharmaceuticals and specialty chemicals. See also ESCHERICHIA COLIFORM (*E. COLI*).

**Bacteriophage** Discovered in 1917 by Felix d'Herelle (fr. bacteria eaters), a bacteriophage is a virus that attaches to, injects its DNA into, and multiplies inside bacteria, which causes bacteria to die. Often abbreviated as simply phage, another name for virus. As an example, bacteriophage lambda is commonly used as a vector in rDNA experiments in *Escherichia coli* and attaches to a specific receptor, which in the bacteria also normally functions in sugar transport across the cell wall. Viruses come in many shapes and sizes. See also ESCHERICHIA COLIFORM (*E. COLI*), RECEPTORS, VIRUS, TRANSDUCTION (GENE), TRANSDUCTION (SIGNAL), TRANSFECTION, LAMBDA PHAGE.

**Bacterium** See BACTERIA.

**Baculovirus** A class of virus that infects lepidopteran insects (e.g., cotton bollworm or gypsy moth larva). Baculoviruses can be modified via genetic engineering to insert new genes into the larva, causing those larva to then produce proteins desired by man (e.g., pharmaceuticals). Baculoviruses are potentially very useful for pharmaceutical production, because the protein molecules produced are glycosylated (i.e., have relevant oligosaccharides attached to them), and baculoviruses cannot infect vertebrate animals. Such pharmaceuticals are thus not even a theoretical risk to humans. See also VIRUS, GENETIC ENGINEERING, GENE, PROTEIN, GLYCOSYLATION, BACULOVIRUS EXPRESSION VECTORS (BEVs).

**Baculovirus Expression Vectors (BEVs)** Vectors (used by researchers to carry new genes into cells) in which the agent is a baculovirus (a virus that infects certain types of insects only). These could conceivably be used to make a genetically engineered insecticide that is specific to a targeted insect (wouldn't harm anything but that insect). For example, a BEV might be used to cause a cotton bollworm adult protein to be expressed when the bollworm is a juvenile, thus killing the bollworm before it has a chance to damage a cotton crop. See also BACULOVIRUS, VIRUS, VECTOR, GENE, PROTEIN, CELL, GENETIC ENGINEERING.

**Bakanae** See *FUSARIUM MONILIFORME*.

**BAR Gene** A dominant gene from the *Streptomyces hygroscopicus* bacterium, which codes for (causes production of) the enzyme phosphinothricin acetyl transferase (PAT). When the BAR gene is inserted into a plant's genome (its DNA), it imparts resistance to glufosinate-ammonium based herbicides. Because the glufosinate-ammonium herbicides act via inhibition of glutamine synthetase (an enzyme that catalyzes the synthesis of glutamine), this inhibition (of enzyme) kills plants (e.g., weeds). That is because glutamine is crucial for plants to synthesize critically needed amino acids. The BAR gene is often utilized by genetic engineers as a marker gene. See also GENE, GENOME, GENETIC ENGINEERING, MARKER (GENETIC MARKER), DOMINANT ALLELE, ESSENTIAL AMINO ACIDS, HERBICIDE-TOLERANT CROP, GTS, SOYBEAN PLANT, CANOLA, CORN, GLUTAMINE, GLUTAMINE SYNTHETASE, PHOSPHINOTHRICIN, PHOSPHINOTHRICIN ACETYLTRANSFERASE (PAT), PAT GENE.

**Barley** The domesticated plant *Hordeum vulgare*, whose grain is utilized by man for various purposes, such as feed barley varieties (for feeding of livestock). Malting barley varieties (containing beta-amylase in their seeds) were created via mutation breeding (i.e., bombardment of the seeds by ionizing radiation to cause random genetic mutations, followed by selection of the particular mutation in which maltose is produced by that barley plant in its seeds). See also TRADITIONAL BREEDING METHODS, MUTATION, MUTATION BREEDING, AMYLASE.

**Barnase** An enzyme that catalyzes destruction of nucleic acids (which thus kills the cell that the barnase is in). When the gene that codes for barnase is inserted via genetic engineering into a given plant and activated only in that plant's pollen (the barnase is produced only in its pollen cells), that plant's male parts become sterile. For crop plants possessing both male and female parts (monoecious plants), such male sterility facilitates the development of hybrids, because self-pollination does not occur. See also ENZYME, NUCLEIC ACIDS, CELL, GENE, GENETIC CODE, GENETIC ENGINEERING, GENETICS, HYBRIDIZATION (PLANT GENETICS), F1 HYBRIDS, MONOECIOUS.

**Base (general)** A substance with a pH in the range 7–14, which will react with an acid to form a salt. Mild bases normally taste bitter and feel slippery to the touch. See also ACID.

**Base (nucleotide)** A segment of the DNA (and RNA) molecules. One of the four (repeating) chemical units that comprise DNA/RNA that, according to their order and pairing (on the parallel strands of DNA/RNA molecules), represent the different amino acids (within the protein molecule that each gene in the DNA codes for). The four bases comprising DNA are adenine (A), cytosine (C), guanine (G), and thymine (T). See also DEOXYRIBONUCLEIC ACID (DNA), RIBONUCLEIC ACID (RNA), POLYMER, CODING SEQUENCE, CONTROL SEQUENCES, EXPRESSION, AMINO ACID, PROTEIN, GENE, ADENINE, CYTOSINE, GUANINE, THYMINE, URACIL, BASE PAIR (bp).

**Base Excision Sequence Scanning (BESS)** A method that can be utilized to detect a "point mutation" in DNA (via rapid DNA sequence scanning). See also BASE PAIR (bp), NUCLEOTIDE, DEOXYRIBONUCLEIC ACID (DNA), MUTATION, POINT MUTATION, EXCISION, SEQUENCING (OF DNA MOLECULES), SEQUENCE (OF A DNA MOLECULE).

**Base Pair (bp)** Two nucleotides that are in different nucleic acid chains and whose bases pair (interact) by hydrogen bonding. In DNA, the nucleotide bases are adenine (which pairs with thymine) and guanine (which pairs with cytosine). See also DEOXYRIBONUCLEIC ACID (DNA), GENETIC CODE, INFORMATIONAL MOLECULES.

**Base Substitution** Replacement of one base (within a DNA molecule) by another base.

See also BASE (NUCLEOTIDE), TRANSITION, TRANSVERSION.

**Basic Fibroblast Growth Factor (BFGF)** See FIBROBLAST GROWTH FACTOR (FGF).

**Basophilic** Staining strongly with basic dye. For example, basophil leukocytes are polymorphonuclear leukocytes which stain strongly with (take up a lot of) basic dyes. See also POLYMORPHONUCLEAR LEUKOCYTES (PMN).

**Basophils** Also called basophilic leukocytes. A type of white blood cell (leukocyte) produced by stem cells within the bone marrow that synthesizes and stores histamine and also contains heparin. When two IgE molecules of the same antibody "dock" at adjacent receptor sites on a basophil cell, the two IgE molecules capture an allergen between them. A chemical signal is sent to the basophil causing the basophil cell to release histamine, serotonin, bradykinin, and "slow-reacting substance." Release of these chemicals into the body causes the blood vessels to become more permeable, which consequently causes the nose to run. These chemicals also cause smooth muscle contraction, resulting in sneezing, coughing, wheezing, etc. See also MAST CELLS, ANTIGEN, ANTIBODY, HISTAMINE, WHITE BLOOD CELLS, BASOPHILIC, LEUKOCYTES, POLYMORPHONUCLEAR LEUKOCYTES (PMN), STEM CELLS.

**BB T.I.** See TRYPSIN INHIBITORS.

**BBB** See BLOOD-BRAIN BARRIER (BBB).

**Bce4** The name of a promoter (region of DNA) that controls/enhances an oilseed plant's gene(s) that code for components (e.g., fatty acids, amino acids, etc.) of that plant's seeds. For example, the Bce4 promoter causes such genes to be expressed during one of the earliest stages of canola plant's seed production. See also PROMOTER, DEOXYRIBONUCLEIC ACID (DNA), GENE, POLYGENIC, PLASTID, EXPRESS, CANOLA, SOYBEAN PLANT, TRANSCRIPTION.

**Bcr-Abl Gene** The gene (SNP) that causes the blood cancer chronic myelocytic leukemia (CML) in humans that possess it. See also GENE, SINGLE-NUCLEOTIDE POLYMORPHISMS (SNPs), CANCER, GLEEVEC™.

**BESS Method** See BASE EXCISION SEQUENCE SCANNING (BESS).

**BESS T-Scan Method** See BASE EXCISION SEQUENCE SCANNING (BESS).

**Best Linear Unbiased Prediction (BLUP)** A statistical (data) technique employed by livestock breeders to determine the breeding (genetic trait) value of animals in a breeding program. See also GENETICS, TRAIT, PHENOTYPE, GENOTYPE, EXPECTED PROGENY DIFFERENCES (EPD).

**Beta Carotene** A phytochemical (vitamin precursor) that is naturally produced in carrots, other orange vegetables, and in the endosperm portion of the corn (maize) kernel. If the corn kernel seed coat is torn (e.g., via insect chewing), the beta carotene inhibits growth of *Aspergillus flavus* fungi in the endosperm region of the kernel. In 1970, an orange (-fruited) cauliflower was discovered growing in a field in Canada. It was the result of a natural mutation that caused beta carotene to be produced in that cauliflower plant, at a level that was several hundred times higher than normal for cauliflower. Beta carotene has been found to aid eyesight in people who consume it, and may help prevent lung cancer and heart disease. Because beta carotene is processed into vitamin A by the human body, consumption of this phytochemical can help avoid human diseases (e.g., in developing countries where vitamin A is scarce) that result from vitamin A deficiency, e.g., coronary heart disease, certain cancers (cancer of prostate, lungs, etc.), childhood blindness, macular degeneration (a leading cause of blindness in older people), and various childhood diseases which often result in death due to a weakened immune system. See also VITAMIN, GOLDEN RICE, AFLATOXIN, FUNGUS, OH43, PHYTOCHEMICALS, NUTRACEUTICALS, CAROTENOIDS, CANCER, CORONARY HEART DISEASE (CHD), ANTIOXIDANTS, DESATURASE.

**Beta Cells** Insulin-producing cells in the pancreas. If these cells are destroyed, childhood (also known as early-onset or Type I) diabetes results. See also ISLETS OF LANGERHANS, INSULIN, TYPE I DIABETES.

**Beta Conformation** An extended, zigzag arrangement of a polypeptide (molecule) chain. See also POLYPEPTIDE (PROTEIN).

**Beta Interferon** One of the interferons, it is a protein that was approved by the U.S. Food and Drug Administration (FDA) in 1993 to be used to treat multiple sclerosis (MS). See also INTERFERONS, FOOD AND DRUG ADMINISTRATION (FDA), PROTEIN.

**Beta Oxidation** See CARNITINE.

**Beta Sitostanol** See SITOSTANOL.

**Beta Sitosterol** See SITOSTEROL.

**Beta-conglycinin** Abbreviated β-conglycinin. One of the (structural) categories of proteins produced in seeds of legumes. In general, β-conglycinin contains one-quarter to one-third as much cysteine (cys) and methionine (met) per unit of protein as does glycinin. β-conglycinin has greater emulsifying capacity (in water) and emulsion stability than does glycinin, so its presence can assist the manufacture of firmer tofu, and better protein-based (emulsion) drinks. See also PROTEIN, CYSTEINE (cys), METHIONINE (met), GLYCININ, EMULSION.

**Beta-D-Glucouronidase** See GUS GENE.

**Beta-Glucan** See WATER SOLUBLE FIBER.

**Beta-lactam Antibiotics** A category of antibiotics (e.g., penicillin G, ampicillin, etc.) that kill targeted bacteria by altering their essential cellular function of enzymatic controls that keep cell wall (peptido-glycan) synthesis (creation/repair) in balance with cell wall degradation. This causes cell wall breakdown and death of those bacteria (pathogens). See also ANTIBIOTIC, PENICILLIN G, BACTERIA, CELL, ENZYME, PATHOGEN, bla GENE.

**Beta-Secretase** An enzyme that (in the human brain) is linked to presence of Alzheimer's disease. See also ENZYME, ALZHEIMER'S DISEASE, AMYLOID β PROTEIN PRECURSOR (AβPP).

**BEVs** See BACULOVIRUS, BACULOVIRUS EXPRESSION VECTORS (BEVs).

**BFGF** Basic Fibroblast Growth Factor. See also FIBROBLAST GROWTH FACTOR (FGF).

**BGYF** See BRIGHT GREENISH-YELLOW FLUORESCENCE (BGYF).

**Bifidobacteria** See BIFIDUS.

*Bifidus* A "family" of bacteria species that live within the digestive systems of certain animals (humans, swine, etc.). Examples include *Bifidobacterium bifidum*, *Bifidobacterium longum*, *Bifidobacterium infantis*, *Bifidobacterium adolescentis*, and *Bifidobacterium acidophilus*. In general, *Bifidus* bacteria help to promote good health of the host animals, by several means.

They produce organic acids (e.g., propionic, acetic, lactic), which make the host animal's digestive system more acidic. Because most pathogens (disease-causing microorganisms) grow best at a neutral pH (neither acidic nor base/caustic), the growth rates of pathogens are thereby inhibited. They "crowd out" enteric pathogens, since *Bifidus* bacteria grow fast in the acidic environment created by those organic acids. Some of the organic acids (e.g., propionic) produced by *Bifidus* bacteria are able to pass through the outer cell membrane of pathogenic bacteria and fungi; once inside those pathogens' cells, these acids dissociate and acidify the cell interior (which disrupts protein synthesis, growth, and replication of that pathogen). They produce bacteriocins, which are proteins that suppress growth of the pathogenic bacteria. They produce certain short-chain fatty acids, which are absorbed by the host animal (e.g., in the colon) and thereby result in a reduction of triglycerides (fat) levels in the host animal's bloodstream. That triglyceride reduction lowers the risk of coronary heart disease and thrombosis. See also BACTERIA, SPECIES, ACID, BASE (GENERAL), PATHOGEN, CELL, PLASMA MEMBRANE, MICROORGANISM, FUNGUS, PROTEIN, RIBOSOMES, GROWTH (MICROBIAL), FRUCTOSE OLIGOSACCHARIDES, FATTY ACID, TRIGLYCERIDES, CORONARY HEART DISEASE (CHD), THROMBOSIS, PREBIOTICS, BACTERIOCINS, INSULIN, TRANSGALACTO-OLIGOSACCHARIDES.

**Bile** A liquid (mixture) made by the liver to help digest fats (in the intestine) and facilitate intestinal absorption of certain vitamins and minerals. Bile consists primarily of water, cholesterol, lipids (fat), "natural detergents" (i.e., salts of bile acids) that help break up fat globules in the intestines, and bilirubin. See also BILE ACIDS, BILIRUBIN, FATS, DIGESTION (WITHIN ORGANISMS).

**Bile Acids** A "family" of acids derived by the human liver from cholesterol (i.e., from foods), and excreted into the bile by the liver. They help to emulsify (food-source) fats in the small intestine, as part of the crucial first step in the digestion of fats. See also CHOLESTEROL, DIGESTION (WITHIN ORGANISMS), LECITHIN, FATS, LIPIDS.

**Bilirubin** A component (pigment) of red blood cells (i.e., erythrocytes), that is recovered (from old red blood cells) and recycled into making bile (a liquid that aids the digestive process) by the liver. See also ERYTHROCYTES, BILE, DIGESTION (WITHIN ORGANISMS), ENDOTHELIUM.

**BIO** See BIOTECHNOLOGY INDUSTRY ORGANIZATION (BIO).

**Bioassay** Determination of the relative strength or bioactivity of a substance (e.g., a drug). A biological system (such as living cells, organs, tissues, or whole animals) is exposed to the substance in question and the effect on the living test system is measured. See also BIOLOGICAL ACTIVITY, ASSAY, BIOCHIP.

**Biochemistry** The study of chemical processes that comprise living things (systems); the chemistry of life and living matter. Despite the dramatic differences in the appearances of living things, the basic chemistry of all organisms is strikingly similar. Even tiny one-celled creatures carry out essentially the same chemical reactions that each cell of a complex organism (such as man) carries out. See also MOLECULAR BIOLOGY, MOLECULAR DIVERSITY.

**Biochip** A term first used with regard to an electronic device that utilizes biological molecules as the "framework" for other molecules acting as semiconductors and functioning as an integrated circuit.

1. During the 1990s, this term also became commonly used to refer to various "laboratories on a chip" to:
   - Analyze very small samples of DNA
   - Assess the impact of pharmaceuticals — or pharmaceutical drug candidate molecules — on specific cells (i.e., attached to the biochip's surface) or on specific cellular receptors (ligand-receptor response of cell)
   - Size and sort DNA fragments (genes) via the (proportional) fluorescence of dyes intercalated in the DNA molecules
   - Detect presence of specific DNA fragments (genes) via hybridization

to a probe (that was fabricated onto the chip)

- Size and sort protein molecules (via various cells fabricated onto the chip)
- Assess pharmaceuticals via adhesion molecules attached to the chip
- Detect specific pathogens or cancerous cells in a blood sample (e.g., by applying controlled electrical fields to cause those cells to collect at electrodes on the chip)
- Screen for compounds that act against a disease (e.g., by applying antibodies linked to fluorescent molecules, then measuring electronically the fluorescence triggered by antibody-binding)
- Conduct gene expression analysis by measuring the fluorescence of messenger RNA (specific to which particular gene is "turned on") when that mRNA hybridizes with DNA (from genome) on hybridization surface on the chip

2. Shortly after the 1990s, several companies manufactured biochips capable of sequencing (determining the sequence of) DNA samples. Such biochips have, attached to their surfaces, all possible "DNA probes" (short sequences of DNA). The sample (i.e., the unknown DNA molecule) is passed over the probe-covered surface of the biochip, where each relevant segment (within the large unknown DNA molecule) hybridizes ("pairs") with the short "DNA probe" attached to a known location on the surface of the biochip. Because the sequence of each DNA probe — at each specified location on the biochip — is known, that information (i.e., the probes' sequences to which the unknown DNA molecule hybridized) is then used to "assemble the complete sequence" of the unknown DNA molecule.

3. Sometimes refers to an electronic device that uses biological molecules as the framework for other molecules that act as semiconductors and function as an integrated circuit. The future working

parts of the science of bioelectronics, biochips may consist of two- or three-dimensional arrays of organic molecules used as switching or memory elements. If biochip technology proves to be feasible, one application will be to shrink currently existing biosensors in size. This would enable the biosensors to be implanted in the body or in organs and tissues for the sake of monitoring and controlling certain bodily functions. A future possibility is to try to provide sight for the blind using light-sensitive (e.g., protein-covered electrode) biochips implanted in the eyes to replace a damaged retina. For example, during 2001, Alan Chow implanted such biochips into several men whose retinas had been damaged by the disease retinitis pigmentosa.

See also BIOELECTRONICS, BIONICS, BIOSENSORS (ELECTRONIC), DEOXYRIBONUCLEIC ACID (DNA), RIBONUCLEIC ACID (RNA), GENE, RECEPTORS, HIGH-THROUGHPUT SCREENING (HTS), BIOINORGANIC, TARGET-LIGAND INTERACTION SCREENING, ANTIBODY, CHARACTERIZATION ASSAY, BIOASSAY, ASSAY, LUMINESCENT ASSAY, PROTEIN, LIGAND (IN BIOCHEMISTRY), MICROFLUIDICS, PROBE, PROTEOMICS, PROTEOME CHIP, BIORECEPTORS, HYBRIDIZATION (MOLECULAR BIOLOGY), FLUORESCENCE, ADHESION MOLECULE, GENE EXPRESSION ANALYSIS, PATHOGEN, BIOINFORMATICS, MICROARRAY (TESTING), HYBRIDIZATION SURFACES, MESSENGER RNA (mRNA), GENOMICS, QUANTUM DOT, QUANTUM WIRE, NANOCOMPOSITES, SEQUENCING (OF DNA MOLECULES).

**Biocide** Any chemical or chemical compound that is toxic to living things (systems). Literally "biokiller" or killer of biological systems. Includes insecticides, bactericides, fungicides, etc. Most bactericides accomplish their task (killing bacteria) via massive lysis (disintegration) of bacteria cell walls (membranes). However, one (triclosan) kills bacteria by inhibiting enoyl-acyl protein reductase; a crucial enzyme utilized by bacteria in their synthesis of fatty acids. See also BACTERICIDE, MICROBICIDE, LYSIS, BACTERIA, CELL, FATTY ACID, ENZYME, PROTEIN, ESSENTIAL FATTY ACIDS, ESSENTIAL NUTRIENTS.

**Biodegradable** Describes any material that can be broken down by biological action (dissimilation, digestion, denitrification, etc.). The breakdown of material (e.g., animal carcasses, dead plants, even manmade chemicals) by microorganisms (bacteria, fungus, etc.). The biodegradation process is often assisted (i.e., first step) by the actions of animals and insects (e.g., feeding on dead carcasses, which breaks down those carcasses to make their materials more available for microorganisms to "feed" upon). For example, vultures and the yellow swallowtail butterfly often are the first to feed on the carcasses of dead alligators in the state of Florida, which helps make the alligator's material (body tissue) more readily available to microorganisms (e.g., in the dung excreted by those "first step" carcass feeders). See also DIGESTION (WITHIN ORGANISMS), MICROORGANISMS, BACTERIA, FUNGUS, GLYCOLYSIS, METABOLISM, NITRIFICATION.

**Biodesulfurization** The removal of organic and inorganic sulfur (a pollution source) from coal by bacterial and soil microorganisms. See also BIOLEACHING, BIORECOVERY, BIOSORBENTS.

**Biodiversity** Defined to be "the variability among living organisms from all sources including terrestrial, marine/aquatic and the complexes of which they are a part" by the Convention on Biological Diversity. See also CONVENTION ON BIOLOGICAL DIVERSITY.

**Bioelectronics** Also called biomolecular electronics. It is the field where biotechnology is crossed with electronics. The branch of biotechnology that deals with the electroactive properties of biological materials, systems, and processes, together with their exploitation in electronic devices. Bioelectronics will attempt to replace traditional semiconductor materials (e.g., silicon or gallium arsenide) with organic materials such as proteins (biochips). See also BIOCHIPS, BIOSENSORS (ELECTRONIC), BIOINORGANIC, BIONICS, QUANTUM WIRE, SELF-ASSEMBLY (OF A LARGE MOLECULAR STRUCTURE).

**Biogenesis** The theory that living organisms are produced only by other living organisms. That is, the theory of generation from preexisting life. It is the opposite of abiogenesis, or spontaneous generation.

**Biogeochemistry** A branch of geochemistry that is concerned with biological materials and their relation to earth's chemicals in an area.

**Bioinformatics** This term refers to the generation/creation, collection, storage (in databases), and efficient utilization of data/information from genomics (functional genomics, structural genomics, etc.), combinatorial chemistry, high-throughput screening, proteomics, and DNA sequencing research efforts in order to accomplish a (research) objective (e.g., to discover a new pharmaceutical or a new herbicide). Examples of the data/information that are manipulated and stored include gene sequences, biological activity/function, pharmacological activity, biological structure, molecular structure, protein-protein interactions, and gene expression products/amounts/timing. See also GENOMICS, FUNCTIONAL GENOMICS, PHARMACOGENOMICS, STRUCTURAL GENOMICS, COMBINATORIAL CHEMISTRY, HIGH-THROUGHPUT SCREENING, PROTEOMICS, BIOCHIP, GENE, GENETIC MAP, GENETIC CODE, SEQUENCING (OF DNA MOLECULES), IN SILICO BIOLOGY, IN SILICO SCREENING, GENE EXPRESSION ANALYSIS, META-MODEL METHODS (OF BIOINFORMATICS).

**Bioinorganic** This term refers to the combination of organic (life) materials with inorganic materials to create (useful materials). For example, Abalone shellfish make their shells via a combination of protein and calcium carbonate. Researchers are working on making semiconductor devices (chips) containing peptides, etc. attached to silicon or gallium arsenide. See also PROTEIN, BIOCHIP, PEPTIDE, BIOSENSORS (ELECTRONIC), NANOCOMPOSITES.

**Bioleaching** The biomediated recovery of precious metals from their ores. In the recovery of gold, for example, the microorganism *T. ferroxidans* may be used to cause the gold to leach out of the ore so it may then be concentrated and smelted. Aluminum may be similarly bioleached from clay ores, using heterotropic bacteria and fungi. See also BIORECOVERY, BIOGEOCHEMISTRY, BACTERIA, BIOSORBENTS.

**Biolistic® Gene Gun** The word "biolistic" was coined from the words "biological" and "ballistic" (pertaining to a projectile fired from a gun). Used to shoot pellets that are coated with genes (for desired traits) into plant seeds or plant tissues, in order to get those plants to then express the new genes. The gun uses an actual explosive (.22 caliber blank) to propel the material. Compressed air or steam may also be used as the propellant. The Biolistic® Gene Gun was invented in 1983–1984 at Cornell University by John Sanford, Edward Wolf, and Nelson Allen. The gun and its registered trademark are now owned by E. I. du Pont de Nemours and Company. See also WHISKERS™, "SHOTGUN" METHOD, GENETIC ENGINEERING, GENE, BIOSEEDS, MICROPARTICLES.

**Biological Activity** The effect (change in metabolic activity upon living cells) caused by specific compounds or agents. For example, the drug aspirin causes the blood to thin, i.e., to clot less easily. See also BIOASSAY, PHARMACOPHORE, RETINOIDS.

**Biological Oxygen Demand (BOD)** The oxygen used in meeting the metabolic needs of aerobic organisms in water containing organic compounds. Numerically, it is expressed in terms of the oxygen consumed in water at a temperature of 68°F (20°C) during a 5-day period. The BOD is used as an indication of the degree of water pollution. See also METABOLISM.

**Biological Vectors** See VECTORS.

**Biology** From the two Greek words *bios* (life) and *logos* (word), it is the field of science encompassing the study of life. See also GENETICS, CLADISTICS, ORGANISM, SPECIES.

**Bioluminescence** The enzyme-catalyzed production of light by living organisms, typically during mating or hunting. This word literally means *living light*. First identified/analyzed in 1947 by William McElroy, bioluminescence results when the enzyme luciferase comes into contact with adenosine triphosphate (ATP)/luciferin, inside the photophores (organs which emit the light) of the organism. Such production of light by living organisms is exemplified by fireflies, South America's railroad worm, and by many deep ocean marine organisms. Bioluminescence has been utilized by man as a genetic marker (e.g., to cause a genetically engineered plant to glow as evidence that a gene was successfully transferred into that plant). Another use of bioluminescence by man is for the rapid detection of foodborne pathogenic bacteria (e.g., in a food processing factory). One rapid-test for bacteria uses two chemical reagents that first break down bacteria cell membranes, then cause the ATP from those broken cells to luminesce. Another rapid-test uses electrophoresis to first separate the sequences of bacteria's DNA (following its extraction from cell and enzymatic fragmentation), then cause those separated sequences to luminesce. A camera is used to record the sequence-pattern light emission and compare that pattern to patterns of pathogenic bacteria previously stored in a database. See also ENZYME, MARKER (GENETIC MARKER), BACTERIA, TOXIN, PATHOGENIC, *ESCHERICHIA COLIFORM* 0157:H7 (*E. COLI* 0157:H7), CELL, LUMINESCENT ASSAY, ADENOSINE TRIPHOSPHATE (ATP), GENETIC ENGINEERING, ELECTROPHORESIS, POLYACRYLAMIDE GEL ELECTROPHORESIS (PAGE), SEQUENCE (OF A DNA MOLECULE), PHOTORHABDUS LUMINESCENS RESTRICTION ENDONUCLEASES, NITRIC OXIDE.

**Biomass** All organic matter grown by the photosynthetic conversion of solar energy (e.g., plants) and organic matter from animals. See also PHOTOSYNTHESIS, LOW-TILLAGE CROP PRODUCTION, NO-TILLAGE CROP PRODUCTION.

**BioMEMS** Refers to MEMS designed to work within biological systems/organisms. Examples include microfluidic cell sorters, or a biochip possessing diverging nanometer-scale etched channels and a fluorescence detector. Via an electrical field that would drive electrophoretic separation of DNA (fragments), samples of DNA could be separated/sorted/identified by fluorescence. See also MEMS (NANOTECHNOLOGY), ORGANISM, ELECTROPHORESIS, MICROFLUIDICS, CELL SORTING, NANOMETERS (nm), FLUORESCENCE, BIOCHIP, NANOTECHNOLOGY.

**Biomimetic Materials** Synthetic (man-made) molecules or systems that are analogues of natural (made by living organisms) materials.

For instance, molecules have been synthesized by man that act chemically like natural proteins, but are not as easily degraded by the digestive system (as are those natural protein molecules). Other systems, such as reverse micelles and/or liposomes, exhibit certain properties that mimic certain aspects of living systems. See also PROTEIN, DIGESTION (WITHIN ORGANISMS), REVERSE MICELLE (RM), LIPOSOMES, ANALOGUE, BIONICS, BIOPOLYMER.

**Biomolecular Electronics** See BIOELECTRONICS.

**Biomotors** Refers to biologically based technologies/techniques used to "power" nanometer-size machines (e.g., "nanobots") in one way or another. For example, during 2000 Bernard Yurke and colleagues created a molecular-machine "tweezers" (grasper) consisting of three separate strands of DNA (two of them were hybridized separately to small complementary sequences near the two ends of the first DNA strand). The "tweezers" can then be closed (or opened) by sequentially adding other DNA strands (to the three) which can hybridize to small complementary sequences on second and third strands, or hybridize to the fourth strand, causing it to unhybridize from the second and third strands. See also NANO-TECHNOLOGY, BIOLOGY, NANOMETERS (NM), MOLECULAR MACHINES, DEOXYRIBONUCLEIC ACID (DNA), HYBRIDIZATION (MOLECULAR GENETICS), SEQUENCE (OF A DNA MOLECULE), COMPLEMENTARY (MOLECULAR GENETICS), SELF-ASSEMBLY (OF A LARGE MOLECULAR STRUCTURE).

**Bionics** An interscience discipline for constructing artificial systems that resemble or have the characteristics of living systems. Bionics can encompass (in whole, or in part) bioelectronics, biosensors, biomimetic materials, biophysics, biomotors, and self-assembly (of a large molecular structure). See also BIOLOGY, BIOELECTRONICS, BIOMIMETIC MATERIALS, BIOSENSORS (ELECTRONIC), BIOPHYSICS, BIOMOTORS.

**Biophysics** An area of scientific study in which physical principles, physical methods, and physical instrumentation are used to study living systems or systems related to life. It overlaps with biophysical chemistry, which is more specialized in scope since it is concerned with the physical study of chemically isolated substances found in living organisms.

**Biopolymer** A high molecular weight organic compound found in nature, whose structure can be represented by a repeated small unit [i.e., monomer (links)]. Common biopolymers include cellulose (long-chain sugars found in most plants and the main constituent of dried woods, jute, flax, hemp, cotton, etc.) and proteins in general, and specifically collagen and gelatin. See also MOLECULAR WEIGHT, PROTEIN, POLYMER.

**Bioreceptors** Refers to fragments of DNA, antibodies, protein molecules, and cellular probes (e.g., adhesion molecule) when those are attached to a man-made surface (e.g., biochip) for purposes of analyzing biological substances. See also HYBRIDIZATION SURFACES, BIOCHIPS, ANTIBODY, DEOXYRIBONUCLEIC ACID (DNA), PROTEIN, ADHESION MOLECULE, ORPHAN RECEPTORS, MICROARRAY (TESTING).

**Biorecovery** The use of organisms (including bacteria, plants, fungi, and algae) in the recovery (collecting) of various metals and/or organic compounds from ores or garbage (other matrices). See also BIOLEACHING, CONSORTIA, BIOSORBENTS, PHYTOREMEDIATION, METABOLIC ENGINEERING, BACTERIA, FUNGUS.

**Bioremediation** The use of organisms (plants, bacteria, fungi, etc.) to consume or otherwise help remove (biorecovery) materials (toxic chemical wastes, metals, etc.) from a contaminated site (e.g., the land and ponds on the site of an old refinery). See also BIORECOVERY, PHYTOREMEDIATION, METABOLIC ENGINEERING, BIOLEACHING, BIODESULFURIZATION, ORGANISM, BACTERIA, FUNGUS.

**Biosafety** See CONVENTION ON BIOLOGICAL DIVERSITY (CBD).

**Biosafety Protocol** See CONVENTION ON BIOLOGICAL DIVERSITY (CBD), INTERNATIONAL PLANT PROTECTION CONVENTION (IPPC).

**Bioseeds** Plant seeds produced via genetic engineering of existing plants. See also GENETIC ENGINEERING, BIOLISTIC® GENE GUN, HERBICIDE-TOLERANT CROP, PAT GENE, EPSP SYNTHASE, ALS GENE, CP4 EPSPS, GLYPHOSATE OXIDASE, CHOLESTEROL OXIDASE, HIGH-LYSINE CORN, ACURON™ GENE, HIGH-METHIONINE CORN, HIGH-PHYTASE CORN AND SOYBEANS, HIGH-STEARATE SOYBEANS, LOW-STACHYOSE SOYBEANS,

LOX NULL, PLANT'S NOVEL TRAI(PNT), "SHOT-GUN" METHOD [TO INTRODUCE FOREIGN (NEW) GENES INTO PLANT CELLS], *BACILLUS THURINGIEN-SIS* (*B.t.*), *B.t. KURSTAKI, B.t. TENEBRIONIS, B.t. ISRAELENSIS*, CRY PROTEINS, CRY1A (b) PROTEIN, CRY1A (c) PROTEIN, CRY9C PROTEIN.

**Biosensors (chemical)** Chemically based devices that are able to detect and/or measure the presence of certain molecules (DNA, antigens, pesticides, etc.). These devices are currently created in the following forms:

1. A two-part diagnostic test that can detect the presence of trace amounts of specific chemicals (e.g., pesticides). The (chemical) biosensor consists of an immobilized enzyme (to bind the trace chemical) combined with a color reagent (to indicate visually the presence of the trace chemical).

2. A one-part test that can detect specific DNA segments in complex ("dirty," multiple component) samples. The biosensor consists of 13-nm gold particles onto which are attached numerous nucleotide molecular chains. Each nucleotide chain contains 28 nucleotides. The 13 nucleotides that are closest to each gold particle serve as a spacer, and solutions containing such (spaced) randomly distributed gold particles appear red in color when illuminated by light.

The 15 nucleotides that are farthest from each gold particle are chosen to be complementary to, and thus bind to, nucleotide sequences in the target (e.g., DNA) molecule. In the presence of the specific target molecule, a closely linked network of gold particles and double-stranded nucleotide molecular chains forms (overcoming the 13-nucleotide "spacer" which previously held the gold particles apart). When double-stranded chains form (i.e., target molecule is present), the distance between gold particles becomes less than the size of those particles, making the solution containing (bound) particles appear blue in color when illuminated by light. See also ENZYME, IMMUNOASSAY,

NANOCRYSTAL MOLECULES, NANOTECHNOLOGY, DEOXYRIBONUCLEIC ACID (DNA), NANOMETERS (nm), ANTIGEN, SEQUENCE (OF A DNA MOLECULE), NUCLEOTIDE, POLYMER, COMPLEMENTARY DNA (c-DNA), DOUBLE HELIX, DUPLEX, SELF-ASSEMBLY.

**Biosensors (electronic)** Electronic sensors that are able to detect and measure the presence of biomolecules such as sugars or DNA segments. Currently created by:

1. Fusing organic matter (e.g., enzymes, antibodies, receptors, or nucleic acids) to tiny electrodes; yielding devices that convert natural chemical reactions into electric current to measure blood levels of certain chemicals (e.g., glucose or insulin), control functions in an artificial organ, monitor some industrial processes, act as a robot's "nose," etc.

2. Fusing organic matter (segment of DNA, antibody, enzyme, etc.) onto the surfaces of etched silicon wafers; yielding devices that convert supramolecular interactions [e.g., nucleotide hybridization, enzyme-substrate binding, lectin-carbohydrate (sugar) interactions, antibody-antigen binding, host-guest complexation, etc.] into electric current via a charge-coupled device (CCD) detector. The CCD detector measures the shift in interference pattern caused by change in refractive index that results when the (sensed) molecule tightly binds to the fused (electronic) organic matter. For such an etched-silicon-wafer biosensor, the nucleotide hybridization (binding) enables the detection of femtomolar ($10$-$15$ mole or $0.000000000000001$) concentrations of DNA. If the (sensed) DNA segment is not complementary to the fused DNA segment, there is no significant change in the interference pattern.

A major goal is to build future generations of biosensors directly into computer chips. (Researchers have discovered that proteins can replace certain metals in semiconductors.) This would enable low-cost mass production via processes similar to those now

**B**

used for existing semiconductor chips, with circuits built right into the sensor to process data picked up by the biological matter on the chip. See also BIOCHIPS, QUARTZ CRYSTAL MICROBALANCES, BIOELECTRONICS, ENZYME, GENOSENSORS, RECEPTORS, ANTIBODY, BIOINORGANIC, INSULIN, COMBINATORIAL CHEMISTRY, SUBSTRATE (CHEMICAL), LECTINS, SUGAR MOLECULES, CARBOHYDRATES (SACCHARIDES), GLUCOSE (GLc), DEOXYRIBONUCLEIC ACID (DNA), NUCLEOTIDE, HYBRIDIZATION (MOLECULAR GENETICS), HYBRIDIZATION SURFACES, ANTIGEN, COMPLEMENTARY DNA (C-DNA), GENE, NANOTECHNOLOGY, TEMPLATE.

**Biosilk** A biomimetic, man-made fiber produced by:

1. Sequencing the "dragline silk" protein that is produced by the orb-weaving spider
2. Synthesizing genes to code for the "dragline silk" protein (components)
3. Expressing those genes in a suitable host (i.e., yeast, bacteria) to cause production of the protein(s)
4. Dissolving the protein in a solvent, and then "spinning" the protein into fiber form by passing the liquid (dissolved protein) through a small orifice, followed by drying to remove the solvent

See also BIOMIMETIC MATERIALS, BIOPOLYMER, PROTEIN, SEQUENCING (OF PROTEIN MOLECULES), GENE, GENE MACHINE, SYNTHESIZING (OF DNA MOLECULES), DEOXYRIBONUCLEIC ACID (DNA), EXPRESS, SUPERCRITICAL CARBON DIOXIDE.

**Biosorbents** Microorganisms which, either by themselves or in conjunction with a support/substrate system (e.g., inert granules) effect the extraction (e.g., from ore) and/or concentration of desired (precious) metals or organic compounds by means of selective retention of those entities. Retention of organic compounds (e.g., gasoline) may be for the purpose of cleaning polluted soil. See also BIORECOVERY, BIOLEACHING, CONSORTIA.

**Biosphere** All the living matter on or in the earth, the oceans and seas, and the atmosphere. The area of the planet in which life is found to occur.

**Biosynthesis** Production of a chemical compound or entity by a living organism.

**Biotechnology** The means or way of manipulating life forms (organisms) to provide desirable products for man's use. For example, beekeeping and cattle breeding could be considered to be biotechnology-related endeavors. The word biotechnology, coined in 1919 by Karl Ereky, applies to the interaction of biology with human technology. However, usage of the word biotechnology in the U.S. has come to mean all parts of an industry that knowingly create, develop, and market a variety of products through the willful manipulation, on a molecular level, of life forms, or utilization of knowledge pertaining to living systems. A common misconception is that biotechnology refers only to recombinant DNA (rDNA) work. However, recombinant DNA is only one of the many techniques used to derive products from organisms, plants, and parts of both for the biotechnology industry. A list of areas covered by the term biotechnology would more properly include: recombinant DNA, plant tissue culture, rDNA or gene splicing, enzyme systems, plant breeding, meristem culture, mammalian cell culture, immunology, molecular biology, fermentation, and others. See also GENETIC ENGINEERING, BIORECOVERY, RECOMBINANT DNA (rDNA), RECOMBINATION, DEOXYRIBONUCLEIC ACID (DNA), BIOLEACHING, GENE SPLICING, MAMMALIAN CELL CULTURE, FERMENTATION.

**Biotechnology Industry Organization (BIO)** An American trade association composed of companies and individuals involved in biotechnology and in services to biotechnology companies (accounting, law, etc.). Formed in 1993, the BIO was created by the merger of its two predecessor trade associations — the Association of Biotechnology Companies (ABC) and the Industrial Biotechnology Association (IBA). The BIO works with the government and the public to promote safe and rational advancement of genetic engineering and biotechnology. See also BIOTECHNOLOGY, ASSOCIATION OF BIOTECHNOLOGY COMPANIES (ABC), INDUSTRIAL BIOTECHNOLOGY ASSOCIATION (IBA), JAPAN BIOINDUSTRY ASSOCIATION, SENIOR ADVISORY GROUP ON BIOTECHNOLOGY (SAGB).

**B**

**Biotic Stresses** The stress (e.g., to crop plants) caused by insects, bacteria, viruses, fungi, nematodes, or other living things that attack plants. See also NEMATODES, FUNGUS, VIRUS, BACTERIA.

**Biotin** A B-complex vitamin, also known as vitamin H, which is essential (required) for life of many grain-eating insects as well as for many of the metabolic pathways (series of chemical reactions) involved in milk production by cattle. All of the predominant cellulolytic bacteria (i.e., those that breakdown cellulose molecules) within the rumen (first stomach) of cattle require biotin for them to be able to grow. Biotin (within certain molecules) acts as a coenzyme in carboxylation reactions, thereby playing a critical role in gluconeogenesis, fatty acid synthesis (manufacture), and protein synthesis reactions occurring within all animals. Biotin enzymes are inhibited (blocked) by the protein avidin. Since insects must have biotin to live, avidin might be a useful ingredient to add to grain in order to protect it from insects such as weevils during storage. See also VITAMIN, METABOLISM, INTERMEDIARY METABOLISM, PATHWAY, BACTERIA, CELLULOSE, LYSIS, ENZYME, COENZYME, WEEVILS, GLUCONEOGENESIS, FATTY ACID, PROTEIN, AVIDIN.

**Biotransformation (of a biosynthesized product)** See POSTTRANSLATIONAL MODIFICATION OF PROTEIN.

**Biotransformation (of an introduced compound)** Biological portion of definition of *persistence*. See also PERSISTENCE.

**bla Gene** A gene that confers resistance to β-lactam (beta-lactam) antibiotics (e.g., ampicillin). See also GENE, BETA-LACTAM ANTIBIOTICS, MARKER (GENETIC MARKER).

**Black-layered (corn)** An indicator of a corn plant's maturity. It refers to a distinctive dark line that forms in each corn kernel at maturity. See also CORN.

**Black-lined (corn)** See BLACK-LAYERED (CORN).

**Blast Cell** A large, rapidly dividing cell that develops from a B cell (B lymphocyte) in response to an antigenic stimulus. The blast cell then becomes an antibody-producing plasma cell. See also ANTIGEN, ANTIBODY, B LYMPHOCYTES, LYMPHOCYTE.

**Blast Transformation** The process through which a B cell (B lymphocyte) becomes a blast cell. See also ANTIBODY, LYMPHOCYTE, BLAST CELL.

**Blood Clotting** See FIBRIN.

**Blood Derivatives Manufacturing Association** A trade organization of firms involved in producing pharmaceuticals from collected blood. See also SERUM, BUFFY COAT (CELLS), SEROLOGY.

**Blood Plasma** See PLASMA.

**Blood Platelets** See PLATELETS.

**Blood Serum** See SERUM.

**Blood-Brain Barrier (BBB)** The specialized layer of endothelial cells that line all blood vessels in the brain. The BBB prevents most organisms (e.g., bacteria) and toxins from entering the brain via the bloodstream while allowing the passage of oxygen and needed nutrients (iron, glucose, tryptophan, etc.). For example, receptors that line BBB cell surfaces (on the bloodstream side of the BBB) "latch onto" transferrin molecules (which contain iron molecules) as those transferrin molecules pass by in the bloodstream. These transferrin receptors first bind to the (passing) transferrin molecules, transport them through the BBB via a process called vaginosis, then release them (in order to supply needed iron to the brain cells). Factors such as aging, trauma, stroke, multiple sclerosis, and some infections will cause an increase in the permeability of the BBB. See also ENDOTHELIAL CELLS, TOXIN, TRANSFERRIN, TRANSFERRIN RECEPTOR, CHELATING AGENT, GLUCOSE, RECEPTORS, VAGINOSIS, HEME, BACTERIA, TRYPTOPHAN (trp), SEROTONIN.

**Blunt-End DNA** A segment of DNA that has both strands terminating at the same base-pair location, that is, fully base-paired DNA. No sticky ends. See also STICKY ENDS.

**Blunt-End Ligation** A method of joining blunt-ended DNA fragments using the enzyme T4 ligase, which can join fully base-paired, double-stranded DNA. See also LIGASE, DEOXYRIBONUCLEIC ACID (DNA), BASE PAIR (bp), BLUNT-END DNA.

**BLUP** See BEST LINEAR UNBIASED PREDICTION (BLUP).

**BOD** See BIOLOGICAL OXYGEN DEMAND (BOD).

**Boletic Acid** See FUMARIC ACID ($C_4H_4O_4$).

**Bollworms** See *HELIOTHIS VIRESCENS* (*H. VIRE-SCENS*), *HELICOVERPA ZEA* (*H. ZEA*), *PECTINOPHORA GOSSYPIELLA*, *B.t. KURSTAKI*.

**Bone Morphogenetic Proteins (BMP)** A family of proteinaceous growth factors (nine identified as of 1994) for bone tissue formation (e.g., at the site where a bone has been broken). BMPs stimulate a "recruitment" of bone-forming cells (to the site of bone injury) which first form cartilage, then mineralize that cartilage to form bone. See also GROWTH FACTOR, PERIODONTIUM, PROTEIN.

**Bovine Somatotropin (BST)** Also called bovine growth hormone. A protein hormone, produced in a cow's pituitary gland, that increases the efficiency of the cow in converting its feed into milk. Increases milk production, and promotes cell growth in healing tissues of all ages of cattle. Promotes body growth of young cattle. See also PROTEIN, GROWTH HORMONE (GH), HORMONE, SOMATOMEDINS, SPECIES SPECIFIC.

**Bowman-Birk Trypsin Inhibitor** See TRYPSIN INHIBITORS.

**bp** Common abbreviation for base pair. See also BASE PAIR (bp).

*Brassica* A fast-growing category of the mustard plant family, which also produces sulfur-based gases (a natural defense against certain fungi, nematodes, and insect pests). For example, Australian CSIRO scientists discovered in 1994 that sulfur-based isothiocyanates emitted by *Brassica* actively combat Wheat Take-All Disease (a fungal disease that attacks the roots of the wheat plant). See also ARABIDOPSIS THALIANA, WHEAT, WHEAT TAKE-ALL DISEASE, CANOLA, ALLELOPATHY, FUNGUS, NEMATODES.

*Brassica campestre* See *BRASSICA*.

*Brassica campestris* See CANOLA, *BRASSICA*.

*Brassica napus* See CANOLA, *BRASSICA*.

**BRCA Genes** Oncogenes that, when mutated, can cause development of breast cancer or ovarian cancer. All humans possess BRCA genes of one sort or another (the acronym BRCA stands for breast cancer). However, the two specific BRCA genes most likely to lead to breast cancer (BRCA 1 and BRCA 2) are present in only two percent of women who are of Northern European ancestry, most Caucasian women in the U.S., and

Askenazi Jews whose ancesters are from Central and Eastern Europe. Those women possessing the BRCA 1 gene in their genome (DNA) have a 20–40% chance of developing ovarian cancer (and a 50–85% chance of developing breast cancer) in their lifetime. Those women possessing the BRCA 2 gene in their genome (DNA) have a 15–20% chance of developing ovarian cancer (and a 55–85% chance of developing breast cancer) in their lifetimes. See also GENE, MUTATION, CANCER, ONCOGENES, HER2 GENE.

**BRCA 1 Gene** See BRCA GENES.

**BRCA 2 Gene** See BRCA GENES.

**Breeder's Rights** See PLANT BREEDER'S RIGHTS.

**Bright Greenish-Yellow Fluorescence (BGYF)** An indication of the presence of fungus (e.g., in a sample of grain), when light of an appropriate wavelength is shone on sample. For example, when the fungus *Aspergillus flavus* infects cottonseed during boll development on the cotton plant, the resultant seed (when harvested) shows BGYF on its lint and linters. That fungus gains entry into the bolls typically via holes made by the pink bollworm (*Pectinophora gossypiella*). See also MYCOTOXINS, AFLATOXIN, FUNGUS, *PECTINOPHORA GOSSYPIELLA*, FLUORESCENCE.

**Broad Spectrum** See GRAM STAIN.

**Bromoxynil** An active ingredient in some herbicides, it kills certain types of plants (weeds). See also NITRILASE.

**Broth** A fluid culture medium (for growing microorganisms). See also MEDIUM, CULTURE MEDIUM.

**Brown Stem Rot (BSR)** A plant disease that can be caused by the soilborne fungus *Phialaphora gregata* in the soybean plant (*Glycine max L.* Merrill). Some soybean varieties are genetically resistant to BSR. See also FUNGUS, SOYBEAN PLANT, GENOTYPE, GENE, PATHOGENIC.

**BSE** Bovine spongiform encephalopathy. A neurodegenerative disease of cattle. See also PRION.

**BSP** Biosafety protocol. See also CONVENTION ON BIOLOGICAL DIVERSITY (CBD).

**BSR** See BROWN STEM ROT (BSR).

**BST** See BOVINE SOMATOTROPIN (BST).

**BtR-4 Gene** See TOXICOGENOMICS.

*B.t.* See *BACILLUS THURINGIENSIS* (*B.t.*).

**B.t.k.** See *B.t. KURSTAKI*.

**B.t. israelensis** One of the approximately 30 subspecies groupings within the approximately 20,000 different strains of the soil bacteria known (collectively) as *Bacillus thuringiensis* (*B.t.*). When eaten (e.g., due to presence on food), the protoxin proteins produced by *B.t. israelensis* are toxic to mosquitoes and black fly (Diptera) larvae. See also *BACILLUS THURINGIENSIS* (*B.t.*), PROTOXIN, ION CHANNELS.

**B.t. kurstaki** One of the approximately 30 subspecies groupings within the approximately 20,000 different strains of the soil bacteria known (collectively) as *Bacillus thuringiensis* (*B.t.*). When eaten (e.g., as part of a genetically engineered plant), the protoxin proteins produced by *B.t. kurstaki* are toxic to certain caterpillars (Lepidoptera larvae), such as the European corn borer (pyralis). See also *BACILLUS THURINGIENSIS* (*B.t.*), PROTOXIN, CRY1A (b) PROTEIN, ION CHANNELS, EUROPEAN CORN BORER (ECB).

**B.t. tenebrionis** One of the approximately 30 subspecies groupings within the approximately 20,000 different strains of the soil bacteria known (collectively) as *Bacillus thuringiensis* (*B.t.*). When eaten (e.g., as part of a genetically engineered plant), the protoxin proteins produced by *B.t. tenebrionis* are toxic to certain insects. See also *BACILLUS THURINGIENSIS* (*B.t.*), PROTOXIN, GENETIC ENGINEERING, ION CHANNELS.

**B.t. tolworthi** One of the approximately 30 subspecies groupings within the approximately 20,000 different strains of the soil bacteria known (collectively) as *Bacillus thuringiensis* (*B.t.*). When eaten (e.g., as part of a genetically engineered crop plant), the protoxin proteins produced by *B.t. tolworthi* are toxic to certain caterpillars (Lepidoptera larvae), such as the European corn borer (pyralis). See also *BACILLUS THURINGIENSIS* (*B.t.*), PROTOXIN, CRY9C PROTEIN, GENETIC ENGINEERING, ION CHANNELS.

**Buffy Coat (cells)** The layer of white blood cells (leukocytes) that separates out when blood is subjected to centrifugation. See also ULTRACENTRIFUGE, LEUKOCYTES, PLASMA, BLOOD DERIVATIVES MANUFACTURING ASSOCIATION.

**Bundesgesundheitsamt (BGA)** German Federal Health Organization. The German government agency that must approve new pharmaceutical products for sale within Germany, it is the equivalent of the U.S. Food and Drug Administration (FDA). See also FOOD AND DRUG ADMINISTRATION (FDA), KOSEISHO, COMMITTEE FOR PROPRIETARY MEDICINAL PRODUCTS (CPMP), COMMITTEE ON SAFETY IN MEDICINES, MEDICINES CONTROL AGENCY (MCA), EUROPEAN MEDICINES EVALUATION AGENCY (EMEA).

**BXN Gene** See NITRILASE.

# C

*C. elegans*  See CAENORHABDITIS ELEGANS.

**C Value**  The total amount of DNA in a haploid genome. See also DEOXYRIBONUCLEIC ACID (DNA), HAPLOID, GENOME.

**C-DNA**  Also known as copy DNA. A helical form of DNA, it occurs when DNA fibers are maintained in 66% relative humidity in the presence of lithium ions. It has fewer base pairs per turn than B-DNA. See also B-DNA, DEOXYRIBONUCLEIC ACID (DNA), BASE PAIR (bp), COMPLEMENTARY DNA (cDNA).

**Cadherins**  A class of (cell surface) adhesion molecules that causes cells (e.g., in the lining of the intestine known as the epithelium) to "stick together" to form a continuous lining; cadherins sometimes function as cellular adhesion receptors. For example, the (food poisoning) pathogenic bacteria *Listeria monocytogenes* is able to infect humans via its use of the E-cadherin receptor located on the surface of intestinal epithelium cells. That bacteria's "key" (a bacterial membrane surface protein known as internaulin) is "inserted" into the E-cadherin ("lock"), which opens up the otherwise closed-to-bacteria intestinal epithelium. The *L. monocytogenes* bacteria then leave the intestine and infect the human body tissues. See also ADHESION MOLECULE, CELL, RECEPTORS, LISTERIA MONOCYTOGENES, EPITHELIUM.

*Caenorhabditis elegans* (*C. elegans*)  The name of a nematode (microscopic roundworm) that is commonly utilized by scientists in genetics experiments. Because of this, a large base of knowledge about *C. elegans* genetics has been accumulated by the world's scientific community. For example, of the nearly 300 "disease-causing genes" in the human genome, more than half of them have an analogous gene within the *C. elegans* genome. *C. elegans* was one of the first animals to have its entire genome sequenced by man. Thus, one of the methodologies utilized by researchers to rapidly screen large numbers of chemical compounds for their potential use as pharmaceuticals is to:

Expose large numbers of *C. elegans* to the various chemical compounds that the researcher wants to investigate for potential pharmaceutical activity.

Pass those large numbers of previously exposed *C. elegans*, suspended in liquid such as water, through a small transparent chamber where a focused laser beam is shone upon the roundworm's side (for its full length, as the roundworm passes by).

Utilize expression-of-fluorescent-protein, autofluorescence, lectin (in the fluid) binding detected via laser reflectance, antibody (in the fluid) binding detected via laser reflectance, etc. as the basis for individual *C. elegans* to be "sorted" via tiny jets of air that blow into a container those *C. elegans* that show thus visible sign(s) of having been changed by the particular chemical compound to which they were exposed.

Evaluate in detail (e.g., via conventional gene expression analysis) the specific impact of that particular chemical compound on those *C. elegans* that had indicated an apparent change, so were sorted into the "likely target" receptacle.

See also NEMATODES, GENETICS, GENE, GENOME, GENE EXPRESSION, GENE EXPRESSION MARKERS, EXPRESSED SEQUENCE TAGS (EST), SEQUENCING

(OF DNA MOLECULES), HIGH-THROUGHPUT SCREENING (HTS), HIGH-THROUGHPUT IDENTIFICATION, GENE EXPRESSION ANALYSIS, TARGET-LIGAND INTERACTION SCREENING, TARGET (OF A THERAPEUTIC AGENT), FLUORESCENCE, LECTINS, MODEL ORGANISM.

**Caffeine** A chemical naturally produced in some plants (e.g., coffee tree) to repel predatory insects. It also acts as a stimulant (when consumed by humans), so is classified as a phytochemical. Research done by Seymour Diamond during 2000 showed that within the human body, caffeine consumption causes interactions with the synthetic chemical painkiller known as Ibuprofen. Consuming both together was shown to be more effective in relieving pain than consuming Ibuprofen alone, and brought pain relief faster than consumption of Ibuprofen alone. See also PHYTOCHEMICALS.

**Calcium Channel-Blockers** Drugs (e.g., verapamil, amlopidine, diltiazem, nifedipine) used to slow down calcium movement through cell membranes. This leads to dilation of the blood vessels and reduces the heart's workload. Blood vessels need calcium to contract (causing flow constriction and hence an increase in blood pressure), so the drug-induced shortage of available calcium causes the body's blood vessels to remain dilated (which results in lower blood pressure). Research in 1996 indicated the possibility that certain types of calcium channel-blockers might lead to increased rates of some cancers. If so, this is likely due to the drug preventing enough calcium availability for normal apoptosis in body cells. See also CELL, ION CHANNELS, CANCER, MEMBRANE TRANSPORT, APOPTOSIS.

**Calcium Oxalate** A crystalline salt normally deposited in the cells of some species of plants. In spinach, the presence of such oxalate inhibits absorption of the calcium (present in spinach) by humans. In many animals, calcium oxalate is excreted in urine or retained by the animal's body in the form of urinary calculi. See also ABSORPTION, OXALATE, CELL.

**Callipyge** (means *beautiful buttocks* in Greek) An inherited trait in livestock (e.g., sheep) that results in thicker, meatier hind quarters.

First identified as a genetic trait in 1983, this desirable trait results in a higher meat yield per animal. See also TRAIT, GENOTYPE, PHENOTYPE, WILD TYPE.

**Callus** An undifferentiated cluster of plant cells that is a first step in regeneration of plants from tissue culture. See also SOMACLONAL VARIATION.

**Calorie** The amount of heat (energy) required to raise the temperature of one gram of water from 14.5°C (58°F) to 15.5°C (60°F) at a constant pressure of one standard atmosphere. This unit measure of energy is also frequently utilized to express the amount of energy contained within certain foods or animal feeds. See also CARBOHYDRATES (SACCHARIDES), FATS, TME(N).

**Calpain-10** A gene that increases the likelihood for development of diabetes disease in humans whose DNA carries that gene (approximately 80% of humans carry the gene). See also DIABETES, INSULIN, INSULIN-DEPENDENT DIABETES MELLITIS (IDDM), GENE, DEOXYRIBONUCLEIC ACID (DNA).

**Campesterol** A phytosterol produced within the seeds of the soybean plant (*Glycine max* L.), among others. Evidence shows human consumption of campesterol helps reduce total serum (blood) cholesterol and low-density lipoproteins (LDLP) levels, thereby lowering risk of coronary heart disease (CHD). Evidence indicates certain phytosterols (including campesterol) interfere with absorption of cholesterol by the intestines, and decrease the body's recovery and reuse of cholesterol-containing bile salts; this causes more (net) cholesterol to be excreted from the body. See also PHYTOSTEROLS, PHYTOCHEMICALS, STEROLS, SOYBEAN PLANT, CHOLESTEROL, STIGMASTEROL, BETA-SITOSTEROL (β-SITOSTEROL), CORONARY HEART DISEASE (CHD).

**Campestrol** See CAMPESTEROL.

**Campsterol** See CAMPESTEROL.

**Camptothecins** See RUBITECAN.

**CaMV** See CAULIFLOWER MOSAIC VIRUS 35S PROMOTER (CaMV 35S).

**CaMV 35S** See CAULIFLOWER MOSAIC VIRUS 35S PROMOTER (CaMV 35S).

**Canavanine** An uncommon amino acid. It is used in biology as an arginine (another amino acid) analogue. It is a potent growth inhibitor of many organisms. See also AMINO ACID, BIOMIMETIC MATERIALS.

**Cancer** The name given to a group of diseases that are characterized by uncontrolled cellular growth (e.g., formation of tumor) without any differentiation of those cells (into specialized and different tissues). See also CARCINOGEN, ONCOGENES, TUMOR-SUPPRESSOR GENES, TUMOR, TELOMERES, RETINOIDS, MUTAGEN, CELL, TELOMERASE, NEOPLASTIC GROWTH, CHEMOTHERAPY, DIFFERENTIATION, ORAL CANCER, MYCOTOXINS.

**CANDA** Computer Assisted New Drug Application. An application to the U.S. Food and Drug Administration (FDA) seeking approval of a drug that has undergone Phase 2 and Phase 3 clinical trials. A CANDA is submitted in the form of computer-readable (clinical) data that provide the FDA with a sophisticated database that allows administration reviewers to evaluate (statistically) the data themselves. See also NDA (to FDA), NDA (to Koseisho), FOOD AND DRUG ADMINISTRATION (FDA), MAA MARKETING AUTHORIZATION APPLICATION, PHASE I CLINICAL TESTING.

**Canola** *Brassica napus* or *Brassica campestris* strains of the rapeseed plant, which were developed by plant breeders after the 1960s. Oil produced from rapeseed grown prior to 1971 contained 30–60% erucic acid (high dietary levels of which were associated with cardiac lesions in experimental animals via toxicology testing). By 1974, canola varieties producing oil containing less than 5% erucic acid constituted virtually all of that year's Canadian rapeseed crop, and Canadian breeders continued to develop new canola varieties with ever-lower erucic acid content.

In 1982, Canada filed with the U.S. Food and Drug Administration (FDA) to have low-erucic-acid rapeseed (LEAR) oil affirmed to be GRAS (Generally Recognized As Safe) which the FDA did. LEAR was one of the first foodstuffs to be determined "substantially equivalent" under the Organization for Economic Cooperation and Development (OECD)-defined criteria for "substantial equivalence" because LEAR was shown (in OECD petition) to be very similar to, and composed of the same basic components as, traditional rapeseed oil (and other commonly

consumed vegetable oils) except for a lower level of erucic acid (the component of concern, per above). See also STRAIN, FATS, LAURATE, FATTY ACID, OLEIC ACID, GRAS LIST, ORGANIZATION FOR ECONOMIC COOPERATION AND DEVELOPMENT (OECD), GLUCOSINOLATES, BRASSICA, HIGH-STEARATE CANOLA.

**CAP** Catabolite gene-activator protein, also known as CRP, catabolite regulator protein (or cyclic AMP receptor protein). The protein mediates the action of cyclic AMP (cAMP) on transcription in that cAMP and CAP must first combine. The cAMP-CAP complex then binds to the promoter regions of *Escherichia coli* and stimulates transcription of its operon. Since a cell component increases rather than inhibits transcription, this type of regulation of gene expression is called positive transcriptional control. See also *ESCHERICHIA COLIFORM* (*E. COLI*), CATABOLITE REPRESSION, TRANSCRIPTION, OPERON.

**Capsid** The external protein coat of a virus particle that surrounds the nucleic acid. The individual proteins that make up the capsid are called capsomers or protein subunits. It has been discovered that resistance to certain viral diseases may be imparted to some plants by inserting the gene for production of the protein coat into the plants. See also TOBACCO MOSAIC VIRUS (TMV), VIRUS, PROTEIN.

**Capsule** An envelope surrounding many types of microorganisms. The capsule is usually composed of polysaccharides, polypeptides, or polysaccharide-protein complexes. These materials are arranged in a compact manner around the cell surface. Capsules are not absolutely essential cellular components. See also MICROORGANISM, POLYSACCHARIDES, POLYPEPTIDE (PROTEIN), PROTEIN, CELL, GRAM-NEGATIVE (G-), MANNANOLIGOSACCHARIDES (MOS), GRAM-POSITIVE (G+).

**CARB** See CENTER FOR ADVANCED RESEARCH IN BIOTECHNOLOGY (CARB).

**Carbetimer** An antineoplastic (i.e., anticancer) low molecular weight polymer that acts against several types of cancer tumors, perhaps via stimulation of the patient's immune system. It has minimal toxicity.

**Carbohydrate Engineering** The selective, deliberate alteration/creation of carbohydrates (and the oligosaccharide side chains of glycoprotein molecules) by man. See also GLUCONEOGENESIS, GLYCOBIOLOGY, GLYCO-FORM, GLYCOLIPID, GLYCOLYSIS, GLYCOPROTEIN, GLYCOSIDASES, RESTRICTION ENDOGLYCOSI-DASES, GLYCOSIDE, GLYCOSYLATION.

**Carbohydrates** (saccharides) A large class of carbon-hydrogen-oxygen compounds. Monosaccharides are called simple sugars, of which the most abundant is D-glucose. It is both the major fuel for most organisms and constitutes the basic building block of the most abundant polysaccharides, such as starch and cellulose. While starch is a fuel source, cellulose is the primary structural material of plants. Carbohydrates are produced by photosynthesis in plants. Most, but not all, carbohydrates are represented chemically by the formula $Cx(H_2O)n$, where $n$ is three or higher. On the basis of their chemical structures, carbohydrates are classified as polyhydroxy aldehydes, polyhydroxy ketones, and their derivatives. See also GLUCOSE (GLc), GLYCOGEN, MONOSACCHARIDES, OLIGOSACCHA-RIDES, POLYSACCHARIDES.

**Carcinogen** A cancer-causing agent. See also MUTAGEN, PROTO-ONCOGENES, AFLATOXINS, ANTIOXIDANTS.

**Carnitine** A vitamin-like nutrient that occurs naturally in animal cells, and which is needed for the body to convert fatty acids to energy (which can then be used by the body's cells). Carnitine is essential to facilitate the transport of Acyl-CoA enzyme (attached to a fatty acid molecule) into the cell's mitochondria, where the beta-oxidation of fatty acids occurs (thereby providing energy to the cell). Before fatty acids can enter the mitochondria, they must be "activated" by a chemical reaction (which occurs on the outer mitochondrial membrane), in which Acyl-CoA is attached to the fatty acid molecule by a chemical reaction driven by adenosine triphosphate (ATP) and catalyzed by Acyl-CoA synthetase. Adenosine monophosphate (AMP) is a byproduct of that chemical reaction. See also FATTY ACIDS, METABOLISM, ACYL-CoA, ENZYME, ACETYL CAR-NITINE, ACETYLCARNITINE TRANSFERASE, MITO-CHONDRIA, PLASMA MEMBRANE, ACTIVATION ENERGY, ADENOSINE TRIPHOSPHATE (ATP), SYN-THASE, ADENOSINE MONOPHOSPHATE (AMP).

**Carotenoids** A general term for a group of plant-produced and microorganism-pro-duced pigments ranging in color from yellow to red and brown, that act as protective antioxidants in photosynthetic plants and animals that consume carotenoids. Approximately 600 carotenoids have been discovered and studied by man. The carotenes and the xanthophylls, orange to yellow in color, are the most common. Carotenoids are responsible for the coloration of certain plants (e.g., the carrot) and of some animals (e.g., the lobster). The carotenoid pigments are transferred to animals as an element in their foods. Carotenoids are composed of isoprene units (usually eight) which may be modified by the addition of other chemical groups on the molecule. The carotenes are of importance to higher animals because they are utilized in the formation of vitamin A. Carotenoids act as antioxidants ("quench-ers" of free radicals), so consumption of car-otenoids apparently thereby reduces the risk of some cancers, coronary heart disease, eyesight loss, and cataracts. See also VITA-MIN, BETA CAROTENE, CANCER, CORONARY HEART DISEASE (CHD), ASTAXANTHIN, LYCOPENE, ANTI-OXIDANTS, FREE RADICAL, OXIDATIVE STRESS, INSULIN, LUTEIN, ZEAXANTHIN, GOLDEN RICE, PHOTOSYNTHESIS, MICROORGANISM.

**Cartilage-Inducing Factors A and B** Com-pounds produced by the body which also have immunosuppressive activity. See also IMMUNOSUPPRESSIVE.

**Cascade** A sequential series of events (chemical reactions, immune responses, etc.) initiated ("set off") by a specific first event (e.g., a signaling molecule "docking" at a receptor molecule, an antibody-antigen complex forming in the body, thrombin cleaving fibrinogen, etc.). See also SIGNALING MOLE-CULE, SIGNAL TRANSDUCTION, RECEPTORS, PRO-TEIN SIGNALING, SYSTEMIC ACQUIRED RESISTANCE (SAR), HARPIN, COMPLEMENT (COM-PONENT OF IMMUNE SYSTEM), COMPLEMENT CAS-CADE, THROMBIN, FIBRIN, GENE EXPRESSION CASCADE, R GENES, VIRAL TRANSACTIVATING PROTEIN.

**Cassette** A "package" of genetic material (containing more than one gene) inserted into the genome of a cell via gene splicing techniques. May include promoter(s), leader sequence, termination codon, etc. See also GENE SPLICING, LEADER SEQUENCE, PROMOTER, GENETIC CODE, TERMINATION CODON (SEQUENCE), GENETIC ENGINEERING, TRANSGENE, GENOME.

**Catabolism** Energy-yielding pathway. The phase of metabolism involved in the energy-yielding degradation of nutrient (food) molecules. See also DISSIMILATION, METABOLISM, PATHWAY, STEROLS.

**Catabolite Activator Protein** See CAP.

**Catabolite Repression** Common in bacteria. The decreased expression of catabolic enzymes as brought about by a catabolite such as glucose. For example, glucose is the preferred fuel source for certain bacteria, and when present in the culture medium, it represses the formation of enzymes required for the utilization of other fuel sugars, such as $\beta$-galactosidase. Since glucose or other catabolites (other molecules derived from glucose) cause the repression, it is known as catabolite repression. See also CAP, OPERON, GLUCOSE (GLc), ADENOSINE MONOPHOSPHATE (AMP), PATHWAY FEEDBACK MECHANISMS.

**Catalase** An enzyme that catalyzes the very rapid decomposition of hydrogen peroxide to water and oxygen. Catalase is in the group of enzymes known as metalloenzymes because it requires the presence of a metal in order to be catalytically active. The metal (known as a cofactor) is, in the case of catalase, iron. Found in both plants and animals. See also HYDROLYSIS, HUMAN SUPEROXIDE DISMUTASE (hSOD), PEG- SOD (POLYETHYLENE GLYCOL SUPEROXIDE DISMUTASE).

**Catalysis** Coined by Jons J. Berzelius in 1838, this term refers to the act of increasing the rate of a given chemical reaction via use of a catalyst. Almost all chemical reactions in biological systems (e.g., within an organism) are catalyzed by molecules known as enzymes. Enzymes typically increase the rate of a given biological/chemical reaction by at least a millionfold. See also CATALYST, CATALYTIC SITE, ENZYME, METALLOENZYME.

**Catalyst** From the Greek word *katalyein*, to dissolve. Any substance (entity), either of protein or of nonproteinaceous nature, that increases the rate of a chemical reaction, without being consumed itself in the reaction. In the biosciences, the term "enzyme" is used for a proteinaceous catalyst. Enzymes catalyze biological reactions. See also ENZYME, CATALYTIC SITE, ACTIVE SITE, CATALYTIC ANTIBODY, SEMISYNTHETIC CATALYTIC ANTIBODY, METALLOENZYME.

**Catalytic Antibody** An antibody produced (e.g., via monoclonal antibody techniques) in response to a carefully selected antigen (e.g., target molecule in bloodstream, or molecule involved in chemical reaction of interest) which itself catalyzes the "splitting" of a molecule in the bloodstream (e.g., heroin into two harmless small molecules) or mimics:

1. Restriction endonucleases that cleave (cut) proteins or DNA molecules precisely at specific locations on those molecules.
2. Restriction endoglycosidases that are capable of cleaving oligosaccharides or polysaccharide molecules precisely at specific locations on those molecules.
3. Transition state chemical complex in the chemical reaction that is to be catalyzed; resultant antibody acts both as an antibody (to the selected transition-state-complex antigen) and as a catalyst (for the chemical reaction possessing that selected transition state chemical complex).

This catalyst (enzyme) thus possesses the remarkable specificity of an antibody (i.e., specific only to the desired transition-state reactant) which holds the potential to yield chemical reaction products of greater purity than those achieved via current (less specific) catalysts. Because the immune system will (in theory) produce an antibody to virtually every molecule of sufficient size to be detected by the immune system (i.e., 6 to 34 Angstroms), it should be possible to raise catalytic antibodies for a large number of industrial chemical reactions that are currently catalyzed via conventional (less specific) catalysts. See also OLIGOSACCHARIDES, CATALYST, ANTIBODY, RESTRICTION ENDONUCLEASES, RESTRICTION ENDOGLYCOSIDASES,

**C**

MONOCLONAL ANTIBODIES (MAb), ANTIGEN, TRANSITION STATE, PROTEIN, ACTIVATION ENERGY, SEMISYNTHETIC CATALYTIC ANTIBODY, ANGSTROM (Å), ABZYMES.

**Catalytic Domain** See DOMAIN (OF A PROTEIN).

**Catalytic RNA** An RNA (ribonucleic acid) molecule that acts to cleave ("cut") any other RNA. See also RIBOZYMES.

**Catalytic Site** The site (geometric area) on an enzyme molecule (or other catalyst) that is actually involved in the catalytic process. The catalytic site usually consists of a small portion of the total area of the enzyme. See also CATALYST, ENZYME, ACTIVE SITE, CATALYTIC ANTIBODY.

**Catecholamines** Hormones (such as adrenalin) that are amino derivatives of a base structure known as catechol. Catecholamines are released into the bloodstream by exercise, and act as natural tranquilizers. See also ENDORPHINS, HORMONE.

**Cation** See ION, CHELATION, CHELATING AGENT.

**Cauliflower Mosaic Virus 35S Promoter (CaMV 35S)** A promoter (sequence of DNA) that is often utilized in genetic engineering to control expression of (inserted) gene; i.e., synthesis of desired protein in a plant. See also VIRUS, PROMOTER, DEOXYRIBONUCLEIC ACID (DNA), GENE, GENETIC ENGINEERING, PROTEIN.

**CBD** See CONVENTION ON BIOLOGICAL DIVERSITY (CBD).

**CBF1** A transcription factor (i.e., special protein) that is synthesized (manufactured) within certain plants (*Arabidopsis thaliana*, etc.) when those plants are exposed to cold temperatures. CBF1 then interacts with certain portions of the plants' DNA (i.e., regulatory sequences) to thus "switch on" the process of cold hardening (via proteins coded for by the plants' genes). See also TRANSCRIPTION FACTORS, PROTEIN, SYNTHESIZING (OF PROTEINS), ARABIDOPSIS THALIANA, GENETIC CODE, CODING SEQUENCE, REGULATORY SEQUENCE, DEOXYRIBONUCLEIC ACID (DNA), COLD HARDENING.

**CCC DNA** A covalently linked circular DNA molecule, such as a plasmid. See also DEOXYRIBONUCLEIC ACID (DNA), PLASMID.

**CD4 EPSP Synthase** See EPSP SYNTHASE, CP4 EPSPS.

**CD4 EPSPS** See EPSP SYNTHASE, CP4 EPSPS.

**CD4 Protein** An adhesion molecule (protein) imbedded in the outer wall (envelope) of human immune system and brain cells that functions as the receptor (door to entry into the cell) for the HIV (AIDS) virus. The gp120 envelope glycoprotein of the HIV (i.e., AIDS) virus directly interacts with the CD4 protein on the surface of helper T cells to enable the virus to invade the helper T cells. See also T CELL RECEPTORS, ADHESION MOLECULE, GP120 PROTEIN, SOLUBLE CD4.

**CD4-PE40** An experimental drug discovered in 1988 by Ira Pastan and Bernard Moss that has indicated potential to combat acquired immune deficiency syndrome (AIDS). CD4-PE40 is a conjugated protein consisting of a CD4 protein (molecule) attached to *Pseudomonas* exotoxin (a substance produced by *Pseudomonas* bacteria that is toxic to certain living cells). The gp 120 glycoprotein on the surface of the HIV (i.e., AIDS) virus attaches preferentially to the CD4 portion of this immunoconjugate, and the virus is inactivated by the *Pseudomonas* exotoxin portion of this immunoconjugate. See also PROTEIN, CD4 PROTEIN, GP120 PROTEIN, SOLUBLE CD4, IMMUNOTOXIN, CONJUGATED PROTEIN, ACQUIRED IMMUNE DEFICIENCY SYNDROME (AIDS), HUMAN IMMUNODEFICIENCY VIRUS TYPE 1 (HIV-1), HUMAN IMMUNODEFICIENCY VIRUS TYPE 2 (HIV-2), RICIN, ABRIN.

**CD44 Protein** One of the adhesion molecules (embedded in the surface of the linings of blood vessels) that assists the neutrophils on their journey from the bloodstream through the walls of blood vessels (e.g., to combat pathogens into adjacent tissues). Tumor cells also exploit CD44 molecules in order to metastasize (spread throughout the body's tissue from a single beginning tumor) via a similar (tumor cell) through-blood vessel-wall adhesion molecule mechanism. See also ADHESION MOLECULE, CD4 PROTEIN, PROTEIN, NEUTROPHILS, PATHOGEN, TUMOR, CANCER, SOLUBLE CD4.

**CD95 Protein** Also called APO-1/Fas, it is a transmembrane protein (embedded within the surface membrane of the cell) that transmits an apoptosis ("programmed" cell death) "signal" into cells. Transduction of that apoptosis signal occurs when certain ligands or antigens (i.e., the APO-1/Fas antigen) bind to the extracellular (portion outside of cell membrane) part (i.e., receptor) of the CD95 protein. See also APOPTOSIS, PROTEIN, CELL, SIGNAL TRANSDUCTION, SIGNALLING, NUCLEAR RECEPTORS, ANTIGEN, RECEPTORS, *FUSARIUM*.

**cDNA** See COMPLEMENTARY DNA (cDNA).

**cDNA Array** See MICROARRAY (TESTING).

**cDNA Clone** A DNA molecule synthesized (made) from an mRNA sequence via sequential use of reverse transcriptase (acting on mRNA) and DNA polymerase. See also DEOXYRIBONUCLEIC ACID (DNA), MESSENGER RNA (mRNA), COMPLEMENTARY DNA (cDNA), SEQUENCE (OF A DNA MOLECULE), REVERSE TRANSCRIPTASE, DNA POLYMERASE, CLONE (A MOLECULE).

**cDNA Microarray** See MICROARRAY (TESTING), COMPLEMENTARY DNA (cDNA).

**Cecrophins** (lytic proteins) Proteins produced by certain white blood cells [called cytotoxic T lymphocytes (CTL) or killer T cells]. The proteins allow lysis (i.e., bursting) of infected cells. Cecrophins are amphopathic (i.e., contain both a hydrophobic region and a hydrophilic region); and work by "worming" the hydrophobic portion into the cell membrane (so the hydrophobic portion of the cecrophin molecule is out of the water). This creates a transmembrane pore (a hole in the membrane) which is lined with the cecrophin's hydrophilic portion. Membranes function simply to separate various components. This separation is required for life to exist. When holes are introduced into cell membranes, water rushes into the targeted cell due to differences in osmotic pressure and the cell ruptures (explodes). The cecrophins are only able to lyse (burst) infected cells because only "sick" cells have a weakened cytoskeleton (located just inside the cell membrane), which cannot prevent the contents of the cell from spilling out through the pores (created by cecrophins). See also HELPER T CELLS (T4 CELLS), PATHOGEN, COMPLEMENT, HYDROPHOBIC, HYDROPHILIC, COMPLEMENT CASCADE, LYSE, LYSIS.

**Cecropin A** See CECROPIN A PEPTIDE.

**Cecropin A Peptide** See CECROPHINS, PEPTIDE.

**Cell** From the Latin *cella*, which means small room. The fundamental self-containing unit of life. The living tissue of every multicelled organism is composed of these fundamental living units. Certain organisms may consist of only one cell, such as yeast or protein bacteria, protozoa, some algae, and gametes (the reproductive stages) of higher organisms. Larger organisms are subdivided into organs that are relatively autonomous but cooperate in the functioning of that plant or animal. Unicellular (single-cell) organisms perform all life functions within the one cell. In a higher organism (a multicellular organism), entire populations of cells (i.e., an organ) may be designated a particular specialized task (e.g., the heart to facilitate circulation). The cells of muscle tissue are specialized for movement, and those of bone and connective tissue for structural support. While most cells are too small to be seen with the unaided eye, the egg yolk of birds is a single cell, so the egg yolk of an ostrich is the world's largest cell. See also PLASMA MEMBRANE, GAMETE, GERM CELL, MICROBIOLOGY, OOCYTES.

**Cell Culture** The *in vitro* (outside of body, in a test tube) propagation of cells isolated from living organisms. See also MAMMALIAN CELL CULTURE, DISSOCIATING ENZYMES, HARVESTING ENZYMES.

**Cell Cytometry** See CELL, CELL SORTING, FLUORESCENCE ACTIVATED CELL SORTER (FACS), MAGNETIC PARTICLES.

**Cell Differentiation** The process whereby descendants of a common parental cell achieve and maintain specialization of structure and function. In humans, for instance, all the different types of cells (muscle cells, bone cells, etc.) differentiate from the zygote (itself formed by union of the simple sperm and egg). In humans, the various blood cell types (red blood cells, white blood cells, etc.) differentiate from stem cells in the bone marrow. Cell differentiation is caused/triggered/assisted by colony stimulating factors (CSFs), growth factors (GFs), and certain other proteins (e.g., hedgehog proteins). See also STEM CELLS, STEM CELL ONE, PROTEIN, HEDGEHOG PROTEINS, ERYTHROCYTES, LEUKOCYTES, COLONY STIMULATING FACTORS, GROWTH FACTORS, DIFFERENTIATION.

**Cell Fusion** The combining of cell contents of two or more cells to become a single cell.

Fertilization is such a process (fusing of gametes' cells). See also GAMETE.

**Cell Recognition** See ADHESION MOLECULE, SIGNAL TRANSDUCTION, RECEPTORS.

**Cell Signaling** See SIGNALING.

**Cell Sorting** A process utilized (e.g., by researchers) to sort/separate different cells (pathogens, cancerous vs. normal cells, sperm that are bearing chromosomes for male vs. female, etc.). Some automated means of cell sorting include biochips (utilizing controlled electrical fields to collect specific cell types onto electrodes in the biochip), fluorescence-activated cell sorter (FACS) machines, magnetic particles (e.g., attached to antibodies), etc. See also CELL, PATHOGEN, CANCER, CHROMOSOME, BIOCHIP, BIOMEMS, FLUORESCENCE ACTIVATED CELL SORTER (FACS), MAGNETIC PARTICLES.

**Cell-Differentiation Proteins** The various growth factors and other proteins which cause/assist in cell differentiation. See also CELL DIFFERENTIATION, HEDGEHOG PROTEINS.

**Cell-Mediated Immunity** See CELLULAR IMMUNE RESPONSE.

**Cellular Adhesion Molecule** See ADHESION MOLECULE.

**Cellular Adhesion Receptors** See ADHESION MOLECULE, RECEPTORS, INTEGRINS, SELECTINS, CADHERINS.

**Cellular Affinity** Tendency of cells to adhere specifically to cells of the same type. This property is lost in some cancer cells. See also CELL, ADHESION MOLECULE, CELL DIFFERENTIATION.

**Cellular Immune Response** Also called cell-mediated immunity. The immune response that is carried out by specialized cells, in contrast to the response carried out by soluble antibodies. The specialized cells that make up this group include cytotoxic T lymphocytes (CTL), helper T lymphocytes, macrophages, and monocytes. This system works in concert with the humoral immune response. See also HUMORAL IMMUNITY, T CELLS, T CELL RECEPTORS, PHAGOCYTE, HELPER T CELLS (T4 CELLS), CYTOKINES, MACROPHAGE.

**Cellular Oncogenes** See PROTO-ONCOGENES.

**Cellulase** The enzyme that digests cellulose to simple sugars such as glucose. See also ENZYME, DIGESTION (WITHIN CHEMICAL PRODUCTION PLANTS).

**Cellulose** A polymer of glucose units found in all plant matter; it comprises 40–55% of the cell wall in plant cells. Because of its presence in all plant cells, cellulose is the most abundant biological compound on earth. See also CARBOHYDRATES, GLUCOSE (GLc), CELL, VAN DER WAALS FORCES.

**Center for Advanced Research in Biotechnology (CARB)** A protein engineering research consortium established in Rockville, MD, during 1989 by the U.S. government, the University of Maryland, and local government. See also PROTEIN ENGINEERING.

**Central Dogma (new)** Coined by Shankar Subramaniam during 1999, it is a restatement of the (old) former "Central Dogma" to include the fact that an organism's environment also impacts when and how some of its genes are expressed (e.g, to cause certain proteins to be "manufactured"). Environmental factors impacting gene expression include temperature, sunlight, humidity, presence of certain bacteria, presence of signal transducers and activators of transcription (STATs), etc. That restatement also expressly includes the fact that more than one protein can result from each gene in an organism's genome [e.g., due to interactions between genes, interactions between genes and their protein products (e.g., STATs), and interactions between genes and some environmental factors]. Mechanistically, this results in (different) proteins via alternative splicing of the mRNA transcript. For example, a single intronic base substitution that is present within the IKAP gene (the allele responsible for the disease known as Familial Dysautonomia) affects the splicing of the IKAP transcript (i.e., the mRNA that determines which protein is subsequently "manufactured" by the cell); varying translation start or stop site (on the gene); or frameshifting (i.e., different set of triplet codons in the mRNA is translated). See also CENTRAL DOGMA (OLD), ORGANISM, MOLECULAR GENETICS, COMPLEMENTARY DNA (cDNA), GENE, ALLELE, PROTEIN, ENZYME, REPLICATION (OF VIRUS), TRANSCRIPTION, TRANSLATION, DEOXYRIBONUCLEIC ACID (DNA), GENOME, RIBONUCLEIC ACID (RNA), MESSENGER RNA (mRNA), TRANSCRIPTION FACTORS, RIBOSOMES, SIGNAL TRANSDUCTION,

SIGNAL TRANSDUCERS AND ACTIVATORS OF TRAN-SCRIPTION (STATs), PHOTOPERIOD, GENE EXPRES-SION, GENE SPLICING, SPLICING, SPLICE VARIANTS, FRAMESHIFT, CODON, INTRON, FRAMESHIFT, PHARMACOENVIROGENETICS.

**Central Dogma (old)** The historical organizing principle of molecular genetics; it states that genetic information flows from DNA to RNA to protein. Stated in another way: DNA makes RNA which makes protein. This principle was first stated by Watson and Crick. It is, however, not rigorously accurate as illustrated by the facts that: DNA (i.e., genes) "information flow" is influenced (timing, amounts, etc.) by some environmental factors (temperature, humidity, etc.). The enzyme reverse transcriptase produces (makes) DNA using an RNA template. Prions do not contain any DNA. See also MOLECULAR GENETICS, COMPLEMENTARY DNA (cDNA), PROTEIN, ENZYME, REPLICATION (OF VIRUS), TRANSCRIPTION, TRANSLATION, DEOXYRI-BONUCLEIC ACID (DNA), RIBONUCLEIC ACID (RNA), MESSENGER RNA (mRNA), PRION, TEMPLATE, CEN-TRAL DOGMA (NEW), REVERSE TRANSCRIPTASES.

**Centrifuge** A machine that is used to separate heavier from lighter molecules and cellular components and structures. See also ULTRA-CENTRIFUGE.

**Centromere** A constricted region of a chromosome that includes the site of attachment to the mitotic or meiotic spindle. See also CHROMOSOMES, MEIOSIS, CHROMATIN, MITOSIS, KARYOTYPE, KARYOTYPER.

**Cerebrose** See GALACTOSE (GAL).

**Cessation Cassette** A three-gene cassette (genetic sequence construct) that, when inserted into a plant and when activated via tetracycline antibiotic, prevents the seeds produced by that plant from germinating. That is because the "cessation cassette" stops those resultant seeds from synthesizing a specific protein needed for seed germination. See also CASSETTE, GENE, GENETIC ENGI-NEERING, PROTEIN, SYNTHESIZING (OF PROTEINS), SEQUENCE (OF A DNA MOLECULE), ANTIBIOTIC.

**CFTR** See CYSTIC FIBROSIS TRANSMEMBRANE REGULATOR PROTEIN (CFTR).

**CGE** Acronym for Control of Gene Expression. See also GENETIC USE RESTRICTION TECHNOLOGIES.

**CGIAR** See CONSULTATIVE GROUP ON INTERNA-TIONAL AGRICULTURAL RESEARCH (CGIAR).

**cGMP** Current Good Manufacturing Practices. The set of current, up-to-date methodologies, practices, and procedures mandated by the Food and Drug Administration (FDA) which are to be followed in the testing and manufacture of pharmaceuticals. The set of rules and regulations promulgated and enforced by the FDA to ensure the manufacture of safe clinical supplies. The cGMP guidelines are more fine-tuned and up to date (technologically speaking) than the more general GMP. See also PHASE I CLINICAL TEST-ING, IND, GOOD MANUFACTURING PRACTICES (GMP).

**Chaconine** A neurotoxin that is naturally present at low levels within potatoes. As a result of that, chaconine is present at detectable levels in the bloodstream of humans that consume potatoes. See also TOXIN, SOLANINE.

**Chakrabarty Decision** Diamond vs. Chakrabarty, U.S. Department of Commerce, 1980; a landmark case in which the U.S. Supreme Court held that the inventor of a new microorganism whose invention otherwise met the legal requirements for obtaining a patent, could not be denied a patent solely because the invention was alive. It essentially allowed the patenting of life forms. See also U.S. PATENT AND TRADEMARK OFFICE (USPTO), MICROORGANISM.

**Channel-Blockers** See CALCIUM CHANNEL-BLOCKERS.

**Chaperone Molecules** See CHAPERONES.

**Chaperone Proteins** See CHAPERONES.

**Chaperones** Protein molecules inside living cells that assist with correct protein folding as the protein molecule emerges from the cell's ribosomes. Also, they help to convey those protein(s) to their ultimate destination(s) in the organism. Later, when cellular protein molecules begin to "unfold" due to age, heat, viruses, or exposure to certain chemicals or ultraviolet light, chaperones often cause those unfolded protein molecules to return to their correct (initial) conformation. Examples of such chaperone molecules include heat-shock protein 70 and heat-shock protein 40. See also PROTEIN

FOLDING, HEAT-SHOCK PROTEINS, PROTEIN, RIBO-SOMES, CELL, CONFORMATION, VIRUS.

**Chaperonins** Protein molecules inside living cells that facilitate proper folding of the (new) protein molecules that are synthesized (manufactured) in the cell's ribosomes. See also PROTEIN, CO-CHAPERONIN, CHAPERONES, MOLECULAR CHAPERONES, PROTEIN FOLDING, RIBOSOMES, CELL, CONFORMATION.

**Characterization Assay** See ASSAY, HIGH-THROUGHPUT SCREENING (HTS), BIOASSAY, BIOCHIPS.

**CHD** Acronym for Coronary Heart Disease. See also CORONARY HEART DISEASE (CHD), ATHEROSCLEROSIS, LOW-DENSITY LIPOPROTEINS (LDLP), CAROTENOIDS.

**Chelating Agent** A molecule capable of "binding" metal atoms. The chelating agent/metal complex is held together by coordination bonds which have a strong polar character. One example of a common chelating agent is ethylenediamine tetra-acetate (EDTA), which tightly and reversibly binds $Mg^{2+}$ and other divalent cations (positively charged ions). If a chelate is allowed to bind to metal ions required for enzyme activity, the enzyme will be inactivated (inhibited). Cobalamin (vitamin $B_{12}$), EDTA and the iron-porphyrin complex of heme (which provides the red color of blood) are other examples of chelates. See also EDTA, PHYTATE, LOW-PHYTATE SOYBEANS, LOW-PHYTATE CORN, CHELATION, HEME, TRANSFERRIN.

**Chelation** The binding of metal cations (metal atoms or molecules possessing a positive electrical charge) by atoms possessing unshared electrons (thus the electrons can be "donated" to a bond with a cation). The binding of the metal (cation) to the (electron-excess) chelator atom (ligand) results in formation of a chelator/metal cation complex. The intraatom bonds thus formed are given the name of coordination bonds. The properties of the chelator/metal cation complex frequently differ markedly from the "parent" cation. Both carboxylate and amino (molecular) groups readily bind metal cations. One of the most widely used chelators is EDTA (ethylenediamine tetraacetate). It has a strong affinity for metal cations possessing two (bi) or more positive (electrical) charges. Each

EDTA molecule binds one metal cation. The EDTA molecule can be visualized as a "hand" (having only four fingers) which grasps the metal cation. Some enzymes (which require metal cations for their activity) are inactivated by EDTA (and other chelators) in that the chelators preferentially remove the metal from the enzyme. See also ION, EDTA, LIGAND (IN BIOCHEMISTRY), CARBOHYDRATES, ENZYME, HEME, CHELATING AGENT, TRANSFERRIN, PHYTATE, LOW-PHYTATE CORN, LOW-PHYTATE SOYBEANS.

**Chemical Genetics** Coined by Rebecca Ward and Tim Mitchison, this term refers to the creation and use of synthetic chemicals that act to either block or enhance the activity of a protein (or gene that codes for protein). This enables scientists to then determine the specific function(s) of specific protein molecules. See also GENOMICS, FUNCTIONAL GENOMICS, PROTEIN, GENE, GENETIC CODE, ZINC FINGER PROTEINS, COMBINATORIAL CHEMISTRY, GENOMIC SCIENCES, GENE FUNCTION ANALYSIS.

**Chemiluminescence** See LUMINESCENT ASSAYS.

**Chemometrics** An empirical methodology utilized to (inexpensively) infer a chemical quantity/value from (indirect) measurement(s) of other physical/chemical values (which can be obtained inexpensively). The term chemometrics was coined in 1975 by Bruce Kowalski. One example of the use of chemometrics is to infer the TME (N) or "true metabolizable energy" of high-oil corn from that corn's protein and oil (fat) content. See also HIGH-OIL CORN, TME (N), PROTEIN, FATS.

**Chemopharmacology** Therapy (to cure disease) by chemically synthesized drugs. See also PHARMACOLOGY, CISPLATIN.

**Chemotaxis** Sensing of, and movement toward or away from, a specific chemical agent by living, freely moving cells (bacteria, macrophages, etc.). See also CELL, BACTERIA, MACROPHAGE, NODULATION.

**Chemotherapy** When this term was first coined by Paul Ehrlich in 1905, it was defined as any therapy (to cure diseases) via chemically synthesized drugs. Over time, the term "chemotherapy" has increasingly been utilized to refer only to application of such therapy to treat cancers. See also CHEMO-PHARMACOLOGY, CANCER, CISPLATIN, TAXOL, PACLITAXEL.

**Chimera** An organism consisting of tissues or parts of diverse genetic constitution. An example of a chimera would be a centaur: the half-man, half-goat figure of Greek mythology. The word "chimera" is from the mythological creature by that name which possessed the head of a lion, the body of a goat, and the tail of a serpent. The word chimera is very general and may be applied to any number of entities. For example, chimeric antibodies may be produced by cell cultures in which the variable, antigen-binding regions are of murine (mouse) origin while the rest of the molecule is of human origin. It is hoped that this combination will lead to an antibody which, when injected, would not elicit "rejection" and would not give rise to a lesser immune response by the host against disease(s) the antibody is "aimed" at. See also DEOXYRIBONUCLEIC ACID (DNA), GENETIC ENGINEERING, CHIMERIC DNA, CHIMERIC PROTEINS, CHIMERAPLASTY, ORGANISM, ANTIBODY, ENGINEERED ANTIBODIES.

**Chimeraplasty** A method utilized by man to introduce a gene (from the same or another species) into the DNA of a living organism or cell, via "gene repair" mechanism. Scientists add the desired DNA (gene) to a cell, along with RNA, in a paired-group known as a chimeraplast. The chimeraplast attaches itself to the cell's DNA at the site of the specific gene (to be changed), and "repairs" it utilizing its (new) chimeraplast-DNA as a "template." See also GENE REPAIR (DONE BY MAN), GENE, SPECIES, DEOXYRIBONUCLEIC ACID (DNA), ORGANISM, CELL, CHIMERA, TEMPLATE, RIBONUCLEIC ACID (RNA).

**Chimeric DNA** (Recombinant) DNA containing spliced genes from two different species. See also DEOXYRIBONUCLEIC ACID (DNA), GENE, GENE SPLICING, SPECIES, RECOMBINANT DNA (rDNA), GENETIC ENGINEERING, GENE FUSION.

**Chimeric Proteins** Fused proteins from different species, produced from the chimeric DNA template. See also CHIMERA, CHIMERIC DNA, DEOXYRIBONUCLEIC ACID (DNA), ANTIBODY, ENGINEERED ANTIBODIES, GENE FUSION.

**Chiral Compound** A chemical compound that contains an asymmetrical center and is capable of occurring in two nonsuperimposable mirror images. This phenomenon was first described by Louis Pasteur. "Chiral" is a word derived from the Greek *cheir* (meaning hand). For example, human hands may be used to illustrate chirality in that when the left and right hands are held one on top of the other, one thumb sticks out on one side while the other thumb sticks out on the other side. The point is that the same number and type of fingers and thumbs exist in both hands, but their arrangement in space may be different. So it is with the arrangement of a given molecule's (e.g., a drug's) atoms in three-dimensional space.

Approximately 40% of drugs on the market today consist of chiral compounds. In many chiral drugs, only one type of the molecule is beneficially biologically active (acts beneficially to control disease, reduce pain, etc.), while the other type of the drug molecule is either inactive or else causes undesired impacts (called "side effects" of the drug mixture). For example, one enantioner of the drug thalidomide is a potent angiogenesis inhibitor, but the other enantiomer causes birth defects in babies of pregnant women taking it. See also STEREOISOMERS, ANGIOGENESIS, OPTICAL ACTIVITY, ENANTIOMERS, *cis/trans* ISOMERISM.

**Chitin** A water-insoluble polysaccharide polymer composed of *N*-acetyl-D-glucosamine molecular units, which forms the exoskeletons of arthropods (insects) and crustacea. Shellac is produced from chitin. See also POLYSACCHARIDES, POLYMER, CHITINASE.

**Chitinase** An enzyme that degrades (breaks down) chitin. It is one of the pathogenesis-related proteins produced by certain plants as a disease-fighting response to entry-into-plant of pathogenic (disease-causing) fungi. Chitinase is also sometimes produced by certain fungi and actinomycetes that destroy the eggs (i.e., chitin-containing shells) of harmful roundworms. See also CHITIN, ENZYME, STRESS PROTEINS, PATHOGENESIS RELATED PROTEINS, FUNGUS, AFLATOXIN.

**Chloroplast Transit Peptide (CTP)** A transit peptide that, when fused to a protein, acts to transport that protein into chloroplast(s) in a plant. Once both are inside the chloroplast, the transit peptide is cleaved off the protein and that protein is then free (to do the task it was designed for). For example, the CP4 EPSPS enzyme in genetically engineered glyphosate-resistant soybean [*Glycine max* (L) Merrill] plant is transported into the soybean plant's chloroplasts by the CTP known as "N-terminal petunia chloroplast transit peptide." After both reach the chloroplast, the CTP is cleaved and degraded, so the CP4 EPSPS is then free to do its task (i.e., confer resistance to glyphosate). See also PEPTIDE, CHLOROPLASTS, GATED TRANSPORT, VESICULAR TRANSPORT, TRANSIT PEPTIDE, FUSION PROTEIN, PROTEIN, SOYBEAN PLANT, CP4 EPSPS, EPSP SYNTHASE, HERBICIDE-TOLERANT CROP.

**Chloroplasts** Specialized chlorophyll-containing photosynthetic organelles (plastids) in eucaryotic cells (the sites where photosynthesis takes place in plants). See also EUCARYOTE, ORGANELLES, CELL, PHOTOSYNTHESIS, CHLOROPLAST TRANSIT PEPTIDE (CTP), TRANSIT PEPTIDE.

**Cholera Toxin** The toxin produced by the *Vibrio cholerae* (Latin America) bacteria, a source of food/water-borne gastrointestinal disease. The cholera toxin has a strong affinity for certain receptors that are present on the surface of gastrointestinal cells. See also TOXIN, ENTEROTOXIN, CONJUGATE, IMMUNOCONJUGATE, RECEPTORS, G-PROTEINS.

**Cholesterol** From the Greek *chole* (bile), it is a sterol (sterol-lipid) that is an essential material for creation of cell membranes, and a "building block" for certain hormones and acids used by the body. For example, the bile acids are made in the liver from cholesterol. Cholesterol is also vital for normal embryonic development (e.g., of humans in the uterus) because it comprises a crucial portion of the "hedgehog proteins" that direct tissue differentiation (of the mammal embryo into various organs, limbs, etc.). However, deposition of (excess) oxidized cholesterol on the interior walls of blood vessels [in the form of plaque] can result in atherosclerosis and/or coronary heart disease (CHD) — two often fatal diseases. See also HIGH-DENSITY LIPOPROTEINS (HDLPs), LOW-DENSITY LIPOPROTEINS (LDLPs), CELL, STEROLS, PHYTOSTEROLS, HORMONE, SITOSTANOL, FRUCTOSE OLIGOSACCHARIDES, CHOLESTEROL OXIDASE, CORONARY HEART DISEASE (CHD), HIGH-OLEIC OIL SOYBEANS, STEROID, LIPIDS, HEDGEHOG PROTEINS, CAMPESTEROL, STIGMASTEROL, SITOSTEROL, SITOSTANOL, RESVERATROL, BILE ACIDS, ATHEROSCLEROSIS, PLAQUE.

**Cholesterol Oxidase** An enzyme that catalyzes the breakdown of cholesterol molecules (causing oxygen consumption in the breakdown process). Because cholesterol molecules are essential for creation and maintenance of cell membranes and some hormones, an excess of cholesterol oxidase can be harmful (e.g., to certain insects). When the gene (that codes) for cholesterol oxidase is inserted into the genome of the corn (maize) plant, it can enable that plant to resist many of the worm pests (corn earworm, European corn borer, corn rootworm, black cutworm, armyworm, etc.) that attack corn (maize) in the field. When the gene (that codes) for cholesterol oxidase is inserted into the cotton plant, it can enable that plant to resist weevils and other sucking insects that attack cotton plants in the field. See also ENZYME, GENE, GENETIC ENGINEERING, GENOME, CORN, CHOLESTEROL, *HELICOVERPA ZEA* (*H. ZEA*), CORN ROOTWORM.

**Choline** Formerly known as vitamin $B_4$, choline is a nutrient that takes part in many of the metabolism processes in the human body. Naturally present in egg yolks, organ meats, dairy products, soybean lecithin, spinach, and nuts. Choline promotes fat metabolism in the liver and the synthesis of high-density lipoproteins (HDLP, also known as "good" cholesterol) by the liver. It is also utilized by the body in order to synthesize (manufacture) acetylcholine, an important neurotransmitter (substance that transmits nerve impulses). Because significant choline deficiency can cause liver carcinogenesis, cirrhosis, and can impair cell signaling, the U.S. government has defined choline to be an essential nutrient. One active metabolite of choline is Platelet Activating Factor (PAF), which is involved in the body's hormonal and reproductive functions. Choline is so important in proper infant development/growth that it is included in manufactured infant formula at the rate of at least 7 mg per 100 kcal. See also LECITHIN, METABOLISM, METABOLITE, HIGH-DENSITY LIPOPROTEINS (HDLPs),

HORMONE, SOYBEAN OIL, VITAMIN, ACETYLCHO-
LINE, CHOLINESTERASE, NEUROTRANSMITTER,
FATS, CANCER, SIGNALING.

**Cholinesterase** An enzyme that catalyzes the chemical reaction in which the neurotransmitter (substance that transmits nerve impulses) molecule acetylcholine is synthesized (manufactured) from Ac-CoA and choline. See also ENZYME, NEUROTRANSMITTER, AC-CoA, CHOLINE, LECITHIN, ALZHEIMER'S DISEASE.

**Chromatids** Copies of a chromosome produced by replication within a living eucaryotic cell during the prophase (the first stage of mitosis). They are compact cylinders consisting of DNA coiled around flexible rods of histone protein. See also CHROMATIN, EUCARYOTE, MITOSIS, CHROMOSOMES, REPLICATION (OF VIRUS), HISTONES, PROTEIN.

**Chromatin** From the Greek word *chroma* for color. Named by Walter Flemming in 1879 due to the fact that chromatin's band-like structures stained darkly, chromatin is the complex of DNA and (histone) protein of which the chromosomes are composed. Consisting of fibrous swirls of unraveled DNA molecules in the nucleus of the interphase (i.e., the prolonged period of cell growth between cell division phases) eucaryote cell. Chromatin DNA gradually coils itself around flexible rods of histone protein during the prophase (the first stage of mitosis), forming two parallel compact cylinders (called chromatids) connected by a knot-like structure (called a centromere) at their middles. In appearance they resemble two rolls of carpeting standing side-by-side that are tied together with rope at their middles.

These (recently replicated) cylinders (that are joined at their middles) are homologous chromosomes (i.e., the genes of the two chromosomes are linked in the same linear order within the DNA strands of both chromosomes). While they are still joined at their middles, these paired chromosomes appear X-shaped when photographed by a karyotyper. Chromatin is usually not visible during the interphase of a cell but can be made more visible during all phases by reaction with basic stains (dyes) specific for DNA. See also BASOPHILIC, DEOXYRIBONUCLEIC ACID (DNA), PROTEIN, HISTONES, CHROMATIDS, CHROMOSOMES,

MITOSIS, REPLICATION (OF VIRUS), CENTROMERE, KARYOTYPE, EUCARYOTE, KARYOTYPER.

**Chromatography** Coined by Mikhail S. Tswett in 1906, this word refers to a process by which complex mixtures of different molecules may be separated from each other. During the process, the mixture is subjected to many repeated partitionings between a flowing phase and a stationary phase. Chromatography constitutes one of, if not *the* most fundamental, separation techniques used in the biochemistry/biotechnology arena to date. See also POLYACRYLAMIDE GEL ELECTROPHORESIS (PAGE), SUBSTRATE (IN CHROMATOGRAPHY), AFFINITY CHROMATOGRAPHY, BIOTECHNOLOGY, AGAROSE, GEL FILTRATION.

**Chromosome Map** See LINKAGE MAP.

**Chromosomes** Discrete units of the genome carrying many genes, consisting of (histone) proteins and a very long molecule of DNA. Found in the nucleus of every plant and animal cell. See also GENOME, GENE, GENETIC CODE, CHROMATIN, CHROMATIDS, KARYOTYPE, KARYOTYPER, PHILADELPHIA CHROMOSOME.

**Chronic Heart Disease** See CORONARY HEART DISEASE (CHD).

**Chymosin** Also known as rennin. It is an enzyme used to make cheeses (from milk). Chymosin occurs naturally in the stomachs of calves and is one of the oldest commercially used enzymes. Chymosin (rennin) is chemically similar to renin, an enzyme that plays an important role in regulating blood pressure in humans. See also RENIN.

**Cilia** Protein-based structures that occur in certain cells of both the plant and animal world. Cilia are very tiny hair-like structures occurring in large numbers on the outside of certain cells. In higher organisms such as man, they usually function to move extracellular material along the cell surface. An example is the "sweeping-out-of-foreign matter" action of cilia in the bronchial tubes in which very small particles are moved into the throat to be expelled or swallowed. Lower organisms may use cilia for locomotion (swimming). Cilia are used in the swimming motion of bacteria toward sources of nutrients in a process called chemotaxis. Cilia are shorter and occur in larger numbers

**C**

per cell than flagella. Singular: cilium. See also CHEMOTAXIS, FLAGELLA.

**Ciliary Neurotrophic Factor (CNTF)** A human protein shown to help the survival of those cells in the nervous system that act to convey sensation and control the function of muscles and organs. CNTF was approved by the U.S. Food and Drug Administration to treat amyotrophic lateral sclerosis (also known as ALS or Lou Gehrig's disease) in 1992. ALS causes a victim's muscles to degenerate severely. It affects approximately 30,000 people per year in the U.S. CNTF might prove useful for treating Alzheimer's disease and/or other human neurological diseases. See also PROTEIN, CELL, NERVE GROWTH FACTOR (NGF), FOOD AND DRUG ADMINISTRATION (FDA).

*cis*-**Acting Protein** A *cis*-acting protein has the exceptional property of acting only on the molecule of DNA from which it was expressed. See also TRANS-ACTING PROTEIN, DEOXYRIBONUCLEIC ACID (DNA).

**Cisplatin** A drug used in chemotherapy regimens against certain types of cancer tumors. Cisplatin works against (tumor) cells by binding to the cell's DNA and generating intrastrand cross-links (between the two strands of the DNA molecule). These intrastrand cross-links prevent replication and cause cell death. See also CHEMOPHARMACOLOGY, CHEMOTHERAPY, CANCER, DEOXYRIBONUCLEIC ACID (DNA), REPLICATION FORK, REPLICATION (OF DNA).

*cis/trans* **Isomerism** A type of geometrical isomerism found in alkenic systems in which it is possible for each of the doubly bonded carbons to carry two different atoms or groups. Two similar atoms or groups may be on the same side (*cis*) or on opposite sides (*trans*) of a plane bisecting the alkenic carbons and perpendicular to the plane of the alkenic systems. See also ISOMER, CHIRAL COMPOUND, TRANS FATTY ACIDS.

*cis/trans* **Test** Assays (determines) the effect of relative configuration on expression of two (gene) mutations. In a double heterozygote, two mutations in the same gene show mutant phenotype in *trans* configuration, wild (phenotype) in *cis* configuration. The phenotypic distinction is referred to as the position effect. See also GENE, PHENOTYPE, *cis*-ACTING PROTEIN, POSITION EFFECT, HETEROZYGOTE, MUTATION.

**Cistron** Synonymous with gene. See also GENE.

**Citrate Synthase** The enzyme utilized (by plants) to synthesize (create) citric acid. See also ENZYME, CITRIC ACID.

**Citrate Synthase (CSb) Gene** A bacterial gene utilized by certain bacteria (*Pseudomonas*) to code for (i.e., cause to be produced by bacterium possessing that gene) the enzyme known as citrate synthase. That enzyme is used to synthesize (create) citric acid. In 1996, Luis Herrera-Estrella discovered that inserting the CSb gene from *Pseudomonas aeruginosa* into certain plants caused those plants to produce up to ten times more citrate in their roots, and to release up to four times more citric acid from those roots into the surrounding soil (thus decreasing aluminum toxicity, via chemically "binding" aluminum ions that are present in some soils). Such soil aluminum, which slows plant growth and decreases crop yields, is present to a certain degree in approximately one-third of the Earth's arable land (e.g., in the country of Colombia, it affects 70%). See also GENE, ENZYME, EXPRESS, CITRATE SYNTHASE, ION, CITRIC ACID.

**Citrate Synthase Gene** A gene that codes for (i.e., causes to be produced by an organism possessing that gene) the enzyme known as citrate synthase. See also GENE, ENZYME, EXPRESS, CITRATE SYNTHASE, CITRIC ACID.

**Citric Acid** A tricarboxylic acid occurring naturally in plants, especially citrus fruits. It is used as a flavoring agent, as an antioxidant in foods, as an animal feed ingredient, and as a sequestering agent. The commercially produced form of citric acid melts at 153°C (307°F). Citric acid is found in all cells, its central role is in the metabolic process. Some plants naturally release citric acid from their roots into the surrounding soil, in order for that citric acid to chemically "bind" aluminum ions that are present in some soils. Such aluminum, which slows plant growth and decreases crop yields, is present to a certain degree (which causes at least some crop yield reduction) in approximately one-third of the world's arable land. For example, 70% of the agricultural land in the country of Colombia possesses harmful amounts/conditions of aluminum to damage crops. Corn (maize) yields are reduced up to 80% by such aluminum in soils. Soybeans, cotton, and field bean yields are also reduced. See also METABOLISM, ACID, CELL, CITRATE SYNTHASE, CITRATE SYNTHASE GENE, CITRATE SYNTHASE (Csb) GENE, CITRIC ACID CYCLE, METABOLITE, CELL, ION, SOYBEAN PLANT, CORN, PROBIOTICS.

**Citric Acid Cycle** Also known as the tricarboxylic acid cycle [TCA cycle because the citric acid molecule contains three (tri) carboxyl (acid) groups]. Also known as the Krebs cycle after H. A. Krebs, who first postulated the existence of the cycle in 1937 under its original name of "citric acid cycle." A cyclic sequence of chemical reactions that occurs in almost all aerobic (air-requiring) organisms. A system of enzymatic reactions in which acetyl residues are oxidized to carbon dioxide and hydrogen atoms, and in which formation of citrate is the first step. See also CITRIC ACID, CITRATE SYNTHASE, CITRATE SYNTHASE GENE, CITRATE SYNTHASE (Csb) GENE, ACID, AEROBIC, METABOLISM, ENZYME, OXIDATION.

**CKR-5 Proteins** See HUMAN IMMUNODEFICIENCY VIRUS TYPE 1 (HIV-1), HUMAN IMMUNODEFICIENCY VIRUS TYPE 2 (HIV-2), RECEPTORS, PROTEIN.

**CLA** Abbreviation for Conjugated Linoleic Acid. See also CONJUGATED LINOLEIC ACID.

**Clades** The taxonomic subgroups within cladistics. See also CLADISTICS.

**Cladistics** Initially popularized by Willi Hennig's 1950 book, *Phylogenetic Systematics*, cladistics is a system of taxonomic classification of organisms (and/or their specimens) based upon (determined) similar lines of selected shared traits. See also CLADES, TYPE SPECIMEN, GENETICS, BIOLOGY, SPECIES, SYSTEMATICS, AMERICAN TYPE CULTURE COLLECTION (ATCC), TRAIT.

**Clinical Trial** One of the final stages in the collection of data (for drug approval prior to commercialization) in which the new drug is tested in human subjects. Used to collect data on effectiveness, safety, and required dosage. See also PHASE I CLINICAL TESTING, FOOD AND DRUG ADMINISTRATION (FDA), KOSEISHO, BUNDESGESUNDHEITSAMT (BGA), COMMITTEE ON SAFETY IN MEDICINES, COMMITTEE FOR PROPRIETARY MEDICINAL PRODUCTS (CPMP).

**Clone (a molecule)** To create copies of a given molecule via various methods. See also POLYMERASE CHAIN REACTION (PCR), MONOCLONAL ANTIBODIES (MAb), COCLONING, ANTIBODY, cDNA CLONE.

**Clone (an organism)** A group of individual organisms (or cells) produced from one individual cell through asexual processes that do not involve the interchange or combination of genetic material. As a result, members of a clone have identical genetic compositions. For example, many plants reproduce asexually (without sex) via a process known as apomixis. Protozoa, bacteria, and some animals (e.g., the anemone *Anthopleura elegantissima*) can reproduce asexually by binary fission, a process in which a single-celled organism undergoes cell division. The result is two cells with identical genetic composition. When these two identical cells divide, the result is four cells with identical genetic composition. These identical offspring are all members of a clone. The word "clone" may be used either as a noun or a verb. Scientists have cloned some adult mammals via nuclear transfer. In that process, the nucleus of an oocyte is removed and replaced with a nucleus taken from another conventional somatic (adult's body) cell. That oocyte can then grow up to become a clone of the (adult) animal. See also ORGANISM, APOMIXIS, BACTERIA, CELL, OOCYTES, SOMATIC CELLS.

*Clostridium* A genus of bacteria. Most are obligate anaerobes, and form endospores. See also ANAEROBE, ENDOSPORE.

**CMC** See CRITICAL MICELLE CONCENTRATION.

**CML** Abbreviation for Chronic Myelogenous Leukemia (also known as Chronic Myeloid Leukemia, or Chronic Myelocytic Leukemia). See also GLEEVEC™.

**CMV** Acronym for Cucumber Mosaic Virus.

**CNTF** See CILIARY NEUROTROPHIC FACTOR (CNTF).

**Co-chaperonin** A protein molecule inside living cells that "works together" with applicable chaperonin(s) to help ensure proper folding of the (new) protein molecules that are synthesized (manufactured) in the cell's ribosomes. See also CHAPERONINS, PROTEIN, PROTEIN FOLDING, CELL, RIBOSOMES, CONFORMATION.

**CoA** See COENZYME A.

**Coccus** A spherical-shaped bacterium. See also BACILLUS.

**Cocloning** (of molecules) The additional (accidental) cloning (i.e., copying) of extra molecular fragments, other than the desired one, that sometimes occurs when a scientist is attempting to clone a molecule. See also CLONE (A MOLECULE), POLYMERASE CHAIN REACTION (PCR), Q-BETA REPLICASE TECHNIQUE.

**Codex Alimentarius** See CODEX ALIMENTARIUS COMMISSION.

**Codex Alimentarius Commission** An international regulatory body that is part of the United Nations' Food and Agriculture Organization (FAO), it is one of the three international SPS (sanitary and phytosanitary) standard-setting organizations recognized by the World Trade Organization (WTO). Created in 1962 by the UN's FAO and the World Health Organization (WHO), the commission has 165 member nations.

In Latin, *codex alimentarius* means food law or food code. Responsible for execution of the Joint FAO/WHO Food Standards Program, the Codex Alimentarius standards are a set of international food mandates adopted by the organization. With delegates from member country governmental agencies, the Codex Secretariat is headquartered in Rome, Italy. The commission periodically determines, then publishes, a list of food ingredients and maximum allowable levels that it deems safe for human consumption (known as the *codex alimentarius*). See also MAXIMUM RESIDUE LEVEL (MRL), SPS, INTERNATIONAL PLANT PROTECTION CONVENTION (IPPC), INTERNATIONAL OFFICE OF EPIZOOTICS (OIE), WORLD TRADE ORGANIZATION (WTO).

**Coding Sequence** The region of a gene (DNA) that encodes the amino acid sequence of a protein. See also GENETIC CODE, INFORMATIONAL MOLECULES, GENE, MESSENGER RNA (mRNA), BASE (NUCLEOTIDE), CONTROL SEQUENCES.

**Codon** A triplet of nucleotides [three nucleic acid units (residues) in a row] that code for

an amino acid (triplet code) or a termination signal. See also GENETIC CODE, TERMINATION CODON (SEQUENCE), AMINO ACID, NUCLEOTIDE, INFORMATIONAL MOLECULES, MESSENGER RNA (mRNA).

**Coenzyme** A nonproteinaceous organic molecule required for the action of certain enzymes. The coenzyme contains as part of its structure one of the vitamins. This is why vitamins are so critically important to living organisms. Sometimes the same coenzyme is required by different enzymes involved in the catalysis of different reactions. By analogy, a coenzyme is like a part of a car, such as a tire, which can be identified in and of itself and which can, furthermore, be removed from the car. The car (enzyme), however, must of necessity have the tire in order to carry out its prescribed function. Coenzymes have been classified into two large groups: fat soluble and water soluble. Examples of a few water-soluble vitamins are: thiamin, biotin, folic acid, vitamin C, and vitamin $B_{12}$. Examples of fat-soluble vitamins are: vitamins A, D, E, and K. See also ENZYME, CATALYST, HOLOENZYME, VITAMIN, POLYPEPTIDE (PROTEIN), BIOTIN.

**Coenzyme A** A water-soluble vitamin known as pantothenic acid. A coenzyme in all living cells, it is required by certain condensing enzymes and functions in acyl-group transfer and in fatty-acid metabolism. Abbreviated CoA. See also ENZYME, FATS, FATTY ACID.

**Cofactor** A nonprotein component required by some enzymes for activity. The cofactor may be a metal ion or an organic molecule called a coenzyme. The term cofactor is a general term. Cofactors are generally heat stable. See also COENZYME, HOLOENZYME, MOLECULAR WEIGHT.

**Cofactor Recycle** The regeneration of a spent cofactor by an auxiliary reaction such that it may be reused many times over by a cofactor-requiring enzyme during a reaction. See also COFACTOR, HOLOENZYME, ENZYME.

**Cohesive Termini** See STICKY ENDS.

**Cold Acclimation** See COLD HARDENING.

**Cold Acclimatization** See COLD HARDENING.

**Cold Hardening** A process of acclimatization in which certain organisms produce specific proteins that protect them from freezing to death during the winter. Among other organisms, the common housefly, the *Arabidopsis thaliana* plant, the fruit fly *Drosophila*, and "no-see-ems" (*Culicoides variipennis*) can produce these proteins (during the gradually decreasing temperatures of a typical autumn season in North America). The amount of such proteins produced within their bodies is proportional to the severity and duration of the cold experienced. For example, prior to cold hardening, *Culicoides variipennis* insects usually die after exposure for two hours to a temperature of 14°F (–10°C). If those insects are first exposed for one hour to a temperature of 41°F (5° C), approximately 98% of these insects can then survive exposure for three days to a temperature of 14° F (–10°C). See also ACCLIMATIZATION, PROTEIN, LOW-TILLAGE CROP PRODUCTION, NO-TILLAGE CROP PRODUCTION, *DROSOPHILA*, *ARABIDOPSIS THALIANA*, CBF1, TRANSCRIPTION FACTORS, LINOLEIC ACID.

**Cold Tolerance** See COLD HARDENING.

**Colicins** Proteins produced by *Escherichia coli* (*E. coli*) , that are toxic (primarily) to other closely-related strains of bacteria. The particular *E. coli* that produce a given colicin are generally unaffected by the colicin that they produce. See also BACTERIOCINS, BACTERIOLOGY, STRAIN, BACTERIA, PROTEIN, TOXIN, *ESCHERICHIA COLIFORM* (*E. COLI*).

**Collagen** The major structural protein in connective tissue. It is instrumental in wound healing [stimulated by fibroblast growth factor (FGF), platelet-derived growth factor, and insulin-like growth factor-1]. See also PROTEIN, FIBROBLAST GROWTH FACTOR (FGF), PLATELET-DERIVED GROWTH FACTOR (PDGF), INSULIN-LIKE GROWTH FACTOR-1 (IGF-1).

**Collagenase** An enzyme that catalyzes the cleavage of collagen, such as when bacteria in the mouth cause production of collagenase that then cleaves (breaks down) the collagen that holds teeth in place. Some cancers use collagenase to break down connective tissues in the body they inhabit, enabling the cancers to form the (new) blood vessels that nourish those cancers and help those cancers spread through the body. Collagenase may also be responsible indirectly for certain autoimmune diseases such as arthritis, by

breaking down the protective proteoglycan coat that covers cartilage in the body. See also STROMELYSIN (MMP-3), PROTEOLYTIC ENZYMES, ENZYME, COLLAGEN, CANCER, AUTOIMMUNE DISEASE.

**Colony** A growth of a group of microorganisms derived from one original organism. After a sufficient growth period, the growth is visible to the eye without magnification.

**Colony Hybridization** A technique using *in situ* hybridization to identify bacterial colonies carrying inserted DNA that is homologous with some particular sequence (probe). See also DNA PROBE, HOMOLOGY, *IN SITU*, REGULATORY SEQUENCE.

**Colony Stimulating Factors (CSFs)** Specific glycoprotein growth factors required for the proliferation and differentiation of hematopoietic progenitor cells. Different CSFs stimulate the growth of different cells. See also MACROPHAGE COLONY STIMULATING FACTOR (M-CSF), GRANULOCYTE COLONY STIMULATING FACTOR (G-CSF), GRANULOCYTE-MACROPHAGE COLONY STIMULATING FACTOR (GM-CSF), EPIDERMAL GROWTH FACTOR (EGF), FIBROBLAST GROWTH FACTOR (FGF), HEMATOLOGIC GROWTH FACTORS (HGF), INSULIN-LIKE GROWTH FACTOR-1 (IGF-1), MEGAKARYOCYTE STIMULATING FACTOR (MSF), NERVE GROWTH FACTOR (NGF), PLATELET-DERIVED GROWTH FACTOR (PDGF), TRANSFORMING GROWTH FACTOR-ALPHA (TGF-ALPHA), TRANSFORMING GROWTH FACTOR-BETA (TGF-BETA).

**Combinatorial Biology** A term used to describe the set of DNA technologies used to generate a large number of samples of new chemicals (metabolites) via creation of non-natural metabolic pathways. The collection of samples thus generated is called a "library," and the samples are then tested for potential use (e.g., for therapeutic effect, in the case of a pharmaceutical). These technologies enable greater efficiency in a pharmaceutical researcher's screening process for drug discovery. See also COMBINATORIAL CHEMISTRY, TARGET, MOLECULAR DIVERSITY, METABOLISM, INTERMEDIARY METABOLISM, METABOLITE, RECEPTORS.

**Combinatorial Chemistry** A term used to describe the set of technologies utilized to generate a large number of samples of (new) chemicals, which are then tested (screened) for potential use (e.g., for therapeutic effect, in the case of a pharmaceutical). These large numbers of chemical samples, thus generated, are called a "library" and are screened (e.g., for therapeutic effect) via a variety of laboratory, biosensor, computational, receptor, or animal tests. Combinatorial chemistry was made feasible by H. Mario Geysen, who, during the 1980s, developed a methodology to synthesize arrays of peptides on pin-shaped solid supports. In addition, Richard A. Houghten developed a technique for creation of peptide libraries in small mesh "bags" by solid-phase parallel synthesis; thereby enabling automation of the process.

For a library that is used for new drug (candidate) screening, high diversity in molecular structure among the chemicals in the library is desired to increase the efficiency of the screening process. One method used to measure diversity of the molecular structure among samples in a library is called "molecular fingerprinting." If two samples are identical in molecular structure, the "fingerprint" coefficient is 1.0. If two samples are totally dissimilar in molecular structure, the coefficient is 0. The diversity of a library is measured by comparing each sample's molecular structure to that of all the others in the library. See also COMBINATORIAL BIOLOGY, TARGET, MOLECULAR DIVERSITY, RECEPTORS, BIOSENSORS (ELECTRONIC), PEPTIDE, SYNTHESIZING (OF PROTEINS), BIOCHIPS, HIGH-THROUGHPUT SCREENING, TARGET-LIGAND INTERACTION SCREENING.

**Combinatorics** See COMBINATORIAL CHEMISTRY.

**Combining Site** The site on an antibody molecule that locks (binds) onto an epitope (hapten). See also ANTIBODY, EPITOPE, ENGINEERED ANTIBODIES, HAPTEN, CATALYTIC ANTIBODY.

**Commensal** A term that literally means eating at the same table; it is used to refer to organisms such as the house mouse (*Mus musculus*), that tend to thrive alongside/among humans. For example, the numerous strains of *Salmonella* bacteria can live within the intestine of an adult cow without harming that cow, but would be pathogenic (disease-causing) in a human's intestine. Similarly, the *E. coli* 0157:H7 strain of *Escherichia coliform* bacteria can live within the digestive system of an adult cow without harm, but is pathogenic in a human's digestive system. However, hundreds of other strains of *E. coli* bacteria live within the digestive system of humans, without causing harm to the human body. See also ORGANISM, MICROORGANISM, BACTERIA, SALMONELLA TYPHIMURIUM, SALMONELLA ENTERITIDIS, PATHOGEN, PATHOGENIC, STRAIN, ESCHERICHIA COLIFORM (*E. COLI*), ESCHERICHIA COLIFORM 0157:H7.

**Commission E Monographs** Documents published by the government of Germany, which detail the proven safety and efficacy of certain phytochemical-containing herbs (approved by the German government). For example, consumption of St. John's Wort (a plant native to Europe) is approved in Germany for treatment of depressive mood disorders, anxiety, and nervous unrest. See also PHYTOCHEMICALS.

**Commission of Biomolecular Engineering** An agency of the French government, established to oversee and regulate all genetic engineering activities in France. See also GENETIC ENGINEERING, IOGTR, RECOMBINANT DNA ADVISORY COMMITTEE (RAC), ZKBS (CENTRAL COMMITTEE ON BIOLOGICAL SAFETY), INDIAN DEPARTMENT OF BIOTECHNOLOGY, GENE TECHNOLOGY REGULATOR (GTR), GENE TECHNOLOGY OFFICE.

**Committee for Proprietary Medicinal Products (CPMP)** The European Union's (EU's) scientific advisory organization dealing with new human pharmaceuticals approval. Its recommendations (e.g., to either approve or not approve a new product) are usually adopted by the European Medicines Evaluation Agency (EMEA), to which the CPMP reports. Within 60 days of a CPMP "approval for recommendation" being adopted by the EMEA, each of the EU's member countries must advise the EMEA of its progress toward a regulatory decision on that pharmaceutical's submission for approvals. See also FOOD AND DRUG ADMINISTRATION (FDA), KOSEISHO, EUROPEAN MEDICINES EVALUATION AGENCY (EMEA), COMMITTEE ON SAFETY IN MEDICINES, BUNDESGESUNDHEITSAMT (BGA).

**Committee for Veterinary Medicinal Products (CVMP)** The European Union's (EU's) scientific advisory organization dealing with approvals of new medicinal products intended for use in animals. Its recommendations (e.g., to either approve or not approve a new product) are usually adopted by the European Medicines Evaluation Agency (EMEA). See also COMMITTEE FOR PROPRIETARY MEDICINAL PRODUCTS (CPMP), FOOD AND DRUG ADMINISTRATION (FDA), KOSEISHO, COMMITTEE ON SAFETY IN MEDICINES, MEDICINES CONTROL AGENCY (MCA), EMEA, BUNDESGESUNDHEITSAMT (BGA).

**Committee on Safety in Medicines** The British government agency that must approve new pharmaceutical products for sale within the United Kingdom. In concert with the Medicines Control Agency (MCA), it regulates all pharmaceutical products in the U.K. It is the equivalent of the U.S. Food and Drug Administration. See also FOOD AND DRUG ADMINISTRATION (FDA), MEDICINES CONTROL AGENCY (MCA), COMMITTEE FOR PROPRIETARY MEDICINAL PRODUCTS (CPMP), KOSEISHO, NDA (TO KOSEISHO), IND, BUNDESGESUNDHEITSAMT (BGA), EMEA.

**Community Plant Variety Office** An agency of the European Union established by Council Regulation 2100/94; and located in Angers, France. It applies UPOV rules across all countries of the European Union when a plant breeder registers a new plant variety at the Community Plant Variety Office. Thus, it confers and protects plant breeder's rights (PBR) across the entire European Union in a manner analogous to the way the European Patent Office (EPO) confers patent rights (for patented inventions) across the entire European Union. See also UNION FOR PROTECTION OF NEW VARIETIES OF PLANTS (UPOV), PLANT BREEDER'S RIGHTS (PBR), EUROPEAN PATENT OFFICE (EPO), PLANT VARIETY PROTECTION ACT (PVP).

**Comparative Analysis** See HOMOLOGOUS (CHROMOSOMES OR GENES).

**Competence Factor** See PLATELET-DERIVED GROWTH FACTOR (PDGF).

**Complement** (component of immune system) A group of more than 15 soluble proteins found in blood serum that interacts in a sequential fashion, in which a precursor molecule is converted into an active enzyme. Each enzyme uses the next molecule in the system as a substrate and converts it into its active (enzyme) form. This cascade of events and reactions leads ultimately to the formation of an attack complex that forms a transmembrane channel in the cell membrane. It is the presence of the channel that leads to lysis (rupturing) of the cell. See also PLASMA MEMBRANE, CELL, CASCADE, COMPLEMENT CASCADE, CECROPHINS, HUMORAL IMMUNITY, LYSE, LYSIS.

**Complement Cascade** The precisely regulated, sequential interaction of proteins (in the blood) triggered by a complex of antibody and antigen to cause lysis of infected cells. The triggering of lysis by multivalent antibody-antigen complexes is mediated by the classical pathway, beginning with the activation of C1, the first component (protein) of the pathway. This activation step, in which C1 undergoes conversion from a zymogen to an active protease, results in sequential cleavage of the C4, C2, C3, and C5 components (proteins). C5b, a fragment of C5, then joins C6, C7, and C8 to penetrate the (cell) membrane bearing the antigen. Finally, the binding of some 16 molecules of C9 to this "bridgehead" produces large pores in the (cell) membrane, which cause the lysis and destruction of the target cell. See also ANTIBODY, ANTIGEN, LYSIS, CELL, PLASMA MEMBRANE, COMPLEMENT, ZYMOGENS, CECROPHINS, CASCADE, PATHWAY.

**Complementary (molecular genetics)** Refers to strands of DNA that will hybridize (bind) to each other, due to one-for-one matchup of each strand's sequence of nucleotides. Any sequence (within the two strands) that does not match up one-for-one will not hybridize to the respective sequence (in the adjacent strand). See also MOLECULAR GENETICS, HYBRIDIZATION (MOLECULAR GENETICS), DEOXYRIBONUCLEIC ACID (DNA), DOUBLE HELIX, NUCLEOTIDE, MICROARRAY (TESTING), BIOMOTORS, SOUTHERN BLOT ANALYSIS.

**Complementary DNA (cDNA)** A single-stranded DNA that is complementary to a strand of mRNA. The DNA is synthesized *in vitro* by an enzyme known as reverse transcriptase. Then, a second DNA strand is synthesized via the enzyme known as DNA polymerase. Complementary DNA is often utilized in hybridization studies and in microarrays (e.g., to detect/identify genes) because cDNAs usually don't contain regulatory sequences of DNA, since the cDNA was copied from mRNA (messenger RNA). This "rebukes" the (old) Central Dogma. See also cDNA, DEOXYRIBONUCLEIC ACID (DNA), MESSENGER RNA (mRNA), CENTRAL DOGMA (OLD), DNA POLYMERASE, HYBRIDIZATION (MOLECULAR GENETICS), MICROARRAY (TESTING), REGULATORY SEQUENCE.

**Compound Q** See TRICHOSANTHIN.

**Computer Assisted New Drug Application** (also called Computer Assisted NDA). See also CANDA.

**Con-Till** An abbreviation that refers to conservation tillage farming practices. See also CONSERVATION TILLAGE, LOW-TILLAGE CROP PRODUCTION, NO-TILLAGE CROP PRODUCTION, GLOMALIN.

**Configuration** The three-dimensional arrangement in space of substituent groups in stereoisomers.

**Conformation** The three-dimensional arrangement of substituent groups in a protein or other molecular structure free to assume different positions. The geometric form or shape of a protein in three-dimensional space. See also NATIVE CONFORMATION, TERTIARY STRUCTURE, EFFECTOR, PROTEIN FOLDING, PROTEOMICS, TRANSCRIPTOME.

**Conjugate** A molecule created by fusing together (via recombination or chemically) two unlike (different) molecules. The purpose is to create a molecule in which one of the original molecules has one function, i.e., a toxic, cell-killing function, while the other original molecule has another function, such as targeting the toxin to a specific site which might include cancerous cells. For example, molecules of interleukin-2 (IL-2) have been fused with molecules of diphtheria toxin to create a conjugate that does the following:

1. It enters leukemia and lymphoma cells. Because these two types of cancer cells possess IL-2 receptors on their surfaces, the IL-2 (targeting function) binds to that receptor and is internalized.
2. The diphtheria toxin (killing function) then shuts down protein synthesis within the cancer cells.
3. It then kills the cancerous cells.

This type of approach is widespread and there are many different types of conjugates. One consists of enzymes used in the treatment of certain molecular diseases attached covalently to polyethylene glycol (PEG). In this case the PEG greatly diminishes both the immunogenicity (the ability to induce an immune reaction) and the antigenicity (the ability to react with preformed antibodies). Antibodies may be used as vectors to carry both relatively small molecules of destructive chemicals or proteins to specific sites (cells) within the body. Antibodies may be coupled to enzymes, toxins, and/or ribosome-inhibiting proteins, as well as to radioisotopes. These conjugates are known collectively as immunoconjugates. See also IMMUNOCONJUGATE, CONJUGATED PROTEIN, "MAGIC BULLET", FUSION PROTEIN, RECOMBINATION, TOXIN, INTERLEUKIN-2 (IL-2), RICIN, ABRIN, RECEPTORS, RIBOSOMES, MESSENGER RNA (mRNA), DIPHTHERIA TOXIN.

**Conjugated Linoleic Acid (CLA)** A naturally occurring $n$-6 polyunsaturated fatty acid (PUFA) discovered in 1979, whose consumption by humans has been linked to reduction in risk for atherosclerosis, reduction in blood triglyceride levels, reduction in body fat (adipose tissue) in obese humans, and reduction in risk for breast cancer, skin cancer, and some other types of cancer. CLA exhibits powerful antioxidant properties (i.e., it "quenches" free radicals). Chemically, CLA consists of two linoleic acid molecules linked together by a chemical bond, so it is a dimer.

Foods that are naturally highest in CLA content include beef, lamb, full-fat milk, butter, cheese, some creams, and full-fat yogurt. Feeding of soybean oil (in feed rations) to livestock has been proven to increase CLA content in the resultant meat. In 1998, T.R. Dhiman showed that feeding of soybean oil containing (i.e., whole) soybeans to dairy cattle also increased the content of CLA in their milk. Research conducted during the 1990s indicated that consumption of CLA (by humans, swine, rats, etc.) causes the bodies of those animals to change the way they utilize and store energy. Thus, the body requires less food to perform at the same level. The body also tends to produce less body fat (adipose tissue) and more lean protein (muscle) tissue. See also POLYUNSATU- RATED FATTY ACIDS (PUFA), FATS, LINOLEIC ACID, ATHEROSCLEROSIS, OXIDATIVE STRESS, ANTIOXI- DANTS, SOYBEAN OIL, ADIPOSE, CANCER, VOLIC- ITIN, OLIGOMER.

**Conjugated Protein** A protein containing a metal or an organic prosthetic group, or both. For example, a glycoprotein is a conjugated protein bearing at least one oligosaccharide group. See also PROSTHETIC GROUP, GLYCOPRO- TEIN, PROTEIN, OLIGOSACCHARIDES, CONJUGATE, CD4-PE40.

**Conjugation** A process akin to sexual repro- duction occurring in bacteria; mating in bac- teria. A process that involves cell-to-cell contact and the one-way transfer of DNA from the donor to the recipient. In contrast to some other DNA-transfer processes of bacteria, conjugation may involve the trans- fer of large portions of the genome. The discovery caused considerable controversy at the time. See also TRANSFORMATION, BAC- TERIA, TRANSDUCTION (GENE), TRANSDUCTION (SIGNAL), DEOXYRIBONUCLEIC ACID (DNA), GENOME, SEXUAL CONJUGATION.

**Consensus Sequence** The nucleotide sequence (within a DNA molecule) which gives the most common nucleotide at each position (along that sequence of that DNA molecule), for those instances (in certain organisms) where a (usually small) number of variations in nucleotide sequences can occur (e.g., for a given nucleotide sequence such as a pro- moter sequence). See also NUCLEOTIDE, DEOXY- RIBONUCLEIC ACID (DNA), SEQUENCE (OF A DNA MOLECULE), GENETIC CODE, GENE, PROMOTER, PHARMACOGENOMICS.

**Conservation Tillage** Refers to crop produc- tion (farming) techniques/practices such as low-tillage crop production, no-tillage crop production, etc. that avoid or minimize the disturbance of topsoil. See also LOW-TILLAGE CROP PRODUCTION, NO-TILLAGE CROP PRODUC- TION, GLOMALIN.

**Conserved** A term used to describe:

1. The number of genes present within the DNA of more than one species. For example, approximately 25% of the genes found within the human genome (DNA) are also found within the DNA of plants.
2. A particular domain (region) of a mol- ecule on the surface of a rapidly mutat- ing microorganism (e.g., the influenza virus, the AIDS virus) that remains the same in all, or most, variations of that microorganism.

If that conserved region is suitable to act as an antigen (hapten, epitope), it may be pos- sible to create a successful vaccine against that microorganism, that would otherwise be unsuccessful due to the fact that the rapid mutation would cause it (e.g., the AIDS virus) to appear to be different than the one (antigen) the vaccine was designed against. See also DOMAIN (OF A PROTEIN), GP120 PROTEIN, SUPERANTIGENS, MUTATION, ACQUIRED IMMUNE DEFICIENCY SYNDROME (AIDS), ANTIGEN, HAPTEN, EPITOPE, VIRUS, GENE, DEOXYRIBONUCLEIC ACID (DNA), HIV-1 AND HIV-2.

**Consortia** Microorganisms that interact with each other (or at least "coexist peacefully") when growing together. An example of such interaction/coexistence would be bioleach- ing. See also BIOLEACHING, BIORECOVERY, BIODESULFURIZATION, BIOSORBENTS.

**Constitutive Enzymes** Enzymes that are part of the basic, permanent enzymatic machin- ery of the cell. They are formed at a constant rate and in constant amounts regardless of the metabolic state of the organism. For example, enzymes that function in the pro- duction of cell-usable energy (such as ATP) might be good candidates. And this, in fact, is the case with the enzymes of the glycolytic sequence, which is the most ancient energy- yielding catabolic pathway. See also ENZYME, METABOLISM.

**Constitutive Genes** Expressed as a function of the interaction of RNA polymerase with the promoter, without additional regulation.

C

**C**

They are sometimes also called "household genes" in the context of describing functions expressed in all cells at a low level. See also GENE, RNA POLYMERASE, PROMOTER.

**Constitutive Heterochromatin** The inert state of permanently nonexpressed sequences, usually satellite DNA. See also EXPRESS, COD-ING SEQUENCE, DEOXYRIBONUCLEIC ACID (DNA), CHROMATIN.

**Constitutive Mutations** Mutations (changes in DNA) that cause genes which are noncon-stitutive (have controlled protein expression) to become constitutive (in which state the protein is expressed all of the time). See also CONSTITUTIVE GENES, MUTATION, REGULATORY SEQUENCE, PROTEIN.

**Construct** See CASSETTE, TRANSGENE.

**Consultative Group on International Agri-cultural Research (CGIAR)** An organiza-tion that is cosponsored by the Rome-based United Nations Food and Agriculture Orga-nization (FAO), the United Nations Devel-opment Programme (UNDP), and the World Bank. The CGIAR is an association of 58 public and private donors that jointly support 16 international agricultural research centers located primarily in developing countries. Twelve of the research centers have collec-tively assembled 500,000 different preserved samples (i.e., germplasm) of major food, for-age, and forest plant species into a gene bank. This, the world's largest internation-ally held collection of genetic resources, was legally placed under the auspices of the FAO in 1994 in order "to hold the collection in trust for the international community." Since 1970, CGIAR's collection has supported research efforts to develop better varieties of staple foods consumed primarily in develop-ing countries of the world. See also AMERICAN TYPE CULTURE COLLECTION (ATCC), TYPE SPECI-MEN, GERMPLASM.

**Contaminant** By definition, any unwanted or undesired organism, compound, or molecule present in a controlled environment. Unwanted presence of an entity in an other-wise clean or pure environment.

**Continuous Perfusion** A type of cell culture in which the cells (either mammalian or oth-erwise) are immobilized in part of the sys-tem, and nutrients/oxygen are allowed to flow through the stationary cells, thus effect-ing nutrient-waste exchange. Ideally the sys-tem incorporates features that retard the activity of proteolytic enzymes, and reduce the need for anti-infective agents (e.g., anti-biotics) and fetal bovine serum, which are required by most other cell culture systems. Continuous perfusion is used because, among other things, it eliminates the need to separate the cells from the culture medium when fresh medium is exchanged for old. See also MAMMALIAN CELL CULTURE, ENZYME, PROTEOLYTIC ENZYMES.

**Control Sequences** Those sequences of DNA adjacent to a gene (in genome) and "turn on" and/or "turn off" that gene. See also SEQUENCE (OF A DNA MOLECULE), GENE, GENOME, PROMOTER, TERMINATION CODON (TERMINATOR SEQUENCE), BASE (NUCLEOTIDE), CODING SEQUENCE.

**Convention on Biological Diversity (CBD)** The international treaty governing the con-servation and use of biological resources around the world that was signed by more than 150 countries at the 1992 United Nations Conference on Environment and Development. Article 19.4 of the CBD called for establishment of a "protocol on biosafety" to govern the transnational-boundary move-ment of nonindigenous living organisms. See also MEA, CONSULTATIVE GROUP ON INTERNA-TIONAL AGRICULTURAL RESEARCH (CGIAR), INTERNATIONAL PLANT PROTECTION CONVENTION (IPPC), BIODIVERSITY.

**Convergent Improvement** See TRANSGRESSIVE SEGREGATION.

**Coordinated Framework for Regulation of Biotechnology** The regulatory "frame-work" through which the U.S. evaluates and approves new products derived via biotech-nology. The Coordinated Framework assigns specific regulatory tasks to each of the U.S. government's applicable agencies (see below). For example, the U.S. Environmen-tal Protection Agency (EPA) is assigned to evaluate and regulate all genetically modi-fied pest protected (GMPP) new plants, in terms of their impact on pests. The U.S. Food and Drug Administration (FDA) is assigned to evaluate and regulate all new food crops derived via biotechnology, in terms of their potential impact on food safety (allergenicity, toxicity, etc.). The U.S. Department of Agriculture (USDA) is assigned to evaluate and regulate all new plants derived via biotechnology, in terms of field (outdoor) testing and of potential impact on the environment such as weediness. See also BIOTECHNOLOGY, FOOD AND DRUG ADMINISTRATION (FDA), GENETICALLY MODIFIED PEST PROTECTED (GMPP) PLANTS, ALLERGIES (FOODBORNE), APHIS.

**Coordination Chemistry** See CHELATION.

**Copy DNA (C-DNA)** See c-DNA.

**Copy Number** The number of molecules (copies) of an individual plasmid or plastid typically present in a single (e.g., bacterial for plasmid, plant for plastid) cell. Each plasmid has a characteristic copy number value ranging from 1 to 50 or more. Higher copy numbers result in a higher yield of the protein encoded for by the plasmid gene in each cell. See also PLASMID, PLASTID, PROTEIN, GENE, EXTRANUCLEAR GENES, GENETIC CODE, MULTI-COPY PLASMIDS.

**Corepressor** A small molecule that combines with the repressor to trigger the shutting down of transcription. See also TRANSCRIPTION.

**Corn** The domesticated plant *Zea mays* L. also known as maize. A green, leafy (grain) plant that is one of the world's largest providers of edible starch and fructose (sugar) for mankind's use. This summer annual varies in height from 2 feet (0.5 meter) to more than 20 feet (6 meters) tall. The seeds (kernels) are borne in cobs, ranging in size from 2 feet long to smaller than a man's thumb. Due to genetic variation (of different hybrids/varieties), the fraction of kernel that consists of recoverable starch varies between 42 and 73% for different corn varieties. Due to genetic variation (of different hybrids/varieties), the fraction of kernel that consists of protein varies between 8 and 10%, but that protein content can be increased by 10% by inserting the glutamate hydrogenase (GDH) gene into the corn plant. Due to genetic variation, the fraction of kernel that consists of oil varies between 3.5 and 8.5% for different varieties.

Grown widely in the world's temperate zones, corn is grown as far north as latitude 58° in Canada and Russia and as far south as latitude 40° in the Southern Hemisphere.

During the 1980s, scientists were able to insert genes from *Bacillus thuringiensis* (*B.t.*) bacteria into the corn plant to make that plant resistant to certain insects. During the 1990s, scientists were able to insert genes into the corn plant to make it tolerant to certain herbicides and to cause the corn plant to produce monoclonal antibodies (MAb). Some of the major economic pests of corn include the European corn borer (*Ostrinia nubilalis*), corn earworm (*Helicoverpa zea*), corn rootworm (*Diabrotica virgifera virgifera*), and beet armyworm (*Pseudaletia unipuncta*). See also HYBRIDIZATION (PLANT GENETICS), *BACILLUS THURINGIENSIS* (*B.t.*), PROTEIN, STRESS PROTEINS, CRY PROTEINS, CRY1A (b) PROTEIN, CRY1A (c) PROTEIN, CRY9C PROTEIN, GENE, "STACKED" GENES, OPAGUE-2, HIGH-METHIONINE CORN, HIGH-LYSINE CORN, *B.t. KURSTAKI*, VALUE-ENHANCED GRAINS, *HELICOVERPA ZEA* (*H. ZEA*), CHLOROPLAST TRANSIT PEPTIDE (CTP), HERBICIDE-TOLERANT CROP, HIGH-OIL CORN, EUROPEAN CORN BORER (ECB), AFLATOXIN, *FUSARIUM*, CORN ROOTWORM, VOLICITIN, GA21, TRANSPOSON, GLUTAMATE DEHYDROGENASE, BLACK-LAYERED (CORN), MONOCLONAL ANTIBODIES (MAb), PHOTORHABDUS LUMINESCENS, CHOLESTEROL OXIDASE.

**Corn Borer** See EUROPEAN CORN BORER (ECB), ASIAN CORN BORER.

**Corn Earworm** See *HELICOVERPA ZEA* (*H. ZEA*), CORN.

**Corn Rootworm** A complex of several strains of beetles referring to the larva stage of the corn rootworm beetle (*Diabrotica virgifera virgifera*), which historically has laid its eggs on corn/maize (*Zea mays* L.) plants. When they hatch, the larva must feed on the roots of the corn/maize plant in order to live. Some strains of *Bacillus thuringiensis* (*B.t.*) have proven to be effective against the corn rootworm, when sprayed onto them or genetically engineered into the corn/maize plant. In 1992, a new genetic variant of corn rootworm known as the "western phenotype" or Western corn rootworm (*Diabrotica virgifera virgifera* LeConte) was discovered in the U.S. It prefers to lay its eggs on soybean plants instead of corn plants. See also CORN, PHENOTYPE, SOYBEAN PLANT, STRAIN, *BACILLUS*

*THURINGIENSIS* (*B.t.*), GENETIC ENGINEERING, CRY3BB PROTEIN, CRW, ANTIBIOSIS.

**Coronary Heart Disease (CHD)** A disease of the heart and arteries, in which (among other effects) cholesterol is deposited on the interior walls (lumen endothelium), where it can sometimes later break off and cause death (via heart attack). Risk factors (increased risk) for CHD include high blood levels of triglycerides, high levels of apolipoprotein B, high levels of LDLPs/VLDLs (the two lipoproteins that are most likely to deposit cholesterol on artery walls), and/or low levels of HDLPs (the lipoproteins that help to clear away cholesterol deposits from artery walls). A human diet containing a large amount of certain phytosterols (e.g., campesterol, beta-sitosterol, and stigmasterol) has been shown to lower total serum (blood) cholesterol and low-density lipoproteins (LDLP) levels by approximately 10%; and thereby lower the risk of CHD. A human diet containing a large amount of oleic acid causes lower blood cholesterol levels and thus lower risk of CHD and atheroslerosis. See also CHOLESTEROL, LOW-DENSITY LIPOPROTEINS (LDLP), SITOSTEROL, VERY LOW-DENSITY LIPOPROTEINS (VLDL), HIGH-OLEIC OIL SOYBEANS, PHYTOSTEROLS, STEROLS, CAMPESTEROL, HIGH-DENSITY LIPOPROTEINS (HDLP), BETA-SITOSTEROL (β-SITOSTEROL), STIGMASTEROL, SERUM LIFETIME, LYCOPENE, ATHEROSCLEROSIS, RESVERATROL, LUMEN, ENDOTHELIUM, TRIGLYCERIDES.

**Corticotropin** See ACTH.

**Cosuppression** A significant decrease ("silencing") in the expression of a gene (within an organism's genome/DNA) that (often) results when man inserts and causes a homologous gene to be expressed. For example, high-oleic oil soybeans result when the GmFad2-1 gene (which codes for native Δ12 desaturase enzyme) is inserted and expressed in traditional varieties of soybeans. That is because the inserted gene silences itself and the endogenous GmFad2-1 gene (i.e., the one naturally/originally present in the soybean plant), prevents formation of the Δ12 desaturase enzyme (which normally causes most oleic acid within soybeans to be converted into polyunsaturated acid/linoleic acid). See also GENE SILENCING, OLEIC ACID, LINOLEIC ACID, EXPRESS, GENE GENOME, HOMOLOGOUS (CHROMOSOMES OR GENES), SOYBEAN PLANT, HIGH-OLEIC OIL SOYBEANS, Δ12 DESATURASE, ANTISENSE (DNA SEQUENCE).

**Cowpea Mosaic Virus (CpMV)** A virus that infects cowpea (*Vigna unguiculata*) plants (known as black-eyed peas in the U.S.), but does not infect animals. Researchers have discovered how to cause CpMV to express certain animal virus proteins (i.e., antigens) on its surface, through genetic engineering. These virus antigens hold potential to replace the antigens currently used in vaccines, which are fraught with problems due to their production in animal cells, bacterial cells, or yeast cells. In addition, CpMV acts as an intrinsic, natural adjuvant to the (animal virus) antigens, since it provokes an immune response itself. See also VIRUS, COWPEA TRYPSIN INHIBITOR (CpTI), EXPRESS, PROTEIN, ADJUVANT (TO A PHARMACEUTICAL), IMMUNE RESPONSE, ANTIGEN.

**Cowpea Trypsin Inhibitor (CpTI)** A chemical that is naturally coded for by a certain cowpea (*Vigna unguiculata*) plant gene. It kills certain insect larvae by inhibiting digestion of ingested trypsin, thereby starving the larvae to death. See also TRYPSIN, TRYPSIN INHIBITORS, GENE, CODING SEQUENCE.

**COX** See CYCLOOXYGENASE.

**COX-1** See CYCLOOXYGENASE.

**COX-2** See CYCLOOXYGENASE.

**CP4 EPSP Synthase** See CP4 EPSPS.

**CP4 EPSPS** The enzyme 5-enolpyruvyl-shikimate-3-phosphate synthase, which is naturally produced by an Agrobacterium species (strain CP4) of soil bacteria. CP4 EPSPS is essential for the functioning of that bacterium's metabolism biochemical pathway. CP4 EPSPS happens to be unaffected by glyphosate-containing or sulfosate-containing herbicides, so introduction of the CP4 EPSPS gene into crop plants (e.g., soybeans) makes those plants essentially impervious to glyphosate-containing or sulfosate-containing herbicides. See also ENZYME, METABOLISM, GENE, GENETIC ENGINEERING, EPSP SYNTHASE, GLYPHOSATE, SULFOSATE, SOYBEAN PLANT, GLYPHOSATE OXIDASE, BACTERIA, CHLOROPLAST

TRANSIT PEPTIDE (CTP), HERBICIDE-TOLERANT CROP, PATHWAY.

**CPMP** See COMMITTEE FOR PROPRIETARY MEDICINAL PRODUCTS (CPMP).

**CpMV** See COWPEA MOSAIC VIRUS (CpMV).

**CpTI** See COWPEA TRYPSIN INHIBITOR (CpTI).

**Critical Micelle Concentration** Also known as the CMC of a surfactant, it is the lowest surfactant concentration at which micelles are formed. That is, the CMC represents that concentration of surfactant at which individual surfactant molecules aggregate into distinct, high molecular weight spherical entities called micelles. Or from another viewpoint, it represents the concentration of a surfactant, above which micelles or reverse micelles will spontaneously form through the process of selfaggregation (selfassembly). See also MICELLE, REVERSE MICELLE (RM).

**Cross Reaction** When an antibody molecule (against one antigen) can combine with (bind to) a different (second) antigen. The combination sometimes occurs because the second antigen's molecular structure (shape) is very similar to that of the first antigen. See also ANTIBODY, ANTIGEN.

**Crossing Over** The reciprocal exchange of material between chromosomes that occurs during meiosis. The event is responsible for genetic recombination. The process involves the natural breaking of chromosomes, the exchange of chromosome pieces, and the reuniting of DNA molecules. See also LINKAGE, DEOXYRIBONUCLEIC ACID (DNA), CHROMOSOMES, RECOMBINATION.

**Crown Gall** See AGROBACTERIUM TUMEFACIENS.

**CRP** See CAP.

**CRTL Gene** See GOLDEN RICE, GENE.

*Cruciferae* A taxonomic group ("family") of plants that includes canola, mustard, oilseed rape, etc. See also BRASSICA.

**CRW** Refers to one type of corn (maize) that has been made resistant to the depradations of corn rootworm larvae (*Diabrotica virgifera virgifera*) via genetic engineering. See also CRY PROTEINS, GENETIC ENGINEERING, CORN ROOTWORM, ION CHANNELS, CRY3B(b) PROTEIN

**Cry Proteins** A class of proteins produced by *Bacillus thuringiensis* (*B.t.*) bacteria (or plants into which a *B.t.* gene has been inserted). Cry ("crystal like") proteins are toxic to certain categories of insects (corn borers, corn rootworms, mosquitoes, black flies, armyworm, tobacco hornworm, some types of beetles, etc.), but harmless to mammals and most beneficial insects. See also *BACILLUS THURINGIENSIS* (*B.t.*), PROTEIN, BACTERIA, GENE, PROTOXIN, CORN, EUROPEAN CORN BORER (ECB), CORN ROOTWORM, ARMYWORM, TOBACCO HORNWORM, CRY1A (b) PROTEIN, CRY1A(c) PROTEIN, CRY3B (b) PROTEIN, CRY9C PROTEIN, ION CHANNELS, COTTON, TOXICOGENOMICS.

**Cry1A (b) Protein** One of the cry ("crystal-like") proteins, it is a protoxin that, when eaten by certain insects (e.g., *Lepidoptera* larvae such as the armyworm or tobacco hornworm or European corn borer), is toxic to those crop pest insects. However, if eaten by a mammal, the Cry1A(b) protein is digested harmlessly within one minute. See also CRY PROTEINS, PROTEIN, *B.t. KURSTAKI*, PROTOXIN, EUROPEAN CORN BORER (ECB), ARMYWORM, TOBACCO HORNWORM, ION CHANNELS.

**Cry1A (c) Protein** One of the cry ("crystal-like") proteins. See also CRY PROTEINS, ION CHANNELS.

**Cry1F Protein** One of the cry ("crystal like") proteins, it is a protoxin that, when eaten by the European corn borer, southwestern corn borer, black cutworm, and fall armyworm, is toxic to those insects. See also CRY PROTEINS, *BACILLUS THURINGIENSIS* (*B.t.*), PROTOXIN, PROTEIN, EUROPEAN CORN BORER (ECB), ARMYWORM, ION CHANNELS.

**Cry3B(b) Protein** One of the cry ("crystal-like") proteins, it is a protoxin that, when eaten by certain insects (e.g., larvae of corn rootworm *Diabrotica virgifera virgifera*), is toxic to those insects. See also PROTEIN, CRY PROTEINS, PROTOXIN, CORN ROOTWORM, ION CHANNELS.

**Cry9C Protein** One of the cry ("crystal-like") proteins, it is a protoxin that, when eaten by the European corn borer, southwestern corn-borer, black cutworm, and some species of armyworm, is toxic to those insects. See also CRY PROTEINS, *BACILLUS THURINGIENSIS* (*B.t.*), *B.t. TOLWORTHI*, PROTOXIN, PROTEIN, EUROPEAN CORN BORER (ECB), ARMYWORM, ION CHANNELS.

**CryX Protein** One of the cry ("crystal-like") proteins, it is a protein that, when eaten by corn rootworm larvae (*Diabrotica virgifera*

**C**

*virgifera*), is toxic to those insects. See also CRY PROTEINS, PROTEIN, PROTOXIN, CORN, CORN ROOTWORM, ION CHANNELS.

**CSF** See COLONY STIMULATING FACTORS (CSFs).

**CT** Refers to Conservation Tillage practices of crop production. See also LOW-TILLAGE CROP PRODUCTION, NO-TILLAGE CROP PRODUCTION, GLOMALIN.

**CTAB** See HEXADECYLTRIMETHYLAMMONIUM BROMIDE (CTAB).

**CTNBio** Acronym for Brazil's National Technical Commission on Biosafety, the Brazilian government's regulatory body for granting formal approval to a new genetically engineered plant (e.g., a genetically engineered crop to be planted). CTNBio is analogous to Germany's ZKBS (Central Commission on Biological Safety), Australia's GMAC (Genetic Manipulation Advisory Committee), Kenya's Biosafety Council, and India's Department of Biotechnology. See also GMAC, RECOMBINANT DNA ADVISORY COMMITTEE (RAC), ZKBS (CENTRAL COMMISSION ON BIOLOGICAL SAFETY), GENETIC ENGINEERING, KENYA BIOSAFETY COUNCIL, INDIAN DEPARTMENT OF BIOTECHNOLOGY.

**CTP** See CHLOROPLAST TRANSIT PEPTIDE (CTP).

**Culture** Any population of cells (bacteria, algae, protozoa, virus, yeasts, plant cells, mammalian cells, etc.) growing on, or in, a medium that supports their growth. Typically used to refer to a population of the cells of a single species or a single strain. A medium which contains only one specific organism (e.g., *E. coli* bacteria) is known as a pure culture. A culture may be preserved (stored alive) by freezing, drying (in which the cells go dormant), subculturing on an agar medium, or other methods. See also CULTURE MEDIUM, TYPE SPECIMEN, LYOPHILIZATION, AMERICAN TYPE CULTURE COLLECTION (ATCC), SPECIES, STRAIN, CELL CULTURE, MAMMALIAN CELL CULTURE.

**Culture Medium** Any nutrient system for the artificial cultivation of bacteria or other cells. The medium usually consists of a complex mixture of organic and inorganic materials. For example, the classic culture (growth) medium used for bacteria consists of nutrients (required by that bacteria) plus agar to solidify or semisolidify the nutrient containing

mass. See also MEDIUM, AGAR, CELL CULTURE, MAMMALIAN CELL CULTURE.

**Curing Agent** A substance that increases the rate of loss of plasmids during bacterial growth. See also GROWTH (MICROBIAL), PLASMID.

**Current Good Manufacturing Practices** See cGMP.

**Cut** An enzyme-induced, highly specific break in both strands of a DNA molecule (opposite one another). The enzymes involved are called restriction enzymes. See also RESTRICTION ENDONUCLEASES, ENZYME, DEOXYRIBONUCLEIC ACID (DNA).

**Cyclic AMP** A molecule of AMP (adenosine monophosphate) in which the phosphate group is joined to both the 3′ and the 5′ positions of the ribose, forming a cyclic (ring) structure. When cAMP binds to CAP, the complex is a positive regulator of procaryotic transcription. See also ADENOSINE MONOPHOSPHATE (AMP), CAP, PROCARYOTES, TRANSCRIPTION, ADENILATE CYCLASE.

**Cyclic Phosphorylation** Synthesis (manufacturing) of adenosine triphosphate (chemical reaction) that occurs during photosynthesis in plants. Also called PHOTOSYNTHETIC PHOSPHORYLATION (photophosphorylation). See also ATP SYNTHASE, ADENOSINE TRIPHOSPHATE (ATP), PHOTOSYNTHESIS, PHOTOSYNTHETIC PHOSPHORYLATION.

**Cyclodextrin** A macrocyclic (doughnut-shaped) carbohydrate ring produced enzymatically from starch. The external surface is hydrophobic while the interior is hydrophilic in nature. The hole of the doughnut is large enough to accommodate guest molecules. Uses include solubilization, separation, and stabilization of molecules in the interior cavity of, or in association with, the cyclodextrin molecules.

**Cycloheximide** Also called actidione. A chemical that inhibits protein synthesis by the 80S eucaryotic ribosomes; it does not, however, inhibit the 70S ribosomes of procaryotes. The chemical blocks peptide bond formation by binding to the large ribosomal subunits. See also PROTEIN, RIBOSOMES.

**Cyclooxygenase** Abbreviated COX, it is an enzyme that converts arachidonic acid to prostaglandins in the human body. There are two forms of cyclooxygenase: COX-1, which converts arachidonic acid to constitutive prostaglandins, which help to maintain the tissues of the stomach, kidneys, and intestines, and COX-2, which converts arachidonic acid to inducible prostaglandins, which can cause pain and inflammation in the body's joints when they accumulate in those joints. Aspirin and some other pain-relieving drugs chemically block the above-described activity of COX-1 and/or COX-2. See also ENZYME, ARACHIDONIC ACID, PLATELETS, INDUCIBLE ENZYMES, SELECTIVE APOPTOTIC ANTI-NEOPLASTIC DRUG (SAAND), EICOSANOIDS.

**Cyclosporin** An immune-system-supressing drug isolated from a mold in the mid-1970s by the Swiss firm of F. Hoffmann-LaRoche & Co. AG. The drug is used to prevent an (organ recipient's) immune system from rejecting a transplanted organ and typically must be taken by the organ recipient for the duration of his or her lifetime. Cyclosporin's mechanism of action is to prevent the divalent calcium cation (Ca 2+) from entering T lymphocytes to activate certain genes within those T lymphocytes (that trigger the rejection process). In 1996, Thomas Eisner reported that the mold *Tolypocladium inflatum*, from which cyclosporin is harvested, prefers a natural (wild) substrate of a deceased dung beetle. During 2000, it was discovered that cyclosporin inhibits growth of the parasitic microorganism *Toxoplasma gondii* (which can cause loss of sight, and neurological disease in humans). See also T LYMPHOCYTES, FUNGUS, XENOGENEIC ORGANS, CATION, GENE, GRAFT-VERSUS-HOST DISEASE (GVHD), HUMAN LEUKOCYTE ANTIGENS (HLA), MAJOR HISTOCOMPATIBILITY COMPLEX (MHC), MICROORGANISM, GROWTH (MICROBIAL).

**Cyclosporine** See CYCLOSPORIN.

**Cysteine (cys)** An amino acid of molecular weight (mol wt) 121 Daltons. Incorporated in many proteins, it possesses a sulfhydryl group (SH) that makes cysteine a mild reducing agent. Cysteine can cross-link with another cysteine located on the same or on a different polypeptide chain to form disulfide bridges. The "free" cysteine group is called a thiol group. High levels of cysteine content in certain genetically engineered corn (maize) kernels have been shown to inhibit in-field production of mycotoxins in corn (e.g., by several species of fungi that can be carried into corn plants by insects). See also AMINO ACID, CYSTINE, DISULFIDE BOND, POLYPEPTIDE (PROTEIN), PROTEIN, MYCOTOXINS.

**Cystic Fibrosis** See CYSTIC FIBROSIS TRANSMEMBRANE REGULATOR PROTEIN (CFTR).

**Cystic Fibrosis Transmembrane Regulator Protein (CFTR)** A protein that regulates proper chloride ion transport across the cell membranes of human lung airway epithelial cells. When the gene that codes for CFTR protein is damaged or mutated, the (mutant) CFTR protein fails to function properly, causing mucous (and bacteria) to accumulate in the lungs. This lung disease is known as cystic fibrosis. See also PROTEIN, GENE, ION, DEOXYRIBONUCLEIC ACID (DNA), INFORMATIONAL MOLECULES, GENOME, GENETIC CODE, RIBOSOMES, TRANSCRIPTION, SINGLE-NUCLEOTIDE POLYMORPHISMS (SNPs).

**Cystine** Two cysteine amino acids covalently linked by a disulfide bond. These units are important in biochemistry in that disulfide bridges represent one important way in which the conformation of a protein is maintained in the active form. Cystine bridges lock the structure of the proteins in which they occur in place by disallowing certain types of (molecule) chain movement. When the disulfide bond is with a free cysteine (i.e., one that is not a part of the same protein molecule's amino-acid backbone), the free cysteine is known as a thiol group. Cystine can be metabolized from methionine by certain animals (e.g., swine), but not vice versa. See also CYSTEINE (cys), AMINO ACID, CONFORMATION, PROTEIN, METHIONINE (met), METABOLISM, DISULFIDE BOND.

**CystX** Refers to a naturally occurring gene present in the genome (DNA) in some varieties of soybean plant, that confers on those particular soybean varieties (some) resistance to the soybean cyst nematode. Discovered and developed during the 1990s by Jamal Faghihi, John Ferris, Virginia Ferris, and Rick Vierling. See also SOYBEAN PLANT, SOYBEAN CYST NEMATODES (SCN), GENE.

**Cytochrome** Any of the complex protein respiratory pigments (enzymes) occurring within plant and animal cells. They usually occur in mitochondria and function as electron carriers in biological oxidation. Cytochromes are involved in the "handing off" of electrons to each other in a stepwise fashion. In the process of "handing off," other events take place which result in the production of energy that the cell needs and is able to use. See also PROTEIN, ENZYME, MITOCHONDRIA, CELL.

**Cytochrome P450** An enzyme within the liver that contains an iron-heme cofactor. It catalyzes many different biological hydroxylation reactions. Essentially, the enzyme renders fat-soluble (hydrophobic) molecules water soluble or more water soluble (by introduction of the hydrophilic hydroxyl group), so that the molecules may be removed (washed) from the body via the kidneys. This enzyme is being investigated for its potential as a catalyst in the hydroxylation of specific (valuable) industrial chemicals. See also CYTOCHROME, ENZYME, COFACTOR, HEME, HYDROXYLATION REACTION, CYTOCHROME P4503A4.

**Cytochrome P4503A4** An enzyme within the liver that, in humans, catalyzes reactions involved in the metabolism (breakdown) of certain pharmaceuticals. Those pharmaceuticals include some sedatives, antihypertensives, the antihistamine terfenadine, and the immunosuppressant cyclosporin. See also ENZYME, CYTOCHROME P450, METABOLISM, HISTAMINE, CYCLOSPORIN, METABOLIC PATHWAY, CYTOCHROME.

**Cytokines** A large class of glycoproteins similar to lymphokines but produced by non-lymphocytic cells such as normal macrophages, fibroblasts, keratinocytes, and a variety of transformed cell lines. They participate in regulating immunological and inflammatory processes, and can contribute to repair processes and to the regulation of normal cell growth and differentiation. Although cytokines are not produced by glands, they are hormone-like in their intercellular regulatory functions. They are active at very low concentrations and for the most part appear to function nonspecifically.

For example, the cytokines stimulate the endothelial cells to express (synthesize and present) P-selectins and E-selectins on the internal surfaces (of blood vessels). These selectins protrude into the bloodstream, which causes passing white blood cells (leukocytes) to adhere to the selectins, then leave the bloodstream by "squeezing" between adjacent endothelial cells. Cytokines are exemplified by the interferons. See also INTERLEUKIN-1 (IL-1), LYMPHOKINES, INTERFERONS, GLYCOPROTEIN, PROTEIN, T CELLS, INTERLEUKIN-6 (IL-6), MACROPHAGE, LECTINS, FIBROBLASTS, HORMONE, ENDOTHELIAL CELLS, ENDOTHELIUM, SELECTINS, P-SELECTIN, ELAM-1, LEUKOCYTES, ADHESION MOLECULE.

**Cytolysis** The dissolution of cells, particularly by destruction of their surface membranes. See also LYSIS, CECROPHINS, LYSOZYME, MAGAININS, COMPLEMENT, COMPLEMENT CASCADE.

**Cytomegalovirus (CMV)** A virus that infects different groups of people in varying amounts, depending on their behavior. For example, 40–90% of American heterosexuals, and about 95% of homosexuals are infected with CMV. CMV normally produces a latent (nonclinical, nonobvious) infection, but with AIDS or other events can cause immune system suppression. CMV produces a febrile (fever-causing) illness that is usually mild in nature but can become retinitis (eye infection). CMV can be treated (to halt life- and sight-threatening infection) in immunocompromised patients (i.e., transplant patients and AIDS victims) with Ganciclovir™, an antiviral compound developed by Syntex or Foscarnet™, a compound developed by Astra Pharmaceuticals. In 1996, Stephen E. Epstein found that latent CMV may cause changes in artery wall cells that aid clogging of arteries in adults (especially following balloon angioplasty). See also VIRUS, ACQUIRED IMMUNE DEFICIENCY SYNDROME (AIDS).

**Cytopathic** Damaging to cells.

**Cytoplasm** The protoplasmic contents of the cell not including the nucleus. See also NUCLEUS, CELL, PROTOPLASM, CYTOPLASMIC DNA, PLASMA MEMBRANE, MITOCHONDRIA, CHLOROPLASTS.

**Cytoplasmic DNA** The DNA within an organism (e.g., plant) that is not inside the cell's nucleus. Cytoplasmic DNA (i.e., located in the cell's mitochondria and the chloroplasts) is not transferred from plant to plant via pollen, as nuclear DNA is. See also DEOXYRIBONUCLEIC ACID (DNA), ORGANISM, CELL, CYTOPLASM,

NUCLEUS, MITOCHONDRIA, MITOCHONDRIAL DNA, CHLOROPLASTS.

**Cytoplasmic Membrane** See PLASMA MEMBRANE.

**Cytosine** A pyrimidine occurring as a fundamental unit (one of the bases) of nucleic acids. See also NUCLEIC ACIDS, BASE (NUCLEOTIDE).

**Cytotoxic** Poisonous to cells.

**Cytotoxic Killer Lymphocyte** See CYTOTOXIC T CELLS.

**Cytotoxic T Cells** Also called killer T cells. T cells that have been created by stimulated helper T cells. The T refers to cells of the cellular system rather than to cells of the humoral system (B cells). Cytotoxic T cells detect and destroy infected body cells by use of a special type of protein. The protein attaches to the infected cell's membrane and forms holes in it. This allows the uncontrolled leakage of ions out of, and water into, the cell, causing cell death. In general, the loss of the integrity of the cell membrane leads to death. The cytotoxic T cells also transmit a signal to the (leaking) infected cells that causes the cell to "chew up" its DNA. This includes its own DNA as well as that of the virus. See also CECROPHINS, MAGAININS, INTERLEUKIN-4 (IL-4), HELPER T CELLS (T4 CELLS), VIRUS, T CELLS, SUPPRESSOR T CELLS, PROTEIN, INTERLEUKIN-2 (IL-2), DEOXYRIBONUCLEIC ACID (DNA), PLASMA MEMBRANE, INSULIN-DEPENDENT DIABETES MELLITIS.

C

# D

**Δ 12 Desaturase** One of the desaturases (enzymes). See also DELTA 12 DESATURASE, COSUPPRESSION, ENZYME, DESATURASE.

**δ Endotoxins** See DELTA ENDOTOXINS.

**Δ 15 Desaturase** One of the desaturases (enzymes). See also ENZYME, DESATURASE, DELTA 12 DESATURASE.

**D Loop** A region within mitochondrial DNA in which a short stretch of RNA is paired with one strand of DNA, displacing the original partner DNA strand in this region. The same term is used to describe the displacement of a region of one strand of duplex DNA by a single-stranded invader in the reaction catalyzed by RecA protein.

**Daffodil Rice** See GOLDEN RICE.

**Daffodils** Refers to the approximately 80 species of flowering plants within the genus *Narcissus*. Native to southern Europe and northern Africa, they are the source of "golden rice" and the Alzheimer's disease treatment compound galantamine hydrobromide. See also GOLDEN RICE, ALZHEIMER'S DISEASE.

**Daidzein** See ISOFLAVONES.

**Daidzen** See ISOFLAVONES.

**Daidzin** The β-glycoside form (isomer in which glucose is attached to the molecule at the seven position of the A ring) of the isoflavone known as daidzein (aglycone form). See also ISOFLAVONES, ISOMER, DAIDZEIN.

**Dalton** A unit of mass very nearly equal to that of a hydrogen atom (precisely equal to 1.0000 on the atomic mass scale). Named after John Dalton (1766–1844), who developed the atomic theory of matter. It is $1.660 \times 10^{-24}$ gram. See also KILODALTON (Kd).

**DBT** An acronym used by some to designate the Indian Department of Biotechnology. See also INDIAN DEPARTMENT OF BIOTECHNOLOGY.

**Deamination** The removal of amino groups from molecules (e.g., in an animal's food) via the energy-consuming metabolism of excess amino acids eaten by that animal. For example, when livestock are fed more lysine (amino acid) than their body needs in a given day (animals' bodies can only utilize the essential amino acids in precise amounts/ratios of their daily diet), the excess lysine is metabolized to urea and then excreted. See also METABOLISM, AMINO ACID, ESSENTIAL AMINO ACIDS, LYSINE (lys), IDEAL PROTEIN, "IDEAL PROTEIN" CONCEPT, PDCAAS, ACC SYNTHASE.

**Defective Virus** A virus that, by itself, is unable to reproduce when infecting its host (cell), but that can grow in the presence of another virus. The other virus provides the necessary molecular machinery that the first virus lacks.

**Defensins** A class of peptides that inhibits certain fungal diseases. These are produced as a natural defense by some plants. For example, the alfalfa plant produces a defensin known as alfAFP (alfalfa antifungal peptide). In addition to protecting the plant from certain diseases, the alfAFP also inhibits a fungal disease known as potato early dying complex (also called *Verticillium* wilt), which is caused by the fungus *Verticillium dahliae*. See also PEPTIDE, FUNGUS.

**Degenerate Codons** Two or more codons that code for the same amino acid. For example, isoleucine is specified by the AUU, AUC, and AUA triplets. Since in this case more than one triplet codes for isoleucine, the codons are called degenerate. See also GENETIC CODE, CODON.

**Dehydrogenases** Enzymes that catalyze the removal of pairs of hydrogen atoms from

D

their substrates. See also SUBSTRATE (CHEMI-CAL), GLUTAMATE DEHYDROGENASE, ENZYME, DEHYDROGENATION.

**Dehydrogenation** The removal of hydrogen atoms from molecules. When those molecules are the components of vegetable oils/fats, a lower content percentage of saturated fats results. See also FATS, MONOUN-SATURATED FATS, SATURATED FATTY ACIDS (SAFA), FATTY ACID.

**Deinococcus radiodurans** A species of bacteria capable of surviving 1.5 million rads of gamma radiation (3000 times the lethal radiation dose for humans), surviving long periods of dehydration, and surviving high doses of ultraviolet radiation. *Deinococcus radiodurans* was discovered in 1956 in some canned meat. See also BACTERIA, EXTREMO-PHILIC BACTERIA.

**Delaney Clause** Formerly part of American federal law (1959 Delaney amendment to the Food, Drug and Cosmetic Act), it was eliminated in 1996. The Delaney Clause had set a zero-risk tolerance level for carcinogenic pesticide residues in processed foods. See also CARCINOGEN.

**Deletions** Loss of a section of the genetic material from a chromosome. The size of a deleted material can vary from a single nucleotide to sections containing a number of genes. See also GENE, CHROMOSOMES.

**Delta 12 Desaturase** An enzyme present within the soybean plant and in other oilseed crops (canola, maize/corn, etc.). Delta 12 desaturase ($\Delta12$) is involved in the synthesis "pathway" utilized by oilseed crops to synthesize (manufacture) polyunsaturated fatty acids (e.g., linoleic acid) from monounsaturated fatty acids (e.g., oleic acid) in seeds (while those seeds are developing). See also ENZYME, DESATURASE, FATTY ACID, UNSATUR-ATED FATTY ACID, MONOUNSATURATED FATTY ACIDS (MUFA), POLYUNSATURATED FATTY ACIDS (PUFA), PATHWAY, OLEIC ACID, LINOLEIC ACID, SOYBEAN PLANT, CORN, CANOLA.

**Delta Endotoxins** See CRY PROTEINS, PROTEIN.

**Denaturation** The loss of the native conformation of a macromolecule resulting, for instance, from heat, extreme pH (i.e., by acidity or basicity) changes, chemical treatment, etc. It is accompanied by loss of biological

activity. See also CONFORMATION, CONFIGURA-TION, MACROMOLECULES.

**Denatured DNA** DNA converted from double-stranded to single-stranded form by a denaturation process such as heating the DNA solution. In the case of heat denaturation, the solution becomes very gelatinous and viscous. See also DENATURATION, DEOXY-RIBONUCLEIC ACID (DNA), DUPLEX.

**Denaturing Gradient Gel Electrophoresis** See DENATURING POLYACRYLAMIDE GEL ELECTROPHORESIS.

**Denaturing Polyacrylamide Gel Electrophoresis** The use of PAGE (polyacrylamide gel electrophoresis) in order to separate and analyze DNA fragments (sequences) after that DNA is first denatured. This methodology can be employed to scan DNA in order to detect point mutations. See also POLYACRYLAMIDE GEL ELECTROPHORESIS (PAGE), POINT MUTATION, DENATURING GRADIENT GEL ELECTROPHORESIS, DEOXYRIBONUCLEIC ACID (DNA), DENATURED DNA, BASE EXCISION SEQUENCE SCANNING (BESS).

**Dendrimers** Polymers (i.e., molecules composed of repeating atomic units within the molecule) that repeatedly branch (while "growing" due to addition of more atoms in a repeating pattern) until that branching is stopped by the physical constraint of contacting itself (i.e., having formed a complete, hollow sphere). Discovered during the 1970s by Donald Tomalia, dendrimers possess sites on their exterior surface to which genetic material (e.g., genes or other portions of DNA) can be "attached." Dendrimers bearing such genetic material have shown the capacity to successfully transfer that genetic material into more than thirty types of living animal cells. See also POLYMER, DENDRITIC POLYMERS, GENE, GENETIC ENGINEERING, GENE DELIVERY, INFORMATIONAL MOLECULES, CODING SEQUENCE, TUMOR-SUPPRESSOR GENES, DEOXY-RIBONUCLEIC ACID (DNA), GENETIC TARGETING, GENETICS.

**Dendrites** Highly branched structures that extend from the (nucleus of) neurons to (synapse junctions with) other neurons (e.g., in human brain tissue). The primary purpose of dendrites is to process signals that are generated/received at the synapses (e.g., from the dendrites of adjoining neurons). Neuron ribosomes are located in the dendritic spines, the dendrite projections that form synapses (the junctions between dendrites where "signal transfer" between neurons takes place). Thus, those ribosomes make the proteins that are crucial to learning and memory (e.g., accomplished via growth/changes of dendrites). Messenger RNAs are synthesized (manufactured) in the nucleus of the neuron, then transported on microtubules (filaments within the neuron cell) to the ribosomes in the dendrites, where they cause manufacture of proteins (e.g., enzymes) in response to synapse activity (i.e., signals). See also NEURON, CELL, NEUROTRANSMITTER, RIBOSOMES, PROTEIN, ENZYME, MESSENGER RNA (mRNA), MICROTUBULES.

**Dendritic Cells** These are rare white blood cells, which act to stimulate the human immune system (T cells) to combat certain types of cancer. See also CELL, IMMUNE RESPONSE, CANCER, LEUKOCYTES.

**Dendritic Langerhans Cells** A type of cell, located in the mucous membranes of the mouth and genital areas, that permits the human immunodeficiency virus (the virus that causes AIDS) to enter and infect the body, even when there are no cuts or abrasions through those mucous membranes. See also HUMAN IMMUNODEFICIENCY VIRUS TYPE 1 (HIV-1), HUMAN IMMUNODEFICIENCY VIRUS TYPE 2 (HIV-2), ACQUIRED IMMUNE DEFICIENCY SYNDROME (AIDS), ADHESION MOLECULE, DENDRITIC POLYMERS.

**Dendritic Polymers** Polymers (i.e., molecules composed of repeating atomic units within the molecule) that repeatedly branch (while "growing" due to the addition of more atoms in a repeating pattern) until that branching is stopped (e.g., by physical constraints, for those polymers within living tissues). In the absence of physical constraints, dendritic polymers can continue branching (and growing) until they form a complete (hollow) sphere. Such spheres are potentially useful for protecting and "delivering" a fragile pharmaceutical molecule to specific tissue(s) within the body. See also POLYMER, DENDRIMERS.

**Denitrification** The process (i.e., internal respiration) by which denitrifying bacteria (e.g., in soil) convert nitrates to gaseous nitrogen/nitrous oxide, which then enters the atmosphere. See also NITRATES, BACTERIA, RESPIRATION.

**Denitrification** Reduction of nitrate to nitrites or into gaseous oxides of nitrogen, or even into free nitrogen by organisms. See also REDUCTION (IN A CHEMICAL REACTION).

**Denitrifying Bacteria** See DENITRIFICATION.

**Deoxynivalenol** A mycotoxin (toxin that is naturally produced by a fungus under certain conditions) which, under specific temperature and moisture conditions, is sometimes produced by certain fungi (e.g., some *Fusarium*) growing in some grains (e.g., corn/maize). Deoxynivalenol is also known as DON, and/or "vomitoxin," because certain animals (especially swine) will often vomit after they have consumed grain that contains deoxynivalenol due to its toxicity. See also TOXIN, DON, MYCOTOXINS, FUNGUS, *FUSARIUM*.

**Deoxyribonucleic Acid (DNA)** Discovered by Frederick Miescher in 1869, DNA is the chemical basis for genes. The chemical building blocks (molecules) of which genes (i.e., paired nucleotide units that code for a protein to be produced by a cell's machinery, such as its ribosomes) are constructed. Every inherited characteristic has its origin somewhere in the code of the organism's complement of DNA. The code is made up of subunits, called nucleic acids. The sequence of the four nucleic acids is interpreted by certain molecular machines (systems) to produce the required proteins of which the organism is composed.

The structure of the DNA molecule was elucidated in 1953 by James Watson, Francis Crick, and Maurice Wilkins. The DNA molecule is a linear polymer made up of deoxyribonucleotide repeating units (composed of the sugar 2-deoxyribose, phosphate, and a purine or pyrimidine base). The bases are linked by a phosphate group, joining the 3' position of one sugar to the 5' position of the next sugar. Most molecules are double-stranded and anti-parallel, resulting in a right-handed helix structure that is held together by hydrogen bonds between a purine on one chain and pyrimidine on the other chain. DNA is the carrier of genetic

**D**

information, which is encoded in the sequence of bases; it is present in chromosomes and chromosomal material of cell organelles such as mitochondria and chloroplasts, and also present in some viruses. See also A-DNA, B-DNA, cDNA, Z-DNA, TRANSCRIPTION, ANTIPARALLEL, DOUBLE HELIX, MESSENGER RNA (mRNA), NUCLEOTIDE, PROTEIN, RIBOSOMES, GENETIC CODE, GENE, CHROMOSOMES, CHROMATIDS, CHROMATIN, MITOCHONDRIAL DNA, CYTOPLASMIC DNA, NUCLEAR DNA.

**Deprotection** (of a peptide) See also HF CLEAVAGE.

**Desaturase** An enzyme (group) family that is present within the soybean plant and other oilseed crops (e.g., canola, corn/maize). One or more desaturases is involved in the synthesis "pathway" through which oilseed crops produce unsaturated fatty acids (e.g., linoleic acid). A desaturase is also involved in production of beta carotene (in some plants). See also ENZYME, FATS, STEAROYL-ACP DESATURASE, DELTA 12 DESATURASE, SOYBEAN PLANT, PATHWAY, LINOLEIC ACID, FATTY ACID, UNSATURATED FATTY ACID, GOLDEN RICE, BETA CAROTENE.

**Desferroxamine Manganese** An iron chelating agent (i.e., it chemically binds to iron atoms in the blood, thus trapping the iron atoms). The molecule also acts as an hSOD mimic by capturing harmful oxygen free radicals in the blood before they damage the walls of blood vessels. Recent research indicates that desferroxamine manganese may be useful in blocking the onset of cataracts. See also HUMAN SUPEROXIDE DISMUTASE (hSOD), XANTHINE OXIDASE, LAZAROIDS.

*Desulfovibrio* A genus of bacteria that reduces sulfate to $H_2S$ (hydrogen sulfide). Energy is obtained by oxidation of $H_2$ or organic molecules. Not a strict autotroph because $CO_2$ cannot be used as a sole carbon source. See also REDUCTION (IN A CHEMICAL REACTION), AUTOTROPH.

**Dextran** A polysaccharide produced by yeasts and bacteria as an energy storage reservoir (analogous to fat in humans). Consists of glucose residues, joined almost exclusively by alpha-1,6 linkages. Occasional branches (in the molecule) are formed by alpha 1,2, alpha 1,3, or alpha 1,4 linkages. Which linkage is

used depends on the species of yeast or bacteria producing the dextran. See also POLYSACCHARIDES.

**Dextrorotary (D) Isomer** A stereoisomer that rotates the plane of plane-polarized light to the right. *Dextro* means right. See also STEREOISOMERS, LEVOROTARY (L) ISOMER, POLARIMETER.

**DHA** See DOCOSAHEXANOIC ACID (DHA).

**Diabetes** A grouping of diseases in which the body either does not synthesize (manufacture) insulin, or else its tissues are insensitive to the insulin that it does synthesize. Approximately 5–10% of all people with diabetes are unable to synthesize insulin (e.g., because their insulin-making tissue was destroyed by autoimmune disease). Approximately 90–95% of all people with diabetes are insensitive to the insulin their body synthesizes. See also PANCREAS, INSULIN, INSULIN-DEPENDENT DIABETES MELLITIS (IDDM), AUTOIMMUNE DISEASE, BETA CELLS, N-3 FATTY ACIDS, CALPAIN-10, TYPE I DIABETES, TYPE II DIABETES, HAPTOGLOBIN.

**Diacylglycerols** Molecules that consist of two fatty acids attached to a glycerol "backbone." Research during the 1990s indicated that consumption of vegetable oils (e.g., used in frying foods) containing primarily diacylglycerols (versus typical triacylglycerols), is less likely to result in it being deposited as body fat (adipose tissue). See also FATTY ACID, SATURATED FATTY ACIDS (SAFA), UNSATURATED FATTY ACID, ADIPOSE, TRIACYLGLYCEROLS.

**Diadzein** See DAIDZEIN, ISOFLAVONES.

**Dialysis** The separation of low molecular weight compounds from high molecular weight components in solution by diffusion through a semipermeable membrane. Frequently utilized to remove salts and biological effectors (such as nicotinamide adenine dinucleotides, nucleotide phosphates, etc.) from polymeric molecules such as protein, DNA, or RNA. Commonly used membranes have a molecular weight cutoff (threshold) of around 10,000 Daltons, but other membrane pore sizes are available. See also HOLLOW FIBER SEPARATION, ACTIVE TRANSPORT.

*Diamond vs. Chakrabarty* See CHAKRABARTY DECISION.

**Diastereoisomers** Four variations of a given molecule, consisting of a pair of stereoisomers about a second asymmetric carbon atom for each of the two isomers of the first asymmetric carbon atom. See also STEREOISOMERS, CHIRAL COMPOUND.

**Differentiation** Refers to processes by which a single type of cells (stem cells, embryonic stem cells, etc.) become multiple, different types of (specialized) cells. See also CELL, STEM CELLS, STEM CELL ONE, STEM CELL GROWTH FACTOR (SCF), TOTIPOTENT STEM CELLS, COLONY STIMULATING FACTORS (CSFs), EMBRYONIC STEM CELLS, HUMAN EMBRYONIC STEM CELLS, GRANULOCYTE-MACROPHAGE COLONY STIMULATING FACTOR (GM-CSF), HEDGEHOG PROTEINS.

**Digestion (within chemical production plants)** Breakdown of feed stocks by various processes (chemical, mechanical, and biological) to yield their desired building-block components for inclusion as raw materials in subsequent chemical or biological processes.

**Digestion (within organisms)** The enzyme-enhanced hydrolysis (breakdown) of major nutrients (food) in the gastrointestinal system to yield their building-block components (to the organism), such as amino acids, fatty acids, or other essential nutrients. See also HYDROLYSIS, FATS, PROTEIN, AMINO ACID, ESSENTIAL AMINO ACIDS, ESSENTIAL NUTRIENTS, FATTY ACID, ESSENTIAL FATTY ACIDS, LIPASE, "IDEAL PROTEIN" CONCEPT, ENZYME, PROTEASES, PROTEOLYTIC ENZYMES, ABSORPTION, TRYPSIN, LECITHIN, PROTEIN DIGESTIBILITY-CORRECTED AMINO ACID SCORING (PDCAAS).

**Diglycerides** See TRIGLYCERIDES.

**Diphtheria Antitoxin** Discovered by Emil von Behring in 1900. See also ANTITOXIN, ENTEROTOXIN.

**Diploid** The state of a cell in which each of the chromosomes, except for the sex chromosomes, is always represented twice (46 chromosomes in humans). In contrast to the haploid state in which each chromosome is represented only once. See also DIPLOPHASE, CHROMOSOMES, HOMOZYGOUS, TRIPLOID.

**Diplophase** A phase in the life cycle of an organism in which the cells of the organism have two copies of each gene. When this state exists the organism is said to be diploid. See also DIPLOID, GENE, HOMOZYGOUS, CELL.

**Direct Transfer** Refers to methods of inserting a gene directly into a cell's DNA without the use of a vector. One example of direct transfer is electroporation. See also GENE, GENETIC ENGINEERING, VECTORS, CELL, DEOXYRIBONUCLEIC ACID (DNA), ELECTROPORATION.

**Directed Self-Assembly** See SELF-ASSEMBLY (OF A LARGE MOLECULAR STRUCTURE).

**Disaccharides** Carbohydrates consisting of two covalently linked monosaccharide units; hence *di* for two. See also OLIGOSACCHARIDES, MONOSACCHARIDES, POLYSACCHARIDES.

**Dissimilation** The breakdown of food material to yield energy and building blocks for cellular synthesis. See also DIGESTION (WITHIN ORGANISMS).

**Dissociating Enzymes** See HARVESTING ENZYMES.

**Distribution** See "ADME" TESTS, PHARMACOKINETICS.

**Disulfide Bond** An important type of covalent bond formed between two sulfur atoms of different cysteines in a protein. Disulfide bonds (linkages, bridges) contribute to holding proteins together and also help provide the internal structure (conformation) of the protein. See also PROTEIN, CYSTEINE (cys), CYSTINE.

**Diversity (within a species)** Refers to the genetic variation that exists within a population (of organisms) in a species. For example, black cattle and white cattle; or both toxic and nontoxic strains/serotypes of *Escherichia coliform* (*E. coli*) bacteria. This diversity is due to one or more single-nucleotide polymorphisms (SNPs) in each individual's genome (DNA) within the population of organisms. See also SPECIES, SINGLE-NUCLEOTIDE POLYMORPHISMS (SNPs), POLYMORPHISM (GENETIC), NUCLEOTIDE, ORGANISM, STRAIN, SEROTYPES, *ESCHERICHIA COLIFORM* (*E. COLI*), *ESCHERICHIA COLIFORM* 0157:H7 (*E. COLI* 0157:H7).

**Diversity Biotechnology Consortium** A non-profit U.S. organization formed in August of 1994 by a group of research institutions and companies. The consortium's first president was Stuart A. Kauffman of the Santa Fe Institute. The consortium's purpose is to further the use of molecular diversity as a tool in drug design, and in the study of mutating viruses. See also MOLECULAR DIVERSITY, RATIONAL DRUG DESIGN, DIVERSITY ESTIMATION (OF MOLECULES), MOLECULAR BIOLOGY, VIRUS, MUTATION, MUTANT, SITE-DIRECTED MUTAGENESIS (SDM), COMBINATORIAL CHEMISTRY, COMBINATORIAL BIOLOGY.

**Diversity Estimation (of molecules)** See COMBINATORIAL CHEMISTRY.

**DNA** See DEOXYRIBONUCLEIC ACID (DNA).

**DNA Analysis** See DNA PROFILING.

**DNA Bridges** Large segments of DNA whose sequence (i.e., composition) is known and mapped in total. Those sequences are then utilized by scientists to piece together (bridging the DNA segments) and assemble a (more) complete map (e.g., of an organism's chromosome or genome). See also DEOXYRIBONUCLEIC ACID (DNA), GENETIC MAP, SEQUENCE (OF A DNA MOLECULE), CHROMOSOME, GENOME, SEQUENCE MAP, SHOTGUN SEQUENCING.

**DNA Chimera** One DNA molecule composed of DNA from two different species. See also CHIMERA.

**DNA Chip** See BIOCHIPS, GENE EXPRESSION ANALYSIS, PROTEOMICS.

**DNA Fingerprinting** See DNA PROFILING.

**DNA Ligase** An enzyme that creates a phosphodiester bond between the 3′ end of one DNA segment and the 5′ end of another, while they are base-paired to a template strand. The enzyme seals (joins) the ends of single-stranded DNA in a duplex DNA chain. DNA ligase constitutes a part of the DNA repair mechanism available to the cell. See also NICK, LIGASE, DEOXYRIBONUCLEIC ACID (DNA), GENE REPAIR (NATURAL), DUPLEX.

**DNA Marker** See MARKER (DNA MARKER).

**DNA Methylation** Refers to a DNA molecule that is saturated with methyl groups (i.e., methyl submolecule groups $CH_3$ have attached themselves to the DNA molecule's "backbone" at all possible locations on that DNA molecule). DNA methylation is used by healthy cells to turn off certain genes when those particular genes are no longer needed (e.g., turn off genes involved in juvenile development after organism reaches adulthood). DNA methylation (of cell genes that would normally prevent inappropriate cell division/proliferation) also occurs in some cancers. See also DEOXYRIBONUCLEIC ACID (DNA), METHYLATED, CELL, GENE, CANCER, TRANSCRIPTION, GENETIC CODE, MESSENGER RNA (mRNA), p53 GENE, TUMOR-SUPPRESSOR GENES.

**DNA Microarray** Initially developed by Patrick Brown during the 1980s, these microarrays enable analysis of the levels of expression of genes in an organism, or comparison of gene expression levels (e.g., between diseased and nondiseased tissues) via hybridization of messenger RNA (mRNA) to its counterpart DNA sequence, when biological samples containing DNA (e.g., in liquid) are passed over the array surface. To manufacture the DNA microarray, cellular mRNA is used to make segments of complementary DNA (cDNA) in lengths of approximately 500–5000 base pairs long, using the reverse transcriptase polymerase chain reaction (RT-PCR). These cDNA segments are then attached to a nylon or glass surface at known spots, so when hybridization-of-sample-DNA occurs, the location of the spot tells what DNA was in the sample. Another way to manufacture a type of DNA microarray is to similarly attach oligonucleotides or peptide nucleic acids of known sequence (composition) at known spots on the nylon or glass surface, and pass the biological sample containing DNA (e.g., in liquid) over that surface to identify the DNA in the sample by the spot to which it hybridizes. See also GENE, ORGANISM, BIOCHIPS, MICROFLUIDICS, DEOXYRIBONUCLEIC ACID (DNA), MESSENGER RNA (mRNA), HYBRIDIZATION (MOLECULAR GENETICS), EXPRESS, GENE EXPRESSION ANALYSIS, PROTEOMICS, MICROARRAY (TESTING), OLIGONUCLEOTIDE, NUCLEIC ACIDS, SEQUENCE (OF A DNA MOLECULE), BIOINFORMATICS.

**DNA Polymerase** An enzyme that catalyzes the synthesis of DNA. The process is accomplished by catalyzing the addition of deoxyribonucleotide residues to the free 3′-hydroxyl end of a DNA chain, starting from a mixture of the appropriate triphosphorylated bases, which are dATP, dGTP, dCTP, and dTTP. This chemical reaction is reversible and, hence, DNA polymerase also functions as an exonuclease. See also ENZYME, EXONUCLEASE, TAQ DNA POLYMERASE, DEOXYRIBONUCLEIC ACID (DNA), SYNTHESIZING (OF DNA MOLECULES).

**DNA Probe** Also called gene probe or genetic probe. Short, specific (complementary to desired gene) artificially-produced segments of DNA are used to combine with and detect the presence of specific genes (or shorter DNA segments) within a chromosome. If a DNA probe of known composition and length is mingled with pieces of DNA (genes) from a chromosome, the probe will cling to its exact counterpart in the chromosomal DNA pieces (genes), forming a stable double-stranded hybrid. The presence of this (now) "labeled" probe is detected visually or with the aid of another detection instrument. Because the composition of the DNA probes is known, scientists can riffle through a chromosome, spotting segments of DNA (i.e., genes) that seem to be linked to genetic diseases. See also MUSCULAR DYSTROPHY (MD), PROBE, POLYMERASE CHAIN REACTION (PCR), GENE, POLYMERASE CHAIN REACTION (PCR) TECHNIQUE, CHROMOSOMES, DOUBLE HELIX, DUPLEX, HYBRIDIZATION (MOLECULAR GENETICS), HYBRIDIZATION SURFACES, DEOXYRIBONUCLEIC ACID (DNA), HOMEOBOX, RAPID MICROBIAL DETECTION (RMD), SOUTHERN BLOT ANALYSIS.

**DNA Profiling** Invented in 1985 by Alec Jeffreys, this technique is used by forensic (i.e., crime-solving) chemists to match biological evidence (e.g., a blood stain) from a crime scene to the person (e.g., the assailant) involved in that particular crime. DNA profiling involves the use of RFLP (restriction fragment length polymorphism) analysis or ASO/PCR (allele-specific oligonucleotide/polymerase chain reaction) analysis to identify the specific sequence of bases (i.e., nucleotides) in a piece of DNA taken from the biological evidence. Since the specific sequence of bases in DNA molecules is different for each individual (due to DNA polymorphism), a criminal's DNA can be matched to that of the evidence to prove guilt or innocence. Biological evidence may include, among other things, blood, hair, nail fragments, skin, and sperm. See also DEOXYRIBONUCLEIC ACID (DNA), RESTRICTION FRAGMENT LENGTH POLYMORPHISM (RFLP) TECHNIQUE, POLYMORPHISM (CHEMICAL), POLYMERASE CHAIN REACTION (PCR) TECHNIQUE, ALLELE, NUCLEOTIDE, NUCLEIC ACIDS, OLIGOMER, GENETIC CODE, INFORMATIONAL MOLECULES, OLIGIONUCLEOTIDE, CODON.

**DNA Synthesis** See SYNTHESIZING (OF DNA MOLECULES).

**DNA Typing** See DNA PROFILING.

**DNA Vaccines** Products in which "naked" genes (i.e., pieces of bare DNA) are used to stimulate an immune response (e.g., either a cellular immune response, humoral immune response, or otherwise raise antibodies against the pathogen from which the naked genes have arisen or been derived). See also DEOXYRIBONUCLEIC ACID (DNA), IMMUNE RESPONSE, CELLULAR IMMUNE RESPONSE, HUMORAL IMMUNITY, ANTIBODY, "NAKED" GENE, PATHOGEN, DNA VECTOR.

**DNA Vector** A vehicle (such as a virus) for transferring genetic information (DNA) from one cell to another. See also BACTERIOPHAGE, RETROVIRUSES, VECTOR.

**DNA-Dependent RNA Polymerase** See RNA POLYMERASE.

**DNA-RNA Hybrid** A double helix that consists of one chain of DNA hydrogen bonded to a chain of RNA by means of complementary base pairs. See also HYBRIDIZATION (MOLECULAR GENETICS), HYBRIDIZATION (PLANT GENETICS), DOUBLE HELIX.

**DNAse** Deoxyribonuclease, an endonuclease enzyme family that degrades (cuts up) DNA molecules. DNase I is produced and secreted by the salivary glands, intestines, liver, and pancreas of animals. It has optimal activity (i.e., greatest ability to cut up DNA molecules) at neutral pH (neither acidic nor basic). DNase II has optimal activity between pH 4.6 and 5.5 (i.e., in slightly acidic solutions). See also ENZYME, DEOXYRIBONUCLEIC ACID (DNA), ENDONUCLEASES, PANCREAS, ACID, BASE (GENERAL).

**Docosahexanoic Acid (DHA)** One of the omega-3 (n-3) highly unsaturated fatty acids (HUFA), DHA is important in the development of the human infant's brain, spinal cord, and retina tissues. DHA aids optimal brain and nervous system development in human infants, and is required for optimal brain function throughout life. Naturally present in human breast milk and fish oil. The human body converts linolenic acid (e.g., from consumption of soybean oil) to the two highly unsaturated fatty acids (HUFA) docosahexanoic acid (DHA) and eicosapentanoic acid (EPA). Research indicates that consumption of docosahexanoic acid also helps to reduce the risk of heart disease (by lowering blood pressure) and depression (via its effect in the brain). See also POLYUNSATURATED FATTY ACIDS (PUFA), HIGHLY UNSATURATED FATTY ACIDS (HUFA), N-3 FATTY ACIDS, FATTY ACIDS, UNSATURATED FATTY ACIDS, ESSENTIAL FATTY ACIDS, LINOLENIC ACID, SOYBEAN OIL, EICOSANOIDS, EICOSAPENTANOIC ACID (EPA).

**Domain (of a chromosome)** May refer either to a discrete structural entity defined as a region within which supercoiling is independent of other domains, or to an extensive region, including an expressed gene that has heightened sensitivity to degradation by the enzyme DNAse I. See also GENE, EXPRESS, ENZYME.

**Domain (of a protein)** A discrete continuous part of the amino acid sequence that can be equated with a particular function. See also COMBINING SITE, EPITOPE, IDIOTYPE, PROTEIN, p53 PROTEIN, MINIMIZED PROTEINS.

**Dominant (gene)** (gene) See also DOMINANT ALLELE.

**Dominant Allele** Discovered by Gregor Mendel in the 1860s, this gene produces the same phenotype when it is heterozygous as it does when it is homozygous (i.e., trait, or protein, is expressed even if only one copy of the gene is present in the genome). See also GENETICS, RECESSIVE ALLELE, HETEROZYGOTE, HOMOZYGOUS, PHENOTYPE, GENOTYPE, GENOME.

**DON** Abbreviation for the mycotoxin deoxynivalenol produced by *Fusarium* fungi DON. Also known as "vomitoxin," because it can cause some animals to vomit if they consume it. See also MYCOTOXINS, DEOXYNIVALENOL, *FUSARIUM*, FUNGUS, VOMITOXIN.

**Donor Junction** The junction between the left 5′ end of an exon and the right 3′ end of an intron. See also EXON, INTRON.

**Double Helix** The natural coiled conformation of two complementary, antiparallel DNA chains. This structure was first put forward by Watson and Crick in 1953. See also DEOXYRIBONUCLEIC ACID (DNA).

**Down Promoter Mutations** Those mutations that decrease the frequency of initiation of transcription. Down promoter mutations lead to the production of less mRNA than is the case in the nonmutated state. See also mRNA, MUTATION, TRANSCRIPTION, DOWN REGULATING.

**Down Regulating** Phrase referring to regulatory sequences, chemical compounds (e.g., transcription factors), mutations (e.g., down promoter mutations), etc. that cause a given gene to express less of the protein that it normally codes for. See also GENE, REGULATORY SEQUENCE, TRANSCRIPTION FACTORS, DOWN PROMOTER MUTATIONS, PROTEIN, CODING SEQUENCE.

**Drosophila** The name of a type of fly (*Drosophila melanogaster*) that reproduces rapidly, and that is commonly utilized in genetics experiments due to its short life cycle (14 days) and simple genome (four chromosome pairs). Because of these factors, a large base of knowledge about *Drosophila* genetics has been accumulated by the world's scientific community. For example, of the nearly 300 "disease-causing" genes in the human genome, more than half have an analogous gene in the *Drosophila* genome. *Drosophila* was one of the first organisms to have its entire genome sequenced by man. See also GENETICS, GENOME, GENETIC CODE, GENETIC MAP, CHROMOSOMES, COLD HARDENING, HOMEOBOX, SEQUENCING (OF DNA MOLECULES), GENE.

**Duchenne Muscular Dystrophy (DMD) Gene** See MUSCULAR DYSTROPHY (MD).

**Duplex** The double-helical structure of DNA (deoxyribonucleic acid). See also DOUBLE HELIX, DEOXYRIBONUCLEIC ACID (DNA).

# E

**E-Selectin** See ELAM-1.

**EAA** See ESSENTIAL AMINO ACIDS.

**EAA** See EXCITATORY AMINO ACIDS (EAAs).

**Early Development** The period of a phage infection before the start of DNA replication. See also PHAGE, BACTERIOPHAGE, DEOXYRIBO-NUCLEIC ACID (DNA).

**Early vs. Late Genes** Those genes transcribed early in a bacteriophage-mediated infection process as compared to those genes transcribed some time later. May require different "p factors" (sigma) for recognition of promotors. See also GENE, PROMOTER.

**Early vs. Late Proteins** During viral infection, viral-specific proteins are synthesized at characteristic times after infection. They are called "early" and "late." Often under positive control of bacterial and viral sigma factors. See also EARLY VS. LATE GENES, PROTEIN.

**Earthworms** (*Eisenia foetida*) These worms live in the soil and consume up to ten tons of organic matter (old crop plant stalks, husks, etc.) per acre (approximately 0.4 hectare) per year. In so doing, earthworms make the soil more fertile, since the process breaks down that organic matter into soil (when excreted by those earthworms). Earthworm tunnels also help aerate soil, which encourages healthy plant root systems. See also LOW-TILLAGE CROP PRODUCTION, GLOMALIN, NO-TILLAGE CROP PRODUCTION.

**E. coli** See ESCHERICHIA COLIFORM (*E. COLI*).

**E. coli 0157:H7** See ESCHERICHIA COLIFORM 0157:H7 (*E. COLI* 0157:H7).

**ECB** See EUROPEAN CORN BORER (ECB).

**Ecology** The study of the interrelationships between organisms and their environments. See also HABITAT.

**Ectodermal Adult Stem Cells** Certain stem cells present within (adult) bodies of organisms, that can be differentiated (via chemical signals) to give rise to cells of skin, hair, tooth enamel, mucous membranes, and some glandular tissues. See also STEM CELLS, MULTIPOTENT ADULT STEM CELLS, CELL, ORGANISM, SIGNALING.

**Edible Vaccines** Edible substances, bearing antigens, that cause activation of an animal's immune system via that animal's GALT (gut-associated lymphoid tissues). These "edible vaccines" are derived from transgenic plants (grains, tubers, fruits, etc.) or eggs (i.e., via the activation of the hen's immune system to cause that hen to secrete desired molecule(s) into the eggs it lays). See also GUT-ASSOCIATED LYMPHOID TISSUES (GALT), ANTIGEN, CELLULAR IMMUNE RESPONSE, MOLECULAR PHARMING™, HUMORAL IMMUNITY, PLANTIGENS.

**EDTA** Ethylenediamine tetraacetate. An organic molecule which, due to the chemical groups it contains and their juxtaposition within that molecule, is able to chelate (bind) certain other molecules such as divalent metal cations. EDTA thus inhibits some enzymes requiring such ions for activity. See also CHELATION, COFACTOR, CHELATING AGENT.

**EFA** See ESSENTIAL FATTY ACIDS.

**Effector** A class of (usually small) molecules that regulates the activity of a specific protein (e.g., enzyme) molecule by binding to a specific site on the protein. Control of (existing) enzyme molecules may be achieved by combination of the effector with the enzyme. The effector molecule may either physically block the active site on the enzyme molecule, or alter the three-dimensional conformation of the enzyme molecule. That conformation change results in a change in the enzyme's catalytic activity. Effector is a general term. Effector molecules may be activators (cause an increase in the enzyme's catalytic activity) or inhibitors (cause a decrease in the enzyme's catalytic activity). A special class of effector, known as an allosteric effector, binds to the enzyme molecule at a site other than the enzyme's active site (thereby activating or inhibiting). See also PROTEIN, ENZYME, CONFORMATION, ALLOSTERIC ENZYMES, ALLOSTERIC SITE, ACTIVE SITE, FEEDBACK INHIBITION, CATALYTIC SITE.

**EGF** See EPIDERMAL GROWTH FACTOR (EGF).

**EGF Receptor** A protein embedded in the surface of the membranes of skin cells. The receptor consists of (1) an outside (of the cell membrane) enzyme that recognizes epidermal growth factor (EGF) and binds to it, and (2) an enzyme on the inside of the cell membrane, which is of the tyrosine kinase class. When free EGF comes in contact with an EGF receptor, they bind (in a lock-and-key fashion) and then enter the cell together (through the cell membrane. There EGF stimulates growth or division of the cell via ras protein and ras gene). The EGF receptor (and receptors in general) is like a butler who allows the EGF (a guest) to enter the cell (home). See also ONCOGENES, PROTEIN, PLASMA MEMBRANE, TRANSMEMBRANE PROTEINS, ras GENE, ras PROTEIN, RECEPTORS, SIGNAL TRANSDUCTION.

**EGFR** See EGF RECEPTOR.

**EHEC** See ENTEROHEMORRHAGIC E. COLI.

**EIA** See ENZYME IMMUNOASSAY (EIA).

**Eicosanoids** A group of chemical compounds which the human body synthesizes (manufactures) from arachidonic acid, docosahexanoic acid, and other starting materials. One subgroup of eicosanoids is that of the prostaglandins (cyclic fatty acids that act as hormones in the body). For example, the COX-1 enzyme converts arachidonic acid to constitutive prostaglandins, and the COX-2 enzyme converts arachidonic acid to inducible prostaglandins. See also ARACHIDONIC ACID (AA), CYCLOOXYGENASE, CONSTITUTIVE ENZYMES, INDUCIBLE ENZYMES, PROSTAGLANDINS, HORMONE, COX-1, COX-2, LEUKOTRIENES.

**Eicosapentaenoic Acid (EPA)** See EICOSAPENTANOIC ACID (EPA).

**Eicosapentanoic Acid (EPA)** One of the omega-3 (n-3) polyunsaturated fatty acids (PUFA), EPA is important for the development of the human brain, retina tissue, prevention of high blood pressure, coronary heart disease (CHD), and some cancers. The human body converts linolenic acid (e.g., from consumption of soybean oil) to the two highly unsaturated fatty acids (HUFA) eicosapentanoic acid (EPA) and docosahexanoic acid (DHA). See also N-3 FATTY ACIDS, POLYUNSATURATED FATTY ACIDS (PUFA), UNSATURATED FATTY ACIDS, ESSENTIAL FATTY ACIDS, CORONARY HEART DISEASE (CHD), CANCER, HIGHLY UNSATURATED FATTY ACIDS (HUFA), LINOLENIC ACID, SOYBEAN OIL.

**ELAM-1** Also known as E-selectin, it is a selectin molecule that is synthesized by endothelial cells after (adjacent) tissue is infected. ELAM-1 molecules then help leukocytes leave the bloodstream to fight the infection. See also SELECTINS, LECTINS, ADHESION MOLECULES, LEUKOCYTES.

**Elastase** An enzyme secreted by neutrophils (white blood cells that engulf pathogens) which catalyzes the cleavage (breakdown) of specific proteins that function to provide elasticity to certain tissues. May be indirectly responsible for some autoimmune diseases, such as arthritis (which results from breakdown of cartilage tissue). Elastase may also be indirectly responsible for the emphysema (caused by loss of lung elasticity) that results from prolonged smoke inhalation. When a-1 antitrypsin (anti-elastase) efficacy is reduced (via smoke), the now-unrestrained excess elastase destroys alveolar walls in the lungs by digesting elastic fibers and other connective tissue proteins. See also LEUKOCYTES, NEUTROPHILS, PROTEOLYTIC ENZYMES.

**Electrolyte** Any compound (salt, acid, base, etc.) which in aqueous solution dissociates into ions (charged atom-sized particles). Electrolytes may either be strong (completely or nearly completely dissociated) or weak (only partially dissociated). See also ION.

**Electron Carrier** A protein, such as flavoprotein or a cytochrome, that can gain and lose electrons reversibly and function in the transfer of electrons from one carrier to another until the electron is taken up by a final molecule or atom such as oxygen. See also PROTEIN, CYTOCHROME.

**Electron Microscopy (EM)** A technique for greatly magnifying and visualizing very small entities such as viruses and even large molecules. The technique uses beams of electrons instead of light rays. Because of the physics involved, beams of electrons permit much greater magnification than is possible with a light microscope. Electron microscopes have been used to examine the structures of viruses, bacteria, pollen grains, molecules, etc.

**Electropermeabilization** See ELECTROPORATION.

**Electrophoresis** A technique for separating molecules based on the differential movement of charged particles through a matrix when subjected to an electric field. The term is usually applied to large ions of colloidal particles dispersed in water. The most important use of electrophoresis (currently) is in the analysis of proteins, and then a technique known as gel electrophoresis is used. Since the proportion of proteins varies widely in different diseases, electrophoresis can be used for diagnostic purposes. Electrophoresis, through agarose or other gel matrices, is a common way to separate, identify, and purify plasmid DNA, DNA fragments resulting from digestion (of DNA) with restriction endonucleases, and RNA. Electrophoresis is also used to study bacteria and viruses, nucleic acids, and some types of molecules, including amino acids. See also PROTEIN, AMINO ACID, BIOLUMINESCENCE, POLYACRYLAMIDE GEL ELECTROPHORESIS (PAGE), TWO-DIMENSIONAL (2D) GEL ELECTROPHORESIS, CHROMATOGRAPHY, GEL, AGAROSE, PLASMID, DEOXYRIBONUCLEIC ACID (DNA), RESTRICTION ENDONUCLEASES, RIBONUCLEIC ACID (RNA), BACTERIA, VIRUS, BIOMEMS.

**Electroporation** A process utilized to introduce a foreign gene into the genome of an organism. In 1995, the U.S. company Dekalb Genetics Corp. received a patent for producing genetically engineered corn via introduction of a foreign gene into corn cells via electroporation. Electroporation, also called electroporesis or electropermeabilization, uses a brief direct-current (dc) electrical pulse to cause formation of "micropores" (tiny holes) in the surface of cells or protoplasts suspended in a solution (water) containing DNA sequences (genes). After the gene(s) enter the cell via the temporarily created micropores, the electrical pulse

ceases, and the micropores close so that the gene(s) cannot depart the cell. The cell then incorporates (some) of the new genetic material (genes) into its genetic complement (genome), and creates whatever product (i.e., a protein) the newly-introduced gene codes for. See also CODING SEQUENCE, GENETIC ENGINEERING, VECTOR, BIOLISTIC® GENE GUN, "EXPLOSION" METHOD, AGROBACTERIUM TUMEFACIENS, GENE, GENOME, CELL, CORN, PROTOPLAST, DEOXYRIBONUCLEIC ACID (DNA), PROTEIN.

**Electroporesis** See ELECTROPORATION.

**ELISA** (test for proteins) An enzyme-linked immunosorbent assay (hence the acronym) which can readily measure less than a nanogram ($10^{-9}$ g) of a protein. This assay is more sensitive than simple immunoassay (tests) because one of the two antibodies used to bind and quantitate (measure) the protein's antigen, based on two concurrent epitopes within the protein, is attached to an enzyme. The enzyme can rapidly convert an added colorless substrate into a colored product or a nonfluorescent substrate into an intensely fluorescent product (thus enabling finer quantitation). See also ABSORBANCE (A), IMMUNOASSAY, PROTEIN, ANTIGEN, ENZYME, NANOGRAM (ng), FLUORESCENCE.

**Elite Germplasm** Refers to germplasm that is adapted (selectively bred) and optimized to new surroundings (i.e., environment). For example, corn/maize (*Zea mays* L.), which is native to Mexico, has been adapted and optimized to grow in field conditions in many of the world's countries. See also GERMPLASM, INTROGRESSION, MARKER ASSISTED SELECTION, CORN.

**Ellagic Acid** A naturally occurring plant phenol (phytochemical) that, when consumed by humans, has been shown to help inhibit some cancers. Ellagic acid is naturally present in strawberries, the pomegranate (*Punica granatum*), etc. See also PHYTOCHEMICALS, POLYPHENOLS, CANCER.

**EMAS** Eco-Management and Audit Scheme.

**Embryo Rescue** Refers to the tissue culture techniques/technologies utilized to enable the fertilized embryo resulting from a "wide cross" (between two nonsexually compatible plant species) to grow and mature into a seed producing plant. See also TRADITIONAL BREEDING METHODS, WIDE CROSS, TISSUE CULTURE.

**Embryology** The study of the early stages in the development of an organism. In these stages a single highly specialized cell, the egg, is transformed into a complex many-celled organism resembling its parents. See also CELL, ANTIANGIOGENESIS, GAMETE.

**Embryonic Stem Cells** See HUMAN EMBRYONIC STEM CELLS.

**EMEA** See EUROPEAN MEDICINES EVALUATION AGENCY (EMEA).

**Emulsion** A stable dispersion of one liquid in a second, immiscible (i.e., nonmixable) liquid. For example, milk is an emulsion of oil (fat) in water, and latex paint is an emulsion of paint resin in water. Certain ingredients (e.g., β-conglycinin protein) help enable a greater content of the first liquid to be dispersed in the second liquid. Certain ingredients (e.g., β-conglycinin protein) make a given emulsion more stable (i.e., prevent the two liquids from separating over an extended period of time). See also PROTEIN, β-CONGLYCININ.

**Enantiomers** From the Greek word *enantios*, which means opposite. Enantiomers are a pair of nonidentical, mirror-image molecules. This means that both molecules are made up of the same atoms, i.e., they have the same molecular formula, but the constituent groups that are attached to a carbon atom can be arranged in two different ways (forms) around the carbon atom. This gives rise to an asymmetric molecule that can exist in either of two mirror-image forms whose mirror images are not superimposable. A pair of these molecules is known as enantiomers. The four attached groups are all different from each other. See also RACEMATE, OPTICAL ACTIVITY, CHIRAL COMPOUND, ENANTIOPURE.

**Enantiopure** Refers to a compound (e.g., a pharmaceutical) that consists of only one of that compound's two possible enantiomers. Sometimes expressed in relative terms. For example, 98% enantiopure would refer to a compound that consists of 98% (of) desired enantiomer. See also ENANTIOMERS, CHIRAL COMPOUND, RACEMATE, OPTICAL ACTIVITY.

**Endergonic Reaction** A chemical reaction with a positive standard free energy change (i.e., an "uphill" reaction). An (heat) energy-requiring reaction. A nonspontaneous reaction at ambient temperature. See also EXERGONIC REACTION, FREE ENERGY.

**Endocrine Glands** Glands that secrete their products (hormones) into the blood, which then carries them to their specific target organs. For example, adrenalin, produced in the adrenal glands, is carried to the heart (and other muscles) when needed during periods of stress. The endocrine glands are: the pituitary, thyroid, adrenals, pancreas, ovaries (in females), and testes (in males). Endocrine glands are found in some invertebrates as well as in vertebrates. See also HORMONE, ENDOCRINE HORMONES.

**Endocrine Hormones** The products secreted by the endocrine glands. These help control long-term bodily processes, such as growth, lactation, sex cycles, and metabolic adjustment. The endocrine system and the nervous system are interdependent and often referred to collectively as the neuroendocrine system. For example, the juvenile hormone, found in insects and annelids, affects sexual maturation. There is currently great interest among scientists in the potential use of such hormones in the control of destructive insects. See also ENDOCRINE GLANDS, HORMONE, PHEROMONES.

**Endocrinology** The branch of science that studies the endocrine glands, hormones, and hormone-like substances. See also ENDOCRINE GLANDS, HORMONE, ENDOCRINE HORMONES.

**Endocytosis** Also called receptor-mediated endocytosis. The import of substances (e.g., hormones, viruses, and toxins) into a cell via specific receptor/ligand binding. The entity under consideration binds to a receptor(s) located in the plasma (cell) membrane, which then invaginates (infolds), hence taking up the entity via "endosomes" (formed by pinching off an infold to form a "bag") into vesicles located within the cell. It is one route to deliver essential metabolites to cells (e.g., low-density lipoprotein), and it is a means to modulate the cell's responses to many protein hormones and growth factors (e.g., insulin, epidermal growth factor, and nerve growth factor).

It is a route by which certain proteins targeted for destruction can be taken up and delivered to the cell's lysosomes. For example, phagocytic cells have receptors enabling them to take up antigen-antibody complexes for subsequent destruction by the phagocytic cell. This route is also a means exploited by certain viruses and toxins to gain entry into cells through the otherwise impervious cell membranes (e.g., used by the AIDS virus and the Semliki Forest Virus). Disorders of endocytosis can lead to disease states (e.g., high cholesterol levels in the blood of people whose low-density lipoprotein receptors are impaired). Drugs (e.g., certain painkillers) can be targeted to specific receptors via receptor mapping (RM) and receptor fitting (RF) for greater efficacy. See also INVASIN, ADHESION MOLECULE, CD4 PROTEIN, EXOCYTOSIS, T CELL RECEPTORS, SIGNAL TRANSDUCTION, VAGINOSIS, RECEPTORS, RECEPTOR FITTING (RF), HIGH-DENSITY LIPOPROTEINS (HDLPs), LOW-DENSITY LIPOPROTEINS (LDLPs), RECEPTOR MAPPING (RM), SIGNALING, NUCLEAR RECEPTORS.

**Endodermal Adult Stem Cells** Certain stem cells present within (adult) bodies of organisms, that can be differentiated (via chemical signals) to give rise to cells of tongue, tonsils, the bladder/urethra, digestive tract, liver, pancreas, lung tissues, etc. See also STEM CELLS, MULTIPOTENT ADULT STEM CELLS, CELL, ORGANISM, SIGNALING.

**Endoglycosidase** An enzyme capable of hydrolyzing (breaking) interior bonds in the oligosaccharide molecular branches of a glycoprotein molecule. That is, the enzyme is capable of cutting a sugar-to-sugar bond anywhere within the sugar polymer molecule (depending, of course, on the specificity of the enzyme). This is in contrast to an exoglycosidase, which must cut away at the polymer from the outside, i.e., from the free end, one unit (or section, as the case may be) at a time. See also EXOGLYCOSIDASE, GLYCOPROTEIN, ENZYME, OLIGOSACCHARIDES, RESTRICTION ENDOGLYCOSIDASES, HYDROXYLATION REACTION.

**Endometrium** The lining of the uterus.

**Endonucleases** A class of enzymes capable of hydrolyzing (breaking) the interior phosphodiester bonds of DNA or RNA chains. As opposed to cleavage (by exonucleases) at the terminal bonds (ends) of a chain. See also ENZYME, DNase 1, DNase 2, EXONUCLEASE, ENDOGLYCOSIDASE.

**Endophyte** A microrganism (fungus or bacterium) that lives inside vascular tissues of plants (in spaces between plant cells). At least one company has incorporated the gene for a protein toxic to insects (taken from *Bacillus thuringiensis*) into an endophyte to confer insect resistance to a crop plant. When endophyte-infested fescue grass is fed to cattle, sheep, horses, or rabbits, it is generally toxic to those animals, due to mycotoxin(s) or alkaloids produced by that endophyte. See also MICROORGANISM, BACTERIA, *BACILLUS THURINGIENSIS* (*B.t.*), FUNGUS, PROTEIN, MYCOTOXINS, TREMORGENIC INDOLE ALKALOIDS.

**Endoplasmic Reticulum (ER)** A highly specialized, complex network of branching, intercommunicating tubules (surrounded by membranes) found in the cytoplasm of most animal and plant cells. The two types of ER recognized are: rough ER and smooth ER. Rough ER is covered with many ribosomes; ER without or with fewer ribosomes attached is called smooth. This nomenclature comes about because of the appearance of the ER under high magnification. The rough ER is very well developed to facilitate cells carrying on abundant protein synthesis, because proteins are synthesized (manufactured) in ribosomes. See also CELL, CYTOPLASM, RIBOSOMES, FATS, LIPIDS, PLASMA MEMBRANE, PROTEIN, PHOSPHOLIPIDS.

**Endorphins** Discovered during the 1970s by U.S. and Scottish scientists, these hormones are produced in the brain, and act as natural painkillers. For example, runners and long-distance walkers achieve something of a "high" due to endorphins released during long runs or walks. See also ENKEPHALINS, CATECHOLAMINES, HORMONE.

**Endosome** See ENDOCYTOSIS.

**Endosperm** The interior portion of a plant seed, beneath the outer hull (the portion that people tend to eat, in food crops). In grains (e.g., rice or corn/maize), the endosperm consists primarily of starch (carbohydrate). In legumes (e.g., beans), the endosperm contains mainly protein, a small amount of carbohydrates, and sometimes vegetable oil. See also STARCH, CORN, SOYBEAN PLANT,

CARBOHYDRATES (SACCHARIDES), SOYBEAN OIL, ALEURONE.

**Endospore** A highly resistant, dormant inclusion body formed within certain bacteria. To kill spores, temperatures above boiling are usually needed. For this, pressure cookers and autoclaves are required. Endospores have survival value since the spore can remain for long periods of time in a nongrowing state and then, under appropriate conditions, can be induced to germinate and regenerate the original cell. Endospore formation may be viewed as being akin to hibernation, i.e., a kind of "bacterial hibernation."

**Endostatin** An antiangiogenesis human protein discovered by Judah Folkman. In concert with angiostatin, it causes certain cancer tumors in mice to shrink. See also PROTEIN, ANTIANGIOGENESIS, ANGIOSTATIN, CANCER.

**Endothelial Cells** These are the flat, sort of plate-shaped cells that line the surface of all blood vessels, heart, and lymphatics within the body. Endothelial cells possess transmembrane (through the cell membrane) molecules known as adhesion molecules, which selectively allow the passage (from bloodstream to tissues) of some molecules (leukocytes, monocytes, hormones, etc.). Endothelial cells are packed much tighter together in the capillaries that provide blood to the brain. This tighter packing limits the size and kind of molecules that can pass into the brain. This blood-brain barrier serves to protect the sensitive brain tissue from pathogens or harmful molecules (e.g., toxins). See also ENDOTHELIUM, VASCULAR ENDOTHELIAL GROWTH FACTOR (VEGF), ADHESION MOLECULES, MONOCYTES, MITOGEN, SELECTINS, BLOOD-BRAIN BARRIER (BBB), LECTINS, ELAM-1, ATP SYNTHASE, OXIDATIVE STRESS, CYCLOOXYGENASE.

**Endothelin** A peptide that causes arteries to contract (which consequently causes blood pressure to increase). See also PEPTIDE, ATRIAL PEPTIDES.

**Endothelium** The layer of epithelial cells that line blood vessels throughout the body. The layer selectively allows the passage (from bloodstream to tissues) of nutrients, hormones, and other molecules essential for tissue growth and function. The endothelium is involved in the recovery and recycling of old red blood cells. It also produces nitric oxide, which causes neighboring smooth-muscle (blood vessel) cells to relax so that those (neighboring ) blood vessels dilate and the body's blood pressure is lowered, and two compounds, prostacyclin and Von Willebrand factor, that prevent blood clotting. See also ENDOTHELIAL CELLS, VASCULAR ENDOTHELIAL GROWTH FACTOR (VEGF), SELECTINS, LECTINS, ADHESION MOLECULES, NITRIC OXIDE, NITRIC OXIDE SYNTHASE, BILIRUBIN.

**Endotoxin** A lipopolysaccharide (fat/sugar complex; poison, also known as LPS) which forms an integral part of the cell wall of gram negative bacteria. It is only released when the cell is ruptured. It can cause, among other things, septic shock and tissue damage. Pharmaceutical preparations are routinely tested for the presence of endotoxins. This is one reason why pharmaceuticals must be prepared in a sterile environment. See also SEPSIS, BACTERIA, LIPIDS, POLYSACCHARIDES, TOXIN, CHOLERA TOXIN, GRAM-NEGATIVE (G-), GOOD MANUFACTURING PRACTICES (GMP).

**Engineered Antibodies** Chimeric monoclonal antibodies, produced via genetic engineering of human antibody-producing cells (clones). For example, the genes coding for antilymphoma binding sites from a rat have been inserted into human antibody-producing cells to yield rat (antigen) binding sites mounted on human antibody "stems." See also CHIMERIC PROTEINS, MONOCLONAL ANTIBODIES (MAb), ANTIBODY, GENETIC ENGINEERING, COMBINING SITE, LYMPHOCYTE, SEMISYNTHETIC CATALYTIC ANTIBODY.

**Enhanced Nutrition Crops** See NUTRIENT ENHANCED™.

**Enkephalins** A class of hormones produced in the brain that act as natural painkillers. Discovered by John Hughes and Hans Kosterlitz in 1975, they are some of the endorphins. See also ENDORPHINS.

**Enolpiruvil Shikimate** See EPSP SYNTHASE.

**Enolpyruvil Shikimate** See EPSP SYNTHASE.

**Enoyl-acyl Protein Reductase** An enzyme that is utilized by bacteria in their synthesis (manufacture) of fatty acids. See also ENZYME, PROTEIN, BACTERIA, FATTY ACID, ESSENTIAL FATTY ACIDS.

**Ensiling** The fermentation of (usually chopped up) agricultural vegetation in order to preserve it. It is carried out for 1–2 weeks, using either indigenous microorganisms (e.g., *Lactobacillus* spp.) or introduced microorganisms (to speed up the process, yield product containing more nutrients for livestock, etc.), in the absence of oxygen (to prevent the growth of aerobic mold fungi). When indigenous microorganisms are used, *Lactobacillus* spp. become the dominant microorganisms present, and heat is generated by the microorganisms within the vegetative mass (optimum temperature is 25–30°C, which is 77–86°F). Lactic acid produced by the microorganisms inhibits the growth of bacteria that would normally putrefy the vegetation. See also FERMENTATION, MICROORGANISM, AEROBIC, FUNGUS, OPTIMUM TEMPERATURE.

**Enterohemorrhagic *E. coli*** The several dozen (approximately 60 known) serotypes (strains) of *E. coli* bacteria that cause internal hemorrhaging in humans that ingest those bacteria. The toxin produced by these particular *E. coli* bacteria attacks the human kidney, which often leads to kidney failure and/or death. See also *ESCHERICHIA COLIFORM* 0157:H7 (*E. COLI* 0157:H7), TOXIN, SEROTYPES, ENTEROTOXIN.

**Enterotoxin** The category (i.e., intestinally active) of toxins, produced by certain bacterial strains and/or serotypes, which attack the body's internal organs. For example, the serotype of *Escherichia coliform* bacteria known as *E. coli* 0157:H7 attacks the kidneys and other internal organs of humans, also causing internal bleeding and sometimes death. See also TOXIN, BACTERIA, *ESCHERICHIA COLIFORM* 0157:H7, ENTERPHEMORHAGIC *E. COLI*, SEROTYPES, CHOLERA TOXIN.

**Enzyme** An organic, protein-based catalyst that is not itself used up in the reaction. It is naturally produced by living cells to catalyze biochemical reactions. Each enzyme is highly specific with regard to the type of chemical reaction that it catalyzes, and to the substances (called substrates) upon which it acts. This specific catalytic activity and its control by other biochemical constituents are of primary importance in the physiological functions of all organisms. Although all enzymes are proteins, they may, and usually do, contain additional nonprotein components called coenzymes that are essential for catalytic activity. See also APOENZYME, CATALYST, COENZYME, HOLOENZYME, SUBSTRATE (CHEMICAL), PROTEIN, HORMONE, EXTREMOZYMES, TURNOVER NUMBER.

**Enzyme Denaturation** The loss of enzyme (catalytic) activity due to loss of the correct functional structure of the protein. Denaturation may be caused by factors such as exposure to heat and organic solvents, degradation of the enzyme molecule by proteases, oxygen, and acid or alkaline pH. See also ENZYME, CONFORMATION, DENATURATION, EXTREMOZYMES.

**Enzyme Derepression** Commonly known as induction (of an enzyme). Initially a repressor protein is bound to a specific region of DNA. This binding inhibits transcription to mRNA, thus blocking the synthesis of the protein (enzyme) specified by the mRNA. When present, the inducer molecule binds to the repressor protein and inactivates it. Thus the inhibition caused by the repressor protein is overcome and mRNA can be synthesized, which consequently leads to synthesis of the mRNA-specified protein (enzyme). The word derepression is sometimes used because the repressor protein is, by itself, active in repressing protein (enzyme) synthesis. Its repressive action is mitigated (derepressed) by the inducer molecule. Hence, derepression (or unrepression) of repression equals induction. See also CONTINUOUS PERFUSION, ENZYME REPRESSION, ENZYME, REPRESSION (OF AN ENZYME).

**Enzyme Immunoassay (EIA)** See ELISA.

**Enzyme Repression** Inhibition of enzyme synthesis caused by the availability of the product of that enzyme. On a molecular level a repressor molecule (which could be, e.g., the amino acid arginine) combines with a specific repressor protein that is present in the cell. This repressor molecule/repressor protein complex is then able to bind to a specific region of DNA at the initial end of the gene which is called the operator region. It is in this region where the synthesis of mRNA is initiated. The repressor "roadblock" thus stops the synthesis of mRNA, and therefore the synthesis of the protein is also blocked. See also ENZYME, REPRESSION (OF AN ENZYME), ENZYME DEPRESSION.

**Enzyme-Linked Immunosorbent Assay** See ELISA.

**Eosinophils** Polymorphonuclear leukocytes made in the bone marrow. They circulate in the blood for a number of hours (three to eight) and then migrate into the tissue where they reside. They kill parasites too large to be phagocytized by secreting substances that kill the parasites (hookworms, trichinosis, etc.), inhibit histamine release from mast cells, and secrete chemicals that neutralize histamine. Allergy causes an increase in eosinophils. GM-CSF stimulates eosinophil production. See also POLYMORPHONUCLEAR LEUKOCYTES (PMN), BASOPHILS, ANTIGEN, CELLULAR IMMUNE RESPONSE.

**EPD** See EXPECTED PROGENY DIFFERENCES.

**Epidermal Growth Factor (EGF)** A protein of 53 amino acids that greatly increases growth/reproduction of epidermal (skin) cells. This protein also increases growth of wool in sheep and growth in more than 50% of human tumors. High concentrations of epidermal growth factor are found in human tears. EGF was discovered by Stanley Cohen. See also PROTEIN, EGF RECEPTOR, GROWTH FACTOR, NERVE GROWTH FACTOR (NGF), AMINO ACID, FILLER EPITHELIAL CELLS, TUMOR.

**Epidermal Growth Factor Receptor** See EGF RECEPTOR, HER-2 RECEPTOR, HER-2 GENE.

**Epimerase** An enzyme capable of the reversible interconversion of two epimers. See also ENZYME, EPIMERS.

**Epimers** Two stereoisomers differing in configuration. See also CONFIGURATION, STEREOISOMERS.

**Episome** (of a bacterium) An independent genetic element (DNA) that occurs inside bacterium in addition to the normal bacterial cell genome. The episome can replicate either as an autonomous unit or as one integrated into the host genome. The F (fertility) factor is an episome. See also GENOME, PLASMID, BACTERIA, DEOXYRIBONUCLEIC ACID (DNA).

**Epistasis** Interaction between nonallelic genes in which the presence of a certain allele at one locus prevents expression of an allele at a different locus. See also ALLELE, GENE, EXPRESS, LOCUS.

**Epithelial Projections** Projections that anchor the epidermis (surface skin) to the dermis (subsurface tissue). Growth of these projections is increased by epidermal growth factor during the wound healing process. See also EPIDERMAL GROWTH FACTOR (EGF).

**Epithelium** The prefix *epi-* means on, above, or upon. The membranous cellular tissue that covers a free surface or lines a tube or cavity of an animal body. It serves to enclose and protect the other tissues, to produce secretions and excretions, and to function in assimilation. See also ASSIMILATION, CADHERINS, ION CHANNELS.

**Epitope** Also called antigenic determinant. The specific group of atoms (on an antigen molecule) that is recognized by (that antigen's) antibodies (thereby causing an immune response). See also ANTIBODY, ANTIGEN, IDIOTYPE, HUMORAL IMMUNE RESPONSE.

**EPO** See ERYTHROPOIETIN, EUROPEAN PATENT OFFICE.

**EPPO** See EUROPEAN PLANT PROTECTION ORGANIZATION.

**EPSP Synthase** Enolpyruvyl-shikimate phosphate synthase. An enzyme produced by virtually all plants and internally transported into their cells' chloroplasts, it is essential in a plant's metabolism biochemical pathway and for the biosynthesis (creation) of the aromatic (ring-shaped molecule) amino acids tyrosine, phenylalanine, and tryptophan, which are needed for plants to live. Some (glyphosate-containing and sulfosate-containing) herbicides kill unwanted plants (e.g., weeds) by inhibiting EPSP synthase. By incorporating a gene that causes (over-) production of CP4 EPSP synthase into several crops (soybeans, cotton, etc.), scientists have been able to help those crops survive post-emergence application(s) of glyphosate-containing herbicide. Additional resistance to glyphosate-containing and sulfosate-containing herbicides can be conferred to plants by incorporating into plants a gene (GO) which causes those plants to produce glyphosate oxidase. See also ENZYME, METABOLISM, GENE, PAT GENE, BAR GENE, GENETIC ENGINEERING, SOYBEAN PLANT,

CORN, GLYPHOSATE, GLYPHOSATE OXIDASE, CP4 EPSPS, HERBICIDE-TOLERANT CROP, SULFOSATE, mEPSPS, CHLOROPLASTS, CHLOROPLAST TRANSIT PEPTIDE (CTP), TARGET (OF A HERBICIDE OR INSECTICIDE).

**EPSPS**  See EPSP SYNTHASE, CP4 EPSPS, mEPSPS.

**ER**  See ENDOPLASMIC RETICULUM.

**Ergotamine**  A mycotoxin (i.e., metabolite produced by a fungus, that is toxic to animals and humans) produced by the fungus (*Claviceps* spp.) known as ergot. Ergotamine is an alkaloid vasoconstrictor, whose consumption can lead to severe constriction of blood vessels in the brain and extremities, causing hallucinations and dry gangrene. Humans whose bodies are deficient in vitamin A are especially vulnerable to ergotism ("ergot poisoning"). See also MYCOTOXINS, TOXIN, FUNGUS, VITAMIN.

*Erwinia caratovora*  A species of bacteria that can cause significant postharvest losses to potato farmers, when it infects potatoes and causes "soft rot" (spoilage). See also BACTERIA, SPECIES.

*Erwinia uredovora*  See GOLDEN RICE.

**Erythrocytes**  (red blood cells) Hemoglobin-containing cells (manufactured in the bone marrow) that transport the oxygen from the lungs to the body tissues where it is needed.

**Erythropoiesis**  The formation of red blood cells from certain stem cells. Stimulated by the protein erythropoietin. See also STEM CELLS, ERYTHROPOIETIN (EPO).

**Erythropoietin (EPO)**  A glycoprotein hormone produced in the kidneys that stimulates stem cells in the bone marrow to increase the number of red blood cells. Erythropoietin can be used to help correct a variety of anemias. See also GLYCOPROTEIN, HORMONE, ERYTHROCYTES, STEM CELLS.

*Escherichia coli*  See ESCHERICHIA COLIFORM (*E. COLI*).

*Escherichia coli* **0157:H7** (*E. coli* **0157:H7**)  See ESCHERICHIA COLIFORM 0157:H7.

*Escherichia coliform* (*E. coli*)  A bacterium that commonly inhabits the human intestine as well as the intestine of other vertebrates (animals possessing a skeleton). The most thoroughly studied of all bacteria, *Escherichia coli* is used in many microbiological experiments. It has historically been considered the workhorse of genetic engineering research, and genetically engineered versions have been used to produce human proteins (e.g., insulin). One of the more exotic uses of genetically engineered *E. coli* was to make indigo dye (originally discovered in 1983, using indole or tryptophan as starting materials). In 1993, Burt D. Ensley and coworkers at Amgen discovered a way to genetically engineer *E. coli* to produce indigo from glucose starting material. *E. coli* has 4,288 genes. See also TRYPTOPHAN (trp), BACTERIA, GENETIC ENGINEERING, GENE, RECOMBINANT DNA (rDNA), *ESCHERICHIA COLIFORM* 0157:H7.

*Escherichia coliform* **0157:H7**  The particular strain (serotype) of *Escherichia coliform* (*E. coli*) bacteria that causes often-fatal diarrhea, internal bleeding, and kidney damage in humans. Children are more susceptible to *E. coli* 0157:H7 than adults, because children possess more of the receptors (on cells inside the digestive tract) that are utilized by *E. coli* 0157:H7 to enter the body from the digestive tract. Although cattle were susceptible to *E. coli* 0157:H7's toxins prior to the 1980s, they eventually developed resistance. That meant that the cattle could carry these bacteria without getting sick, and transmit *E. coli* 0157:H7 to humans whenever conditions allow (e.g., when *E. coli* 0157:H7-infected cattle are slaughtered and people consume the meat without first heating it to a high enough temperature to kill the *E. coli* 0157:H7). Some varieties of *E. coli* 0157:H7 are resistant to the antibiotics tetracycline and streptomycin. In 1996, researchers at Cornell University in New York state, U.S.A., discovered that nonambulatory cows (that could not walk) were approximately four times as likely as other cows to test positive for *E. coli* 0157:H7. Other research in Canada indicates that fasting of cattle (common occurrence for nonambulatory cows) tends to alter the pH inside the cow's rumen (stomach) in a way that encourages the proliferation of *E. coli* 0157:H7 instead of the bacteria that normally populate the rumen. See also *ESCHERICHIA COLIFORM* (*E. COLI*), BACTERIA, SEROTYPES, TOXIN, RECEPTORS, BIOLUMINESCENCE, STRAIN, ENTEROTOXIN, COMMENSAL.

**Essential Amino Acids** Those amino acids that cannot be synthesized by humans and most other vertebrates, and therefore must be obtained from the diet. They are phenylalanine, valine, threonine, tryptophan, isoleucine, methionine, histidine, arginine, leucine, and lysine (glycine and proline for poultry). See also AMINO ACID, LYSINE (lys), METHIONINE (met), SOY PROTEIN, OPAGUE-2, PROTEIN DIGESTIBILITY-CORRECTED AMINO ACID SCORING (PDCAAS).

**Essential Fatty Acids** The group of polyunsaturated fatty acids of plants that are required in the human diet, because the human body cannot synthesize (manufacture) them, yet must have them for proper functioning (of the body's metabolism, immune system function, etc.). These include linoleic acid, linolenic acid, arachidonic acid, and docosahexanoic acid. If humans and other higher animals do not consume enough essential fatty acids per day, they suffer decreased growth rates, increased susceptibility to infection, impaired reproduction, kidney damage, and other adverse physiological effects. See also FATTY ACID, SOYBEAN OIL, LECITHIN, FATS, ESSENTIAL NUTRIENTS, POLYUNSATURATED FATTY ACIDS (PUFA), LINOLEIC ACID, LINOLENIC ACID, DOCOSAHEXANOIC ACID (DHA), ARACHIDONIC ACID (AA).

**Essential Nutrients** Chemical compounds in foods required for (consuming organism's) life, growth, or tissue repair, and cannot be synthesized by that organism. See also ESSENTIAL AMINO ACIDS, ESSENTIAL FATTY ACIDS, ESSENTIAL POLYUNSATURATED FATTY ACIDS, VITAMIN.

**Essential Polyunsaturated Fatty Acids** See ESSENTIAL FATTY ACIDS.

**EST** See EXPRESSED SEQUENCE TAGS (EST).

**Estrogen** A female sex hormone, secreted by the ovaries, that promotes estrus and helps to regulate the pituitary gland's production of luteinizing hormone (LH) and follicle-stimulating hormone (FSH). Estrogen causes proliferation of breast tissue (cells) and is also responsible for the development of female secondary sex characteristics (e.g., smaller body size, lack of facial hair, higher pitch voice in humans). Research indicates that lack of estrogen (e.g., in post-menopausal women) makes humans more prone to colon cancer and heart disease, but less prone to the "hormone dependent" cancers (ovarian cancer, uterine cancer, etc.). See also HORMONE, PITUITARY GLAND, FOLLICLE STIMULATING HORMONE (FSH), SELECTIVE ESTROGEN EFFECT, TESTOSTERONE, LUTEINIZING HORMONE (LH), HYPOTHALAMUS, CANCER, CELL.

**Ethylene** A plant hormone synthesized (manufactured) by some plants to induce ripening (of their fruit). See also PLANT HORMONE, ACC SYNTHASE, ACC, SAM-K GENE.

**Etiological Agent** (of a disease) The microorganism (or other agent) that causes the disease. See also PATHOGEN, ETIOLOGY.

**Etiology** The science (study) of the cause (source) of a disease. See also PATHOGEN, ETIOLOGICAL AGENT.

**Eucaryote** Also spelled eukaryote. A cell characterized by compartmentalization (by membranes) of its extensive internal structures; or an organism made up of such cells. For example, eucaryotes possess a distinct membrane-surrounded nucleus containing the DNA. Eucaryotic cells (e.g., human cells) are much larger and more complex than procaryotic cells (e.g., bacteria). The cells of all higher organisms, both plant and animal, are eucaryotic, so those higher (complex) organisms are often referred to as eucaryotes. Most eucaryotic organisms cannot survive temperatures greater than 131°F (55°C). However, one called the Pompeii worm (*Alvinella pompejana*) can withstand long-term exposure in water up to a temperature of 176°F (80°C). See also PROCARYOTES, CELL, THERMOPHILE, DEOXYRIBONUCLEIC ACID (DNA), PLASMA, MEMBRANE, MICROTUBULES.

**Eugenics** First formulated by Francis Galton, who was a contemporary of Gregor Mendel in the 19th century, eugenics is the concept that a species can be "improved" by encouraging reproduction of only those organisms in that species that possess "desired" traits. This belief became popular in a number of countries during the early 20th century. Margaret Sanger, founder of America's Planned Parenthood organization, referred to African-Americans as "human weeds" and called for "more children from the fit, less from the unfit." Based upon Charles Darwin's written assertion that "the civilized races of man will almost certainly exterminate and replace the savage races," a number of large genocides were committed by national governments. See also GENETICS, GENE, TRAIT, GENOTYPE, HEREDITY, HERITABILITY, GENOME.

**Eukaryote**  See EUCARYOTE.

**Euploid**  A cell carrying an exact multiple of the haploid chromosome number. For example, a diploid possesses twice the haploid number of chromosomes. See also HAPLOID, DIPLOID, CHROMOSOMES.

**European Corn Borer (ECB)**  Also known as pyralis. Latin name *Ostrinia nubilalis*, it is an insect whose larvae (caterpillars) eat and bore into the corn/maize plant (*Zea mays* L.). In doing so, they can act as vectors (i.e., carriers) of the fungi known as *Aspergillus flavus* (source of aflatoxin) or *Fusarium moniliforme* (source of fumonisin) or *Aspergillus parasiticus* (source of aflatoxin). Full-grown ECB larvae winter by sheltering inside a variety of vegetative materials (e.g., plant stalks lying on top of soil in some fields). ECB control can be effected by some of the following methods:

1. Spraying of conventional synthetic chemical pesticides
2. Spraying of pesticides produced via promulgation of *Bacillus thuringiensis* (*B.t.*) bacteria
3. Incorporating a (protoxin) gene from *Bacillus thuringiensis* (*B.t.*) into the DNA of the corn plant, so that the plant itself produces *B.t.* protoxin

As part of Integrated Pest Management (IPM), farmers can utilize:

1. Corn possessing *Bacillus thuringiensis* (*B.t.*) gene(s) to control populations of ECB without applying insecticides
2. The parasitic *Euplectrus comstockki* wasp to help control the ECB. (When that wasp's venom is injected into ECB larva, it stops the larva from molting and thus maturing)
3. Additional methods, alone or in concert with above

See also CORN, FUNGUS, AFLATOXIN, INTEGRATED PEST MANAGEMENT (IPM), *BACILLUS THURINGIENSIS* (*B.t.*), *B.t.* KURSTAKI, *FUSARIUM*, *FUSARIUM MONILIFORME*, ASIAN CORN BORER, PROTOXIN, VOLICITIN.

**European Medicines Evaluation Agency (EMEA)**  A London-based agency of the European Union (EU) that began operation in 1995. It coordinates drug licensing and safety matters throughout the nations of the EU. Its licensing/approval process is compulsory throughout the EU. See also COMMITTEE FOR PROPRIETARY MEDICINAL PRODUCTS (CPMP), MEDICINES CONTROL AGENCY (MCA), FOOD AND DRUG ADMINISTRATION (FDA), KOSEISHO, BUNDESGESUNDHEITSAMT (BGA), COMMITTEE ON SAFETY IN MEDICINES, COMMITTEE FOR VETERINARY MEDICINAL PRODUCTS (CVMP).

**European Patent Convention**  An international patent treaty signed in 1973, by which the countries of Europe agreed to recognize and honor the patents granted by each country, plus those patents granted by the European Patent Office (EPO). Plant varieties or animal breeds were initially excluded from patentability by the European Patent Convention. In 1998, the European Parliament removed that exclusion. See also EUROPEAN PATENT OFFICE (EPO), U.S. PATENT AND TRADEMARK OFFICE (USPTO), PLANT'S NOVEL TRAIT (PNT), PLANT BREEDER'S RIGHTS (PBR), UNION FOR PROTECTION OF NEW VARIETIES OF PLANTS (UPOV).

**European Patent Office (EPO)**  The Munich, Germany-based agency of the European Union (EU) — established in 1977 — that is responsible for common patent protection matters for all of the (EU) member countries, plus the non-EU countries of Switzerland and Liechtenstein. The European Patent Office originally did not allow a "plant or animal breed" to be patented, whereas its U.S. counterpart — the U.S. Patent and Trademark Office (USPTO) — does allow patenting of microbes, plants, and animals (e.g., those which have been genetically engineered by man). In 1998, the European Parliament removed that exclusion, and in 1999, the European Patent Court issued a ruling which caused the European Patent Convention to allow patents on novel plants, thus making the two patent systems compatible.

E

See also EUROPEAN PATENT CONVENTION, MICROBE, GENETIC ENGINEERING, BIOTECHNOLOGY, AMERICAN TYPE CULTURE COLLECTION (ATCC), U.S. PATENT AND TRADEMARK OFFICE (USPTO), PLANT'S NOVEL TRAIT (PNT), PLANT BREEDER'S RIGHTS (PBR), UNION FOR PROTECTION OF NEW VARIETIES OF PLANTS (UPOV), COMMUNITY PLANT VARIETY OFFICE.

**European Plant Protection Organization (EPPO)** One of the international SPS standard-setting organizations that develops plant health standards, guidelines, and recommendations (e.g., to prevent transfer of a plant disease or plant pest from one country to another). Its secretariat is in Paris, France. EPPO, one of the organizations within the International Plant Protection Convention (IPPC), covers the countries of Europe. See also INTERNATIONAL PLANT PROTECTION CONVENTION (IPPC), NORTH AMERICAN PLANT PROTECTION ORGANIZATION (NAPPO), SPS, PLANT'S NOVEL TRAIT (PNT), PLANT BREEDER'S RIGHTS (PBR).

**Event** Refers to each instance of a genetically engineered organism. For example, the same gene inserted by man into a given plant genome at two different locations (*loci*) along that plant's DNA would be considered two different events. Alternatively, two different genes inserted into the same *locus* of two same-species plants would also be considered two different events. Generally speaking, the world's regulatory agencies confer new biotech-derived product approvals in terms of events. See also GENETIC ENGINEERING, GENETICALLY ENGINEERED ORGANISM (GEO), GENE, DEOXYRIBONUCLEIC ACID (DNA), LOCUS, LOCI, GENOME, MUTUAL RECOGNITION AGREEMENTS (MRAS).

**Excision** The cutting out of a piece of damaged or defective DNA by enzymes. DNA damage might be constituted by the presence of a thymine dimer which inactivates that part of the DNA. The region of the dimer is cut out and then repaired. See also RECOMBINATION, GENOME, INFORMATIONAL MOLECULES.

**Excitatory Amino Acids (EAAs)** Amino acids present in the brain (when released by certain immune system cells) that can kill brain cells when in excess (e.g., results from strokes, which cause the release of too many EAAs in the brain). Another source of harmful EAAs (e.g., glutamate) is the disease known as multiple sclerosis. Some spiders paralyze their prey with venom that contains a substance that blocks the action of EAAs; thus, pharmaceuticals based on an active ingredient in that venom may someday be used to prevent brain damage in stroke and in multiple sclerosis victims. See also AMINO ACID, MULTIPLE SCLEROSIS, CELL, IMMUNE RESPONSE.

**Exclusion Chromatography** See GEL FILTRATION.

**Exergonic Reaction** A chemical reaction with a negative standard free energy change (i.e., a "downhill" reaction). A reaction which releases energy (exothermic; in the form of heat). See also ENDERGONIC REACTION, FREE ENERGY.

**Exobiology** Extraterrestrial biology.

**Exocytosis** The releasing of an entity that was bound inside an "endosome" (e.g., inside a cell). See also ENDOCYTOSIS.

**Exoglycosidase** An enzyme that hydrolyzes (cuts) only a terminal (end) bond in the oligosaccharide (molecular) branch(es) of a glycoprotein. See also ENDOGLYCOSIDASE, GLYCOPROTEIN, RESTRICTION ENDOGLYCOSIDASES.

**Exon** The segment of a eucaryotic gene that is transcribed into an mRNA (messenger RNA) molecule; it codes for a specific domain of a protein. See also PROTEIN, EUCARYOTE, MESSENGER RNA (mRNA), GENE, HOMEOBOX.

**Exonuclease** An enzyme that hydrolyzes (cuts) only a terminal phosphodiester bond of a nucleic acid. See also HYDROLYZE.

**Exotic Germplasm** Germ plasm that has not been adapted (selectively bred) to the environment intended (for its offspring, via selective breeding by man). See also GERM PLASM, INTROGRESSION, HYBRIDIZATION (PLANT GENETICS).

**Exotoxin** Proteins (toxins) produced by certain bacteria that are released by the bacteria into their surroundings (growth medium). Produced by primarily Gram-positive bacteria. Diphtheria toxin was the first one discovered. Other exotoxins cause botulism, tetanus, gas gangrene, and scarlet fever. Exotoxins are generally more potent and specific in their actions than endotoxins. See also ENDOTOXIN, TOXIN, GRAM-POSITIVE (G+).

**Expected Progeny Differences (EPD)** Numerical rankings of (livestock) parental genetics, in terms of an animal's genetic impact on progeny's four following commercial traits:

1. Number of progeny born alive
2. Weight of progeny at weaning age
3. Number of days required to reach slaughter weight, when fed adequately
4. Carcass lean meat vs. fat percentages

EPDs allow a farmer to estimate differences in performance of future offspring (of a given parent) vs. offspring produced by parents of average genetic value. For example, a boar (male pig) possessing an EPD of –4 for "number of days required to reach slaughter weight" produces offspring that reach slaughter weight in four fewer days (of feeding time) than offspring that are sired by a boar possessing an EPD of 0. See also GENETICS, TRAIT, PHENOTYPE, GENOTYPE, BEST LINEAR UNBIASED PREDICTION (BLUP).

**"Explosion" Method** [to introduce foreign (new) genes into plant cells] A technique for gene-into-cell introduction in which the gene (genetic material) is driven into plant cells by the force of an explosion (vaporization) of a drop of water (to which the gene and gold particles have been added). The explosion is caused by application of high-voltage electricity to the drop of gene-laden water; the water is then vaporized explosively, driving the "shot" (gold particles) and genetic material through the cell membrane. The plant cell then heals itself (reseals the hole where the gene entered), incorporates the new gene into its genetic complement, and produces whatever product (e.g., a protein) for which the newly introduced gene codes. See also *AGROBACTERIUM TUMEFACIENS*, CODING SEQUENCE, GENETIC ENGINEERING, VECTOR, "SHOTGUN" METHOD, GENE, GENOME, RIBOSOMES.

**Express** To translate the cell's genetic information stored in the DNA (gene) into a specific protein (synthesized by the cell's ribosome system). Certain proteins (i.e., when present in relevant cells) regulate the

expression (e.g., increase/decrease/timing) of some genes. See also GENE EXPRESSION CASCADE, RIBOSOMES, GENE, DEOXYRIBONUCLEIC ACID (DNA), CELL, TRANSCRIPTION, TRANSLATION, MESSENGER RNA (mRNA), TRANSCRIPTION UNIT, PROTEIN, COSUPPRESSION, GENE EXPRESSION ANALYSIS, FUNCTIONAL GENOMICS.

**Expressed Sequence Tags (EST)** Molecular tags utilized to "label" a given gene (i.e., in terms of that gene's function/protein). Physically, the EST is composed of cRNA [i.e., the gene's "message" after the "junk DNA" (introns) have been edited out], produced by the analogous gene in (simple) model organisms such as (traditionally) *Caenorhabditis elegans* nematode, which has been sequenced/mapped. Functions of the "labeled" genes are (at least initially) inferred from (known function) *C. elegans* genes. See also GENE, INTRON, PROTEIN, COMPLEMENTARY DNA (cDNA), DEOXYRIBONUCLEIC ACID (DNA), JUNK DNA, BEST, *CAENORHABDITIS ELEGANS* (*C. ELEGANS*), SEQUENCING (OF DNA MOLECULES), SEQUENCE (OF A DNA MOLECULE), MAPPING, MODEL ORGANISM, BACTERIAL EXPRESSED SEQUENCE TAGS (BEST).

**Expression Analysis** See GENE EXPRESSION ANALYSIS, MICROARRAY (TESTING).

**Expression Array** See MICROARRAY (TESTING).

**Expression Profiling** See GENE EXPRESSION ANALYSIS.

**Expressivity** The intensity with which the effect of a gene is realized in the phenotype. The degree to which a particular effect is expressed by individuals. See also PHENOTYPE, EXPRESS, RIBOSOMES.

**Extension** (in nucleic acids) The nucleic acid strand elongation (lengthening) that occurs in a polymerization reaction. See also NUCLEIC ACIDS, POLYMER.

**Extranuclear Genes** Genes that reside within the cell, but outside the nucleus. Generally, extranuclear genes reside in the organelles such as mitochondria and chloroplasts. See also GENE, CELL, NUCLEUS, COPY NUMBER, ORGANELLES, CHLOROPLASTS, MITOCHONDRIA.

**Extremophilic Bacteria** Bacteria that live and reproduce outside (either colder or hotter) the typical temperature range of 40°F (4°C) to 140°F (60°C) that bacteria tend to be found in, on earth. Other extremes are high

E

pressure (e.g., at the ocean bottom), salt saturation, (e.g., the Dead Sea), pH lower than 2 (e.g., coal deposits), pH higher than 11 (e.g., sewage sludge), high levels of radiation, etc. See also BACTERIA, THERMOPHILIC BACTERIA, THERMOPHILE, THERMODURIC, *DEINOCOCCUS RADIODURANS*.

**Extremozymes** Enzymes within the microorganisms (e.g., extremophilic bacteria) that populate extreme environments. Because extremozymes can catalyze reactions under high pressure, high temperatures, etc., they are increasingly being used as catalysts for industrial processes. See also EXTREMOPHILIC BACTERIA, ENZYME, ARCHAEA, PHYTO-MANUFACTURING.

***Ex vivo* (testing)** The testing of a substance by exposing it to (excised) living cells (but not to the whole, multicelled organism) in order to ascertain the effect of the substance (e.g., pharmaceutical) on the biochemistry of the cell. See also IN VITRO, IN VIVO.

***Ex vivo* (therapy)** Removal of cells (e.g., certain blood cells) from a patient's body, alteration of those cells in one or more therapeutic ways, followed by reinsertion of the altered cells into the patient's body. See also IN VITRO, IN VIVO.

# F

**F-Box Proteins** Proteins produced (manufactured) within some eucaryotic cells, that play an essential role in the degradation (i.e., breakdown) of cellular regulatory proteins, after those regulatory proteins have "completed ther job" in the cell. See also PROTEIN, CELL, EUCARYOTE.

**F1 Hybrids** The first-generation offspring of crossbreeding; also known as first filial hybrids. They tend to be more healthy, productive, and uniform than their parents. See also GENETICS, HYBRIDIZATION (PLANT GENETICS).

**FACS** See FLUORESCENCE ACTIVATED CELL SORTER (FACS).

**Factor IX** A protein factor in the blood serum that is instrumental in the cascade of chemical reactions (involving 17 blood components) that leads to clot formation, following a cut or other wound to body tissue. A deficiency of Factor IX is the cause of the disease known as hemophilia B (approximately 15% of all hemophilia patients). See also FIBRIN, FIBRONECTIN, PROTEIN, CASCADE, FACTOR VIII.

**Factor VIII** Also known as antihemophilic globulin (AHG) or antihemophilic Factor VIII. A protein factor in the blood serum that is instrumental in the "cascade" of chemical reactions (involving 17 blood components in the intrinsic pathway) that leads to clot formation following a cut or other wound to body tissue. Also, a deficiency of AHG is the cause of the classical type of hemophilia sometimes known as hemophilia AM (approximately 85% of all hemophilia patients). See also FIBRIN, PROTEIN, FIBRONECTIN, CASCADE, PATHWAY, FACTOR IX.

**Facultative Anaerobe** An organism that will grow under either aerobic or anaerobic conditions. See also AEROBE, ANAEROBE, ORGANISM.

**Facultative Cells** Cells that can live either in the presence or absence of oxygen. See also AEROBE, ANAEROBE.

**FAD** See FLAVIN ADENINE DINUCLEOTIDE (FAD).

**FAO** Food and Agriculture Organization of the United Nations. See also CONSULTATIVE GROUP ON INTERNATIONAL AGRICULTURAL RESEARCH, CODEX ALIMENTARIUS COMMISSION.

**Farnesyl Transferase** An enzyme utilized by the ras gene (to help "signal" certain cells to divide/grow). See also RAS GENE, GENE, ENZYME, CELL, SIGNALING MOLECULE.

**Fats** Energy storage substances produced by animals and some plants (e.g., soybeans), which consist of a combination of fatty acids and glycerol that form predominantly triglyceride molecules (although some diglyceride molecules are also often present in fats). The structure of triglyceride molecules consists of three fatty acids attached to a glycerol molecular backbone, so "triglyceride" molecules are more accurately called "triacylglycerides," but the triglyceride term is most often used.

Two separate components of plant cells are involved in the synthesis (manufacturing) of plant fats (lipids); the plastid and the endoplasmic reticulum. Synthesis of fatty acids begins in the plastid, where Ac-CoA is first carboxylated (thereby becoming Malonyl CoA) via the enzyme Acetyl-CoA carboxylase. Next, a group of seven related enzymes (known as "fatty acid synthetases") catalyzes synthesis of palmityl-CoA (which is a long molecule possessing 18 carbon atoms in its "molecular backbone"); although shorter-length molecules result when a specific ACP (acyl carrier protein) thioesterase enzyme is present in plastid (e.g., C16:0ACP), which results in fatty acids of various "carbon chain" length. After the palmityl-CoA is elongated (i.e., made a longer molecule via addition of carbons to its molecular backbone) to become the (stearate-like) molecule oleoyl-ACP in a chemical reaction catalyzed by a palmioyl elongase enzyme, the oleoyl-ACP is transported to the plant's endoplasmic reticulum.

In the endoplasmic reticulum, the oleoyl-ACP is either further elongated (via the addition of more carbon atoms to the fatty acid's molecular carbon chain "backbone") or it is further desaturated (i.e., via desaturase-catalyzed removal of hydrogen atoms from that fatty acid molecule). Stearic acid (also known as stearate) is desaturated to become oleic acid, which can be desaturated to become linoleic acid, which can be desaturated to become linolenic acid. Three of the resultant fatty acid molecules are then chemically attached to a glycerol-3-phosphate molecule (with the cleaved-off phosphate atom "recycled" in the endoplasmic reticulum, for further utilization in the energy cycle of the cell).

The content levels of individual fatty acids vary somewhat with the diet of the animal (i.e., for animal fat) and vary somewhat with the plant's growing conditions (i.e., for plant fat also known as vegetable oil). No natural fat is either totally saturated or unsaturated. When eaten, fats are generally not absorbed directly through the intestinal wall. They are first emulsified, then hydrolyzed by the lipase enzyme. The components (fatty acids, cholesterol, monoacylglycerol, phospholipids, etc.) form micelles that pass through the intestinal wall and are absorbed by the body. Such emulsification/micelle formation is aided by the nutrient lecithin (a component in soybeans). When fats are oxidized in cells, they provide energy for the body. Some of the energy is released as heat and some is stored in the form of adenosine triphosphate (ATP), which "fuels" metabolic processes. See also FATTY ACID, HYDROLYSIS, HYDROLYTIC CLEAVAGE, HYDROLYZE, LIPASE, MONOUNSATURATED FATS, SATURATED FATTY ACIDS, TRIGLYCERIDES, TRIACYLGLYCEROLS, DIACYLGLYCEROLS, MICELLE, CELL, METABOLISM, DIGESTION (WITHIN ORGANISMS), CHOLESTEROL, LIPIDS, LECITHIN, SOYBEAN OIL, FREE FATTY ACIDS, OXIDATIVE STRESS, PLASTID, ACP, OXIDATION (of fats/oils/lipids), PLASMA MEMBRANE, ENZYME, Ac-CoA, ENDOPLASMIC RETICULUM, FATTY ACID SYNTHETASE, THIOESTERASE, DESATURASE, MITOCHONDRIA, LAUROYL-ACP THIOESTERASE, STEAROYL-ACP DESATURASE, ADIPOCYTES, ADENOSINE TRIPHOSPHATE (ATP), BILE ACIDS, PHOSPHATE TRANSPORTER GENES, PHOTOSYNTHESIS, OLEOSOMES, STEARATE (STEARIC ACID), OLEIC ACID, LINOLEIC ACID, LINOLENIC ACID (a-linolenic acid), CONJUGATED LINOLEIC ACID (CLA).

**Fatty Acid** A long-chain aliphatic acid found in natural fats and oils. Fatty acids are abundant in cell membranes and (after extraction/purification) are widely used as industrial emulsifiers, e.g., phosphatidylcholine (lecithin).

In general, fats possessing the highest levels of saturated fatty acids tend to be solid at room temperature, and those fats possessing the highest levels of unsaturated fatty acids tend to be liquid at room temperature. That rule of thumb was the original "dividing line" between compounds called fats and oils, respectively. In general, saturated fatty acids tend to be more stable (resistant to oxidation and thermal breakdown) than unsaturated fatty acids. Fatty acids in biological systems (e.g., produced by plants in oilseeds) tend to contain an even number of carbon atoms in their molecular "backbone," typically between 14 and 24 carbon atoms. The molecular backbone (alkyl chain) may be saturated (no double bonds) or it may contain one or more double bonds. The configuration of the double bonds in most unsaturated fatty acids is CIS. See also ESSENTIAL FATTY ACIDS, LAURATE, PHYTOCHEMICALS, SATURATED FATTY ACIDS, LECITHIN, SOYBEAN OIL, UNSATURATED FATTY ACID, MONOUNSATURATED FATS, POLYUNSATURATED FATTY ACIDS (PUFA), LPAAT PROTEIN, STEAROYL-ACP DESATURASE, SOYBEAN OIL, CANOLA, FATS, OLEIC ACID, TRANS FATTY ACIDS, ENOYL-ACYL PROTEIN REDUCTASE, OXIDATION (of fats/oils/lipids), LIPIDS, MITOCHONDRIA, ADIPOCYTES, OLEOSOMES, DELTA 12 DESATURASE, LINOLEIC ACID, LINOLENIC ACID, FATTY ACID SYNTHETASE, CARNITINE, BIOTIN.

**Fatty Acid Synthetase** A group of seven related enzymes that catalyze synthesis (manufacturing) of fatty acids within the soybean plant (*Glycine max* (L.) Merrill). See also ENZYME, CATALYZE, FATTY ACID, SOYBEAN PLANT, DESATURASE, FATS, OLEOSOMES, PATHWAY, DELTA 12 DESATURASE.

**Federal Coordinated Framework for Regulation of Biotechnology** The legal framework created by the U.S. government in 1986, which divided regulation of biotechnology among the U.S. Department of Agriculture, the U.S. Environmental Protection Agency, and the U.S. Food and Drug Administration. See also FOOD AND DRUG ADMINISTRATION.

**Federal Insecticide Fungicide and Rodenticide Act (FIFRA)** A law enacted by the U.S. Congress in 1972. During 1994, the U.S. Environmental Protection Agency (EPA) proposed that the substances produced by plants (e.g., genetically engineered crops) for their defense against pests and diseases would be regulated by EPA under FIFRA. See also TOXIC SUBSTANCES CONTROL ACT (TSCA), GENETICALLY ENGINEERED MICROBIAL PESTICIDES (GEMP), WHEAT TAKE-ALL DISEASE, *BACILLUS THURINGIENSIS* (B.t.).

**Feedback Inhibition** Inhibition of the first enzyme in a metabolic pathway by the end product of that pathway. This is a method of shutting down a metabolic pathway that is producing a product that is no longer needed. See also METABOLISM, ENZYME, EFFECTOR.

**Feedstock** Raw material(s) used for the production of chemicals; or growth substrates of microbes (e.g., yeasts or bacteria that require a solid phase on which to attach themselves).

**Fermentation** A term first used with regard to the foaming that occurs during the manufacture of wine and beer. The process dates back to at least 6,000 B.C. when the Egyptians made wine and beer by fermentation. From the Latin word *fermentare*, to cause to rise. The term "fermentation" is now used to refer to so many different processes that fermentation is no longer accepted for use in most scientific publications. Three typical definitions are given below:

1. A process in which chemical changes are brought about in an organic substrate through the actions of enzymes elaborated (produced) by microorganisms.

2. The enzyme-catalyzed, energy-yielded pathway in cells by which "fuel" molecules such as glucose are broken down anaerobically (in the absence of oxygen). One product of the pathway is always the energy-rich compound adenosine triphosphate (ATP). The other products are of many types: alcohol, glycerol, and carbon dioxide from yeast fermentation of various sugars; butyl alcohol, acetone, lactic acid, and acetic acid from various bacteria; citric acid, gluconic acid, antibiotics, vitamin $B_{12}$ and $B_2$ from mold fermentation. The Japanese utilize a bacterial fermentation process to make the amino acid, L-glutamic acid, a derivative of which is widely used as a flavoring agent.

3. An enzymatic transformation of organic substrates (feedstocks), especially carbohydrates, generally accompanied by the evolution of gas. A physiological counterpart of oxidation, permitting certain organisms to live and grow in the absence of air; used in various industrial processes for the manufacture of products such as alcohols, acids, and cheese by the action of yeasts, molds, and bacteria. Alcoholic fermentation is the best known example. Also known as zymosis. The leavening of bread depends on the alcoholic fermentation of sugars. The dough rises due to production of carbon dioxide gas that remains trapped within the viscous dough.

See also ZYMOGENS, SUBSTRATE (CHEMICAL), ADENOSINE TRIPHOSPHATE (ATP), MICROORGANISM, ENZYME, FEEDSTOCK, CARBOHYDRATES (SACCHARIDES).

**Ferritin** An iron-protein complex (a metalloprotein) that occurs in living tissues. Functions in iron storage in the spleen. See also HEMOGLOBIN.

**Ferrobacteria** Also called iron bacteria. Any of a group of bacteria that oxidize iron as a source of energy. The oxidized iron in the form $Fe(OH)_3$ is then deposited in the environment by secretion from the bacterium.

The energy obtained from these reactions is used to carry on processes in which the basic substances needed by the bacterium are manufactured. These bacteria are commonly found in seepage waters of coal and iron mining areas where iron compounds abound. Ferrobacteria are not disease producers (i.e., pathogenic), but they are important as scavengers. Sometimes they create a nuisance by multiplying so profusely in iron water pipes that they stop the flow of water. Ferrobacteria have been active through long periods of geologic time. For example, the great Mesabi iron (ore) seam of America's Lake Superior region is thought to be a product of ferrobacteria activity. See also PATHOGEN.

**Ferrochelatase** A mitochondrial enzyme that catalyzes the incorporation of iron into the protoporphyria molecule. See also MITO-CHONDRIA, ENZYME, CATALYST, PORPHYRINS.

**Ferrodoxin** An iron- and sulfur-containing protein important in the electron-transfer processes of photosynthesis in plants. It also plays a role in the metabolism of some bacteria and was first found in an anaerobic bacterium. See also PHOTOSYNTHESIS, METABOLISM.

**Fertility Factor (F)** A type of transmissible (i.e., can enter other cells) plasmid that is often found in *Escherichia coli* (*E. coli*). See also PLASMID, VECTOR, *ESCHERICHIA COLIFORM* (*E. COLI*).

**Fertilization** The union of the (haploid) male and (haploid) female germ cells (sex cells or gametes) to produce a diploid zygote. Fertilization marks the start of development of a new individual (organism), the beginning of cell differentiation. See also GERM CELL.

**FFA** Acronym for Free Fatty Acids. See also FREE FATTY ACIDS.

**FGF** See FIBROBLAST GROWTH FACTOR (FGF).

**FGMP** See FOOD GOOD MANUFACTURING PRACTICE (FGMP).

**FIA** Refers to immunodiagnostic tests that are based on fluorescence tracers (labels). See also IMMUNOASSAY, FLUORESCENCE, RADIO-IMMUNOASSAY.

**Fibrin** The ordered fibrous array of fibrin monomers, called a fibrin-platelet clot (blood clot), which spontaneously assembles from fibrin monomers (which themselves are formed by the thrombin-catalyzed conversion of fibrinogen into fibrin). Fibrinogen itself is the product of a controlled series of zymogen activation steps (enzymatic cascade) triggered initially by substances released from body tissues as a consequence of trauma (harm). See also FIBRONECTIN, ZYMOGENS, CASCADE, LIPOPROTEIN-ASSOCIATED COAGULATION (CLOT) INHIBITOR (LACI).

**Fibrinogen** See FIBRIN, LIPOPROTEIN-ASSOCIATED COAGULATION (CLOT) INHIBITOR (LACI).

**Fibrinolytic Agents** Bloodborne compounds that activate fibrin in order to dissolve blood clots. See also TISSUE PLASMINOGEN ACTIVATOR (tPA), THROMBOLYTIC AGENTS, FIBRIN.

**Fibroblast Growth Factor (FGF)** First described in the mid-1970s by Dr. Gospodarowicz and fellow researchers at the University of California, San Francisco. It is a protein that stimulates the formation/development of blood vessels and fibroblasts (precursors to collagen, the connective tissue "glue" that holds cells together). FGF also is mitogenic (causes cells to divide and multiply) for both fibroblasts and endothelial cells, and attracts those two cell types (i.e., is chemotactic). Dr. Gospodarowicz named the FGF originally derived from bovine (cow) brain tissue to be Acidic FGF. Dr. Gospodarowicz named the FGF originally derived from bovine pituitary tissue to be Basic FGF. This was due to their identical biological activity, but differing isoelectric points (the former being acidic, and the latter being basic). Basic FGF is, however, ten times more "potent" than acidic FGF in most bioassays. See also ANGIOGENIC GROWTH FACTORS, PROTEIN, FIBROBLASTS, PITUITARY GLAND, COLLAGEN, MITOGEN, ENDOTHELIAL CELLS, CHEMOTAXIS, BIOLOGICAL ACTIVITY, BIOASSAY, ACID, BASE.

**Fibroblasts** Cells that are precursors to the connective tissue cells found in the skin. They make structural proteins like collagen, which gives skin its strength. Because fibroblasts do not express antigens on their cell surfaces (free standing, separated), fibroblasts possess potential for use in making artificial organs (e.g., artificial pancreas for diabetics), since recipent immune systems cannot recognize the fibroblast cells as foreign. See also CELLULAR IMMUNE RESPONSE,

HUMORAL IMMUNITY, GRAFT-VERSUS-HOST DISEASE (GVHD), XENOGENEIC ORGANS, CELL, FIBROBLAST GROWTH FACTOR (FGF), COLLAGEN.

**Fibronectin** An adhesive glycoprotein that forms a link between the epithelial cells and the connective tissue matrix (essential for blood clotting). Research has indicated that fibronectin may solve the problem of getting new cells to stick to existing tissue, once a growth factor has caused them to grow (e.g., when growth factor is administered after a serious wound to tissue). See also FIBRIN, GLYCOPROTEIN, GROWTH FACTOR, ORGANOGENESIS.

**Field Inversion Gel Electrophoresis (FIGE)** A chromatographic procedure for the separation of a mixture of molecules by means of a two-dimensional electrical field, applied across a gel matrix containing those molecules. For example, FIGE is commonly used to separate mixtures of large DNA molecules by their size and (electrical) charge. FIGE can be used to separate (resolve) DNA molecules up to 2000 Kbp in length. See also TWO-DIMENSIONAL (2D) GEL ELECTROPHORESIS, CHROMATOGRAPHY, ELECTROPHORESIS, KILOBASE PAIRS (Kbp), POLYACRYLAMIDE GEL ELECTROPHORESIS (PAGE), DEOXYRIBONUCLEIC ACID (DNA).

**FIFRA** See FEDERAL INSECTICIDE FUNGICIDE AND RODENTICIDE ACT (FIFRA).

**Filler Epithelial Cells** Skin cells that initially form under a scab in the wound healing process, in response to stimulation by epidermal growth factor (EGF). See also EPIDERMAL GROWTH FACTOR (EGF).

**Finger Proteins** See ZINC FINGER PROTEINS.

**Fingerprinting** See PEPTIDE MAPPING ("FINGERPRINTING"), COMBINATORIAL CHEMISTRY.

**First Filial Hybrids** See F1 HYBRIDS.

**Flagella** A protein-based, flexible, whip-like organ of locomotion found on some microorganisms. With these, microorganisms are able to swim. Flagella are usually very long and there are usually only one or two per cell. The tails of sperm cells are examples of flagella. Flagella are used in the swimming motion of bacteria toward sources of nutrients in a process called chemotaxis. Singular: flagellum. See also CILIA, CHEMOTAXIS, BACTERIA, PROTEIN.

**Flanking Sequence** A segment of DNA molecule that either precedes or follows the region of interest on the molecule. See also DEOXYRIBONUCLEIC ACID (DNA).

**Flavin** Also known as lyochrome. One of a group of pale yellow, greenly fluorescing biological pigments widely distributed in small quantities in plant and animal tissues. Flavins are synthesized only by bacteria, yeast, and green plants; for this reason, animals are dependent on plant sources for riboflavin (vitamin $B_2$), the most prevalent member of the group.

**Flavin Adenine Dinucleotide (FAD)** The coenzyme of some Adenine Dinucleotide (FAD) oxidation-reduction enzymes; it contains riboflavin. See also FLAVIN, ENZYME, COENZYME, OXIDATION-REDUCTION REACTION.

**Flavin Mononucleotide (FMN)** Riboflavin phosphate, a coenzyme of certain oxidoreduction enzymes. See also COENZYME.

**Flavin Nucleotides** Nucelotide coenzymes (FMN and FAD) containing riboflavin. See also FLAVIN MONONUCLEOTIDE (FMN), FLAVIN ADENINE DINUCLEOTIDE (FAD).

**Flavin-Linked Dehydrogenases** Dehydrogenases are enzymes (involved in removing hydrogen atoms from their substrate) which require one of the riboflavin coenzymes, FMN or FAD, in order to function. See also DEHYDROGENASES, FLAVIN MONONUCLEOTIDE (FMN), FLAVIN ADENINE DINUCLEOTIDE (FAD), SUBSTRATE (CHEMICAL).

**Flavinoids** See FLAVONOIDS.

**Flavonoids** A category of phytochemicals, that are typically beneficial to the health of humans that consume them. Hundreds of flavonoids are naturally produced (by plants) in common human foods. For example, the three isoflavones (genistein, daidzein, and glycitein) produced in seeds of the soybean plant (*Glycine max* (L.) Merrill) are flavonoids, and they confer several health benefits to humans that consume them. Coffee, tea, and chocolate products contain a number of antioxidant flavonoids (i.e., polyphenols). Because oxidation of lipids (low-density lipoproteins) in the bloodstream is the initial step in atherosclerosis disease, consumption of large amounts of coffee may help to prevent atherosclerosis. Research conducted by Joe Vinson in 1999 indicated that high coffee consumption by humans reduced oxidation of lipids in the bloodstream by 30%. Cranberries (*Vaccinium macrocarpon*) contain a number of antioxidant flavonoids, and research indicates that consumption of large amounts on a regular basis may inhibit development of breast cancer. Blueberries (genus *vaccinium*) contain a number of flavonoids, and research indicates that consumption of large amounts on a regular basis helps to strengthen eyesight, improve memory, and inhibit some physical aspects of the aging process. Other subcategories of flavonoids are flavones, flavonols, flavanols, aurones, chalcones, etc. One example of a not-very-beneficial flavonoid is quercetin, a nonnutritive antioxidant produced in almonds. See also PHYTOCHEMICALS, ISOFLAVONES, SOYBEAN PLANT, ATHEROSCLEROSIS, OXIDATION, ANTIOXIDANTS, OXIDATIVE STRESS, CANCER, LIPIDS, ANTHOCYANIDINS, PROANTHOCYANIDINS, FLAVONOLS.

**Flavonols** A group of phytochemicals, consisting of a subcategory of the flavonoid "family" of phytochemicals. Flavonols are typically beneficial to the health of humans that consume them, and are found in citrus fruits such as grapefruit, oranges, etc. However, at least one flavonol (quercitin glycoside) is found in tomato peels. See also PHYTOCHEMICALS, FLAVONOIDS.

**Flavoprotein** An enzyme containing a flavin nucleotide as a prosthetic group. See also PROSTHETIC GROUP.

**FLK-2 Receptors** See TOTIPOTENT STEM CELLS.

**Flora** The microorganisms found in a given situation, e.g., reservoir flora (the microorganisms present in a given municipal water reservoir) or intestinal flora (the microorganisms found in the intestines).

**Floury-2** A gene in corn/maize (*Zea mays* L.) that (when present in the DNA of a given plant) causes that plant to produce seed that contains higher-than-traditional levels of the amino acids methionine and tryptophan. See also GENE, CORN, METHIONINE (met), HIGH-METHIONINE CORN, ESSENTIAL AMINO ACIDS, VALUE-ENHANCED GRAINS, DEOXYRIBONUCLEIC ACID (DNA).

**Flow Cytometry** See CELL SORTING, FLUORESCENCE ACTIVATED CELL SORTER (FACS), MAGNETIC PARTICLES.

**Fluorescence** The reaction of certain molecules upon absorption of specific amount/wavelength of light; in which those molecules emit (reradiate) light energy possessing a longer wavelength than the original light absorbed. All cells will naturally fluoresce, at least a bit. Human colon cancer cells, and precursor cells, fluoresce much more (and emit much more red light when they fluoresce) than noncancerous cells; which may lead to a new and better means of early detection. See also CELL, CANCER, FIA, BRIGHT GREENISH-YELLOW FLUORESCENCE (BGYF), BIOCHIP, NEAR-INFRARED SPECTROSCOPY (NIR).

**Fluorescence *In Situ* Hybridization (FISH)** A method for detecting the presence of a particular gene (e.g., in a biological sample), which utilizes a fluorescein-"tagged" DNA probe. When the DNA probe hybridizes to that particular gene, the "tag" fluoresces (thereby indicating positively the presence in sample of that particular gene). See also GENE, FLUORESCENCE, PROBE, DNA PROBE.

**Fluorescence Activated Cell Sorter (FACS)** A machine used to sort cells from a mixed group of cells (e.g., to remove only the cells into which a new gene has been inserted via genetic engineering techniques). The desired cells are first labeled with a specific fluorescent dye, then passed through a flow chamber that is illuminated by a laser beam, which causes the labeled cells to fluoresce (glow). The molecules of the fluorescent dye, which "stick" to only one type of cell in the mixture, contain chromophores that can be elevated to an excited, unstable state via irradiation with specific wavelength(s) of light. Those chromophores remain in that excited state for a maximum of $10^{-9}$ seconds before releasing their energy by emitting light, and returning to their unexcited "ground" state. This fluorescence (glow) is a measurable property and the FACS machine utilizes it to separate the desired cells from the rest of the mixture. See also BASOPHILIC, GENETIC ENGINEERING, CELL, FLUORESCENCE, CELL SORTING.

**Fluorogenic Probe** See MOLECULAR BEACON.

**Follicle Stimulating Hormone (FSH)** A protein hormone used in conventional medical therapy in an attempt to increase production of sperm in men (inside the follicles of the testes). See also THYROID STIMULATING HORMONE (TSH), GRAVE'S DISEASE, PROTEIN, HORMONE, PITUITARY GLAND.

**Food and Drug Administration (FDA)** The federal agency charged with approving all pharmaceutical and food ingredient products sold within the U.S. In 1992, prior to approval of any of the biotechnology-derived food crop plants, the FDA decided that food crops produced via "biotechnological (i.e., recombinant) technologies" must meet the same rigorous safety standards as those created via "traditional breeding methods," both categories of which are regulated by the FDA. Historically, new food crops created via "traditional breeding technologies" (e.g., crossing with wild type in order to confer disease resistance, increased yield, etc. on the resultant domesticated plant varieties/strains) have sometimes contained unexpectedly high levels of known (and naturally occurring) toxins (e.g., solanine, a naturally occurring toxin in potatoes and some other plants, psoralene, a naturally occurring toxin in celery, etc.). See also KOSEISHO, COMMITTEE FOR PROPRIETARY MEDICINAL PRODUCTS (CPMP), COMMITTEE FOR VETERINARY MEDICINAL PRODUCTS (CVMP), COMMITTEE ON SAFETY IN MEDICINES, WILD TYPE, STRAIN, "TREATMENT" IND REGULATIONS, KEFAUVER RULE, IND, IND EXEMPTION, RECOMBINANT DNA (rDNA), PHASE I CLINICAL TESTING, EUROPEAN MEDICINES EVALUATION AGENCY (EMEA), MEDICINES CONTROL AGENCY (MCA), BUNDESGESUNDHEITSAMT (BGA), TRADITIONAL BREEDING METHODS, SOLANINE, PSORALENE.

**Food Good Manufacturing Practice (FGMP)** The Food and Drug Administration's (FDA's) approval mechanism for a process to manufacture a given food or food additive. It is implemented instead of specific regulations (such as those used to dictate processes in simple food manufacture, as in beef packing), due to the newness of the technology, and may later be superceded due to further advances in the technology. See also FOOD AND DRUG ADMINISTRATION (FDA).

**Footprinting** A technique used by researchers to determine precisely where (on a DNA molecule) certain DNA-binding proteins

make specific contact with that DNA molecule. For example, certain types of drugs act by binding tightly to certain DNA molecules in specific locations (e.g., in order to halt cancerous growth of cells). See also DEOXYRIBONUCLEIC ACID (DNA), PROTEIN, GENOTOXIC.

**For Treatment IND** See "TREATMENT" IND REGULATIONS.

**Formaldehyde Dehydrogenase** An enzyme which catalyzes the oxidation of formaldehyde to formic acid (formate at intracellular pH). It requires NAD (nicotinamide-adenine dinucleotide) as an electron acceptor. It is important in the metabolism of methanol. See also METABOLISM, ENZYME, NAD (NADH, NADP, NADPH), CATALYST.

**Forward Mutation** A mutation from the wild (natural) type to the mutant type. See also MUTATION, WILD TYPE.

**FOS** See FRUCTOSE OLIGOSACCHARIDES.

**FOSHU** A Japanese government designation meaning "Foods of Specified Health Use." Introduced in the early 1980s, these are foods or food ingredients that meet the following specific criteria:

1. Must improve human nutrition and health. A benefit to human health and nutrition must be proven for that food/ingredient.
2. An appropriate daily dose (amount to be consumed) must be confirmed by doctors or dieticians.
3. The food/ingredient must guarantee balanced nourishment.
4. The active component (e.g., phytochemical) must be scientifically confirmed regarding its quantitative and qualitative definition, and its chemical and/or physical features.
5. The active component must not lower nutritional value (e.g., of the food it is added to).
6. The food/ingredient must be consumed in a normal fashion (i.e., eaten or drank, not as pill or powder form).
7. The active component must be of natural origin. Some of the foods/ingredients designated "FOSHU" have been those containing polyphenols, anthocyanins, and diacylglycerols.

F

See also NUTRACEUTICALS, PHYTOCHEMICALS, MANNANOLIGOSACCHARIDES, FRUCTOSE OLIGO-SACCHARIDES, ANTHOCYANINS, POLYPHENOLS, DIACYLGLYCEROLS.

**Foundation on Economic Trends** A small organization that lobbies against agricultural biotechnology. See also BIOTECHNOLOGY.

**Frameshift** A shift (displacement) of the reading frame in a DNA or RNA molecule. Frameshifts generally result from the addition or deletion of one or more nucleotides to or from the DNA or RNA molecule. See also READING FRAME, CODON, GENETIC CODE, MUTATION, DEOXYRIBONUCLEIC ACID (DNA), NUCLEOTIDE, RIBONUCLEIC ACID (RNA), CENTRAL DOGMA (NEW).

**Free Energy** The component of the total energy of a system that can do work at a constant temperature and pressure. Also known as Gibbs free energy. Free energy is a key variable calculated and monitored for different (proposed) drug molecules or drug/target interactions during rational drug design activities (e.g., molecular modeling). See also RATIONAL DRUG DESIGN, TARGET (of a therapeutic agent), ACTIVATION ENERGY.

**Free Fatty Acids (F.F.A.)** Individual fatty acid molecules within a vegetable oil, which exist in an uncombined-with-glycerine molecular state. The presence of F.F.A. can be caused by naturally occurring noncombination (e.g., in some varieties of oilseeds), sprouting of the oilseeds prior to processing into vegetable oil, or breakdown of the fat (oil) during processing or usage. See also FATS, FATTY ACID, SATURATED FATTY ACIDS, UNSATURATED FATTY ACID.

**Free Radical** Sometimes called Reactive Oxygen Species, Singlet Oxygen, or Oxygen Free Radical. Term utilized to refer to an oxygen (atom) bearing an "extra" electron. Because of that, it possesses a large amount of energy, and in a biological system (i.e., inside the body of an organism), it can damage body tissues when it "discharges" that energy. See also OXIDATIVE STRESS, ANTIOXI-DANTS, HUMAN SUPEROXIDE DISMUTASE (hSOD), CAROTENOIDS, CONJUGATED LINOLEIC ACID (CLA), INSULIN.

**Fructan** A general term utilized to refer to any carbohydrate in which fructosyl-fructose

(molecule) linkages constitute the majority of the molecule's glycosidic bonds (i.e., between atoms in the molecule). See also CARBOHYDRATES (saccharides), OLIGOSACCHA-RIDES, FRUCTOSE OLIGOSACCHARIDES, GLYCOSIDE.

**Fructo Oligosaccharides** See FRUCTOSE OLIGO-SACCHARIDES.

**Fructose Oligosaccharides** A "family" of oligosaccharides, some of which help foster the growth of bifidobacteria in the lower colon of monogastric animals (humans, swine, etc.). Those bifidobacteria generate certain short-chain fatty acids, which are absorbed by the colon and result in a reduction of triglyceride (fat) and cholesterol levels in the bloodstream, thereby lowering risk of coronary heart disease and thrombosis. Research indicates they also promote absorption of calcium from foods (in the large intestine). Fructose oligosaccharides are classifed as a "water soluble fiber" (by the European Union's government food regulatory agencies), because humans cannot digest them. See also BIFIDOBACTERIA, *BIFIDUS*, INULIN, FOSHU, OLIGOSACCHARIDES, NUTRACEUTICALS, CHOLESTEROL, HIGH-DENSITY LIPOPROTEINS (HDLPs), LOW-DENSITY LIPOPROTEINS (LDLPs), BACTERIA, FATTY ACID, PREBIOTICS, MANNANOLI-GOSACCHARIDES (MOS), CORONARY HEART DISEASE (CHD), TRIGLYCERIDES, THROMBOSIS.

**Fumarase (fum)** An enzyme that catalyzes the hydration (addition of hydrogen atoms) of fumaric acid to maleic acid, as well as the reverse dehydration reaction (removal of hydrogen atoms). See also ENZYME, CATALYST.

**Fumaric Acid ($C_4H_4O_4$)** A dicarboxylic organic acid produced commercially by chemical synthesis and fermentation; the trans isomer of maleic acid; colorless crystals, melting point 87°C (191°F); used to make resins, paints, varnishes and inks, in food as a mordant (dye fixer/stabilizer), and as a chemical intermediate. Also known as BOLETIC ACID. See also ACID, ISOMER, BOLETIC ACID.

**Fumonisins** Mycotoxins that are primarily produced by the fungus *Fusarium monili-forme* (e.g., in insect-damaged corn/maize). Consumption of fumonisins by horses and swine can be fatal to those animals. Consumption of fumonisins by other animals (including humans) can result in tumors (e.g., cancer of the esophagus, in humans). See also MYCOTOXINS, FUNGUS, *FUSARIUM*, *FUSARIUM MONILIFORME*, EUROPEAN CORN BORER (ECB).

**Functional Foods** Refers to foods that provide health benefits beyond basic nutrition. See also NUTRACEUTICALS, PHYTOCHEMICALS, FOSHU.

**Functional Genomics** Study of, or discovery of, what traits/functions (generally via proteins expressed) are conferred to an organism by given (gene) sequences. The timing and location of the expression of those genes is also impacted by external/environmental factors sometimes, i.e., temperature, sunlight, humidity, the presence of signal transducers and activators of transcription (STATs), etc. Also impacting the functions/traits are interactions among genes, signaling cascades, and response/reaction mechanisms within the body of that organism. Typically, functional genomic study follows after discovery of gene sequences found via structural genomics study. Some methods utilized to determine which traits/functions result from which gene(s) are:

1. Site-directed mutagenesis (SDM), to compare two same-species organisms possessing two different genes at the same site on the genome.
2. Antisense DNA sequence, to compare two same-species organisms (one of which has gene at same site "turned off" via antisense DNA).
3. Reporter gene, to compare two same-species organisms (with two different genes at same site on genome) via a "reporter" gene adjacent to gene/site, to detect presence of desired trait/function.
4. Chemical genetics, to compare two same-species organisms (one of which has gene at same site on DNA molecule at least partially inactivated by a specifical chemical).
5. "Silencing" or "knocking out" a particular gene via other methods than antisense or chemical genetics, to compare.

See also GENOMICS, TRAIT, GENE, GENOTYPE, PHENOTYPE, POLYGENIC, EXPRESS, STRUCTURAL GENE, STRUCTURAL GENOMICS, DEOXYRIBONUCLEIC ACID (DNA), SEQUENCE (OF A DNA MOLECULE), PLEIOTROPIC, GENETIC CODE, EXPRESSED SEQUENCE TAGS, INFORMATIONAL MOLECULES, POINT MUTATION, SITE-DIRECTED MUTAGENESIS (SDM), ANTISENSE (DNA SEQUENCE), REPORTER GENE, METHYLATION, ZINC FINGER PROTEINS, DNA METHYLATION, POSITIONAL CLONING, CHEMICAL GENETICS, GENE SILENCING, DROSOPHILA, *CAENORHABDITIS ELEGANS*, CENTRAL DOGMA (NEW), TRANSCRIPTION FACTORS, SIGNAL TRANSDUCERS AND ACTIVATORS OF TRANSCRIPTION (STATs), GENE EXPRESSION ANALYSIS, GENE FUNCTION ANALYSIS, PATHWAY, PATHWAY FEEDBACK MECHANISMS, CASCADE.

**Functional Group** A molecule, or portion of a molecule, that will react with other molecule(s). For example, "hedgehog proteins" must first add a cholesterol molecule (to themselves) before they can carry out their task of directing/controlling tissue differentiation during mammal embryo development (into various organs, limbs, etc.). An "acetyl (functional) group" must be added to a choline molecule in order for the body to have the critical neurotransmitter acetylcholine. See also PROTEIN, PEPTIDE, HEDGEHOG PROTEINS, CHOLESTEROL, ACETYL CHOLINE, NEUROTRANSMITTER, SIGNAL TRANSDUCTION.

**Fungicide** Any chemical compound toxic to fungi. See also BIOCIDE, FUNGUS.

**Fungus** (plural: fungi) Any of a major group of saprophytic and parasitic plants that lack chlorophyll and flowers, including molds, toadstools, rusts, mildews, smuts, ergot, mushrooms *Aqaricus bisporus*, and yeasts. Under certain conditions (temperature, humidity, etc.), some fungi can produce mycotoxins via their metabolism. See also RUSTS, *ASPERGILLUS FLAVUS*, MYCOTOXINS, *FUSARIUM*, AFLATOXIN, FUMONISINS, VOMITOXIN, DON, ERGOTAMINE, METABOLISM.

**Furanose** A sugar molecule containing the five-membered furan ring. See also SUGAR MOLECULES.

**Fusaric Acids** See *FUSARIUM MONILIFORME*.

***Fusarium*** A genus of fungus, also known as "scab," that infests certain grains (e.g., wheat *Triticum aestivum*, corn or maize *Zea mays* L., etc.) during growing seasons in which climate (e.g., high humidity, cool weather) and other conditions combine to

**F**

enable rapid growth/proliferation of the fungus. In wheat, fungus infestation (*Fusarium* head blight) causes the wheat plant to weaken and to produce empty seed heads, which reduces yield.

As a by-product of their metabolism, some of the *Fusarium* types (species) produce deoxynivalenol (also known as DON or "vomitoxin"), zearalenone, and fumonisins (a group of very potent mycotoxins that are produced by *Fusarium moniliforme* and *Fusarium proliferatum* fungi). Fumonisin B1 is the most prevalent *Fusarium*-produced mycotoxin in corn (maize). Its presence can cause livestock to refuse to eat infested feed, decrease reproductive efficiency in swine, and even kill horses (via equine leukoencephalomalacia).

When consumed by humans, fumonisin B1 induces cell death via apoptosis; the tissues that are adjacent to killed cells respond with cell replication/proliferation to replace the lost cells.

Fumonisin B1 inhibits the enzyme ceramide synthetase (which is crucial to the biosynthetic pathway for the creation of sphingolipids in cells), resulting in accumulation of sphinganine in cells, and decreases ceramides and complex sphingolipids. These internal changes signal the cells to die via apoptosis ("programmed cell death"), especially liver and kidney cells.

Maximum fumonisin content allowed in flour (for U.S. bread) is one part per million. Maximum fumonisin content allowed in U.S. malting barley (*Hordeum vulgare*) is zero.

In 1997, Iowa State University research showed that *B.t.* corn varieties (which express the *B.t.* protoxin in the corn ears) have significantly less ear mold caused by *Fusarium* fungi. That is because the European corn borer (ECB) is a vector (carrier) of *Fusarium*. See also FUNGUS, MYCOTOXINS, TOXIN, METABOLISM, APOPTOSIS, ENZYME INHIBITION, LIPIDS, VOMITOXIN, DON, DEOXYNIVALENOL, *BACILLUS THURINGIENSIS* (*B.t.*), EUROPEAN CORN BORER (ECB), CD95 PROTEIN, SOYBEAN CYST NEMATODES (SCN), ZEARALENONE, *FUSARIUM MONILIFORME*.

***Fusarium moniliforme***  One of the Fusarium fungi; therefore it can produce one or more fumonisins (a group of mycotoxins) under certain environmental conditions, when it grows in some grains (see the entry for *Fusarium*). When *Fusarium moniliforme* grows within growing plants of domesticated rice (*Oryza sativa*), it can cause the plant disease known as Bakanae (also known as "foolish seedling" disease). Symptoms of Bakanae include rice plants that are much taller than normal rice plants, and leaves that are much longer than normal. That abnormal growth (of rice plant/leaves) is caused by a gibberellin compound excreted by the *Fusarium moniliforme* fungus. The fungus also excretes fusaric acids, which can stunt or kill rice plants. See also FUSARIUM, MYCOTOXINS, FUMONISINS, FUNGUS, GIBBERELLINS.

**Fusion Protein**  A protein consisting of all or part of the amino acid sequences (known as the "domain") of two or more proteins. Formed by fusing the two protein-encoding genes (which causes the ribosome to subsequently produce the fusion protein). This fusion is often done deliberately, either to put the expression of one of the (fused) genes under the control of the strong promoter for the first gene, or to allow the gene of interest (which is difficult to assay) to be more easily studied via substituting some of the (gene) protein with a more easily measured (assayed) function. For example, by fusing a difficult-to-study gene with the β-galactosidase gene, the (protein) product that results can easily be measured (assayed) using chromatography. See also PROTEIN, AMINO ACID, SEQUENCE (OF A PROTEIN MOLECULE), GENE, RIBOSOMES, PROMOTER, ASSAY, CODING SEQUENCE, DOMAIN (OF A PROTEIN), GENE FUSION.

**Fusion Toxin**  A fusion protein that consists of a toxic protein (domain) plus a cell receptor binding region (protein domain). The cell receptor portion (of the total fusion toxin molecule) delivers the toxin directly to the (diseased) cell, thus sparing other healthy tissues from the effect of the toxin. See also FUSION PROTEIN, TOXIN, RICIN, PROTEIN, PROTEIN ENGINEERING, DOMAIN (OF A PROTEIN), RECEPTORS, ENDOCYTOSIS.

**Fusogenic Agent**  Any compound, virus, etc., that causes cells to fuse together. For example, one of the effects of the HIV (i.e., AIDS-causing) viruses is to cause the T cells of the human immune system to fuse (causing collapse of the immune system). See also ACQUIRED IMMUNE DEFICIENCY SYNDROME (AIDS), HUMAN IMMUNODEFICIENCY VIRUS TYPE 1 (HIV-1), HUMAN IMMUNODEFICIENCY VIRUS TYPE 2 (HIV-2), HELPER T CELLS (T4 CELLS), ADHESION MOLECULE.

**Futile Cycle** An enzyme-catalyzed set of cyclic reactions that results in release of thermal energy (heat) through the hydrolysis of ATP (adenosine triphosphate). The hydrolysis of ATP is normally coupled to other cycles and reactions in which the energy released is metabolically used. However, futile cycles would appear to waste the energy of ATP as heat, except when one is shivering to keep warm. The production of heat by shivering is an example of the futile cycle. See also ADENOSINE TRIPHOSPHATE (ATP), ENZYME, HYDROLYSIS.

F

# G

**G-** See GRAM-NEGATIVE (G-).

**G+** See GRAM-POSITIVE (G+).

**G-Protein-coupled Receptors** See G-PROTEINS.

**G Proteins** See G-PROTEINS.

**G-Proteins** (Guanyl-Nucleotide Binding Proteins) Discovered by Rodbell and co-workers at America's National Institutes of Health (NIH), and Alfred G. Gilman and co-workers at the American University of Virginia-Charlottesville, during the 1970s–1980s. These are proteins embedded in the surface membrane of cells. G-proteins "receive chemical signals" from outside the cell (e.g., hormones) and "pass the signal" into the cell, so that cell can "respond to the signal." For example, a hormone, drug, neurotransmitter, or other "signal" binds to a receptor molecule on the surface of the cell's exterior membrane. That receptor then activates the G-protein, which causes an effector inside cell to produce a second "signal" chemical inside the cell, which causes the cell to react to the original external chemical signal. The G-proteins are called thus, because they become GTP and GDP forms alternately, as part of their reaction cycle (i.e., in "passing the signal"). Dysfunction of G-proteins in humans causes the salt and water losses inherent in cholera (the body's compromised immune defense inherent in pertussis), and is believed responsible for some symptoms of diabetes and alcoholism. Dysfunction of G-proteins in plants causes rapid water loss (wilting). See also PROTEIN, SIGNALING, SIGNAL TRANSDUCTION, HORMONE, CELL, BETA CELLS, GTPases, GPA1, INSULIN, RECEPTORS, NATIONAL INSTITUTES OF HEALTH (NIH), NEUROTRANSMITTERS, TRANSMEMBRANE PROTEINS, ION CHANNELS, CHOLERA TOXIN.

**GA21** A naturally occurring gene (i.e., expressed at low levels in some plants) which confers resistance to glyphosate-containing herbicides. When the GA21 gene is inserted by man into crop plants (e.g., maize/corn) in a way that causes high expression, those crop plants are subsequently unaffected when glyphosate-containing herbicides are applied to fields to cotnrol weeds in those crops. See also GENE, EXPRESS, EXPRESSIVITY, PROTEIN, GENETIC ENGINEERING, CORN, HERBICIDE-TOLERANT CROP, GLYPHOSATE.

**Galactose (gal)** A monosaccharide occurring in both levo (L) and dextro (D) forms as a constituent of plant and animal oligosaccharides (lactose and raffinose) and polysaccharides (agar and pectin). Galactose is also known as cerebrose. See also STEREOISOMERS, DEXTROROTARY (D) ISOMER, LEVOROTARY (L) ISOMER.

**Gall** See Ti PLASMID.

**GalNAc** *N*-acetyl-D-galactosamine.

**GALT** See GUT-ASSOCIATED LYMPHOID TISSUE.

**Gamete** A germ or reproductive cell. In animals (and humans) the functional, mature, male gamete is called a spermatozoon; in plants it is called a spermatozoid. In both animals and plants the female gamete is called the ovum, or egg. See also OOCYTES.

**Gamma Globulin** A type of blood protein that plays a major role in the process of immunity (immune system response). Sometimes the term "gamma globulin" refers to a whole group of blood proteins that are known as antibodies or immunoglobulins (Ig). Most often, however, it applies to a particular immunoglobulin, designated as IgG, believed to be the most abundant type of antibody in the body. See also ANTIBODY, G GUT-ASSOCIATED LYMPHOID TISSUE (GALT), PROTEIN, IMMUNOGLOBULIN.

**Gamma Interferon** Produced by T lymphocytes. See also INTERFERONS, T LYMPHOCYTES.

**GAP** A double-stranded DNA is said to be
"gapped" when one strand is missing over a
short region of the molecule. See also DEOXY-
RIBONUCLEIC ACID (DNA).

**Gated Transport** (of a protein) One of three
means for a protein molecule to pass
between compartments within eucaryotic
cells. The compartment "wall" (membrane)
possesses a "sensor" (receptor) that detects
the presence of a correct protein (e.g., after
that protein has been synthesized in the cell's
ribosomes), then opens a "gate" (pore) in the
membrane to allow that protein to pass from
the first compartment to the second compartment. See also PROTEIN, EUCARYOTE, CELL,
RIBOSOMES, SIGNALING, VESICULAR TRANSPORT.

**GDH Gene** See GLUTAMATE DEHYDROGENASE.

**GDNF** See GLIAL DERIVED NEUROTROPHIC FACTOR.

**GEAC** The country of India's Genetic Engineering Approval Committee. The GEAC
must approve a rDNA product (e.g., a genetically engineered crop plant that earlier
received its "bio safety clearance" from the
Indian Department of Biotechnology) before
that rDNA product is allowed to be commercially planted. See also GENETIC ENGINEERING,
rDNA, INDIAN DEPARTMENT OF BIOTECHNOLOGY.

**Gel** A colloid, where the dispersed phase is
liquid and the dispersion medium is solid.

**Gel Electrophoresis** See TWO-DIMENSIONAL (2D)
GEL ELECTROPHORESIS, POLYACRYLAMIDE GEL
ELECTROPHORESIS (PAGE), ELECTROPHORESIS.

**Gel Filtration** Also known as exclusion chromatography. An effective technique for separating molecules (such as peptide mixtures)
on the basis of size. This is accomplished by
passing a solution of the molecules to be
separated over a column of Sephadex®, for
example, which is a polymerized carbohydrate derivative that contains tiny holes.
The holes are of such a size that some of the
smaller molecules diffuse into them and are
in this way retained (held back) while the
larger molecules are not able to get into the
holes and pass on by the solid phase (Sepha-
dex®, in this example). This, simplistically,
is how separation is effected. See also
ELECTROPHORESIS, CHROMATOGRAPHY, FIELD
INVERSION GEL ELECTROPHORESIS.

**GEM (Germ plasm Enhancement for
Maize)** A project conducted under the auspices of the U.S. Department of Agriculture,
in concert with 16 American universities and
20 corn (maize) seed companies. GEM's
intent is to cross exotic (not in current use)
germ plasm with commercial maize lines in
order to increase corn yield. See also CORN,
GERM PLASM, HYBRIDIZATION (PLANT GENETICS),
PLEIOTROPIC.

**GEMP (Genetically Engineered Microbial
Pesticide)** See GENETICALLY ENGINEERED
MICROBIAL PESTICIDE, INTEGRATED PEST MAN-
AGEMENT (IPM).

**Gene** A natural unit of the hereditary material,
which is the physical basis for the transmission of the characteristics of living organisms
from one generation to another. The basic
genetic material is fundamentally the same
in all living organisms: it consists of chain-
like molecules of nucleic acids — deoxyribonucleic acid (DNA) in most organisms and
ribonucleic acid (RNA) in certain viruses —
and is usually associated in a linear arrangement that (in part) constitutes a chromosome.

The segment of DNA that is involved in
producing a polypeptide chain. It includes
regions preceding and following the coding
region (leader and trailer) as well as intervening sequences (introns) between individual coding segments (exons). More than one
protein can be expressed (made) from a
given gene (i.e., the particular protein
expressed is determined by factors such as
the cell's temperature or other environmental
variable, or the presence of STATs, some of
which themselves are proteins). See also
INFORMATIONAL MOLECULES, DEOXYRIBO-
NUCLEIC ACID (DNA), RIBONUCLEIC ACID (RNA),
GENE EXPRESSION, CHROMOSOMES, EXPRESS,
MESSENGER RNA (mRNA), CODON, INTRON, EXON,
CODING SEQUENCE, GENE EXPRESSION CASCADE,
CENTRAL DOGMA (NEW), SIGNAL TRANSDUCERS
AND ACTIVATORS OF TRANSCRIPTION (STATs).

**Gene "Stacking"** See "STACKED" GENES.

**Gene Amplification** The copying of segments
(e.g., genes) within the DNA or RNA
molecule. This can be done by man (e.g.,

polymerase chain reaction), can be caused by certain chemical carcinogens (e.g., phorbol ester), or occur naturally (e.g., in procaryotes and certain lower eucaryotes). The five primary techniques used by man to perform gene amplification are:

1. Polymerase Chain Reaction (PCR)
2. Ligase Chain Reaction (LCR)
3. Self-sustained Sequence Replication (SSR)
4. Q-beta Replicase Technique
5. Strand Displacement Amplification (SDA)

See also GENE, Q-BETA REPLICASE TECHNIQUE, POLYMERASE CHAIN REACTION (PCR), CARCINOGEN, PROCARYOTE, EUCARYOTE.

**Gene Array Systems** See BIOCHIPS, PROTEOMICS, GENE EXPRESSION ANALYSIS.

**Gene Chips** See BIOCHIPS, GENE EXPRESSION ANALYSIS, PROTEOMICS.

**Gene Delivery** (gene therapy) The insertion of genes (e.g., via retroviral vectors) into selected cells in the body in order to:

1. cause those cells to produce specific therapeutic agents (growth hormone in livestock, factor VIII in hemophiliacs, insulin in diabetics, etc.). A potential way of curing some genetic diseases, in that the inserted gene will produce the protein and/or enzyme that is missing in the body due to a defective gene (thus causing the genetic disease). Approximately 3,000 genetic diseases are known to man. Examples of genetic diseases include cystic fibrosis, sickle cell anemia, Huntington's disease, phenylketonuria (PKU), Tay-Sach's disease, ADA deficiency (adenosine deaminase enzyme deficiency), and thalassemia.
2. cause those cells to become (more) susceptible to a conventional therapeutic agent that previously was ineffective against that particular condition/disease (e.g., insertion of Hstk gene into brain tumor cells to make those tumor cells susceptible to the Syntex drug Ganciclovir).
3. cause those cells to become less susceptible to a conventional therapeutic agent (e.g., insert genes into healthy tissue in order to enable that healthy tissue to resist the harmful effects of such conventional chemotherapy agents as vincristine).
4. counter the effects of abnormal (damaged) tumor suppressor genes via insertion of normal tumor suppressor genes.
5. cause expression of ribozymes that cleave oncogenes (cancer-causing genes).
6. be used for other therapeutic uses of genes in cells.

See also TUMOR SUPPRESSOR GENES, ONCOGENES, CANCER, p53 GENE, TUMOR, PROTO-ONCOGENES, RETROVIRAL VECTORS, RETROVIRUSES, HUNTINGTON'S DISEASE, GENETIC CODE, INFORMATIONAL MOLECULES, DEOXYRIBONUCLEIC ACID (DNA), CHROMOSOMES, HORMONE, ENZYME, PROTEIN, GENETIC TARGETING.

**Gene Expression** Conversion of the genetic information within a gene, into an actual protein (or cell process). Note that many genes are only expressed at specific times during the lifetime of an organism. Some genes are expressed in a "cascade" of related expressions. See also GENE, GENETIC CODE, INFORMATIONAL MOLECULES, EXPRESS, GENE EXPRESSION ANALYSIS, BIOCHIPS, GENE EXPRESSION CASCADE, CENTRAL DOGMA (NEW).

**Gene Expression Analysis** Generally done via use of "biochips" (which have numerous detection/analysis devices fabricated onto their silicon surface) or "microarrays," gene expression analysis involves evaluation of the expression (and expression levels) of numerous genes in a biological sample, to analyze/compare any differences between gene expression/products in:

1. Normal cells vs. diseased cells.
2. Normal cells vs. those responding to a stimulus.
3. Cells from the same organism, at different stages of development (e.g., embryo versus adult).

G

4. Normal (historic wild type) cells vs. genetically engineered cells (those that have been engineered to cure a disease, resist an herbicide, etc.).
5. normal cells vs. those same cells treated with a given pharmaceutical (candidate).

Analysis generally involves measurement of gene expression markers (i.e., molecules synthesized, or cellular consequences such as apoptosis) to determine which genes are expressed (and when/how much, etc.). See also GENE, GENE EXPRESSION, GENE EXPRESSION PROFILING, MICROARRAY (TESTING), GENOMICS, FUNCTIONAL GENOMICS, EXPRESS, EXPRESSED SEQUENCE TAGS (EST), ZINC FINGER PROTEINS, BIOCHIPS, HIGH-THROUGHPUT SCREENING (HTS), MICROFLUIDICS, HERBICIDE-TOLERANT CROP, GENE DELIVERY (GENE THERAPY), HORMONE, PROTEOMICS, PROMOTER, GENE EXPRESSION MARKERS, GENE EXPRESSION CASCADE, APOPTOSIS, RT-PCR.

**Gene Expression Cascade** A sequential series of individual gene expressions (i.e., each gene causing a separate/different protein to be "manufactured"), that is initiated ("set off") by the first gene expression. For example, a gene expression cascade is often initiated by the first gene causing expression of a transcription factor (i.e., protein that itself interacts with cell's DNA to either cause or speed up yet another gene expression). The protein resulting from that second gene expression could be yet another transcription factor that triggers another (i.e., third) gene expression, and so on. See also GENE, EXPRESS, GENE EXPRESSION, CASCADE, PROTEIN, CELL, DEOXYRIBONUCLEIC ACID (DNA), PROMOTER, TRANSCRIPTION FACTORS, APOPTOSIS.

**Gene Expression Markers** Refers to molecules (e.g., synthesized due to a specific gene's expression) or consequences (e.g., cell apoptosis due to a specific gene's expression) that can be measured as proof of gene's expression in gene expression analysis. See also GENE EXPRESSION, GENE, GENE EXPRESSION ANALYSIS, EXPRESS, BIOCHIPS, PROTEIN, CELL, APOPTOSIS, GREEN FLUORESCENT PROTEIN.

**Gene Expression Profiling** Determination of specifically which genes are "switched on" (e.g., in a cell), thereby enabling precise definition of the phenotypic condition of that cell (i.e., the phenotype of that cell at that moment). Typical uses (i.e., comparison of such tissue phenotypes) include:

1. Comparing diseased cell with normal cell.
2. Defining quantitatively the "normal" state.
3. Comparing a given drug's impact (i.e., treated cell with normal cell).
4. Comparing old cell with young cell.

In subsequent gene expression analysis, the quantitative amounts of each protein being expressed can be determined via use of such technologies as two-dimensional (2D) gel electrophoresis, Southern blot analysis, fluorescence tagging, radiolabeling, RT-PCR, QPCR, plane polarimetry, etc. See also GENE, GENE EXPRESSION, PROTEIN, CELL, PHENOTYPE, GENE EXPRESSION ANALYSIS, TWO-DIMENSIONAL (2D) GEL ELECTROPHORESIS, SOUTHERN BLOT ANALYSIS, RADIOLABELED, RT-PCR, QPCR, GENE EXPRESSION MARKERS, MICROARRAY (TESTING).

**Gene Function Analysis** The determination of which protein is expressed (i.e., caused to be "manufactured") by each gene in an organism's genome/DNA. Typically, gene function analysis follows after discovery of gene sequences found via structural genomics study. Some methods utilized to determine which proteins result from which gene(s) are:

1. Site-directed mutagenesis (SDM) to compare two same-species organisms possessing two different genes at the same site (SNP) on the genome (i.e., on organism's DNA).
2. Antisense DNA sequences to compare two same-species organisms, one of which has a gene at the same site "turned off" (silenced) via antisense DNA.
3. Reporter gene, to compare two same-species organisms (possessing two different genes at the same site on genome/DNA) via a reporter gene adjacent to the gene/site, to detect

presence or absence of the desired trait/function.

4. Comparison of same organism (e.g., crop plant) when one of the two is "challenged" by a specific plant disease.
5. Chemical genetics, to compare two same-species organisms (one of which has gene at the specific site at least partially inactivated by a specific chemical).
6. "Silencing" or "knocking out" a particular gene via other methods than antisense or chemical genetics, to compare.
7. Use of already-known "model organisms" (e.g., *Drosophila* for comparing insect genes, *Arabidopsis thaliana* for plant genes, *Caenorhabditis elegans* for animal genes).

See also GENE, GENE EXPRESSION, GENETIC CODE, INFORMATIONAL MOLECULES, EXPRESS, PROTEIN, GENOME, GENOMICS, STRUCTURAL GENOMICS, FUNCTIONAL GENOMICS, ZINC FINGER PROTEINS, TRAIT, DEOXYRIBONUCLEIC ACID (DNA), SEQUENCE (OF A DNA MOLECULE), POINT MUTATION, SITE-DIRECTED MUTAGENESIS (SDM), ANTISENSE (DNA SEQUENCE), GENE SILENCING, REPORTER GENE, METHYLATION, POSITIONAL CLONING, DNA METHYLATION, CHEMICAL GENETICS, MODEL ORGANISM, *DROSOPHILA*, *ARABIDOPSIS THALIANA*, *CAENORHABDITIS ELEGANS* (*C. ELEGANS*), CENTRAL DOGMA (OLD), CENTRAL DOGMA (NEW), TRANSCRIPTION FACTORS, TRANSWITCH®, SINGLE-NUCLEOTIDE POLYMORPHISMS (SNPs).

**Gene Fusion** Refers to the technology/methods utilized to fuse together two or more genes. When such a "fused gene" is then inserted into a genome (e.g., the DNA of a plant), it causes production (in plant's ribosomes) of protein(s) consisting of all or part of the amino acid sequences (known as the "domain") of the two proteins typically coded for by those two genes. This fusion is often done in order to put expression of the second (fused) gene under the control of the (strong) promoter of the first gene. During 2001, Rajbir Sangwan and colleagues inserted a fused gene into a potato plant (*Solanum tuberosum*), a major source of plant starch. That fused gene coded for production of the two proteins α-amylase and glucose isomerase; both are enzymes. α-amylase catalyzes the conversion of potato starch into glucose (a sugar), and glucose isomerase catalyzes conversion of glucose to fructose (a more valuable sugar). See also GENE, GENOME, DEOXYRIBONUCLEIC ACID (DNA), GENETIC ENGINEERING, RIBOSOMES, CODING SEQUENCE, PROTEIN, AMINO ACID, SEQUENCE (OF A PROTEIN MOLECULE), FUSION PROTEIN, EXPRESS, PROMOTER, ENZYME, AMYLASE, GLUCOSE, ISOMERASE.

**Gene Machine** An instrument which, when fed information on the amino acid sequence of a protein (usually via a protein sequencer), will automatically produce polynucleotide gene segments to code for that protein. See also SEQUENCING (OF DNA MOLECULES), SYNTHESIZING (OF DNA MOLECULES), GENE, AMINO ACID, PROTEIN.

**Gene Manipulation** See GENETIC ENGINEERING.

**Gene Map** See LINKAGE MAP, GENETIC MAP, PHYSICAL MAP (OF GENOME).

**Gene Mapping** See SEQUENCING (OF DNA MOLECULES), GENETIC MAP, LINKAGE MAP, PHYSICAL MAP (OF GENOME).

**Gene Probe** See DNA PROBE.

**Gene Repair (done by man)** The "repair" of a damaged gene (e.g., mutation) or replacement of a given gene, via a process invented by Eric Kmiec in 1993. The desired DNA (gene) is added to a cell, along with RNA, in a paired-group known as a chimeraplast. The chimeraplast attaches itself to the cell; DNA at the site of the specific gene (i.e., the one that is to be changed), and "repairs" it using its (new) chimeraplast-DNA as a "template." See also GENE, CHIMERAPLASTY, MUTATION, DEOXYRIBONUCLEIC ACID (DNA), RIBONUCLEIC ACID (RNA), CELL, TEMPLATE.

**Gene Repair (natural)** Refers to the natural processes via which all cells in an organism are continually repairing their DNA (which can be damaged by ultraviolet light, various chemicals, etc.). In these natural cell (gene repair) processes, first, an enzyme complex detects the damaged DNA (e.g., on one of the two strands of the DNA molecule). Next, an enzyme cuts out the damaged portion of the DNA (on that one strand, leaving the other — good — strand intact). Then a DNA polymerase enzyme enters the gap and synthesizes (manufactures) the new DNA (to replace the portion that was cut out), using the intact — good — DNA strand as a template.

Finally, the new DNA is joined to the old DNA via the help of DNA ligase enzyme. See also CELL, ENZYME, DEOXYRIBONUCLEIC ACID (DNA), DNA POLYMERASE, DNA LIGASE, TEMPLATE.

**Gene Replacement Therapy** See GENE DELIVERY.

**Gene Silencing** The suppression of gene expression (e.g., of the gene for polygalacturonase which causes fruit to ripen) via a variety of methods (e.g., via chemical genetics, "zinc finger proteins," sense or antisense genes, etc.). See also GENE, EXPRESS, GENE EXPRESSION, GENETIC CODE, INFORMATIONAL MOLECULES, PROTEIN, CHEMICAL GENETICS, ZINC FINGER PROTEINS, GENE FUNCTION ANALYSIS, COSUPPRESSION, ANTISENSE (DNA SEQUENCE), TRANSWITCH®, SENSE, POLYGALACTURONASE (PG), GPA1.

**Gene Splicing** The enzymatic attachment (joining) of one gene (or part of a gene) to another; also removal of introns and splicing of exons during mRNA synthesis. See also SPLICING, CENTRAL DOGMA (NEW), MESSENGER RNA (mRNA), GENE, B LYMPHOCYTES, RECOMBINASE.

**Gene Switching** See GENE, GENETIC CODE, CODING SEQUENCE, DEOXYRIBONUCLEIC ACID (DNA), SEQUENCE (OF A DNA MOLECULE), REGULATORY SEQUENCE, TRANSCRIPTION FACTORS, CBF1, COLD HARDENING, CESSATION CASSETTE, SYSTEMIC ACQUIRED RESISTANCE (SAR).

**Gene Targeting** See GENETIC TARGETING, GENE SPLICING, GENE DELIVERY, GENETIC ENGINEERING.

**Gene Technology Office** An agency of the Australian government, established in 1997, to oversee and regulate all genetic engineering activities conducted in the country of Australia. Replaced/superceded by Australia's newly formed Interim Office of the Gene Technology Regulator (IOGTR) in 1999. See also IOGTR, GENE TECHNOLOGY REGULATOR (GTR), GENETIC ENGINEERING, RECOMBINANT DNA ADVISORY COMMITTEE (RAC), ZKBS (CENTRAL COMMITTEE ON BIOLOGICAL SAFETY), INDIAN DEPARTMENT OF BIOTECHNOLOGY, COMMISSION OF BIOMOLECULAR ENGINEERING.

**Gene Technology Regulator (GTR)** The regulatory body of Australia's government that is responsible for approvals of new rDNA products (e.g., new genetically engineered crops) before they can be introduced into Australia. GTR replaced Australia's IOGTR (Interim Office of the Gene Technology Regulator) in this role on June 21, 2001. See also INTERIM OFFICE OF THE GENE TECHNOLOGY REGULATOR (IOGTR), GENE TECHNOLOGY OFFICE, GENETIC MANIPULATION ADVISORY COMMITTEE (GMAC), rDNA, DEOXYRIBONUCLEIC ACID (DNA), GENETIC ENGINEERING, RECOMBINANT DNA ADVISORY COMMITTEE (RAC), COMMISSION OF BIOMOLECULAR ENGINEERING, INDIAN DEPARTMENT OF BIOTECHNOLOGY.

**Gene Therapy** See GENE DELIVERY.

**Gene Transcript** See TRANSCRIPT.

**Generation Time** The time required for a population of cells to double. The average time required for a round of cell division. See also CELL, MITOSIS.

**Genestein** See GENISTEIN (Gen).

**Genetic Code** The set of triplet code words in DNA coding for all of the amino acids. There are more than 20 different amino acids and only four bases (adenine, thymine, cytosine, and guanine). The mRNA code is a triplet code, that is, each successive "frame" of three nucleotides (sometimes called a codon) of the mRNA corresponds to one amino acid of the protein. This rule of correspondence is the genetic code. The genetic code consists of 64 entries — the 64 triplets possible when there are four possible nucleotides, each of which can be at any of three places ($4 \times 4 \times 4 = 64$). A triplet code was required because a doublet code would have only been able to code for ($4 \times 4 = 16$) 16 amino acids. A triplet code allows for the coding of 64 theoretical amino acids. Since only a little over 20 exist, there is some redundancy in the system. Hence some certain amino acids are coded for by two or three different triplets. See also MESSENGER RNA (mRNA), DEOXYRIBONUCLEIC ACID (DNA), INFORMATIONAL MOLECULES.

**Genetic Engineering** The selective, deliberate alteration of genes (genetic material) by man. This term has come to have a very broad meaning, including the manipulation and alteration of the genetic material (constitution) of an organism in such a way as to allow it to produce endogenous proteins with properties different from those of the traditional (historic/typical), or to produce entirely different (foreign) proteins altogether. Some other words often applicable to the same process are gene splicing, gene manipulation, or recombinant DNA technology (techniques). See also GENE, INFORMATIONAL MOLECULES, CHROMOSOMES, GENE AMPLIFICATION, VECTOR, PLASMID, *AGROBACTERIUM TUMEFACIENS*, GENE SPLICING, DEOXYRIBONUCLEIC ACID (DNA), TRANSGENIC (ORGANISM), BIOLISTIC R GENE GUN, WHISKER™, "SHOTGUN" METHOD, NUCLEAR TRANSFER, GMO, RECOMBINANT DNA (rDNA), RECOMBINATION, HETEROKARYON, HEREDITY, MESSENGER RNA (mRNA), HETERODUPLEX, POSITIVE AND NEGATIVE SELECTION (PNS), POLYMERASE CHAIN REACTION (PCR) TECHNIQUE, BIOTECHNOLOGY, METABOLIC ENGINEERING.

**Genetic Engineering Approval Committee** See GEAC.

**Genetic Event** See EVENT.

**Genetic Linkage** See LINKAGE, LINKAGE GROUP.

**Genetic Manipulation** See GENETIC ENGINEERING.

**Genetic Manipulation Advisory Committee (GMAC)** A body that advises the Australian government on matters pertaining to genetic engineering (e.g., new rDNA product approvals). The GMAC is analogous to Germany's ZKBS (Central Commission on Biological Safety), Brazil's CTNBio (National Technical Biosafety Commission), and the Kenya Biosafety Council. See also GMAC, ZKBS (CENTRAL COMMISSION ON BIOLOGICAL SAFETY), RECOMBINANT DNA ADVISORY COMMITTEE (RAC), GENETIC ENGINEERING, rDNA, DEOXYRIBONUCLEIC ACID (DNA), CTNBio, KENYA BIOSAFETY COUNCIL, GENE TECHNOLOGY OFFICE, GENE TECHNOLOGY REGULATOR (GTR).

**Genetic Map** A diagram showing the relative sequence and position of specific genes along a chromosome (DNA) molecule. Markers utilized as "signposts"/guideposts in such maps include single-nucleotide polymorphisms (SNPs), restriction sites (i.e., the specific locations where each restriction endonuclease "cuts" a DNA strand), and microsatellites. Such markers located in or close to the gene of interest (e.g., a disease-causing gene within a chromosome) to a researcher are more likely to be inherited along with that gene. See also POSITION EFFECT, GENE, GENOME, CHROMOSOMES, DEOXYRIBONUCLEIC ACID (DNA), PHYSICAL MAP (OF GENOME), SINGLE-NUCLEOTIDE POLYMORPHISMS (SNPs), RESTRICTION SITE, MICROSATELLITE DNA, MARKER ASSISTED SELECTION.

**Genetic Marker** See MARKER (GENETIC MARKER).

**Genetic Probe** See DNA PROBE.

**Genetic Targeting** The insertion of antisense DNA molecules *in vivo* into selected cells of the body in order to block the activity of undesirable genes. These genes might include oncogenes, or genes crucial to the life cycle of parasites such as trypanosomes (which cause sleeping sickness). See also ANTISENSE (DNA SEQUENCE), GENE, GENE DELIVERY, ONCOGENES, DENDRIMERS.

**Genetic Use Restriction Technologies (GURTs)** A general term refering to several different technologies intended to control the expression (or nonexpression) of the gene(s) for specific (e.g., valuable) traits. See also CESSATION CASSETTE, GENE, TRAIT, EXPRESS, VALUE-ENHANCED GRAINS.

**Genetically Engineered Microbial Pesticides (GEMP)** One or more microbes that have been genetically engineered to be effective in combatting pest(s) that attack crops or livestock. For example, a microbe that naturally attacks a crop pest could be genetically engineered to make the microbe more potent, or more durable in field environments when applied via selected method of microbe application. See also MICROBE, GENETIC ENGINEERING, WHEAT TAKE-ALL DISEASE, BACULOVIRUS, *BACILLUS THURINGIENSIS* (*B.t.*), FEDERAL INSECTICIDE FUNGICIDE AND RODENTICIDE ACT (FIFRA), TOXIC SUBSTANCES CONTROL ACT (TSCA).

**Genetically Engineered Organism (GEO)** See GEO.

**Genetically Manipulated Organism (GMO)** See GMO.

**Genetically Modified Microorganism (GMM)** See GMM.

**Genetically Modified Organism (GMO)** See GMO.

G

**Genetically Modified Pest Protected (GMPP) Plants** Plants that have been genetically engineered so they resist (or are more tolerant to) attacks by pests (e.g., insects). See also GENETIC ENGINEERING, *BACILLUS THURINGIENSIS* (*B.t.*), COWPEA TRYPSIN INHIBITOR (CpTI), CRY PROTEINS, CRY1A (b) PROTEIN, CRY1A (c) PROTEIN, CRY9C PROTEIN, *B.t. KURSTAKI, B.t. TENEBRIONIS, B.t. ISRAELENSIS,* PATHOGENESIS RELATED PROTEINS, *PHOTORHABDUS LUMINESCENS.*

**Genetics** The branch of biology concerned with heredity, it was literally invented by Gregor Mendel in the 19th century. It is a study of the manner in which genes operate and are transmitted from parents to offspring. In 1865, Mendel defined what gene (alleles) are, and that they can be dominant or recessive (within the offspring's genome/DNA, which has two "copies" of each gene). For example, if a given trait (e.g., black hair) is dominant, and that gene is inherited from only one of the parents (e.g., the father), the offspring will have that trait (black hair). But if a given trait (e.g., red hair) is recessive, the offspring will not have that trait unless the "red hair gene" is inherited from both parents. Genetics also involves the study of the mechanism of gene action — the manner in which the genetic material (DNA) affects physiological reactions within the cell. See also GENE, DEOXYRIBONUCLEIC ACID (DNA), HYBRIDIZATION (PLANT GENETICS), HEREDITY, DOMINANT ALLELE, RECESSIVE ALLELE, CELL, GENE EXPRESSION ANALYIS.

**Genistein (Gen)** One of several phytochemicals produced by the soybean plant as a defense against certain plant diseases; and to signal *Rhizobium japonicum* bacteria to produce nitrogen for the soybean plant via colonization of its roots, followed by nitrogen fixation from the air. Genistein can also be produced as a by-product of mycobacterium fermentation (the process used to produce commercial amounts of certain antibiotics).

Genistein is an isoflavone, a steroid-like compound that can be lethal to certain animal cells via its kinase-inhibiting properties. Genistein fights cancer (tumor cells) by inhibiting protein tyrosine kinase and topoisomerase II. Genistein also exhibits the property of antiangiogenesis (i.e., inhibition of tumor growth via prevention of the formation/development of new blood vessels in tumors). Attached to a pharmaceutical "guided missile" such as a monoclonal antibody or the CD4 protein, genistein is potentially useful for treatment against some tumors and has been investigated as a possible treatment against B-cell precursor leukemia. A human diet containing a large amount of genistein has been shown to increase bone density and to decrease total serum (blood) cholesterol, thereby lowering risk of osteoporosis and coronary heart disease. Research also indicates that human consumption of genistein can help to prevent breast cancer, prevent adverse increases in blood platelet aggregation, and inhibit the proliferation of smooth-muscle cells in plaque deposits (inside blood vessels). See also IMMUNOTOXIN, MONOCLONAL ANTIBODIES (MAb), CD4 PROTEIN, GENETIC ENGINEERING, NITROGEN FIXATION, NODULATION, PHYTOCHEMICALS, FUSION PROTEIN, FUSION TOXIN, SOLUBLE CD4, ISOFLAVONES, SOYBEAN PLANT, RICIN, TYROSINE (tyr), STEROID, CANCER, INHIBITION, STRESS PROTEINS, "MAGIC BULLET", TYROSINE KINASE, CORONARY HEART DISEASE (CHD), CHOLESTEROL, OSTEOPOROSIS, SELECTIVE ESTROGEN EFFECT, ANTIANGIOGENESIS, PROTEIN TYROSINE KINASE INHIBITOR, PLAQUE.

**Genistin** The β-glycoside form (isomer in which glucose is attached to the molecule at the 7 position of the A ring) of the isoflavone known as genistein (aglycone form). See also GENISTEIN (Gen), ISOFLAVONES, ISOMER.

**Genome** The entire hereditary material (which was proven by Oswald Avery in 1944 to be DNA) in a cell. In addition to the DNA contained in cell nucleus (known as nuclear DNA), an organism's cells contain DNA in other locations within those cells: bacteria contain some DNA in plasmids; plants contain some DNA in plastids; animals contain some DNA in mitochondria. An organism's nuclear DNA is composed of one or more chromosomes, depending on the complexity of the organism. See also DEOXYRIBONUCLEIC ACID (DNA), CHROMOSOMES, PLASTID, PLASMID, MITOCHONDRIA, MITOCHONDRIAL DNA.

**Genomic Sciences** An encompassing term utilized to refer to all knowledge of, and attempts to decipher/understand, the structure and function of the genomes of organisms. See also GENOMICS, GENOME, STRUCTURAL GENOMICS, FUNCTIONAL GENOMICS, GENOTYPE, GENE, GENETICS, GENETIC MAP, GENETIC TARGETING, GENETIC CODE, SEQUENCING (OF DNA MOLECULES), INFORMATIONAL MOLECULES, DEOXYRIBONUCLEIC ACID (DNA), GENE AMPLIFICATIONS, CODING SEQUENCE, CHEMICAL GENETICS.

**Genomics** The scientific study of genes and their role in an organism's structure, growth, health, disease, and/or resistance to disease, etc. For example, how the (approximately) 3,000 genes in a given strain of bacteria, or the (approximately) 6,000 genes in a given strain of yeast, contribute to the shape, function, and the development of those whole organisms. Some tools/methods utilized in genomics include:

1. Structural Genomics — The study or discovery of what particular gene sequences are present, and where they are located within an organism's DNA.
2. Gene Function Analysis — The determination of which protein is expressed (i.e., caused to be "manufactured") by each gene in an organism's genome. Typically, gene function analysis follows after structural genomics study.
3. Functional Genomics — The study or discovery of what traits/functions are conferred to an organism by given gene sequence(s).
4. Chemical Genetics — Comparison of two same-species organisms (one of which has a given gene, or genes, inactivated by a specific chemical or site mutation).
5. Gene Expression Analysis — Determination of the product(s) resulting (such as an enzyme or other critical protein) when a given gene is "switched on," by measuring fluorescence of individual messenger RNA (mRNA) molecules (specific to which particular gene is "switched on" at the time), when that mRNA hybridizes (with DNA pieces corresponding to proteins produced/analyzed, that were attached to hybridization surface on biochip).

See also GENOTYPE, GENE, GENETIC MAP, GENETIC TARGETING, GENETICS, GENETIC CODE, SEQUENCING (OF DNA MOLECULES), INFORMATIONAL MOLECULES, DEOXYRIBONUCLEIC ACID (DNA), FUNCTIONAL GENOMICS, GENE AMPLIFICATION, CODING SEQUENCE, STRUCTURAL GENOMICS, GENOMIC SCIENCES, BACTERIA, YEAST, STRAIN, CHEMICAL GENETICS, FLUORESCENCE, ENZYME, PROTEIN, MESSENGER RNA (mRNA), BIOCHIPS, EXPRESS, EXPRESSED SEQUENCE TAGS (EST), HYBRIDIZATION SURFACES, GENE EXPRESSION, GENE EXPRESSION ANALYSIS, GENE FUNCTION ANALYSIS.

**Genosensors** Biosensors (electronic) that can detect the individual nucleotides that comprise a genome (DNA) molecule. Automated genosensors enable rapid, nondestructive sequencing of DNA molecules. See also GENOME, NUCLEOTIDE, DEOXYRIBONUCLEIC ACID (DNA), SEQUENCING (OF DNA MOLECULES), TEMPLATE, BIOSENSORS (ELECTRONIC), FOOTPRINTING, NANOTECHNOLOGY, BIOCHIPS.

**Genotoxic** Refers to compounds that interfere with normal functioning of genetic material (i.e., DNA). For example, the antitumor antibiotic family of duocarmycin drugs. See also DEOXYRIBONUCLEIC ACID (DNA), GENOTOXIC CARCINOGENS, FOOTPRINTING.

**Genotoxic Carcinogens** Compounds that act directly on the genetic material (i.e., DNA) of an organism, thus causing cancer in that organism. Of the numerous chemicals that have been documented to be human carcinogens, the majority of them are genotoxic. See also CARCINOGEN, CANCER, GENE, DEOXYRIBONUCLEIC ACID (DNA).

**Genotype** The total genetic, or hereditary, constitution that individuals receive from their parents. An individual organism's genotype is distinguished from its phenotype, which is its appearance or observable character. See also TRAIT, PHENOTYPE, WILD TYPE.

**Gentechnik Gesetz (Gene Technology Law)** The 1990 law that governs recombinant DNA research and development in the country of Germany. It was amended January 1, 1994, to make it somewhat less restrictive. See also ZKBS (CENTRAL COMMISSION ON BIOLOGICAL SAFETY), RECOMBINANT DNA ADVISORY COMMITTEE (RAC), GENETIC ENGINEERING, RECOMBINANT DNA (rDNA), RECOMBINATION,

BIOTECHNOLOGY, BUNDESGESUNDHEITSAMT (BGA), INDIAN DEPARTMENT OF BIOTECHNOLOGY.

**Genus** A group of closely related species. See also SPECIES, CLADES.

**GEO** Genetically engineered organism. See also GENETIC ENGINEERING, GMO, GENE, GENE SPLICING, GMM.

**Geomicrobiology** Applications of microbiological knowledge to an understanding of geological phenomena. See also FERROBACTERIA.

**Germ Cell** The sex cell (sperm or egg). It differs from other cells in that it contains only half (haploid) the usual number of chromosomes. See also GAMETE, HAPLOID.

**Germ Plasm** The total genetic variability to an organism, represented by the total available pool of germ cells or seed. See also ORGANISM, CELL, GERM CELL, GEM.

**German Gene Law** See GENTECHNIK GESETZ (GENE TECHNOLOGY LAW).

**GFP** Green Fluorescent Protein. See also GREEN FLUORESCENT PROTEIN.

**GH** See GROWTH HORMONE.

**Gibberellins** Plant hormones that, among other functions, regulate the growth of grass species, including rice (after the relevant gibberellin is activated by an enzyme). In 1996, Lew Mander and Richard Pharis discovered an analogue (i.e., a chemical that is similar) to grass gibberellin that does not cause grass to grow. When this analogue is sprayed onto grass, it mixes into the naturally occurring grass gibberellin and significantly slows grass growth (thus potentially reducing the amount of mowing required for lawns, golf courses, etc.). See also PLANT HORMONE, ENZYME, ANALOGUE, *FUSARIUM MONILIFORME*.

**Gleevac™** See GLEEVEC™.

**Gleevec™** A pharmaceutical (imatinib mesylate, also known as STI571), developed and trademarked by Novartis AG, used to treat the blood cancer known as chronic myelogenous leukemia or chronic myeloid leukemia or chronic myelocytic leukemia (CML). CML results from a genetic defect (single-nucleotide polymorphism) that causes excessive production of white blood cells in the body of the affected (human). That excessive production of white blood cells results when the defective gene (i.e., SNP) causes excessive production of the enzyme Bcr-Abl tyrosine kinase. Because Gleevec™ is a protein tyrosine kinase inhibitor, it arrests excessive production of white blood cells (and induces apoptosis — cell death — in the cells that have the Bcr-Abl gene/SNP). See also CANCER, WHITE BLOOD CELLS, GENE, MUTATION, SINGLE-NUCLEOTIDE POLYMORPHISMS (SNPs), ENZYME, APOPTOSIS, PROTEIN TYROSINE KINASE INHIBITOR.

**Glial Derived Neurotrophic Factor (GDNF)** A neurotrophic factor that assists the survival and functional activity of the brain's dopaminergic neurons. Because dopaminergic neurons typically deteriorate and die in brains of the victims of Parkinson's disease, it is possible that GDNF may someday be used in treatment of Parkinson's disease. See also NEUROTRANSMITTER, PARKINSON'S DISEASE.

**Globular Protein** A soluble protein in which the polypeptide chain is tightly folded in three dimensions to yield a globular (roughly oval, circular) shape. See also PROTEIN FOLDING, POLYPEPTIDE (PROTEIN), CONFORMATION, TERTIARY STRUCTURE.

**Glomalin** A "sticky" protein molecule naturally produced by certain fungi which grow on most plant roots (in the soil). Glomalin acts sort of like glue, thereby improving soil stability by "gluing" soil into clumps. Proper soil "clumping" (i.e., glomming together) allows air and water to pass through that soil more easily, increases the amount of carbon contained within the soil (thereby removing that "greenhouse gas" carbon dioxide from the atmosphere), increases the number of ("healthy") bacteria in that soil, and improves that soil's overall fertility (i.e., its ability to produce high-yield crops or a large amount of biomass per hectare/acre). The glomalin (and thus carbon) content of soil in a field is increased by farmer utilization of low-tillage or "no-tillage" methods of crop production. See also PROTEIN, FUNGUS, BACTERIA, BIOMASS, CONSERVATION TILLAGE, LOW-TILLAGE CROP PRODUCTION, NO-TILLAGE CROP PRODUCTION.

**GLP** See GOOD LABORATORY PRACTICES (GLP).

**GLQ223** See TRICHOSANTHIN.

**Glucagon** A hormone produced by the pancreas that causes the breakdown of glycogen in the liver. Glycogen is a form of storage sugar and its breakdown releases glucose for energy production. See also GLYCOGEN, HORMONE, GLUCOSE, PANCREAS.

**Glucan** See WATER SOLUBLE FIBER, POLYPHENOLS.

**Glucocerebrosidase** (trade name Ceredase) An enzyme used in treatment of inherited Gaucher's disease in which there is abnormal deposition of glucocerebrosides (hydrophobic lipid molecules that contain a hydrophilic sugar head group). Gaucher's disease is an enzyme deficiency disease that may be amenable to cure by incorporation of the gene coding for glucocerebrosidase into the patient's genome via gene delivery techniques. See also ENZYME, GENE DELIVERY.

**Glucogenic Amino Acid** Amino acids whose carbon chains can be metabolically converted by cells into glucose or glycogen. See also GLUCONEOGENESIS, CELL, AMINO ACID, METABOLISM.

**Gluconeogenesis** The net biosynthesis (formation) of new glucose from noncarbohydrate precursors such as pyruvate, lactate, glycerol, acetyl-CoA (in plants), certain amino acids, and intermediates of the citric acid cycle. See also CARBOHYDRATES, GLUCOSE (GLc), CITRIC ACID CYCLE, Ac-CoA, BIOTIN.

**Glucose (GLc)** A prime fuel for the generation of energy by organisms. It is broken down (to obtain energy) via a metabolic process called glycolysis. Glucose is a hexose, a sugar possessing six carbon atoms in its molecule. The six carbon atoms are connected to each other to form a closed ring structure known as a hexose (6) ring.

Animal cells store glucose in the form of glycogen (sometimes called animal starch), a large branched polymer of glucose units. Plant cells store glucose in the form of starch, a large polymer of glucose units. Yeasts and bacteria store glucose in the form of dextran, a polymer of glucose units. The difference between the forms of storage glucose is (1) in the size (molecular weight) of the final polymer formed, (2) in the type of linkages that connect the single glucose units together in the branched molecule, and (3) in the degree of branching which occurs in the polymer. Note that a glucose polymer does not consist of just a single long straight chain. The backbone chain has other polymer chains branching off it. The whole molecule may be visualized as looking somewhat like a tree without the trunk. The other very abundant polymer formed by glucose units is structural in nature and is called cellulose. It is the most abundant cell wall and structural polysaccharide in the plant world. Hence, glucose is used not only as an energy source, but also as a structural material. See also AMYLOSE, AMYLOPECTIN, GLYCOLYSIS, GLUCONEOGENESIS, GLYCOGEN, STARCH, DEXTRAN, CELLULOSE.

**Glucose Isomerase** An enzyme that catalyzes the conversion of glucose to fructose. A molecule of fructose contains the same atoms as a molecule of glucose (but in a different arrangement). See also ENZYME, GLUCOSE, GENE FUSION.

**Glucose Oxidase** An enzyme that breaks down sugar molecules (causing oxygen consumption in an organism). Industrial uses include removing dissolved oxygen from certain food products (e.g., sugar-containing drink products). See also ENZYME, GLUCOSE (GLc), GLYCOLYSIS, SUGAR MOLECULES, ORGANISM.

**Glucosinolates** Toxins (neurotoxic phytotoxins) naturally produced in the seeds of some plants (e.g., rapeseed, wild mustard (*Brassica juncea*/*Brassica rapa*), grass pea (*Lathyrus sativus*), etc.) in order to dissuade wild animals from eating those plants' seeds. For example, when large amounts of grass pea (*Lathyrus sativus*) are consumed by humans, the glucosinolates build up in the body and can cause lathyrism (i.e., an irreversible spastic paralysis of the legs). The glucosinolates in rapeseed (*Brassica rapa*) oil have been linked to heart damage in humans who consume rapeseed (high erucic acid) oil; plus those glucosinolates impart a bitter taste to such plant oils.

The rapeseed glucosinolate 5-vinyl oxazolidine I cyano-2-hydroxy-3-butene causes poultry livers to hemorrhage (bleed internally) if it is fed via rapeseed meal or rapeseed oil to poultry for several weeks (at 20% of total diet). Such feeding of rapeseed meal/oil to poultry also predispose those poultry to develop Fatty Liver Syndrome (FLS), a metabolic disease. When glucosinolates from seeds of the wild mustard weed family are mixed into canola meal (e.g., when those weeds grew in a canola field and that resultant canola is processed into canola meal), such canola meal must first be diluted (e.g., via mixing in some soybean meal) in order to reduce glucosinolate concentration (below the legal maximum allowance) before it is fed to livestock in Canada. See also CANOLA, *BRASSICA*, TOXIN, PHYTOTOXINS, METABOLISM, SOYBEAN MEAL.

**Glufosinate** See PAT GENE, BAR GENE, HERBICIDE-TOLERANT CROP, GENE, GLUTAMINE SYNTHETASE.

**Gluphosinate** See PAT GENE, BAR GENE, HERBICIDE-TOLERANT CROP, GENE, GLUTAMINE SYNTHETASE.

**Glutamate Dehydrogenase** An enzyme found naturally in certain soil bacteria, which helps those bacteria to utilize soilborne nitrogen. When its gene (GDH gene) is inserted into corn plant via genetic engineering, the resultant plant production of glutamate dehydrogenase enables that corn plant to better utilize soilborne nitrogen. As a result, such genetically engineered corn (*Zea mays* L.) has a protein yield increase of approximately 10%, according to research begun in 1991 by David Lightfoot. See also ENZYME, BACTERIA, GENE, CORN, NITROGEN CYCLE, DEHYDROGENASES, PROTEIN, GENETIC ENGINEERING.

**Glutamic Acid** A dicarboxylic amino acid of the α-ketoglutaric acid family. See also AMINO ACID.

**Glutamine** An amino acid; the monamide of glutamic acid. Glutamine is of fundamental importance for amino acid biosynthesis in all forms of life. See also GLUTAMINE SYNTHETASE, AMINO ACID, PAT GENE, BAR GENE.

**Glutamine Synthetase** An enzyme that catalyzes the synthesis of glutamine (which is crucial for amino acid biosynthesis). See also GLUTAMINE, ENZYME, PAT GENE, BAR GENE, AMINO ACID.

**Glutathione** A tripeptide that is found in all cells of higher animals, which acts to help protect against oxidative stress. Composed of the amino acids glutamic acid, cysteine, and glycine. The cysteine possesses a sulfhydryl group that makes glutathione a weak reducing agent. See also OXIDATIVE STRESS, REDUCTION (IN A CHEMICAL REACTION).

**Gluten** A term used to refer to a naturally occurring mixture of two different proteins — glutenin and gliadin — in the seeds of bread wheat (*Triticum aestivum*). In flour made from traditional varieties of wheat, glutenin proteins constitute approximately 50% of the total gluten. The relative content of those two proteins determines one of the most commercially important properties of the wheat (strength and elasticity of the flour made from that particular wheat). For example, more of the high molecular weight glutenin (which is "stretchy" and imparts physical strength to a dough made from such flour, so that dough holds together while rising) results in a flour that is better suited to manufacture higher-quality yeast-raised bread products. See also WHEAT, PROTEIN, GLUTENIN, HIGH-GLUTENIN WHEAT, YEAST, MOLECULAR WEIGHT, POLYMER.

**Glutenin** A protein naturally present in the gluten within seeds of wheat (*Triticum aestivum*). See also GLUTEN, WHEAT, PROTEIN.

**Glyceraldehyde (D- and L-)** One of the smallest monosaccharides, it is called an aldose because it contains an aldehyde group. Glyceraldehyde has a single asymmetric carbon atom; thus there are two stereoisomers (D-glyceraldehyde and L-glyceraldehyde). See also MONOSACCHARIDES, STEREOISOMERS.

**Glycetein** See ISOFLAVONES.

**Glycine (gly)** The simplest (and smallest) of the amino acids found in proteins. It is the only amino acid that does not have an asymmetric carbon atom within its molecule. Thus, it is not optically active. See also AMINO ACID, PROTEIN, STEREOISOMERS, OPTICAL ACTIVITY.

*Glycine max* See SOYBEAN PLANT.

**Glycinin** One of the (structural) categories of proteins that are produced within seeds of legumes. In general, glycinins contain 3–4 times more cysteine (cys) and methionine (met) per unit of protein than does β-conglycinin. See also PROTEIN, CYSTEINE (cys), METHIONINE (met).

**Glycitein** See ISOFLAVONES.

**Glycitin** The β-glycoside form (isomer in which glucose is attached to molecule at the 7 position of the A ring) of the isoflavone known as glycitein (aglycone form). See also ISOFLAVONES, ISOMER, GLYCITEIN.

**Glycoalkaloids** See ALKALOIDS.

**Glycobiology** The study of the involvement (function) of sugars in biological processes. See also GLUCOSE (GLc), GLUCOSE OXIDASE, GLYCOGEN, GLYCOLIPID, GLYCOLYSIS, GLYCOPROTEIN, GLYCOSIDASES, GLYCOSIDE, GLYCOSYLATION.

**Glycocalyx** A polysaccharide matrix involved (in some microorganisms) in firm attachment of the organism to a solid surface.

**Glycoform** One of several molecular arrangements that a given glycoprotein can possess [varieties are determined by the attachment of various oligosaccharide(s)]. Some glycoforms of a given glycoprotein may exhibit greater or lesser biological activity (e.g., pharmaceutical effectiveness for biotherapeutic glycoproteins) because the oligosaccharide units of the glycoprotein molecule mediate interactions of the glycoprotein with the cells of the body. See also GLYCOPROTEIN, OLIGOSACCHARIDES.

**Glycogen** A polymer of glucose with a branching, tree-like molecular structure. It is the chief storage form of carbohydrates in animals. In mammals, glycogen is stored mainly in the liver and muscles. Its molecular weight may be several million. See also GLUCOSE (GLc), GLUCAGON, MOLECULAR WEIGHT.

**Glycolipid** A lipid containing at least one carbohydrate group within its molecule. See also LIPIDS, GLYCOPROTEIN, GLYCOSYLATION, GLYCOLYSIS.

**Glycolysis** A metabolic process in which sugars are broken down into smaller compounds with the release of energy. This series of chemical reactions is found in plant and animal cells as well as in many microorganisms. Except for the final reaction in the series, the chemical reaction pathway of glycolysis is the same as that for fermentation. See also GLUCOSE (GLc), METABOLISM, FERMENTATION.

**Glycoprotein** A conjugated protein containing at least one carbohydrate (oligosaccharide) group within its molecule. A commonly occurring category of glycoproteins found in nature is called mucoproteins. These are protein-polysaccharide compounds that occur in the tissues, particularly in mucous secretions. Other glycoproteins include lymphokines (e.g., interleukins), hormones (e.g., somatotropins), receptors (e.g., GP120), enzymes (e.g., tissue plasminogen activator), and some therapeutics (e.g., CD4PE40). See also GLYCOFORM, CONJUGATED PROTEIN, GP120 PROTEIN, CONJUGATE, PROTEIN, OLIGOSACCHARIDES, POLYSACCHARIDES.

**Glycoprotein C** A blood-clot regulating glycoprotein. See also PROTEIN C, GLYCOPROTEIN.

**Glycoprotein Remodeling** The use of restriction endoglycosidases to (enzymatically) remove sugar (i.e., oligosaccharide) "branches" from glycoprotein (i.e., part protein, part oligosaccharide) molecules. One reason to perform such glycoprotein remodeling would be to remove one or more oligosaccharide branches so that the glycoprotein is less or no longer antigenic (i.e., triggers an immune response). This allows the glycoprotein to be injected into the body (e.g., for pharmaceutical purposes) without incurring an unwanted immune response. See also GLYCOPROTEIN, RESTRICTION ENDOGLYCOSIDASES, ENZYME, OLIGOSACCHARIDES, ANTIGEN, CELLULAR IMMUNE RESPONSE, HUMORAL IMMUNITY, ANTIBODY, EPITOPE, HAPTEN.

**Glycosidases** Enzymes that catalyze the cleavage (hydrolysis) of glycosidic molecular bonds. For example, lysozyme (an enzyme found in human tears) lyses (cuts up) certain bacteria by cleaving the (β configuration) glycosidic linkages (bonds) between the monosaccharide units that (when linked) comprise the polysaccharide component of the bacterial cell walls. A bacterial cell devoid of a cell wall usually bursts. See also ENDOGLYCOSIDASE, EXOGLYCOSIDASE, RESTRICTION ENDOGLYCOSIDASES.

**Glycoside** Any of a group of compounds that yield sugar molecules on hydrolysis. All parts of a glycoside compound may be sugar molecules, so that sucrose, raffinose, starch, and cellulose — all of which hydrolyze into sugar molecules — may all be considered to be glycosides. However, the name (glycoside) is usually applied to a compound in which part of the molecule is not a sugar. This nonsugar component is called the aglycon. See also HYDROLYSIS, FRUCTAN.

**Glycosinolates** See GLUCOSINOLATES.

**Glycosylation** (to glycosylate) Addition of oligosaccharide units (e.g., to protein molecules). The oligosaccharide units are linked to either asparagine side chains by *N*-glycosidic bonds or to serine and threonine side chains by *O*-glycosidic bonds. See also OLIGOSACCHARIDES, PROTEIN, GOLGI BODIES, PLANTIBODIES™, BACULOVIRUS.

**Glycosyltransferases** A class of enzymes (transferases) that catalyze the addition (chemical reaction) of specific sugars (molecular groups) to oligosaccharides, glycoproteins, or glycosides. See also OLIGOSACCHARIDES, MONOSACCHARIDES, ENZYME, GLYCOPROTEIN, GLYCOSIDE, TRANSFERASES.

**Glyphosate** An active ingredient in some herbicides, it kills plants (e.g., weeds) by inhibiting the crucial plant enzyme EPSP synthase. See also ENZYME, EPSP SYNTHASE, CP4 EPSPS, GLYPHOSATE OXIDASE, GLYPHOSATE-TRIMESIUM, GLYPHOSATE ISOPROPYLAMINE SALT, GA21.

**Glyphosate Isopropylamine Salt** One of several forms of an active ingredient utilized in some glyphosate-based herbicides. See also GLYPHOSATE, EPSP SYNTHASE, CP4 EPSPS, GLYPHOSATE OXIDASE, GLYPHOSATE-TRIMESIUM.

**Glyphosate Oxidase** An enzyme that (via catalysis) chemically breaks down glyphosate (i.e., the active ingredient in some herbicides). Glyphosate oxidase is produced in nature by acclimated microorganisms. In 1988, Michael Heitkamp discovered a strain of *Pseudomonas* bacteria which possessed a gene (GO) that caused those particular *Pseudomonas* bacteria to produce unusually large amounts of glyphosate oxidase. That GO gene can be incorporated into a variety of crop plants (soybean, cotton, etc.) in order to help enable those plants to survive post-emergence applications of glyphosate-containing herbicides. Additionally, a plant can be genetically engineered to survive post-emergence applications of glyphosate-containing and/or sulfosate-containing herbicides via insertion of gene (cassette) for plant production of the enzyme CP4 EPSPS. See also ENZYME, ACCLIMATIZATION, STRAIN, *PSEUDOMONAS FLUORESCENS*, GENE, GENETIC ENGINEERING, BACTERIA, MICROORGANISM, SOYBEAN PLANT, EPSP SYNTHASE, CP4 EPSPS, CASSETTE, GLYPHOSATE, SULFOSATE, GA21.

**Glyphosate Oxidoreductase** An enzyme naturally produced in one strain of the microorganism *Ochrobactrum anthropi*. That enzyme (by catalysis) chemically breaks down glyphosate (the active ingredient in some herbicides). If a gene (called goxv247) that codes for the production of glyphosate oxidoreductase is inserted via genetic engineering into crop plants, that would help enable such plants to survive post-emergence applications of glyphosate- and/or sulfosate-containing herbicides. Additionally, a plant can be genetically engineered to survive post-emergence applications of glyphosate-and/or sulfosate-containing herbicides via insertion of gene (cassette) for plant production of the enzyme CP4 EPSPS. See also ENZYME, STRAIN, MICROORGANISM, GENE, GENETIC ENGINEERING, EPSP SYNTHASE, CP4 EPSPS, CASSETTE, GLYPHOSATE, SULFOSATE.

**Glyphosate-Trimesium** One of several forms of active ingredient utilized in some glyphosate-based herbicides. See also GLYPHOSATE, EPSP SYNTHASE, CP4 EPSPS, GLYPHOSATE OXIDASE, GLYPHOSATE ISOPROPYLAMINE SALT, GA21.

**Gm Fad2-1** A (plant) gene that codes for delta 12 desaturase ($\Delta$ 12). See also GENE, DELTA 12 DESATURASE, COSUPPRESSION.

**GMAC** Acronym for the Genetic Manipulation Advisory Committee of the country of Australia, which advises the Australian government on matters pertaining to genetic engineering (e.g., new rDNA product approvals). The GMAC is analogous to Germany's ZKBS (Central Commission on Biological Safety), Brazil's CTNBio (National Technical Biosafety Commission), and the Kenya Biosafety Council. See also GENE TECHNOLOGY REGULATOR (GTR), ZKBS (CENTRAL COMMISSION ON BIOLOGICAL SAFETY), RECOMBINANT DNA ADVISORY COMMITTEE (RAC), GENETIC ENGINEERING, rDNA, DEOXYRIBONUCLEIC ACID (DNA), CTNBio, KENYA BIOSAFETY COUNCIL, GENE TECHNOLOGY OFFICE, INTERIM OFFICE OF THE GENE TECHNOLOGY REGULATOR (IOGTR).

**GMO** Genetically manipulated organism, or genetically modified organism. See also GENE, GENE SPLICING, GENETIC ENGINEERING.

**GMP** See GOOD MANUFACTURING PRACTICES (GMP).

**GMP Guanylate** See G-PROTEINS.

**GMPP** See GENETICALLY MODIFIED PEST PRO-
TECTED (GMPP) PLANTS.

**GMS** Genetically modified soya. See also GMO,
SOYBEAN PLANT.

**GNE** Group of National Experts on Safety in
Biotechnology. The group of people within
the OECD that developed OECD's guidelines for nations to utilize in their safety
evaluations of foods derived from biotechnology. See also ORGANIZATION FOR ECONOMIC
COOPERATION AND DEVELOPMENT (OECD), BIO-
TECHNOLOGY, GENETIC ENGINEERING.

**GO Gene** See GLYPHOSATE OXIDASE.

**Golden Rice** A biotechnology-derived rice
(*Oryza sativa*) created in the 1990s by Ingo
Potrykus and Peter Beyer, which contains
large amounts of beta carotene (precursor of
vitamin A) in its seeds. The human body
converts beta carotene into vitamin A. Potrykus and Beyer utilized *Agrobacterium
tumefaciens* bacteria to genetically engineer
rice plants (by inserting the following genes
from daffodil and from the bacterium
*Erwinia uredovora*:

1. Phytoene synthase — from daffodil
   (*narcissus*) which converts geranylger-
   anyl-diphosphate into phytoene.
2. "CRTL" gene — from *Erwinia uredovora*, which codes for phytoene
   desaturase, which causes the rice plant
   to convert phytoene (a "light harvesting" carotenoid involved in photosynthesis) into lycopene (a carotenoid
   which is then utilized by the rice plant
   in the production of beta carotene).
3. Lycopene beta-cyclase — from daffodil, which converts lycopene into beta
   carotene.

The United Nations (UNICEF) estimates
that 1 to 2 million deaths of children aged
1–4 years old could be prevented annually
around the world, if they received a little
more vitamin A daily in their diet (e.g., via
such a rice). Some of the diseases caused by
lack of vitamin A include: childhood blindness (estimated to afflict 350,000–500,000
children per year); coronary heart disease;
certain cancers (cancer of the lungs, prostate,
etc.); macular degeneration, a leading cause
of blindness in older people; and various
childhood diseases which result in death
(due to a weakened immune system).

Research indicates that, when commercialized in the future, "golden rice" will also
contribute more iron (bioavailable) to the
human diet. That will be due to inserted
genes for ferritin (an iron-rich storage protein) and phytase. Because iron deficiency
anemia (IDA) is a major cause of maternal
and childhood illnesses in developing countries, such a reduction in IDA via consumption
of this rice could confer major health benefits
to those countries' populations. See also BIO-
TECHNOLOGY, BETA CAROTENE, VITAMIN, PHYTO-
CHEMICALS, NUTRACEUTICALS, CAROTENOIDS,
GENE, GENETIC ENGINEERING, BACTERIA, *AGRO-
BACTERIUM TUMEFACIENS*, PHOTOSYNTHESIS,
LYCOPENE, CORONARY HEART DISEASE (CHD),
IRON DEFICIENCY ANEMIA (IDA), PROTEIN,
PHYTASE, PATHWAY, METABOLIC PATHWAY, MET-
ABOLIC ENGINEERING.

**GoldenRice™** A registered trademark now
owned by the company Syngenta AG. See
also GOLDEN RICE.

**Golgi Apparatus** See GOLGI BODIES.

**Golgi Bodies** (also known as Golgi complexes)
First described by Camillo Golgi in 1898,
these are the primary "sorting centers" of
cells, and the mechanism for glycosylation
of (i.e., adding oligosaccharide and polysaccharide branches onto) proteins, before those
proteins are then transported by transfer vesicles to lysosomes, secretory vesicles, or the
plasma membrane. In plant cells, Golgi complexes are where complex polysaccharides
are "sorted" and assembled in preparation
for making the cell wall (located just outside
the cell's plasma membrane). Visually, a
Golgi complex is a stack of flattened membranous sacs (usually 6 sacs in mammal cells
and 20 sacs in plant cells). See also GLYCO-
SYLATION, CELL, OLIGOSACCHARIDES, POLYSAC-
CHARIDES, PROTEIN, LYSOSOME, VESICLES, PLASMA
MEMBRANE.

**Golgi Complexes** See GOLGI BODIES.

**Good Laboratory Practice for Nonclinical
Studies (GLPNC)** The Good Laboratory
Practice (GLP) that is required by the U.S.
Food and Drug Administration (FDA) for
studies of the safety and toxicological effects

of new drugs for livestock. See also GOOD LABORATORY PRACTICES (GLP), NADA.

**Good Laboratory Practices (GLP)** A set of rules and regulations issued by the Food and Drug Administration (FDA) that establishes broad methodological guidelines for procedures and record keeping. They are to be followed in laboratories involved in the testing and/or preparation of pharmaceuticals. GLPs also apply to the Environmental Protection Agency (EPA) (e.g., in toxicity testing of new herbicides).

**Good Manufacturing Practices (GMP)** The set of general methodologies, practices, and procedures mandated by the Food and Drug Administration (FDA) which is to be followed in the testing and manufacture of pharmaceuticals. The purpose of GMPs is essentially to provide for record keeping, and in a wider context to protect the public. GMP guidelines exist instead of specific regulations due to the newness of the technology, and may later be superceded (modified) due to further advances in technology and understanding. See also cGMP.

**Gossypol** A yellow pigment produced in glands and seeds of the cotton plant (*Gossypium* spp.), and some other plants. When consumed by monogastric animals (e.g., swine, poultry, etc.), gossypol is somewhat toxic to those animals. See also COTTON, PHYTOTOXIN.

**GP120 Protein** An adhesion molecule (glycoprotein) on the envelope (surface membrane) of HIV (i.e., AIDS-causing) viruses that directly interacts with the CD4 protein on helper T cells; enabling the HIV viruses to bind to and infect helper T cells. In 1994, a group at America's Scripps Research Institute led by Dennis Burton and Carlos Barbas III announced that they had generated a recombinant human antibody to the GP120 protein; which neutralized more than 75% of HIV isolates against which it was tested. This advance holds the potential to someday lead to a vaccine against AIDS. See also MONOCLONAL ANTIBODIES (MAb), HUMAN IMMUNODEFICIENCY VIRUS TYPE 1 (HIV-1), HUMAN IMMUNODEFICIENCY VIRUS TYPE 2 (HIV-2), ACQUIRED IMMUNE DEFICIENCY SYNDROME (AIDS), SOLUBLE CD4, CD4 PROTEIN, HELPER T CELLS (T4 CELLS), CD44 PROTEIN, ADHESION MOLECULE, CONSERVED, GLYCOPROTEIN, SELECTINS, LECTINS, PROTEIN.

**GPA1** A gene, found in most plants, responsible for controlling water retention and cell division in those plants. The GPA1 gene codes for a G-protein, which transmits/regulates signals (light, temperature, phytohormones, nutrients, etc.) controlling the plant's development.

During 2001, Alan Jones and colleagues discovered that "knocking out" (silencing) the GPA1 gene caused the (then-resultant) G-protein to be insensitive to abscisic acid. Because abscisic acid is a phytohormone (plant hormone) utilized by plants to control the size of stomatal pores [i.e., the openings in leaves through which plants exchange oxygen and carbon dioxide (and also water inadvertently) with the atmosphere], the "knocked-out GPA1" plants wilted due to uncontrolled water loss to the atmosphere. See also GENE, CELL, MITOSIS, G-PROTEINS, PLANT HORMONE, ABSCISIC ACID, KNOCKOUT (GENE).

**GPCRs** Acronym for G-Protein-Coupled Receptors. See also G-PROTEIN-COUPLED RECEPTORS.

**Graft-Versus-Host Disease (GVHD)** The rejection of transplanted organs by the recipient's immune system. Also known as hyperacute rejection. It is caused by the attack of the recipient's T lymphocytes (T cells, a certain class of white blood cells) on the transplanted organ. The recipient's T cells are able to distinguish between self and foreign cells, and are hence able to recognize the foreign (nonself) cells of the transplanted organ. They then, naturally, try to destroy the "foreign invaders" in the body. This constitutes rejection of the transplanted organ. From this it should be understood that there is nothing wrong with the body, but that it is behaving exactly as it should. See also CELLULAR IMMUNE RESPONSE, HUMORAL IMMUNITY, XENOGENEIC ORGANS, FIBROBLASTS, CYCLOSPORIN.

**Gram Molecular Weight** The weight in grams of a compound that is numerically equal to its molecular weight; the weight of one mole ($6.02 \times 10^{23}$ molecules). See also MOLECULAR WEIGHT, MOLE.

**Gram Stain** Devised by Hans Christian Joachim Gram in 1884, this is a test that illuminates the composition/makeup of the physical structure of the cell wall of bacteria being tested. It is utilized to judge the effectiveness of a given chemical compound (e.g., an antibiotic) against bacteria types. The test consists of a differential staining procedure, which allows most bacteria to be visually separated into two groups, known as Gram-Positive (G+) and Gram-Negative (G-).

An antibiotic is defined in terms of the group of (pathogenic) bacteria that it is effective against, which is known as that antibiotic's "spectrum of activity." An antibiotic is said to have a spectrum of activity against gram-positive bacteria, gram-negative bacteria, or the bacteria of both groups. An antibiotic that is effective against both groups of bacteria is termed "broad spectrum" or "wide spectrum." See also BACTERIA, GRAM-POSITIVE (G+), GRAM-NEGATIVE (G-), PATHOGENIC, CELL, ANTIBIOTIC.

**Gram-Negative (G-)** Pertaining to one of the most important ways of classifying bacteria by means of the differences in the way they stain. The set of bacteria that are not able to be stained (blue) when treated with the gram staining procedure. Gram negativity (and gram positivity) is conferred not by the chemical constituents of the bacteria, but rather by the physical structure of the bacteria cell wall. The staining procedure involves the staining of all cells in a sample with a blue dye. Gram-negative bacteria have a very thin peptidoglycan cell wall (capsule). Hence, the washing procedure, which is an integral part of the overall staining procedure, washes out the blue dye (known as crystal violet). This leaves the gram-negative bacteria colorless. The cells are then stained with a red acidic counterstain (dye) such as acid fuchsin or safranine. After treatment with counterstain, the gram-negative cells are red and the gram-positive cells are blue. See also GRAM-POSITIVE (G+), BACTERIA, CELL, GRAM STAIN.

**Gram-Positive (G+)** Pertaining to bacteria, holding the color of the primary stain (blue) when treated with Gram's stain (a commercial staining agent), or Gentian violet solution. In contrast to the gram-negative bacteria, the gram-positive bacteria possess a much thicker peptidoglycan cell wall (capsule). Because of this, the blue crystal violet dye (with which the bacteria were stained) does not wash out of the cell and the bacteria appear blue under the microscope. See also GRAM-NEGATIVE (G-), BACTERIA, CELL, GRAM STAIN, CAPSULE.

**Granulation Tissue** A mixture of proteins and cells produced by the fibroblast growth that results from a wound. See also FIBROBLASTS, PROTEIN.

**Granulocidin** A protein produced by white blood cells, which has demonstrated (in the laboratory) an ability to kill a broad spectrum of pathogens. See also PATHOGEN, PROTEIN.

**Granulocyte Colony Stimulating Factor (G-CSF)** A colony stimulating factor (CSF; a protein) that stimulates production of granulocytes, particularly neutrophils. See also COLONY STIMULATING FACTORS, GRANULOCYTES, NEUTROPHILS.

**Granulocyte-Macrophage Colony Stimulating Factor (GM-CSF)** (or Granulocyte-Monocyte Colony Stimulating Factor) A colony stimulating factor (CSF; a protein) that stimulates production of granulocytes/macrophages/monocytes. See also COLONY STIMULATING FACTORS (CSFs), MACROPHAGE, MONOCYTES.

**Granulocytes** (polymorphonuclear granulocytes) Phagocytic (scavenging, ingesting) cells that are part of the immune system. When their cell nucleus is segmented into lobes and they have granule-like inclusions within their cytoplasm (the neutrophils, eosinophils, and basophils), they are collectively known as polymorphonuclear granulocytes. See also PHAGOCYTE.

**GRAS List** A list of food additives/ingredients considered to be Generally Recognized as Safe, by the U.S. Food and Drug Administration (FDA). This list of additives is judged to be safe by a panel of FDA pharmacologists and toxicologists, who base their judgment upon data that is available for each ingredient. In practice, those additives for which extensive experience of common use in foods (without known ill effects) has been accumulated over time (e.g., common table salt) are often approved by the FDA due more to the "common use factor" than to any toxicology data, per se. See also FOOD AND DRUG ADMINISTRATION (FDA), DELANEY CLAUSE, PHARMACOLOGY, CANOLA.

**Grass Pea** See GLUCOSINOLATES.

**Green Fluorescent Protein** A protein that is naturally present within the jellyfish *Aequorea victoria*. Green fluorescent protein (GFP) is utilized by scientists to "mark" certain endpoints in experiments (at which point the green light signals that endpoint was reached). See also FLUORESCENCE, PROTEIN, GENE EXPRESSION MARKERS.

**GRF** See GROWTH HORMONE RELEASING FACTOR.

**GRH** See GROWTH HORMONE RELEASING FACTOR.

**Group of National Experts on Safety in Biotechnology** See GNE.

**Growth (microbial)** An increase in the number of cells. See also GENERATION TIME.

**Growth Curve** The change in the number of cells in a growing culture as a function of time. See also GENERATION TIME.

**Growth Factor** A specific substance that must be present in the organism's tissues (when *in vivo*) or growth medium (when *in vitro*) in order for the growth-factor-specific cells to grow/multiply. See also FIBROBLAST GROWTH FACTOR (FGF), NERVE GROWTH FACTOR (NGF), EPIDERMAL GROWTH FACTOR (EGF), VASCULAR ENDOTHELIAL GROWTH FACTOR (VEGF), ANGIOGENIC GROWTH FACTORS, ANGIOGENIN, BONE MORPHOGENETIC PROTEINS (BMP).

**Growth Hormone (GH)** A hormone produced by the anterior pituitary gland. This hormone is a protein (somatotropin) and can be obtained from the bodies of animals, or produced by genetically engineered microorganisms. Its major action in humans (human growth hormone) is a generalized stimulation of skeletal growth. However, human growth hormone (HGH) is also known to affect the growth of other tissues, to be important in fat, protein, and carbohydrate metabolism, and to enhance the effects of various other hormones. See also BOVINE SOMATOTROPIN (BST), PORCINE SOMATOTROPIN (PST), PITUITARY GLAND.

**Growth Hormone-Releasing Factor (GRF or GHRF)** Also termed growth hormone-releasing hormone (GRH). A factor that causes the release of growth hormone, it is 44 amino acids in length. See also GROWTH HORMONE (GH), GROWTH FACTOR, AMINO ACID, HORMONE.

**GT-AG Rule** Describes the presence of these constant dinucleotides at the first two and last two positions of introns of nuclear genes. See also INTRON, GENE.

**GT/PT Correlation** Abbreviation for Genotype/Phenotype Correlation. See also GENOTYPE, PHENOTYPE.

**GTO** Abbreviation for Gene Technology Office. See also GENE TECHNOLOGY OFFICE.

**GTP** See GMP.

**GTPases** Guanosine triphosphatases. These are G-proteins (enzymes) which are crucial for growth, movement, and maintenance of the cell's shape. When active, GTPases are bound to cell membranes (surfaces) by an isoprene molecule (receptor). See also G-PROTEINS, ENZYME, CELL, PHOSPHORYLATION, RECEPTORS, PROTEIN.

**GTR** See GENE TECHNOLOGY REGULATOR (GTR).

**GTS** Glyphosate tolerant soybean. See also HERBICIDE-TOLERANT CROP, SOYBEAN PLANT, CP4 EPSPS, GLYPHOSATE.

**GTS** Glufosinate-ammonium tolerant soybean. See also HERBICIDE-TOLERANT CROP, SOYBEAN PLANT, PAT GENE, GLUFOSINATE.

**Guanine** A purine base. It occurs naturally as a fundamental component of nucleic acids. See also PURINE, NUCLEIC ACIDS.

**GURTs** See GENETIC USE RESTRICTION TECHNOLOGIES.

**GUS** See GUS GENE.

**GUS Gene** A gene that codes for production of β-glucuronidase (i.e., GUS protein) in *Escherichia coli* bacteria. The GUS gene is commonly utilized as a "marker gene" for genetically engineered plants. β-glucuronidase causes a color change, in the presence of the chemical 5-bromo-4-chloro-3-indoyl-beta-D-glucuronic acid, by cleaving ('cutting') a glucuronic acid molecule off the 5-bromo-4-chloro-3-indoyl-beta-D-glucuronic acid. The (remaining) molecule is an insoluble blue dye. See also GENE, CODING SEQUENCE, *ESCHERICHIA COLIFORM* (*E. COLI*), MARKER (GENETIC MARKER), GENETIC ENGINEERING.

**Gut-Associated Lymphoid Tissues (GALT)**
A variety of specialized lymph-reticular tissues that line the inside of an animal's digestive system. GALT include Peyer's Patches, the appendix, and small solitary lymphoid tissues in the gut. They constitute the intestinal immune system (response to antigens). See also LYMPHOCYTE, PEYER'S PATCHES, ANTIGEN, HUMORAL IMMUNITY, CELLULAR IMMUNE RESPONSE, "EDIBLE VACCINES", PLANTIGENS.

G

# H

***H. pylori*** A bacteria that has been linked (e.g., cause) to gastric ulcers and other gastric problems in humans. That link was first announced by Barry Marshall in the early 1990s. See also BACTERIA, *HELICOBACTER PYLORI*.

***H. virescens*** See *HELIOTHIS VIRESCENS* (*H. VIRESCENS*).

***H. zea*** See *HELICOVERPA ZEA* (*H. ZEA*).

**HA** Abbreviation for the word hemagglutinin. See also HEMAGGLUTININ.

**Habitat** The natural environment of an organism within an ecosystem. The place, in an ecosystem, where an organism lives. See also ECOLOGY.

**HAC** See HUMAN ARTIFICIAL CHROMOSOMES (HAC).

**HACCP** See HAZARD ANALYSIS AND CRITICAL CONTROL POINTS (HACCP).

**Hairpin Loop** A section of highly curving, single-stranded DNA or RNA formed when a long piece (string) of the DNA or RNA bends back on itself and hydrogen-bonds (is able to base pair) in some regions to form double-stranded regions. The structure can be visualized by taking a human hair, bending it back on itself and holding it in such a way as to half its original length. The section where the two ends of hair lie next to each other represents the section of double-stranded DNA or RNA. At one end the hair will have to make a sharp turn and will form a loop. This loop represents the single-stranded hairpin loop. See also RIBONUCLEIC ACID (RNA), DEOXYRIBONUCLEIC ACID (DNA).

**Halophile** Microorganisms that require NaCl (salt) for growth (they are called obligate halophiles). Those that do not require it, but can grow in the presence of high NaCl concentrations, are called facultative halophiles. Natural habitats containing high salt concentrations are, for example, the Great Salt Lake in Utah, the Dead Sea in Israel, and the Caspian Sea in Russia. See also HABITAT.

**HAP Gene** See LOW-PHYTATE CORN.

**Haploid** A cell with one set of chromosomes; half as many chromosomes as the normal somatic body cells contain. A characteristic of sex cells. See also GAMETE.

**Haplophase** A phase in the life cycle of an organism in which it has only one copy of each gene. The organism is then said to be haploid. Yeast can exist as true haploids. Humans are haploid for only a few genes and cannot exist as true haploids. See also HAPLOID.

**Haplotype** A subgroup (e.g., an ethnic minority, all members of a genetically related family group, etc.) of organisms (e.g., humans) whose phenotype results in their body responding in the same way to a physical agent (a certain pharmaceutical, a toxin, a food, etc.). For example, more than 70% of black people in North America are lactose intolerant (their bodies cannot metabolize the lactose sugar in cow's milk), but fewer than 19% of Caucasian people in North America are lactose intolerant. Analogous to that, the drugs acetaminophen, aspirin, and Valium remain in the bodies of women (who constitute a haplotype) longer than in the bodies of men. Haplotypes for the $\beta$2-adrenergic gene are predictive of asthma patients' response to the pharmaceutical albuterol. See also PHARMACOGENOMICS, HERITABILITY, HEREDITY, TRAIT, GENETICS, PHENOTYPE, TOXIN, INSULIN, METABOLISM, SINGLE-NUCLEOTIDE POLYMORPHISMS (SNPs).

**Hapten** A small foreign molecule that will stimulate an immune system response (e.g., antibody production) if the small molecule (now called a haptenic determinant) is attached to a macromolecule (carrier) to make it large enough to be recognized by the immune system. See also EPITOPE, CELLULAR IMMUNE RESPONSE, HUMORAL IMMUNITY.

**Haptoglobin** A protein which is a component in human blood; that can occur in one of two different molecular forms (i.e., a "large" version of that molecule or a "small" version of that molecule). The "small" version of haptoglobin is very effective at capturing and removing free radicals (high-energy oxygen atoms which bear an "extra" electron) from the bloodstream before they damage tissues (e.g., in the eye, kidneys, and/or arteries). The "large" version of haptoglobin, which is the only haptoglobin molecule in the bloodstream of one particular haplotype (genetic subgroup) of people, is not effective at capture/removal of those free radicals (e.g., generated at a high rate in people with diabetes disease), so diabetics within that particular haplotype tend to suffer extreme damage to eyes, kidneys, and arteries (sometimes necessitating limb amputation). See also FREE RADICAL, HAPLOTYPE, INSULIN, OXIDATIVE STRESS.

**Hardening** See COLD HARDENING, HYDROGENATION.

**Harpin** A protein naturally produced by the *Erwinia amylovora* bacteria (which usually causes the plant disease known as fire blight in apple trees, pear trees, and some ornamental plants of the rose family). Discovered in 1992 by Zhong-Min Wei and colleagues, harpin causes numerous species of plants to initiate a protective/defensive response (cascade) against bacteria, viruses, fungi, and some insects and nematodes. Harpin also causes plants (i.e., that it is sprayed onto) to increase their photosynthesis and to have increased root growth/proliferation; which can lead to greater crop yields. See also PROTEIN, BACTERIA, PHYTOALEXINS, PATHOGENESIS RELATED PROTEINS, SIGNALING, SIGNALING MOLECULE, SIGNAL TRANSDUCERS AND ACTIVATORS OF TRANSCRIPTION (STATs), SALICYLIC ACID (SA), JASMONIC ACID, SYSTEMIC ACQUIRED RESISTANCE (SAR), CASCADE, R GENES, NEMATODES.

**Harvesting** A term used to describe the recovery of microorganisms from a liquid culture (in which they have been grown by man). This is usually accomplished by means of filtration or centrifugation. See also MICROORGANISM, CULTURE MEDIUM, ULTRACENTRIFUGE, DIALYSIS.

**Harvesting Enzymes** Enzymes that are used to gently dissociate (break apart) cells in living tissues in order to produce single, separate cells that can then be established and propagated in a cell culture reactor. Harvesting enzymes are also used to dissociate cells that have been grown for some time in a cell culture reactor. See also CELL CULTURE, MAMMALIAN CELL CULTURE, ENZYME, CULTURE MEDIUM.

**Hazard Analysis and Critical Control Points (HACCP)** A quality control program (for food processing) to systematically prevent hazards (e.g., pathogens) from entering the production process. HACCP was initially developed in the 1950s by the Pillsbury Company to supply food products for astronauts in America's space program. Under HACCP, food processors/handlers must analyze and identify in advance the points where hazards are most likely to occur, and eliminate them. For example, because melons lie in pathogen-contaminated dirt while growing, a "critical control point" for restaurants serving sliced melon is cleansing of the knife after each melon is cut (to prevent the knife carrying pathogens from one infected melon to other melons). See also PATHOGEN, RAPID MICROBIAL DETECTION (RMD).

**Heat-Shock Proteins** See STRESS PROTEINS.

**Heavy-Chain Variable (VH) Domains** The regions (domains) of the antibody (molecule's) "heavy chain" that vary in their amino acid sequence. The "chains" (of atoms) comprising the antibody (immunoglobulin) molecule consist of a region of variable (V) amino acid sequence and a region in which the amino acid sequence remains constant (C). An antibody molecule possesses two antigen binding sites, and it is the variable domains of the light (VL) and heavy (VH) chains which contribute to this antigen binding ability. See also ANTIBODY, PROTEIN, IMMUNOGLOBULIN, SEQUENCE (OF A PROTEIN MOLECULE), ANTIGEN, AMINO ACID, COMBINING SITE, DOMAIN (OF A PROTEIN), LIGHT-CHAIN VARIABLE (VL) DOMAINS.

**Hedgehog Proteins** Signaling molecules (consisting of "signaling protein" with cholesterol molecule attached to it), that direct/control tissue differentiation during mammal embryo development (into various organs, limbs, etc.). The signaling protein (within an embryo cell) cleaves itself into two peptides, one of which then acts as a transferase (i.e., enzyme that catalyzes the addition of a functional group to a given molecule — in this case to the other "hedgehog peptide"). When the cell then secretes the cholesterol/peptide molecule, the cholesterol (functional group) "anchors" it to the cell surface, while the "signaling protein" end of the cholesterol/peptide directs differentiation of nearby cells. See also PROTEIN, SIGNALING MOLECULES, SIGNALING, CHOLESTEROL, SIGNAL TRANSDUCTION, PEPTIDE, CELL, TRANSFERASES, ENZYME, FUNCTIONAL GROUP, CELL DIFFERENTIATION.

*Helicobacter pylori* Bacteria. See also *H. PYLORI*.

*Helicoverpa zea (H. zea)* Known as the corn earworm (when it is on corn plants), and known as the tomato fruitworm (when it is on tomato plants), this is one of three insect species that is called "bollworms" (when on cotton plants). *H. zea* chews on those crop plants, and is one of the insects that can act as a vector (carrier) of *Aspergillus flavus* fungus. In 1997, scientists at the U.S. Department of Agriculture created/optimized a monoclonal antibody against *Helicoverpa zea* vitellin, which thus holds potential to be used as a means to control that insect. See also *B.t. KURSTAKI*, *HELIOTHIS VIRESCENS* (*H. VIRESCENS*), FUNGUS, *PECTINOPHORA GOSSYPIELLA*, *ASPERGILLUS FLAVUS*, CORN, MONOCLONAL ANTIBODIES (MAb).

*Heliothis virescens (H. virescens)* Known as the tobacco budworm (when it is on tobacco plants), this is one of three insect species that is called "bollworms" (when they are on cotton plants). As part of Integrated Pest Management (IPM), farmers can utilize the parasitic *Euplectrus comstockki* wasp to help control the tobacco budworm/cotton bollworm. When that wasp's venom is injected into *Heliothis* larva, it stops the larva from molting (and thus maturing). See also *B.t. KURSTAKI*, *HELICOVERPA ZEA* (*H. ZEA*), *PECTINOPHORA GOSSYPIELLA*, *INTEGRATED PEST MANAGEMENT* (*IPM*).

**Helix** A spiral, staircase-like structure with a repeating pattern described by two simultaneous operations (rotation and translation). It is one of the natural conformations exhibited by biological polymers. See also BIOMIMETIC MATERIALS, ANALOGUE.

**Helper T Cells (T4 cells)** T cells (lymphocytes) which bind B cells (upon recognizing a foreign epitope on B cell surface). The binding stimulates B cell proliferation by secreting B cell growth factor. See also B CELLS, CYTOKINES, T CELL, T CELL RECEPTORS, SUPPRESSOR T CELLS.

**Hemagglutinin (HA)** A special protein that some viruses utilize to gain entry into the cells they have "targeted." The HA protein helps the virus adhere to the cell it targets. Hemagglutinin is also utilized to refer to specific plant cell proteins (lectins) that are naturally produced by certain plants such as the soybean plant (*Glycine max* (L) Merrill). The presence of those lectin molecules (e.g., on surfaces of root cells of the soybean plant) help nitrogen-fixing *Rhizobium japonicum* bacteria to adhere to soybean plant roots, where they begin to "fix nitrogen" (i.e., create natural nitrate fertilizer, which improves the soil and helps plants to grow). See also PROTEIN, VIRUS, CELL, LECTINS, SOYBEAN PLANT, NITROGEN FIXATION, BACTERIA, NITRATES, NODULATION.

**Hematologic Growth Factors (HGF)** A class of colony stimulating factors (proteins) that stimulates bone marrow cells to produce certain types of red and white blood cells. Some colony stimulating factors are:

1. Granulocyte-macrophage colony stimulating factor (GM-CSF)
2. Granulocyte-monocyte colony stimulating factor
3. Granulocyte colony stimulating factor (GM-CSF)
4. Erythropoietin (EPO)
5. Interleukin-3 (IL-3)
6. Macrophage colony stimulating factor (M-CSF)

**Hematopoietic Growth Factors** Growth factors that stimulate the body to produce blood cells. See also GROWTH FACTOR, INTERLEUKIN-6 (IL-6).

H

**Hematopoietic Stem Cells** Certain stem cells present (e.g., in infants' bodies, and in the umbilical cords of newborn infants), that can be differentiated (via chemical signals in the growing body) to give rise to red blood cells and the infection-fighting cells of the immune system. See also STEM CELLS, MULTI-POTENT ADULT STEM CELLS, MESODERMAL ADULT STEM CELLS, CELL, ORGANISM, SIGNALING.

**Heme** The iron-porphyrin prosthetic group of a class of proteins called "heme proteins." See also PROSTHETIC GROUP, CHELATING AGENT, PROTEIN, TRANSFERRIN.

**Hemoglobin** An oxygen-transporting respiratory pigment; it is present in humans, animals, and some plants (e.g., land plants that withstand occasional immersion/flooding). In humans, hemoglobin is carried in the red blood cells (erythrocytes), and is responsible for the red color of the blood. It is composed of two pairs of identical polypeptide chains and iron-containing heme groups, comprising the (total) hemoglobin molecule. The molecular structure of hemoglobin was determined by Max Perutz in 1959. A human disease known as sickle-cell anemia is caused by a (genetically induced) small change in the hemoglobin molecule's structure (in victims of that disease). See also HEME, POLYPEPTIDE (PROTEIN), GENETICS, BILIRUBIN, HEREDITY, ERYTHROCYTES, PROTEIN STRUCTURE.

**Hemostasis** See FIBRIN.

**Heparin** A polysaccharide sulfuric acid ester found in the liver, lung, and other tissues that prolongs the clotting time of blood by preventing the formation of fibrin. Used in vascular surgery and in treatment of postoperative thrombosis and embolism. See also FIBRIN, THROMBOSIS.

**HER-2 Gene** Abbreviation for Human Epidermal growth factor Receptor-2 gene, an oncogene that is responsible for approximately 30% of breast cancers (i.e., in those women whose body over-expresses that particular oncogene, and it spreads via metastaticism). In addition to conventional treatments (mastectomy, chemotherapy, etc.), the U.S. Food and Drug Administration (FDA) in 1998 approved use of a humanized monoclonal antibody (trastuzumab) to be utilized alone, or in combination with, certain chemotherapy agents (e.g., paclitaxel) against such metastatic breast cancers. That monoclonal antibody attaches to the extracellular domain (i.e., portion of the Her-2 receptor sticking out of the surface of breast tissue cells) and down-regulates the Her-2 gene, i.e., resulting in fewer Her-2 receptors being produced on the plasma membrane surfaces of that woman's breast tissue cells. See also GENE, RECEPTORS, HER-2 RECEPTOR, RAS GENE, EGF RECEPTOR, ONCO-GENES, CANCER, EXPRESS, EXPRESSIVITY, MONO-CLONAL ANTIBODIES (MAb), BRCA GENES, PACLITAXEL, FOOD AND DRUG ADMINISTRATION (FDA), PLASMA MEMBRANE.

**HER-2 Protein** See HER-2 RECEPTOR.

**HER-2 Receptor** An epidermal growth factor receptor (protein molecule embedded in the surface of cells) that is present in abundance attached to the plasma membrane surface of breast tissue cells in humans possessing the HER-2 gene. See also RECEPTORS, EPIDERMAL GROWTH FACTOR RECEPTOR, PLASMA MEMBRANE, HER-2 GENE.

**Herbicide Resistance** See HERBICIDE-TOLERANT CROP.

**Herbicide-Resistant Crop** See HERBICIDE-TOL-ERANT CROP.

**Herbicide-Tolerant Crop** Crop plants, cultivated by man, that have been altered to be able to survive application(s) of one or more herbicides by the incorporation of certain gene(s), via either genetic engineering, natural mutation, or mutation breeding (i.e., soaking seeds in mutation-causing chemicals, or bombardment of seeds with ionizing radiation, to cause random genetic mutations, followed by selection of the particular mutation in which herbicide-tolerance occurs).

Because it has been utilized for decades, most relevant national laws consider mutation breeding to be one of the so-called "traditional plant breeding" techniques. For example, European laws that require special labeling of food products containing genetically engineered (via rDNA) crops do not require such special labeling for food products that contain crops created via mutation breeding. Several crops (soybean, canola, cotton, etc.) are made tolerant to glyphosate- or sulfosate-containing herbicides by the insertion (via genetic engineering techniques) of the aroA transgene (cassette) for CP4 EPSPS. Corn (maize) is made tolerant to glyphosate-containing herbicides by insertion (via genetic engineering techniques) of the mEPSPS or GA21 transgene (cassette). Some soybean varieties are made tolerant to sulfonylurea-based herbicides by adding (via traditional breeding methods) the ALS gene (which confers the sulfonylurea-tolerance trait). Corn (maize) and rice (*Oryza sativa*) are made tolerant to imidazolinone-containing herbicides by adding (via traditional breeding techniques) the imidazolinone-tolerance trait. That trait is imparted by the T-Gene, IT-Gene, or the IR-Gene. See also GENE, GENETIC ENGINEERING, CASSETTE, TRANSGENIC, DEOXYRIBONUCLEIC ACID (DNA), rDNA, EPSP SYNTHASE, GLYPHOSATE OXIDASE, PAT GENE, BAR GENE, GENETICS, GLYPHOSATE, GA21, SULFOSATE, ALS GENE, EPSP SYNTHASE, CP4 EPSPS, CHLOROPLAST TRANSIT PEPTIDE (CTP), ACURON™ GENE, TRANSGENE, TRAIT, CANOLA, SOYBEAN PLANT, CORN, MUTATION BREEDING, TRADITIONAL BREEDING METHODS.

**Heredity** Transfer of genetic information from parent cells to progeny. See also INFORMATIONAL MOLECULES, GENE, GENETIC CODE, GENOME, GENETICS, GENOTYPE, DEOXYRIBONUCLEIC ACID (DNA), HERITABILITY, QUANTITATIVE TRAIT LOCI (QTL).

**Heritability** The fraction of variation (of an individual's given trait) that is due to genetics. For example, if a pig's trait (e.g., weight at birth) is 30% heritable, that means that 30% of the (birthweight) difference between that individual pig and its (statistically representative) group of contemporaries (pigs) is due to genetics. The other 70% would be due to factors such as nutrition of the mother during pregnancy, etc. See also HEREDITY, TRAIT, GENETICS, INFORMATIONAL MOLECULES, GENE, GENETIC CODE, GENOME, GENOTYPE, DEOXYRIBONUCLEIC ACID (DNA), QUANTITATIVE TRAIT LOCI (QTL).

**Hetero-** A chemical nomenclature prefix meaning "different." For example, a heterocyclic compound is one with a (ring) structure made up of more than one kind of atom. A heterokaryon refers to a cell containing nuclei of different species. See also HETEROCYCLIC, HETERODUPLEX, HETEROGENEOUS (CATALYSIS), HETEROGENEOUS (CHEMICAL REACTION), HETEROGENEOUS (MIXTURE), HETEROKARYON, HETEROLOGOUS PROTEINS, HETEROLOGOUS DNA, HETEROLOGY, HETEROSIS, HETEROTROPH, HETEROZYGOTE.

**Heterocyclic** See HETERO-.

**Heteroduplex** A DNA molecule, the two strands of which come from different individuals so that there may be some base pairs or blocks of base pairs that do not match. Can arise from mutation, recombination, or by annealing DNA single strands *in vitro*. See also DEOXYRIBONUCLEIC ACID (DNA).

**Heterogeneous (catalysis)** Catalysis occurring at a phase boundary, usually a solid-fluid interface. See also HETERO-, HETEROGENEOUS (MIXTURE), CATALYST.

**Heterogeneous (chemical reaction)** A chemical reaction in which the reactants are of different phases: for example, gas with liquid, liquid with solid, or a solid catalyst with liquid or gaseous reactants. See also HETERO-, HETEROGENEOUS (CATALYSIS), CATALYST.

**Heterogeneous (mixture)** One that consists of two or more phases such as liquid-vapor, or liquid-vapor-solid. See also HETERO-.

**Heterokaryon** A fused cell containing nuclei of different species. See also NUCLEOID.

**Heterologous DNA** Refers to a DNA molecule in which each of the (double) strands is from different sources (e.g., different species). See also DEOXYRIBONUCLEIC ACID (DNA), HETERO-, SPECIES.

**Heterologous Proteins** Those proteins produced by an organism that is not the wild type source of those proteins. For example, bacteria have been genetically engineered to produce human growth hormone and bovine (i.e., cow) somatotropin. See also PROTEIN, WILD TYPE, GROWTH HORMONE (GH), BOVINE SOMATOTROPIN (BST), HOMOLOGOUS PROTEIN.

**Heterology** A sequence of amino acids in two or more proteins that are not identical to each other. See also AMINO ACID, PROTEIN, HOMOLOGY.

**Heterosis** Also known as "hybrid vigor." See also F1 HYBRIDS.

**Heterotroph** An organism that obtains nourishment from the ingestion and breakdown of organic matter.

H

**Heterozygote** An individual organism with different alleles at one or more particular loci. See also ALLELE.

**Hexadecyltrimethylammonium Bromide (CTAB)** A solvent that is widely utilized to dissolve plant DNA samples (e.g., when a scientist wants to sequence that sample of plant DNA). CTAB solvent helps the scientist to separate out contaminants that are commonly present in samples from plant tissues (polysaccharides, quinones, etc.) because DNA molecules are much more soluble in CTAB than are the contaminant molecules. See also DEOXYRIBONUCLEIC ACID (DNA), POLYSACCHARIDES, SEQUENCING (OF DNA MOLECULES), SDS.

**Hexose** See GLUCOSE (GLc).

**HF Cleavage** A research process in which hydrofluoric acid is used to sequentially remove side-chain protective groups from peptide chains. Also used to remove the resin support from peptides that have been prepared via solid-phase peptide synthesis. The HF cleavage reaction is a temperature-dependent process. See also PROSTHETIC GROUP, SYNTHESIZING (OF PROTEINS).

**High-Amylose Corn** Refers to those corn (maize) hybrids that produce kernels in which the starch that is contained within those kernels is at least 50% amylose, versus the average of 24–28% amylose in traditional corn starch. See also CORN, STARCH, AMYLOSE.

**High-Density Lipoproteins (HDLPs)** So-called "good" cholesterol, it consists of lipoproteins that can help move excess low-density lipoproteins ("bad" cholesterol, which can clog arteries) out of the human body by binding to the low-density lipoproteins (also known as LDL cholesterol) in the blood and then attaching to special LDLP receptor molecules in the liver. The liver then clears those (bound) low-density lipoproteins out of the body as a part of regular liver functions.

Studies have shown that humans having high bloodstream levels of HDLPs will offset high levels of LDLPs (e.g., the HDLPs can still help lower the risk of developing coronary heart disease). Since cholesterol does not dissolve in water (which constitutes most of the volume of blood), the body makes HDL cholesterol into little "packages" surrounded by a hydrophilic ("water loving") protein. That protein "wrapper" is known as apolipoprotein A-1, or apo A-1, and it enables HDL cholesterol to be transported in the bloodstream because the apolipoprotein A-1 is attracted to water molecules in the blood. See also LOW-DENSITY LIPOPROTEINS (LDLP), RECEPTORS, APOLIPOPROTEINS, WATER SOLUBLE FIBER, CHOLESTEROL, CORONARY HEART DISEASE (CHD).

**High-Glutenin Wheat** See GLUTEN.

**High-Isoflavone Soybeans** Developed in the U.S. in the 1990s, these are soybean varieties which contain greater content of isoflavones than do traditional soybean varieties (i.e., isoflavones constitute 0.15–0.3% of a traditional variety soybean's dry weight). Consumption of isoflavones helps to reduce the blood level of low-density lipoproteins ("bad cholesterol") in humans. A human diet containing a large amount of isoflavones helps prevent osteoporosis, causes reduced risk of certain cancers (breast cancer, prostate cancer, endometrial cancer, etc.), and decreases risk of prostate enlargement. See also ISOFLAVONES, SOYBEAN PLANT, CHOLESTEROL, CANCER, PROSTATE-SPECIFIC ANTIGEN (PSA), LOW-DENSITY LIPOPROTEINS (LDLP), OSTEOPOROSIS.

**High-Lactoferrin Rice** Refers to rice plants (*Oryza sativa*) which have been genetically engineered to produce substantial amounts of lactoferrin in the grain they yield. Lactoferrin is a compound that is naturally produced in human breast milk. Consumption of lactoferrin by infants helps to strengthen their immune system. Consumption of lactoferrin (e.g., from genetically engineered rice) by older humans helps their immune systems to resist some infectious diseases. Lactoferrin "binds" free iron (e.g., in body fluids), thereby denying that iron to pathogenic bacteria (which need free iron to grow/infect). Lactoferrin also promotes intestinal cell growth in humans. See also GENETIC ENGINEERING, PATHOGEN, BACTERIA, VALUE-ENHANCED GRAINS, GROWTH (MICROBIAL), CELL.

**High-Laurate Canola** Refers to canola (*Brassica napus/campesris*) varieties genetically engineered (e.g., via insertion of gene for lauroyl-ACP thioesterase) to produce at least 40% laurate (lauric acid) in their oil (in seed). See also LAURATE, CANOLA, GENETIC ENGINEERING, FATTY ACID, LAUROYL-ACP THIOESTERASE, VALUE-ENHANCED GRAINS.

**High-Lysine Corn** Developed in the U.S. in the mid-1960s, these were initially corn (maize) varieties possessing the opague-2 gene. The opague-2 gene causes such corn to contain 0.30–0.55% lysine (i.e., 50–80% more than traditional No. 2 yellow corn). Other genes have subsequently been discovered that, when inserted into the corn/maize genome (e.g., via genetic engineering techniques), cause production of larger amounts of lysine than in traditional corn/maize varieties. High-lysine corn is particularly useful for feeding of swine, since traditional No. 2 yellow corn does not contain enough lysine for optimal swine growth. See also CORN, LYSINE (lys), GENE, OPAGUE-2, GENETIC ENGINEERING, GENOME, VALUE-ENHANCED GRAINS, "IDEAL PROTEIN" CONCEPT, MAL (MULTIPLE ALEURONE LAYER) GENE.

**High-Methionine Corn** Developed in the U.S. in the mid-1960s, these were initially corn (maize) varieties possessing the floury-2 gene. The floury-2 gene causes such corn to contain slightly higher levels of methionine than traditional No. 2 yellow corn. Other genes have subsequently been discovered that, when inserted into corn/maize genome (e.g., via genetic engineering techniques), cause production of larger amounts of methionine than in traditional corn/maize varieties. High-methionine corn is particularly useful for feeding of poultry, since traditional No. 2 yellow corn does not contain enough methionine for optimal poultry (especially feather) growth. See also METHIONINE (met), CORN, FLOURY-2, GENE, GENOME, GENETIC ENGINEERING, VALUE-ENHANCED GRAINS, OPAGUE-2, "IDEAL PROTEIN" CONCEPT, MAL (MULTIPLE ALEURONE LAYER) GENE.

**High-Oil Corn** Conceived in 1896 at the University of Illinois in the U.S., high-oil corn (HOC) is defined to be corn (maize) possessing a kernel oil content of 5.8% or greater. Traditional No. 2 yellow corn varieties tend

to contain 4.5% or less oil content. See also VALUE-ENHANCED GRAINS, CORN, CHEMOMETRICS.

**High-Oleic Oil Soybeans** Soybeans from plants which have been genetically engineered to produce soybeans bearing oil that contains more than 70% oleic acid, instead of the typical 24% oleic acid content of soybean oil produced from traditional varieties of soybeans. Cosuppression, via inserted gene for $\Delta$ 12 desaturase (an enzyme that normally converts oleic acid to linoleic acid as part of the oil creation process in traditional varieties of soybean plants), causes the higher than traditional amount of oleic acid in the soybean oil. High-oleic soybean oil would tend to have greater oxidative stability (especially at elevated temperatures) than soybean oil from traditional varieties of soybeans. Because of that, nuts that were fried in high-oleic oil have been shown to possess a longer shelf life than nuts fried in traditional oils. A human diet containing a large amount of oleic acid causes lower blood cholesterol level, and thus lower risk of coronary heart disease (CHD). See also SOYBEAN PLANT, SOYBEAN OIL, FATTY ACID, OLEIC ACID, MONOUNSATURATED FATS, GENETIC ENGINEERING, DELTA 12 DESATURASE, CHOLESTEROL, CORONARY HEART DISEASE (CHD), PALMITIC ACID, COSUPPRESSION, ENZYME, LINOLEIC ACID.

**High-Phytase Corn and Soybeans** Crop plants that have been genetically engineered to contain in their grain/seed high(er) levels of the enzyme phytase (which aids digestion and absorption of phosphate in that grain/seed). High-phytase grains or oilseeds are particularly useful for the feeding of swine and poultry, since traditional No. 2 yellow corn (maize) or traditional soybean varieties do not contain phytase in amounts needed for complete digestion/absorption of phosphate naturally contained in those traditional soybeans and corn (maize) in the form of phytate. See also PHYTASE, ENZYME, PHYTATE, VALUE-ENHANCED GRAINS, LOW-PHYTATE CORN, LOW-PHYTATE SOYBEANS.

**High-Stearate Canola** Canola varieties which have been genetically engineered so their seeds contain a higher percentage of stearate (also called stearic acid) in the canola oil than the typical stearate content in canola oil produced from traditional canola varieties. Cosuppression, via inserted gene for D-stearoyl-ACP desaturase (i.e., enzyme that normally converts stearic acid to to oleic acid in the oil creation process in traditional varieties of canola), causes the higher than traditional amount of stearic acid in the canola oil. See also CANOLA, STEARATE, SATURATED FATTY ACIDS (SAFA), GENE, GENETIC ENGINEERING, VALUE-ENHANCED GRAINS, FATTY ACID, COSUPPRESSION, ENZYME, OLEIC ACID, STEAROYL-ACP DESATURASE, CHOLESTEROL, CORONARY HEART DISEASE (CHD).

**High-Stearate Soybeans** Soybean plant varieties which have been bred or genetically engineered so their beans contain at least 12% stearate (also known as stearic acid) within their soybean oil (i.e., more than four times the typical 3% stearic acid content in the soybean oil produced from traditional soybean varieties). Some high-stearate soybeans contain more than 20% stearate. Cosuppression, via inserted gene for D-stearoyl-ACP desaturase (i.e., enzyme that normally converts stearic acid to oleic acid in the oil creation process in traditional varieties of soybeans), is the primary way to cause the higher than traditional amount of stearic acid in the resultant soybean oil. A human diet containing stearate instead of alternative saturated fatty acids, does not cause an increase in blood cholesterol levels (whereas human consumption of the other saturated fatty acids causes bloodstream cholesterol levels to increase, which increases risk of coronary heart disease). See also STEARATE, VALUE-ENHANCED GRAINS, SOYBEAN PLANT, SOYBEAN OIL, GENE, GENETIC ENGINEERING, FATTY ACID, COSUPPRESSION, ENZYME, OLEIC ACID, CHOLESTEROL, SATURATED FATTY ACIDS (SAFA), CORONARY HEART DISEASE (CHD), STEAROYL-ACP DESATURASE.

**High-Sucrose Soybeans** Another name for low-stachyose soybeans because the soybeans replace the (reduced) stachyose with (additional) sucrose. See also LOW-STACHYOSE SOYBEANS, STACHYOSE, VALUE-ENHANCED GRAINS, SOYBEAN PLANT, SUGAR MOLECULES.

**High-Throughput Identification** Determination of the identification of a given chemical compound (e.g., within a mixture), the desired impact (cell apoptosis, etc.), a specific segment (sequence) of DNA (i.e., a specific gene), a specific ligand or receptor (e.g., "attaching" itself to a given molecule), etc. within the overall process known as high-throughput screening. See also HIGH-THROUGHPUT SCREENING (HTS), COMBINATORIAL CHEMISTRY, BIOCHIPS, CELL, APOPTOSIS, GENE, DEOXYRIBONUCLEIC ACID (DNA), GENE EXPRESSION, TARGET-LIGAND INTERACTION SCREENING, RECEPTORS, CHARACTERIZATION ASSAY, SEQUENCE (OF A DNA MOLECULE), GENE EXPRESSION ANALYSIS, *CAENORHABDITIS ELEGANS* (*C. ELEGANS*), MOLECULAR BEACON.

**High-Throughput Screening (HTS)** A methodology utilized to quickly screen large numbers of compounds for use as pharmaceuticals or agrochemicals (e.g., herbicides). For example, when screening chemical compounds for potential use as a pharmaceutical, the goal often is to assess differences between diseased and treated cells; enabling identification of a pharmaceutical candidate that favorably impacts change in protein level (i.e., gene expression) which characterizes a diseased state, or some other gene expression marker (e.g., apoptosis).

When screening compounds for potential use as herbicide active ingredients, the goal is to assess differences between normal and treated weed plant cells; enabling identification of a potential herbicide candidate that imparts desired (fatal) change. Although whole living cells or whole microscopic animals such as nematodes could be utilized in HTS, it is more common to use a proxy (e.g., receptors, enzymes, or STATs from applicable cells) whose interaction with candidate compounds can be inferred to cell (and/or organism) effects. See also COMBINATORIAL CHEMISTRY, BIOCHIP, TARGET-LIGAND INTERACTION SCREENING, CELL, ORGANISM, CHARACTERIZATION ASSAY, PROTEIN, GENE, GENE EXPRESSION, HIGH-THROUGHPUT IDENTIFICATION, RECEPTORS, GENE EXPRESSION ANALYSIS, BIOASSAY, GENE EXPRESSION MARKERS, SIGNAL TRANSDUCERS AND ACTIVATORS OF TRANSCRIPTION (STATs), APOPTOSIS, *IN SILICO* SCREENING, NEMATODES, *CAENORHABDITIS ELEGANS* (*C. ELEGANS*), ENZYME, NORTHERN BLOT ANALYSIS, MOLECULAR BEACON.

**Highly Available Phosphate Corn (maize)**
See LOW-PHYTATE CORN.

**Highly Available Phosphorous (HAP) Gene**
See LOW-PHYTATE CORN.

**Highly Unsaturated Fatty Acids (HUFA)**
Refers to a number of unsaturated fatty acids (e.g., that the human body forms from poly-unsaturated fatty acids it consumes in diet) containing four or more double (molecular) bonds; i.e., arachidonic acid, docosahex-anoic acid, eicosapentanoic acid. These HUFAs are utilized (by the human body) to make prostaglandins and other eicosanoids. See also POLYUNSATURATED FATTY ACIDS (PUFA), UNSATURATED FATTY ACIDS, ESSENTIAL FATTY ACIDS, CORONARY HEART DISEASE (CHD), N-3 FATTY ACIDS, N-6 FATTY ACIDS, DOCOSAHEX-ANOIC ACID (DHA), EICOSAPENTANOIC ACID (EPA), ARACHIDONIC ACID (AA), PROSTAGLANDIN ENDOPEROXIDE SYNTHASE.

**Histamine** A base that is naturally present in ergot (a fungus) and plants; it is also natu-rally produced by basophils (basophilic leu-kocytes) in the human body. It is formed from histidine by decarboxylation, and is held to be responsible for the dilation and increased permeability of blood vessels which play a major role in allergic reactions. See also BASE, HISTIDINE (HIS), BASOPHILS.

**Histidine (his)** A basic amino acid that is essential in the nutrition of the rat. It is formed by the decomposition of most pro-teins (as globin). See also PROTEIN.

**Histiocyte** See MACROPHAGE.

**Histoblasts** See B LYMPHOCYTES.

**Histones** Proteins rich in basic amino acids (e.g., lysine) found complexed with chromo-somes of all eucaryotic cells except sperm, where the DNA is specifically complexed with another group of basic proteins, the protamines. See also CHROMOSOMES, CHROMA-TIDS, CHROMATIN.

**Histopathologic** Refers to changes in tissue caused by a disease. For example, certain diseases (e.g., jaundice) cause the skin to turn yellow. See also PATHOGENIC, VIRUS, CANCER, ADHESION MOLECULE.

**HIV-1 and HIV-2** See HUMAN IMMUNODEFI-CIENCY VIRUS TYPE 1 (HIV-1), HUMAN IMMUNO-DEFICIENCY VIRUS TYPE 2 (HIV-2).

**HLA** See HUMAN LEUKOCYTE ANTIGENS.

**HNE** The common chemical (by-)product of lipid oxidation, known as 4-hydroxy-2-non-enal, which is an aldehyde. See also OXIDA-TIVE STRESS, OXIDATION, PLASMA MEMBRANE, LIPIDS.

**HNGF** Human nerve growth factor. See also NERVE GROWTH FACTOR (NGF).

**HOC** See HIGH-OIL CORN.

**Hollow Fiber Separation** (of proteins). The separation of proteins from a mixture by means of "straining" the mixture through hollow, semipermeable fibers (e.g., polysul-fone fibers) under pressure. The hollow fibers are constructed in such a way that they have very tiny (molecular size) holes in them. In this way, large molecules are retained in the original liquid while smaller molecules, which are able to pass through the holes, are filtered out. See also DIALYSIS, PROTEIN, ULTRAFILTRATION.

**Holoenzyme** The entire, functionally com-plete enzyme. The term is used to designate an enzyme that requires a coenzyme in order for it to function (possess catalytic abilities). The holoenzyme consists of the protein part (apoenzyme) plus a dialyzable, nonprotein coenzyme part that is bound to the apoen-zyme protein. See also COENZYME, APOENZYME, DIALYSIS.

**Homeobox** A short sequence of DNA that is 180 base pairs long and located in the 3' exon of certain genes of the *Drosophila* fly (where they were discovered by Walter Gehring dur-ing the 1970s). In the 1980s, Jani Christian Nusslein-Volhard discovered that one homeobox was attached (in adjacent exon) to each of the genes that are responsible for embryonic development (i.e., "switched on" only in an embryo that is developing into an adult), in a wide variety of species including invertebrates, birds, and mammals. Thus, it is now possible to locate many embryonic-development genes in many species by using a DNA probe (made via a *Drosophila* homeobox DNA sequence) to find homeobox sequences attached to those embryonic-development genes. In such a role, the respective homeobox sequences attached to each gene are known as DNA markers. See also GENE, DEOXYRIBONUCLEIC ACID (DNA), DNA PROBE, DNA MARKER, SEQUENCE

H

(OF A DNA MOLECULE), BASE PAIR (bp), *DROSO-PHILA*, EXON, SPECIES.

**Homeostasis** A tendency toward maintenance of a relatively stable internal environment in the bodies of higher animals through a series of interacting physiological processes. An example is the mammal's maintenance of a constant body temperature despite extremes in weather temperature. See also SELECTINS, LECTINS, ADHESION MOLECULE.

**Homing Receptor** Also known as L-selectin. See also SELECTINS, LECTINS, ADHESION MOLE-CULES.

**Homologous (chemically)** See HOMOLOGY.

**Homologous (chromosomes or genes)** Chromosomes or chromosome segments that are identical with respect to their constituent sequence, genetic loci, and/or their visible structure (in the case of chromosomes). So, for example, a gene of "unknown" function in humans could be compared (in a database) with genes of a simpler organization (e.g., *Caenorhabditus elegans*). If the human gene is homologous, and the function of the *Caenorhabditus elegans* gene is known, the function of the human gene could be inferred by comparison. See also CHROMOSOMES, GENE, SEQUENCE (OF A DNA MOLECULE), LOCUS, *CAENORHABDITUS ELEGANS*.

**Homologous Protein** A protein having identical functions and similar properties in different species. For example, the hemoglobins that perform identical functions in the blood of different species.

**Homology** A sequence of amino acids in two or more proteins that are identical to each other. Nucleic-acids homology refers to complementary strands that can hybridize with each other. See also TATA HOMOLOGY, PROTEIN, HYBRIDIZATION (MOLECULAR GENETICS).

**Homotropic Enzyme** An allosteric enzyme whose own substrate functions as an activity modulator. See also ENZYME.

**Homozygote** An organism in which the corresponding genes (alleles) on the two genomes are identical. An organism which possesses an identical pair of alleles in regard to a given (genetic) characteristic. See also GENE, ALLELE, GENOME, GENOTYPE, PHENOTYPE, HOMOZYGOUS, HETEROZYGOTE.

**Homozygous** In a diploid organism, a state where both alleles of a given gene are the same. See also HETEROZYGOTE, ALLELE, DIPLOID, DIPLOPHASE, HOMOZYGOTE.

**Hormone** A type of chemical messenger (peptide), occurring in both plants and animals, that acts to inhibit or excite metabolic activities (in that plant or animal) by binding to receptors on specific cells to deliver its "message." A hormone's site of production is distant from the site of biological activity (i.e., where the message is delivered). See also PEPTIDE, MINIMIZED PROTEINS, SIGNALING, SIGNALING MOLECULE, ALBUMIN.

**Host Cell** A cell whose metabolism is used for growth and reproduction by a virus. Also the cell into which a plasmid is introduced (in recombinant DNA experiments).

**Host Vector (HV) System** The host is the organism into which a gene from another organism is transplanted. The guest gene is carried by a vector (i.e., a larger DNA molecule, such as a plasmid, or a virus into which that gene is inserted) which then propagates in the host.

**Hot Spots** Sites in genes at which events, such as mutations, occur with unusually high frequency. See also GENE, JUMPING GENES, MUTATION, TRANSLOCATION.

**HPLC** High-performance liquid chromatography. See also CHROMATOGRAPHY.

**HSOD** See HUMAN SUPEROXIDE DISMUTASE (hSOD).

**HTC** See HERBICIDE-TOLERANT CROP, STS, PAT GENE, EPSP SYNTHASE, ALS GENE, BAR GENE, CP4 EPSPS, GLYPHOSATE OXIDASE.

**HTS** Herbicide-Tolerant Soybeans. See also SOYBEAN PLANT, GLYPHOSATE, CP4 EPSPS, EPSP SYNTHASE, GLYPHOSATE OXIDASE, HERBICIDE-TOLERANT CROP, STS, GLUFOSINATE, PAT GENE, BAR GENE.

**HTS** See HIGH-THROUGHPUT SCREENING (HTS).

**Human Artificial Chromosomes (HAC)** Chromosomes that have been synthesized (made) from chemicals that are identical to chromosomes within human cells. See also YEAST ARTIFICIAL CHROMOSOMES (YAC), BACTERIAL ARTIFICIAL CHROMOSOMES (BAC), CHROMOSOMES, *ARABIDOPSIS THALIANA*, SYNTHESIZING (OF DNA MOLECULES).

**Human Chorionic Gonadotropin** A human hormone. In 1986, Mark Bogart discovered that elevated levels of human chorionic gonadotropin in pregnant women are correlated with babies (later) born with Down syndrome. See also HORMONE.

**Human Colon Fibroblast Tissue Plasminogen Activator** A second generation tissue plasminogen activator (tPA), which has the clot-sensitive activation of plasminogen with potentially greater selectivity and (clot) specificity. See also TISSUE PLASMINOGEN ACTIVATOR (tPA).

**Human EGF-Receptor-Related Receptor (HER-2)** A gene that appears to be directly related to human breast cancer mortality. The more copies of the HER-2 gene (in a patient's breast tumor cells), the more dismal that patient's prospects for survival.

**Human Embryonic Stem Cells** Those cells (in the early embryo's inner cell mass) from which each of the human body's 210 different types of tissues arise via differentiation, proliferation, and growth processes. See also PLURIPOTENT, STEM CELL GROWTH FACTOR (SCF), DIFFERENTIATION.

**Human Gamma-Glutamyl Transpeptidase** A glycoprotein that is thought to possess a different oligosaccharide when it is produced by a (liver) tumor cell instead of a healthy cell. Thus, it is a possible early warning marker for liver cancer. See also GLYCOPROTEIN, OLIGOSACCHARIDES.

**Human Growth Hormone (HGH)** See GROWTH HORMONE (GH).

**Human Immunodeficiency Virus Type 1 (HIV-1)** One of the two "families" of the viruses identified (so far) which cause acquired immune deficiency syndrome (AIDS), although not all strains of HIV-2 cause AIDS. HIV-1 and HIV-2 show a preferential tropism (affinity) toward the helper T cells, although other immune system (and nervous system) cells are also infected. The GP120 envelope (surface) protein of HIV-1 and HIV-2 directly interacts (binds) with the CD4 proteins (receptors) on the surface of helper T cells, enabling the viruses to bind (attach to) and infect the helper T cells.

In order to successfully enter and infect cells, the HIV must also bind with CKR-5 proteins (receptors) located on the surface of cells of most humans. In 1996, Nathaniel

Landau and Richard Koup discovered that approximately 1% of humans carry a gene for a version of CKR-5 receptor that resists entry to cells by HIV. As of 1996, a total of nine separate strains (serotypes) of human immunodeficiency virus were known; identified by the letters A, B, C, D, E, F, G, H, I. See also CD4 PROTEIN, TAT, TATA HOMOLOGY, ADHESION MOLECULE, GP120 PROTEIN, ACQUIRED IMMUNE DEFICIENCY SYNDROME (AIDS), RECEPTORS, TROPISM, HELPER T CELLS (T4 CELLS), STRAIN, T CELL RECEPTORS, VIRUS, SEROTYPES, HUMAN IMMUNODEFICIENCY VIRUS TYPE 2 (HIV-2).

**Human Immunodeficiency Virus Type 2 (HIV-2)** See HUMAN IMMUNODEFICIENCY VIRUS TYPE 1 (HIV-1).

**Human Leukocyte Antigens (HLA)** A very complex array of six proteins that cover the surface of leukocytes (and the bone marrow cells that produce leukocytes). These HLA are usually different (i.e., a nonmatch) for individuals that are not genetically related to each other (e.g., a father-son or a father-daughter), so have been used in the past to prove paternity. HLA must also be matched (as nearly as possible) for successful bone marrow transplants, to prevent the donated bone marrow (and the marrow recipient) from "rejecting" each other. See also LEUKOCYTES, ANTIGEN, MAJOR HISTOCOMPATIBILITY COMPLEX (MHC), PROTEIN, GRAFT-VERSUS-HOST DISEASE (GVHD).

**Human Protein Kinase C** An enzyme that is involved in the control of blood coagulation and fibrinolysis. See also FIBRIN.

**Human Superoxide Dismutase (hSOD)** An enzyme that "captures" oxygen free radicals (oxygen atoms bearing an extra electron, thus high in energy: e.g., which are sometimes generated in a biological system such as within the body of an organism). Oxygen free radicals are generated within occluded blood vessels when a blood clot blocks arteries in the heart, causing a heart attack. These oxygen free radicals are highly energized and can cause damage to blood vessel walls after the clot is dissolved (e.g., with tissue plasminogen activator), so hSOD may profitably be administered in conjunction with clot-dissolving pharmaceuticals to minimize

**H**

damage when occluded arteries are reopened.

Research indicates that hSOD may help protect elderly patients from the lethal effects of influenza (the flu), because influenza often causes overproduction of free radicals in the victim's body. Recent research indicates that hSOD may be made more effective when administered in combination with certain copper/zinc compounds to bolster its efficacy. See also FREE RADICAL, PEG-SOD (POLYETHYLENE GLYCOL SUPEROXIDE DISMUTASE), CATALASE, XANTHINE OXIDASE, TISSUE PLASMINOGEN ACTIVATOR (tPA), ANTIOXIDANTS.

**Human Thyroid-Stimulating Hormone (hTSH)** A naturally occurring hormone that causes the thyroid gland to develop. See also HORMONE.

**Humoral Immune Response** Refers to the rapid manufacture and secretion by the body of the soluble blood serum components — e.g., antibodies (by B cells), complement proteins, cecrophins, etc. — in response to an infection. See also ANTIBODY, COMPLEMENT, COMPLEMENT CASCADE, CECROPHINS, HUMORAL IMMUNITY.

**Humoral Immunity** The immune system response consisting of the soluble blood serum components that fight an infection (antibodies, complement proteins, cecrophins, etc.). See also ANTIBODY, COMPLEMENT, COMPLEMENT CASCADE, CECROPHINS, CELLULAR IMMUNE RESPONSE, IMMUNOGLOBULIN.

**HuSNPs** Abbreviation for Human SNPs (single-nucleotide polymorphisms). See also SINGLE-NUCLEOTIDE POLYMORPHISMS (SNPs).

**Hybrid Vigor** See F1 HYBRIDS, HYBRIDIZATION (PLANT GENETICS).

**Hybridization (molecular genetics)** The pairing (tight physical bonding) of two complementary single strands of RNA and/or DNA to give a double-stranded molecule. See also ANNEAL, STICKY ENDS, RIBONUCLEIC ACID (RNA), MESSENGER RNA (mRNA), BIOSENSORS (ELECTRONIC), BIOSENSORS (CHEMICAL), HYBRIDIZATION SURFACES, DNA PROBE, DEOXYRIBONUCLEIC ACID (DNA), ANTISENSE (DNA SEQUENCE), BIOMOTORS.

**Hybridization (plant genetics)** The mating of two plants from different species or genetically very different members of the same species to yield hybrids (first filial hybrids)

possessing some of the characteristics of each parent. Those (hybrid) offspring tend to be more healthy, productive, and uniform than their parents — a phenomenon known as "hybrid vigor." Hybrids can also arise from more than two ("parent") species. Hybrid corn/maize seed was first commercialized (in the U.S.) in 1922. Other recently created crop hybrids include tangelos (produced by crossing grapefruit with tangerines), nectarines (bred from peaches), and brocciflower (produced by crossing broccoli with cauliflower).

Some hybrids have occurrred spontaneously in nature. For example, wheat (*Triticum aestivum*) arose centuries ago from a naturally occurring interbreeding of three Middle East grasses. In the 1980s, sugar beet (*Beta vulgaris*, subspecies *vulgaris*) naturally interbred with the wild native weed known as sea beet (*Beta vulgaris*, subspecies *maritima*) in Europe, resulting in an annual weed (in contrast to sugar beet, which is a biannual). Because that new hybrid weed is closely related to sugar beet, any herbicide that kills the new hybrid weed is likely to harm the sugar beet crop (unless the sugar beet crop is made herbicide-tolerant). See also F1 HYBRIDS, SPECIES, TRANSGRESSIVE SEGREGATION, GENETICS, CORN, WHEAT, GEM, EXOTIC GERMPLASM, BARNASE, HERBICIDE-TOLERANT CROP.

**Hybridization Surfaces** Various physical substrates (surfaces) onto which have been "attached" genetic materials (DNA, RNA, oligonucleotides, etc.). Relevant complementary genetic materials (DNA, RNA, oligonucleotides, etc.) then are hybridized onto those attached-to-surface genetic materials for various specific purposes (e.g., detection of the presence of those unattached genetic materials, in the case of biosensor's hybridization surface). One of the technologies that can be utilized to assay (evaluate) DNA from hybridization surfaces is Matrix-Assisted Laser Desorption Ionization Time of Flight Mass Spectrometry (MALDI-TOF-MS). See also SUBSTRATE (STRUCTURAL), HYBRIDIZATION (MOLECULAR GENETICS), COMPLEMENTARY DNA (c-DNA), DEOXYRIBONUCLEIC ACID (DNA), RIBONUCLEIC ACID (RNA), NANOCRYSTAL MOLECULES, DOUBLE HELIX, BIOSENSORS (ELECTRONIC),

BIOSENSORS (CHEMICAL), BIOCHIPS, OLIGONUCLE-
OTIDE, OLIGONUCLEOTIDE PROBES, MALDI-TOF-MS,
ASSAY, MICROARRAY (TESTING).

**Hybridoma** The cell line produced by fusing a myeloma (tumor cell) with a lymphocyte (which makes antibodies); it continues indefinitely to express the immunoglobulins (antibodies) of both parent cells. See also MONOCLONAL ANTIBODY (MAb), AGING.

**Hydrazine** A chemical with formula $N_2H_4$. Used as a rocket fuel, and in the hydrazinolysis of glycoproteins. See also HYDRAZINOLYSIS (OF GLYCOPROTEINS, TO ISOLATE UNREDUCED OGLIOSACCHARIDE SIDE CHAINS), GLYCOPROTEIN, REDUCTION (IN A CHEMICAL REACTION).

**Hydrazinolysis** (of glycoproteins to isolate unreduced oligosaccharide side chains) A technique that used the chemical hydrazine to separate and isolate the oligosaccharide portion from the protein portion of a glycoprotein. The hydrazine chemically "chews up" the polypeptide (i.e., protein) portion of a glycoprotein molecule, leaving the intact oligosaccharides behind. It can subsequently be analyzed (after chromatographic separation from the peptide pieces and other chemical components). See also REDUCTION (IN A CHEMICAL REACTION), HF CLEAVAGE, POLYPEPTIDE (PROTEIN), GLYCOPROTEIN, SEQUENCING (OF OLIGOSACCHARIDES), HYDRAZINE, CHROMATOGRAPHY.

**Hydrofluoric Acid Cleavage** See HF CLEAVAGE.

**Hydrogenation** A chemical reaction/process in which hydrogen atoms are added to molecules (e.g., of unsaturated fatty acids) in edible oils. In the case of fatty acids, the fraction of each isomeric form (*trans* vs. *cis* fatty acids) and the molecular chain length (of the fatty acids present) have a large impact on the melting characteristics of each (fat or oil), with shorter-chain fats melting at lower temperature.

Hydrogenation is the most common chemical reaction utilized in the edible oils (processing) industry. Hydrogenation increases the solids (i.e., crystalline fat) content of edible fats/oils, and improves their resistance to thermal and atmospheric oxidation (e.g., for frying of foods). Those increases in solids and resistance to oxidation result from the reduction in the fat/oil relative unsaturation, plus increased geometric and positional isomerization of the fat/oil molecules. The edible oil/fat hydrogenation reaction is accomplished by treating fats/oils with pressurized hydrogen gas in the presence of a catalyst. As a result, the (usually) liquid oils are converted to more-saturated fats, which are semisolids at an ambient temperature of 72°F (22°C).

The presence of *trans* fatty acids in hydrogenated edible oils can be reduced significantly via changes in catalyst, temperature, pressure, etc. used in the hydrogenation reaction. In general, natural oils and fats possessing melting points lower than 121°F (50°C) are nearly completely absorbed in the digestive system of typical humans. See also FATTY ACID, MONOUNSATURATED FATS, SATURATED FATTY ACIDS, DEHYDROGENATION, ESSENTIAL FATTY ACIDS, LAURATE, LECITHIN, TRIGLYCERIDES, UNSATURATED FATTY ACID, SOYBEAN OIL, CONJUGATED LINOLEIC ACID (CLA), OXIDATION, ISOMER, STEREOISOMERS, CATALYST, SUBSTRATE (CHEMICAL), *TRANS* FATTY ACIDS.

**Hydrolysis** Literally, means "cleaved by water." It is used for a chemical reaction in which the chemical bond attaching an atom, or group of atoms to the (rest of the) molecule is cleaved, followed by attachment of a hydrogen atom at the same chemical bond.

**Hydrolytic Cleavage** A chemical reaction in which a portion (e.g., an atom or a group of atoms) of a molecule is "cut" off the molecule via hydrolysis. See also HYDROLYSIS.

**Hydrolyze** To "cut" a chemical bond (i.e., with a molecule) via hydrolysis. See also HYDROLYSIS.

**Hydrophilic** This term means water loving or having a great affinity for water. It is used to describe molecules or portions of molecules that have an affinity for water. The property of having an affinity for water at an oil-water interface. For example, ordinary sugar that dissolves readily in water is said to be hydrophilic (i.e., a molecule that is "water loving"). See also AMPHIPHILIC MOLECULES.

**Hydrophobic** This term means water hating or having a great dislike for water. It is used to describe molecules or portions of molecules that have very little or no affinity for water. The property of having an affinity for oil (nonpolar environments) at an oil-water

**H**

interface. For example, a nonpolar hydrocarbon such as butane (as used in lighters) that will not dissolve in water, but which will dissolve (be miscible) in oil is said to be hydrophobic (i.e., a molecule that is "water hating"). See also AMPHIPHILIC MOLECULES.

**Hydroxylation Reaction** A chemical reaction in which one or more hydroxyl groups (i.e., the -OH group) is introduced (i.e., is chemically attached) to a molecule.

**Hyperacute Rejection** See GRAFT-VERSUS-HOST DISEASE (GVHD).

**Hyperchromicity** The increase in optical density that occurs when DNA is denatured. See also DEOXYRIBONUCLEIC ACID (DNA), DENATURED DNA, OPTICAL DENSITY (OD).

**Hypersensitive Response** A protective/defensive response by certain plants to "infection" by plant pathogens (bacteria, fungi, etc.), in which those plant cells that are immediately adjacent (to the infected area of plant) are "instructed" to self-destruct via apoptosis, in order to cordon off the infected area (to prevent further spread of the infection). The initiation of the hypersensitive response is often triggered by signaling molecules that are produced by the pathogens themselves. For example, one particular protein produced by the soil fungus triggers a hypersensitive response that often is so severe that the entire plant dies. See also PATHOGENESIS RELATED PROTEINS, PROTEIN, PATHOGEN, BACTERIA, FUNGUS, CELL, APOPTOSIS, SIGNALING, SIGNALING MOLECULE.

**Hyperthermophilic** (organisms) See also THERMOPHILE, THERMOPHILIC BACTERIA.

**Hypostasis** Interaction between nonallelic genes in which one gene will not be expressed in the presence of a second. See also EPISTASIS, GENE, EXPRESS, ALLELE.

**Hypothalamus** A part of the brain structure, lying near the base of the brain, it regulates a number of hormones. As a part of the brain, it constantly receives (neurochemical) signals from nerve cells (neurons). The hypothalamus monitors those signals, and converts them into hormonal signals [e.g., it generates a "burst" of hormones in response to certain visual stimuli, certain physical (e.g., sexual) stimuli, etc.]. Also, the hypothalamus is able to monitor and detect changes in the blood levels of hormones coming from endocrine glands. For example, the metabolic hormone insulin (from the pancreas) and the reproductive hormone estrogen (from the ovaries) both trigger changes in function in the hypothalamus.

The hypothalamus regulates biological processes (metabolic rate, appetite, etc.). A major function of the hypothalamus is to control reproduction, via secretion of gonadotropin-releasing hormone (GnRH) from the tips of hypothalamic nerve fibers that extend downward toward (into) the pituitary gland. Similarly, the hypothalamus also helps to control the body's growth (from birth until the end of puberty) via secretion of growth hormone-releasing factor (GHRF) to the pituitary gland. See also HORMONE, ENDOCRINE HORMONES, ENDOCRINE GLANDS, ENDOCRINOLOGY, PITUITARY GLAND, GROWTH HORMONE (GH), NEUROTRANSMITTER, GROWTH HORMONE-RELEASING FACTOR (GHRF).

# I

**IBA** See INDUSTRIAL BIOTECHNOLOGY ASSOCIATION.

**IBG** See INTERNATIONAL BIOTECHNOLOGY GROUP.

**ICAM** Intercellular adhesion molecule. See also ADHESION MOLECULE.

**IDA** Acronym for Iron Deficiency Anemia. See also IRON DEFICIENCY ANEMIA (IDA).

**IDE** "Investigational Device Exemption" application to the Food and Drug Administration (FDA) seeking approval to begin clinical studies of a new medical device.

**Ideal Protein Concept** Refers to the protein content in the feed ration (food) eaten by livestock and poultry (and humans). Feed that contains ideal protein contains protein(s) that — when digested by an animal — yields all of the essential amino acids, in proper proportions, for the growth and/or maintenance needs of that animal. "Ideal protein" varies for different species (e.g., pigs require different amino acids/rations than chickens do). "Ideal protein" varies for different stages in the life of a given animal (e.g., poultry require more sulfur-containing amino acids, such as methionine, during life stages when feather growth is at a comparatively high rate). The animal's requirement for one essential amino acid is proportionally linked to the animal's requirements for another. Increasing the supply (when deficient) of one essential amino acid in the animal's diet would improve that animal's (growth) performance if no other amino acids were limiting. Feed rations formulated to contain "ideal protein" have been shown to reduce the amount of nitrogen (nitrates) excreted by livestock and poultry, by as much as 50%. See also AMINO ACID, PROTEIN, ESSENTIAL AMINO ACIDS, ESSENTIAL NUTRIENTS, METHIONINE (met), DIGESTION (WITHIN ORGANISMS), SOY PROTEIN, HIGH-LYSINE CORN, HIGH-METHIONINE CORN.

**Idiotype** The region of the antibody molecule that enables each antibody to recognize a specific foreign structure (i.e., epitope or hapten) is said to have an idiotype (for that epitope or hapten). An identifying characteristic (or property) of the epitope or hapten that one is talking about. See also EPITOPE, HAPTEN, ANTIGEN, ANTIBODY, CATALYTIC ANTIBODY.

**IDM** See INTEGRATED DISEASE MANAGEMENT.

**IFBC** See INTERNATIONAL FOOD BIOTECHNOLOGY COUNCIL.

**IFN-Alpha** Alpha interferon. See also INTERFERONS.

**IFN-Beta** Beta interferon. See also INTERFERONS.

**IGF-1** See INSULIN-LIKE GROWTH FACTOR-1.

**IGF-2** See INSULIN-LIKE GROWTH FACTOR-2.

**IGF-I** See INSULIN-LIKE GROWTH FACTOR-1.

**IGF-II** See INSULIN-LIKE GROWTH FACTOR-2.

**IL-1** See INTERLEUKIN-1.

**IL-Ira** See INTERLEUKIN-1 RECEPTOR ANTAGONIST.

**Immune Response** See CELLULAR IMMUNE RESPONSE, ANTIBODY, HUMORAL IMMUNITY.

**Immunoassay** The use of antibodies to identify and quantify (measure) substances by a variety of methods. The binding of antibodies to antigen (substance being measured) is often followed by tracers, such as fluorescence or (radioactive) radioisotopes, to enable measurement of the substance. See also ANTIBODY, TRACER (RADIOACTIVE ISOTOPIC METHOD), ANTIGEN, ELISA, RADIOIMMUNOASSAY, ASSAY, EIA, FLUORESCENCE, NEAR-INFRARED SPECTROSCOPY (NIR).

**Immunoconjugate** A molecule that has been formed by attachment to each of two originally different molecules. One of these is generally an antibody; hence, the word "immunoconjugate." Classic organic drug molecules such as methotrexate, adriamycin chlorambucil, etc.; radionuclides; enzymes;

toxins; and ribosome-inhibiting proteins may be conjugated to antibodies. The salient point is that the antibody portion of the conjugate is there to "steer" the biologically active molecule to its target. See also CONJUGATE, "MAGIC BULLET", ANTIBODY, MAGNETIC PARTICLES.

**Immunocontraception** Any process or procedure in which an organism's immune system is utilized to attack or inactivate the reproductive cells (e.g., sperm) within the organism. See also CELLULAR IMMUNE RESPONSE, ANTIBODY, HUMORAL IMMUNITY, GERM CELL.

**Immunogen** See ANTIGEN.

**Immunoglobulin** (IgA, IgE, IgG, and IgM) A class of (blood) serum proteins representing antibodies. Often used, along with the more specific monoclonal antibodies, in health diagnostic reagents. In certain people genetically predisposed to foodborne allergies, immunoglobulin-E (IgE) initiates an immune system response to antigen(s) present on protein molecule(s) in the particular food to which that person is allergic. Severe allergic reactions to foods may lead to death. See also PROTEIN, ANTIGEN, ALLERGIES (FOODBORNE), ANTIBODY, IMMUNOASSAY, B LYMPHOCYTES.

**Immunosuppressive** That which suppresses the immune system response (e.g., certain chemicals). See also CELLULAR IMMUNE RESPONSE, HUMORAL IMMUNITY.

**Immunotoxin** A conjugate formed by attaching a toxic molecule (e.g., ricin) to an agent of the immune system (e.g., a monoclonal antibody), that is specific for the pathogen or tumor to be killed. The immune system-agent portion (of the conjugate) delivers the toxic chemical directly to the specified (disease) site, thus sparing other healthy tissues from the effect of the toxin. See also RICIN, MONOCLONAL ANTIBODIES (MAb), "MAGIC BULLET".

**Imprinting** A cellular process in which certain genes within an organism's cells are "disabled" during the earliest stage(s) of the organism's development. For example, the embryo of a female mammal (which receives two copies of the X chromosome — one from each parent) disables one of those copies, at random, in each of its cells, so the female becomes a genetic mixture of its two parents. See also CELL, GENE, CHROMOSOMES, X CHROMOSOME.

**In Silico** See *IN SILICO* BIOLOGY.

**In Silico Biology** A set of computer modeling technologies, via which researchers can:

1. Create computer models of specific cells to
   a. see how a given disease impacts that cell
   b. see how a given pharmaceutical impacts that cell
2. Create computer models of specific organs to
   a. see how a given disease impacts that organ
   b. see how a given pharmaceutical impacts that organ
3. Create computer models of specific organisms to
   a. see how a given disease impacts that organism
   b. see how a given pharmaceutical then impacts that disease within that organism
4. Create computer models of specific organisms that possess a given genome to
   a. see how a given disease impacts that specific organism/phenotype to
   b. see how a given pharmaceutical then impacts that disease within that organism/phenotype
5. Create computer models of protein "digestion" (i.e., breaking apart into constituent peptides), for comparison with the actual peptides (fragments) that are determined (e.g., via MALDI-TOF-MS) to have resulted from chemical digestion of those protein molecules (e.g., via immersion in trypsin).

See also RATIONAL DRUG DESIGN, RECEPTOR MAPPING, CELL, BIOCHIPS, GENOME, GENOMICS, PHARMACOGENOMICS, PROTEIN, PROTEOMICS, PHENOTYPE, MALDI-TOF-MS, PEPTIDE, TRYPSIN.

**In Silico Screening** A set of computer modeling technologies via which researchers can (vicariously) screen chemical compounds for their potential as pharmaceutical candidate compounds, pesticide candidate compounds, etc. The chemical compounds are "generated" (e.g., from data available about compounds actually created in a laboratory in the past), and computer modeling is then utilized to:

1. Assess their impact on "generated" specific cells, tissues, etc. (from data available about that chemical-type of molecule's impact on that type of cell/tissue when actually tested on it in a laboratory in the past).
2. Generate an analogous chemical compound, that is likely to be more efficacious or have fewer undesirable side effects.
3. Repeat the process.

For example, when screening compounds for potential usefulness as a pharmaceutical, the goal is to assess (modeled/predicted) differences between diseased (untreated) and treated cells; thus enabling prediction of (better) pharmaceutuical candidate compounds for eventual actual testing on real cells/tissues. Some of the more sophisticated *in silico* screening software can even model ADME properties for selected pharmaceutical candidate compounds. See also RATIONAL DRUG DESIGN, IN SILICO BIOLOGY, RECEPTOR MAPPING, CELL, BIOCHIPS, HIGH-THROUGHPUT SCREENING (HTS), COMBINATORIAL CHEMISTRY, PHARMACOGENOMICS, PROTEOMICS, QUANTITATIVE STRUCTURE-ACTIVITY RELATIONSHIP (QSAR), ADME TESTS, TARGET (OF A THERAPEUTIC AGENT), TARGET (OF A HERBICIDE OR INSECTICIDE).

*In situ*    In the natural or original position (e.g., inside the body).

*In vitro*    In an unnatural position (e.g., outside the body, in the test tube). *In vitro* is Latin for in glass. For example, the testing of a substance, or the experimentation in (using) a "dead" cell-free system. See also IN VITRO SELECTION.

*In vitro* Selection    A search process (e.g., for a new pharmaceutical) that first involves the construction of a large "pool" of polynucleotide sequences (at least some of which are likely to possess the desired pharmaceutical properties), synthesized by a totally random process. This is followed by repeated cycles of screening (for those sequences possessing desired properties) and/or enriching, and amplification (of the screened/enriched sequences). Common amplification techniques include Polymerase Chain Reaction (PCR), Ligase Chain Reaction (LCR), Self-sustained Sequence Replication (SSR), Q-beta Replicase Technique, and Strand Displacement Amplification (SDA). See also IN VITRO, AMPLIFICATION, GENE AMPLIFICATION, POLYMERASE CHAIN REACTION (PCR), Q-BETA REPLICASE TECHNIQUE, NUCLEOTIDE, DEOXYRIBO-NUCLEIC ACID (DNA), SYNTHESIZING (OF DNA MOLECULES), OLIGONUCLEOTIDE, DNA PROBE, GENE MACHINE, COMBINATORIAL CHEMISTRY.

*In vivo*    Latin for "in living"; e.g., the testing of a new pharmaceutical substance or experimentation in (using) a living, whole organism. An *in vivo* test is one in which an experimental substance is injected into an animal such as a rat in order to ascertain its effect on the organism. See also MODEL ORGANISM.

*In-vitro* Evolution    See IN VITRO SELECTION.

*In-vitro* Selection    See IN VITRO SELECTION.

**Inclusion Bodies**    See REFRACTILE BODIES (RB).

**IND**    "Investigational New Drug" application to the Food and Drug Administration (FDA) seeking approval to begin clinical studies of a new pharmaceutical. See also "TREATMENT" IND, IND EXEMPTION, PHASE I CLINICAL TESTING, FOOD AND DRUG ADMINISTRATION (FDA).

**IND Exemption**    A permit by the Food and Drug Administration (FDA) to begin clinical trials on humans (of a new pharmaceutical) after toxicity data have been reviewed and approved by the FDA. See also KEFAUVER RULE, IND, PHASE I CLINICAL TESTING.

**Indian Department of Biotechnology**    The governmental body in India that regulates all recombinant DNA research. It is the Indian counterpart of the American government's Recombinant DNA Advisory Committee (RAC), the Australian government's Gene Technology Regulator (GTR), and the French government's Commission of Biomolecular Engineering. See also RECOMBINANT DNA ADVISORY COMMITTEE (RAC), ZKBS (CENTRAL COMMISSION ON BIOLOGICAL SAFETY), GENETIC ENGINEERING, RECOMBINANT DNA (rDNA), RECOMBINATION, BIOTECHNOLOGY, GENE TECHNOLOGY

OFFICE, COMMISSION OF BIOMOLECULAR ENGINEERING, GENE TECHNOLOGY REGULATOR (GTR).

**Induced Fit** A substrate-induced change in the shape of an enzyme molecule that causes the catalytically functional groups of the enzyme to assume positions that are optimal for catalytic activity to occur. See also ENZYME.

**Inducers** Molecules that cause the production of larger amounts of the enzymes involved in the uptake and metabolism of the inducer (such as galactose). Inducers may be enzyme substrates. See also ENZYME, INDUCIBLE ENZYMES, SUBSTRATE (CHEMICAL).

**Inducible Enzymes** Enzymes whose rate of production can be increased by the presence of certain chemical molecules.

**Industrial Biotechnology Association (IBA)** An American trade association of companies involved in biotechnology. Formed in 1981, the IBA tended to consist of the larger firms involved in biotechnology. In 1993, the Industrial Biotechnology Association (IBA) was merged with the Association of Biotechnology Companies (ABC) to form the Biotechnology Industry Organization (BIO). See also ASSOCIATION OF BIOTECHNOLOGY COMPANIES (ABC), BIOTECHNOLOGY INDUSTRY ORGANIZATION (BIO), BIOTECHNOLOGY.

**Informational Molecules** Molecules containing information in the form of specific sequences of different building blocks. They include proteins and nucleic acids. See also HEREDITY, GENE, GENETIC CODE, GENOME, GENOTYPE, NUCLEIC ACIDS, MESSENGER RNA (mRNA), DEOXYRIBONUCLEIC ACID (DNA), RIBONUCLEIC ACID (RNA).

**Ingestion** Taking a substance into the body. For example, the amoeba surrounds a food particle, then ingests the particle.

**Inhibition** The suppression of the biological function of an enzyme or system by chemical or physical means. See also APTAMERS, ENZYME, PROTEIN TYROSINE KINASE INHIBITOR.

**Initiation Factors** Specific proteins required to initiate synthesis of a polypeptide on ribosomes. See also RIBOSOMES, PROTEIN, POLYPEPTIDE (PROTEIN).

**Inositol** See PHYTATE.

**Inositol Hexaphosphate (IP-6)** See PHYTATE.

**Insertional Knockout Systems** See GENE SILENCING.

**Insitu** See the link. See also *IN SITU*.

**Insulin** A protein hormone normally secreted by the beta (β) cells of the pancreas (when stimulated by glucose, and the parasympathetic nervous system). Insulin and glucagon are the most important regulators of fuel (food) metabolism. In essence, insulin signals the "fed" state to the body's cells, which stimulates the storage of energy (fuel) in the form of fat; and the synthesis of proteins (i.e., tissue building/repair) in a variety of ways.

The disease known as diabetes results from a body's inability to produce insulin, or its insensitivity to the insulin that is produced. That inability/insensitivity, and thus the disease, can result from several different causes: Type I (also known as childhood or juvenile or early-onset) diabetes results when the body's insulin-making tissue is destroyed by autoimmune disease. See also the entry for INSULIN-DEPENDENT DIABETES MELLITIS (IDDM) below. Type II diabetes results when the body's insulin-utilizing tissues become insensitive to insulin.

The too-high sugar content in bloodstream that results from diabetes, causes creation of free radicals (high-energy oxygen atoms bearing an "extra" electron) which can damage the eyes, kidneys, and extremity arteries (sometimes necessitating limb amputation) in one haplotype (i.e., genetic subgroup) of people (i.e., those possessing the larger-size molecules of haptoglobin — a blood protein). Some research indicates that consumption of amylose (starch only) or inulin (fructose oligosaccharide) in human diet as the primary carbohydrate source, instead of glucose (or other sugars that the human body converts to glucose) can help the human body avoid Type II diabetes by avoiding gluconeogenesis.

In 1922, Canadian scientists Frederick Banting, Charles Best, J. J. R. MacLeod, and J. B. Collip succeeded in extracting insulin from the pancreas of slaughtered livestock (cows, pigs) in a form that could be injected into diabetes patients as a substitute for human insulin. The English biochemist Fred

Sanger was first to determine the complete amino acid sequence of the insulin molecule. In 1977, the American scientist Howard Goodman, collaborating with William Rutter, announced the first cloning of insulin genes. This led to human insulin production by genetically engineered microorganisms (approved by FDA in 1982). See also BETA CELLS, ISLETS OF LANGERHANS, HORMONE, PROTEIN, GLUCOSE (GLc), AMINO ACID, POLYPEPTIDE (PROTEIN), SEQUENCE (OF PROTEIN MOLECULE), GENETIC ENGINEERING, GLUCAGON, INSULIN-DEPENDENT DIABETES MELLITIS (IDDM), G-PROTEINS, CARBOHYDRATES, PANCREAS, AUTOIMMUNE DISEASE, INULIN, FREE RADICAL, HAPLOTYPE, OXIDATIVE STRESS, HAPTOGLOBIN, TYPE I DIABETES, TYPE II DIABETES.

**Insulin-dependent Diabetes Mellitis (IDDM)** An autoimmune disease in which the insulin-producing cells of the pancreas (i.e., beta cells, also known as Islets of Langerhans) are attacked and destroyed by the cytotoxic T cells of the body's immune system. See also AUTOIMMUNE DISEASE, INSULIN, ISLETS OF LANGERHANS, BETA CELLS, CYTOTOXIC T CELLS, HAPTOGLOBIN, DIABETES, TYPE I DIABETES.

**Insulin-Like Growth Factor-1 (IGF-1)** A protein hormone produced by the body's bone cells (when those bone cells have been stimulated by parathyroid hormone and/or estrogen), that is a promoter of bone formation and follicle development (in ovaries). Another function of IGF-1 is to facilitate the transport of amino acids into cells, and further inhibit protein breakdown in cells. If the body is injured, IGF-1 works with platelet-derived growth factor (PDGF) to stimulate fibroblast and collagen cell division/metabolism to cause healing of wounds and bones. IGF-1 also occurs naturally in cow's milk. See also FIBROBLASTS, AMINO ACID, COLLAGEN, ESSENTIAL AMINO ACIDS, DIGESTION (WITHIN ORGANISMS), METABOLISM, PROTEIN, MESSENGER RNA (mRNA), UBIQUITIN.

**Insulin-Like Growth Factor-2 (IGF-2)** See INSULIN-LIKE GROWTH FACTOR-1 (IGF-1).

**Integrated Crop Management** See INTEGRATED PEST MANAGEMENT.

**Integrated Disease Management** See INTEGRATED PEST MANAGEMENT.

**Integrated Pest Management (IPM)** A holistic (system) approach, initially developed as a methodology by Ray Smith and Perry Adkisson, that is utilized by some farmers to try to control agricultural pests (tobacco budworm, European corn borer, soybean cyst nematode, weevils, etc.).

IPM also helps control plant diseases. For example, farmers can plant buckwheat near their cornfields in order to help control European corn borer (ECB), a serious pest of corn (maize) *Zea mays* L. plants. Green lacewing beetles (*Chrysoperla carnea*), which prey on European corn borers, are attracted by the buckwheat and consume ECB in the corn while they live in the buckwheat areas. Because European corn borer is a vector (carrier) of disease and/or mycotoxin-producing microorganisms such as the fungi *Aspergillus flavus*, *Aspergillus parasiticus*, and *Fusarium* spp., this lacewing beetle (IPM) control of ECB also helps reduce those plant diseases and mycotoxins. Often utilized in conjunction with no-tillage crop production. See also WEEVILS, *HELIOTHIS VIRESCENS* (*H. VIRESCENS*), EUROPEAN CORN BORER (ECB), FUNGUS, MYCOTOXINS, AFLATOXIN, LOW-TILLAGE CROP PRODUCTION, NO-TILLAGE CROP PRODUCTION, SOYBEAN CYST NEMATODES (SCN), CORN, SOYBEAN PLANT, *BACILLUS THURINGIENSIS* (*B.t.*).

**Integrins** A class of proteins found on the surface (membranes) of cells, and that function as cellular adhesion receptors. For example, integrin avb3 is a receptor on the surface of endothelial cells in tumors. It binds angiogenic endothelial cells, enabling them to form new blood vessels. See also ADHESION MOLECULES, PROTEIN, GLYCOPROTEINS, CELL, RECEPTORS, LECTINS, SELECTINS, SIGNAL TRANSDUCTION, ANGIOGENESIS, TUMOR, ENDOTHELIAL CELLS, PLASMA MEMBRANE.

**Intercellular Adhesion Molecule (ICAM)** See ADHESION MOLECULE.

**Interferons** A family of small (cytokines) proteins (produced by vertebrate cells following a virus infection) possessing potent antiviral effects. Secreted interferons bind to the plasma membrane of other cells in the organism and induce an antiviral state in them (conferring resistance to a broad spectrum of viruses). Three classes of interferons have been isolated and purified, so far: α-interferon (originally called leukocyte interferon); β-interferon (beta interferon or fibroblast interferon); and γ-interferon (gamma interferon or immune interferon, a lymphokine).

These proteins have been cloned and expressed in *Escherichia coli* (*E. coli*), which has enabled large quantities to be produced for evaluation of the interferons as possible antiviral and anticancer agents. To date, interferons have been used to treat Kaposi's sarcoma, hairy cell leukemia, venereal warts, multiple sclerosis, and hepatitis. See also ALPHA INTERFERON, BETA INTERFERON, CYTOKINES, PROTEIN, LYMPHOKINES, *ESCHERICHIA COLIFORM* (*E. COLI*).

**Interim Office of the Gene Technology Regulator (IOGTR)** The regulatory body of Australia's government that was responsible for approvals of new rDNA products (e.g., new genetically engineered crops) before they could be introduced in Australia, during 1999–2001. IOGTR replaced/superceded Australia's Gene Technology Office (in this role) in 1999, and was itself scheduled to be replaced by the Gene Technology Regulator (GTR) in 2001. See also GENE TECHNOLOGY REGULATOR (GTR), GENE TECHNOLOGY OFFICE, GENETIC MANIPULATION ADVISORY COMMITTEE (GMAC), rDNA, DEOXYRIBONUCLEIC ACID (DNA), GENETIC ENGINEERING, RECOMBINANT DNA ADVISORY COMMITTEE (RAC), COMMISSION OF BIOMOLECULAR ENGINEERING, INDIAN DEPARTMENT OF BIOTECHNOLOGY.

**Interleukin-1 (IL-1)** A cytokine (glycoprotein) released by activated macrophages, during the inflammatory stage of immune system response to an infection, which promotes the growth of epithelial (skin) cells and white blood cells. Recent research has indicated that too much IL-1 is linked to the development of rheumatoid arthritis, diabetes, inflammatory bowel disease, and other autoimmune diseases. See also MACROPHAGE, AUTOIMMUNE DISEASE, ADHESION MOLECULE, TUMOR NECROSIS FACTOR (TNF), CYTOKINES, GLYCOPROTEIN, WHITE BLOOD CELLS, ISLETS OF LANGERHANS, EPITHELIUM, INTERLEUKIN-1 RECEPTOR ANTAGONIST (IL-1ra).

**Interleukin-1 Receptor Antagonist (IL-1ra)** A glycoprotein (produced by macrophages in response to presence of Interleukin-1, and endotoxin in tissues) that preferentially binds to those cell receptors in the body that typically bind the lymphokine, Interleukin-1 (IL-1). When manufactured by man (via genetic engineering) and injected into the body in large quantities, IL-1ra can block the deleterious effects of (too much) Interleukin-1. See also INTERLEUKIN-1 (IL-1), RECEPTORS, RECEPTOR FITTING, GLYCOPROTEIN, MACROPHAGE, ENDOTOXIN, ADHESION MOLECULE, CELLULAR IMMUNE RESPONSE, PROTEIN, LYMPHOKINES, ANTAGONISTS.

**Interleukin-2 (IL-2)** Known as T cell growth factor. A cytokine (glycoprotein) secreted by (immune system response) stimulated helper T cells which promotes the proliferation/differentiation of more helper T cells, and promotes the growth of lymphocytes to combat an infection. Interleukin-2 also stimulates the lymphocytes to produce gamma interferon. It is gamma interferon that prompts the cytotoxic T cells to attack virus-infected cells and kill the virus within them. The structure of the gene that codes for synthesis of IL-2 (by immune system cells) was determined by Tadatsugu Taniguchi in 1983. See also IMMUNE RESPONSE, HUMORAL IMMUNITY, CYTOKINES GLYCOPROTEIN, CYTOTOXIC T CELLS, T CELLS, HELPER T CELLS, T CELL RECEPTORS, INTERFERONS.

**Interleukin-3 (IL-3)** A hematologic growth factor (glycoprotein) cytokine that stimulates the proliferation of a wide range of white blood cells (to combat an infection). See also HEMATOLOGIC GROWTH FACTORS (HGF), CYTOKINES, WHITE BLOOD CELLS.

**Interleukin-4 (IL-4)** A cytokine (glycoprotein) that stimulates production of antibody-producing B cells, Immunoglobulin-E (IgE), and promotes cytotoxic T cell (i.e., killer T cells) growth. See also ANTIBODY, CYTOTOXIC T CELLS, B CELLS, GLYCOPROTEIN, CYTOKINES.

**Interleukin-5 (IL-5)** A cytokine (glycoprotein) that stimulates eosinophil growth. See also EOSINOPHILS, PROTEIN, GLYCOPROTEIN, CYTOKINES, CELLULAR IMMUNE RESPONSE.

**Interleukin-6 (IL-6)** A cytokine (glycoprotein) that is pleiotropic (i.e., stimulates several different types of immune system cells), and is a hematopoietic growth factor. See also HEMATOPOIETIC GROWTH FACTORS (HGF), GROWTH FACTOR, GLYCOPROTEIN, PLEIOTROPIC, MACROPHAGE, CYTOKINES.

**Interleukin-7 (IL-7)** A cytokine (glycoprotein) synthesized in the bone marrow that stimulates early (fetal) proliferation and differentiation of B cells and T cells. May be useful in regenerating lymphoid cells in patients whose immune systems have been devastated by cancer chemotherapy. See also CYTOKINES, GLYCOPROTEIN, STEM CELL ONE, T CELLS, CANCER.

**Interleukin-8 (IL-8)** A basic polypeptide (glycoprotein) with heparin-binding activity. Endogenous endothelial IL-8 appears to regulate transvenular traffic during acute inflammatory responses. See also POLYPEPTIDE (PROTEIN), GLYCOPROTEIN, HEPARIN, ENDOTHELIAL CELLS, ENDOTHELIUM, POLYMORPHONUCLEAR LEUKOCYTES (PMN), CELLULAR IMMUNE RESPONSE.

**Interleukin-9 (IL-9)** A cytokine (glycoprotein) that is released at sites in the body where inflammation has occurred. See also CYTOKINES, GLYCOPROTEIN, CELLULAR IMMUNE RESPONSE.

**Interleukin-12 (IL-12)** A cytokine (glycoprotein) produced by the body, which serves to activate the immune system against certain tumors and pathogens. See also CYTOKINES, GLYCOPROTEIN, TUMOR, TUMOR-ASSOCIATED ANTIGENS, MAJOR HISTOCOMPATIBILITY COMPLEX (MHC), T CELL RECEPTORS, CYTOTOXIC T CELLS, PATHOGEN.

**Intermediary Metabolism** The chemical reactions that take place in the cell that transform the complex molecules derived from food into the small molecules needed for the growth and maintenance of the cell. See also METABOLISM, CELL, DIGESTION (WITHIN ORGANISMS), METABOLIC PATHWAY.

**International Food Biotechnology Council (IFBC)** An organization that was established in 1988 by the Industrial Biotechnology Association (IBA) and the International Life Sciences Institute (ILSI), in order to "produce a (recommended) set of guidelines that could be used to assess the safety of genetically altered foods." See also GNE,

INDUSTRIAL BIOTECHNOLOGY ASSOCIATION (IBA), INTERNATIONAL LIFE SCIENCES INSTITUTE (ILSI), SENIOR ADVISORY GROUP ON BIOTECHNOLOGY, BIOTECHNOLOGY INDUSTRY ORGANIZATION (BIO), GENETIC ENGINEERING, POLYGALACTURONASE, ANTISENSE (DNA SEQUENCE), BIOTECHNOLOGY, BACTERIOCINS.

**International Life Sciences Institute (ILSI)** A nonprofit foundation established in 1978 to advance the understanding of scientific issues relating to nutrition, food safety, toxicology, risk assessment, and the environment. ILSI is headquartered in Washington, D.C. and has branches in Argentina, Australasia, Brazil, Europe, India, Japan, Korea, Mexico, Africa, Thailand, Singapore, China, and other nations.

**International Office of Epizootics (OIE)** One of the three international SPS standard-setting organizations recognized by the World Trade Organization (WTO), the OIE is an international veterinary organization headquartered in Paris. The OIE was established in 1924, originally as part of the League of Nations, and is the worldwide authority for development of animal health and zoonoses standards, guidelines, and recommendations. See also SPS, INTERNATIONAL PLANT PROTECTION CONVENTION (IPPC), ZOONOSES, WORLD TRADE ORGANIZATION (WTO).

**International Plant Protection Convention (IPPC)** One of the three international SPS standard-setting organizations recognized by the World Trade Organization (WTO), the IPPC is the worldwide authority for development of plant health standards, guidelines, and recommendations (e.g., to prevent transfer of a plant disease or plant pest from one country to another). The treaty establishing the IPPC was signed in 1952 (amended in 1979 and 1997), and currently has 107 member countries [i.e., signatories to the 1979 text]. The IPPC Secretariat is within the United Nations' Food and Agriculture Organization (FAO). IPPC standards are set (and enforced) via regional SPS institutions such as the North American Plant Protection Organization (NAPPO), European Plant Protection Organization (EPPO), etc. There are currently nine RPPOs (i.e., regional plant protection organizations) under Article VIII of the 1979 IPPC text. See also SPS, EUROPEAN PLANT PROTECTION ORGANIZATION (EPPO), INTERNATIONAL OFFICE OF EPIZOOTICS (OIE), WORLD TRADE ORGANIZATION (WTO), NORTH AMERICAN PLANT PROTECTION ORGANIZATION (NAPPO).

**International Society for the Advancement of Biotechnology (ISAB)** A nonprofit organization of individuals that was started in 1994 "to advance and promote the general welfare of the science and commercialization of genetic engineering and industrial biotechnology." See also GENETIC ENGINEERING, BIOTECHNOLOGY, AMERICAN SOCIETY FOR BIOTECHNOLOGY (ASB), BIOTECHNOLOGY INDUSTRY ORGANIZATION (BIO).

**International Union for Protection of New Varieties of Plants (UPOV)** See UNION FOR PROTECTION OF NEW VARIETIES OF PLANTS (UPOV).

**Internaulin** See CADHERINS.

**Introgression** The incorporation of exotic (i.e., wild type) genes into elite germplasm (i.e., domesticated breeding lines), or of transgenes (i.e., genes from transgenic organisms) into a wild type's genome. See also TRANSGENIC, OUTCROSSING, WILD TYPE, GENOME, GENE, TRANSLOCATION.

**Intron** A (intervening sequence) segment of deoxyribonucleic acid (DNA) that is transcribed, but is removed from within the transcript by splicing together the sequences (exons) on either side of it (in the molecule). It is generally considered a nonfunctioning portion of the molecule. See also TRANSCRIPTION, DEOXYRIBONUCLEIC ACID (DNA), EXON.

**Inulin** A fructose oligosaccharide (FOS) that is naturally produced in more than 30,000 plants. Like many other FOS, consumption of inulin by humans results in several health benefits (helps prevent coronary heart disease, promote growth of bifidobacteria in the intestines, reduce likelihood of developing diabetes, promote absorption of calcium from foods, etc.). During 2000, the European Union's government regulatory agencies agreed to classify inulin as a water-soluble fiber (because humans cannot digest inulin). See also FRUCTOSE OLIGOSACCHARIDES, WATER SOLUBLE FIBER, BIFIDOBACTERIA, CORONARY HEART DISEASE (CHD), DIABETES.

**Invasin** A transmembrane (through the membrane of the cell) protein that enables bacterial cells to invade normal (body) cells. See also CD4 PROTEIN, RECEPTORS, CELL, T CELL RECEPTORS, ENDOCYTOSIS, PLASMA MEMBRANE.

**Inverted Micelle** See REVERSE MICELLE (RM), MICELLE.

**Investigational New Drug** See IND.

**Invitro** See *IN VITRO*.

**Invivo** See *IN VIVO*.

**IOGTR** See INTERIM OFFICE OF THE GENE TECHNOLOGY REGULATOR (IOGTR).

**Ion** From the Greek *ion*, something that goes. An ion is an atom or molecule possessing a positive or a negative electrical charge. Ions are produced by the dissociation (coming apart) of a (electrolyte) molecule resulting from an electrolyte dissolving in a solution. One example is the dissociation of common table salt (sodium chloride) in water, which results in positively charged sodium ions (called cations) and negatively charged chloride ions (called anions). Ions play critically important roles in many biological processes such as nerve activity. See also CHELATION, CHELATING AGENT, ION CHANNELS, CITRIC ACID, CITRATE SYNTHASE (CSb) GENE.

**Ion Channels** Refers to specialized proteins that act as "pores" (through the plasma membrane of a cell) through which certain ions (atoms or molecules bearing an electrical charge) are allowed to pass. The selectivity of ion channels can be altered when specific molecules (e.g., in the blood or digestive fluids) come in contact with the plasma membrane (i.e., G-protein receptors coupled to the ion channel). For example, the group of pharmaceuticals known as CALCIUM CHANNEL BLOCKERS (verapamil, amlopidine, diltiazem, nifedipine, etc.) act to "block" or hinder the movement of calcium ions through calcium ion channels: "pores" which had previously allowed calcium ions to enter relevant cells (i.e., in blood vessel walls) easily.

Another example is the mode of action of the "cry" (crystal-like) proteins that are naturally present within *Bacillus thuringiensis* (*B.t.*) bacteria. When eaten by certain insects (possessing alkaline digestive fluids in their stomach or gut), cry proteins are hydrolyzed (i.e., chemically "cut") into fragments. One of those fragments — 60 Kd in size — attaches to specific receptors located on the surface (membrane) of certain cells which line the inside (epithelium) of the insect's mid-gut. That attachment to those receptors triggers ion channels in the (epithelium) cell's membrane to suddenly allow cations (atoms or molecules with positive electrical charge) to quickly flow out of the cell (which leads to death of all gut cells that the cry protein piece attached to). See also CELL, PLASMA MEMBRANE, ION, CALCIUM CHANNEL-BLOCKERS, MEMBRANE TRANSPORT, PROTEIN, CRY PROTEINS, G-PROTEINS, *BACILLUS THURINGIENSIS* (*B.t.*), BACTERIA, PROTOXIN, HYDROLYZE, KILO-DALTON (Kd), RECEPTORS, EPITHELIUM.

**Ion-Exchange Chromatography** Separation of ionic compounds (which include nucleic acids and proteins) in a chromatographic column containing a polymeric resin (i.e., the stationary phase) having fixed charge groups. The process works in that the charges of the column (stationary phase) interact with the opposite charges of the material dissolved in the solution that is flowing through the column (mobile phase). The charge interaction between the column material and, i.e., the protein has the effect of slowing down the rate of movement of the protein through the column. The other molecules, meanwhile, which do not interact with the column, flow right on through. This constitutes the separation process. See also CHROMATOGRAPHY.

**IP-6** Inositol hexaphosphate. See also PHYTATE.

**IPM** See INTEGRATED PEST MANAGEMENT (IPM).

**IPPC** See INTERNATIONAL PLANT PROTECTION CONVENTION.

**Iron Bacteria** See FERROBACTERIA.

**Iron Deficiency Anemia (IDA)** A disease caused by lack of iron in an organism's body, due to shortfall in diet or due to dietary iron not being bioavailable (digestible) to that organism's body. For example, the phytate naturally present in traditional varieties of corn (maize) inhibits absorption of the iron in that corn (maize) by humans, swine, and poultry. IDA is a major cause of childhood diseases and maternal death (i.e., death of the mother following childbirth) in many

developing countries. IDA also makes people more susceptible to diphtheria. See also GOLDEN RICE, PHYTATE, LOW-PHYTATE CORN, LOW-PHYTATE SOYBEANS, ORGANISM.

**Islets of Langerhans** (also called beta cells) Cells in the pancreas that produce insulin in response to the presence of glucose (sugar) in the bloodstream. The failure of insulin production results in the disease called diabetes. See also GLUCOSE (GLc), GLYCOLYSIS, AUTOIMMUNE DISEASE, INSULIN, INSULIN-DEPENDENT DIABETES MELLITIS (IDDM).

**Isoenzymes** See ISOZYMES.

**Isoflavins** See ISOFLAVONES.

**Isoflavones** A group of phytochemicals (including genistein, glycitein, and daidzein) that are produced within the seeds of the soybean plant [*Glycine max* (L.) Merrill] at a typical concentration of approximately 0.04–0.24%. Isoflavones are also produced within other types of tissues of the soybean plant (e.g., to ward off infection by plant diseases such as *Phytophthera* ones) and the soybean plant's roots (e.g., to signal and attract the *Rhizobium japonicum* bacteria which live symbiotically among the soybean plant's roots and "fix" nitrogen from the air, thereby providing natural fertilizer for the plant). Much smaller amounts of isoflavones are produced in some wheat, lentils, chickpeas, and edible bean plants.

Evidence shows that consumption of soybean isoflavones by humans can help lower blood content of low-density lipoproteins (LDLP), help prevent osteoporosis, help prevent prostate enlargement, and help prevent certain types of cancer (breast cancer, colon cancer, lung cancer, prostate cancer, uterine cancer, etc.). A human diet containing a large amount of isoflavones has been shown to increase bone density and to decrease total serum cholesterol, thereby lowering risk of osteoporosis and coronary heart disease. Isoflavones also exhibit antioxidant properties. See also GENISTEIN (Gen), SOYBEAN PLANT, PHYTOALEXINS, PHYTOCHEMICALS, LOW-DENSITY LIPOPROTEINS (LDLP), OSTEOPOROSIS, PROSTATE-SPECIFIC ANTIGEN (PSA), CANCER, SELECTIVE ESTROGEN EFFECT, STRESS PROTEINS, CHOLESTEROL, NITROGEN FIXATION, NODULATION, CORONARY HEART DISEASE (CHD), OSTEOPOROSIS,

I

RHIZOBIUM (bacteria), *PHYTOPHTHERA MEGASPERMA* F. SP. *GLYCINEA*, *PHYTOPHTHERA* ROOT ROT, SIGNALING, SIGNALING MOLECULES, HIGH-ISOFLAVONE SOYBEANS, ANTIOXIDANTS, OXIDATIVE STRESS.

**Isoflavonoids** See ISOFLAVONES.

**Isoleucine (ile)** A monocarboxylic amino acid occurring within most dietary proteins. See also AMINO ACID, PROTEIN, ALS GENE.

**Isomer** One of the two or more chemical substances having the same elementary percentage composition (i.e., same atoms) and molecular weight, but differing in structure and therefore in properties. There are many ways in which such structural differences (between the two or more isomeric molecules) occur. One example is n-butane [$CH_3(CH_2)_2CH_3$] and isobutane [$CH_3CH(CH_3)_2$]. See also STEREOISOMERS.

**Isomerase** An enzyme-catalyzing transformation of a compound into its positional isomer. See also ISOMER.

**Isoprene** The five-carbon hydrocarbon molecule 2-methyl-1,3 butadiene. It is a recurring structural unit of the terpenoid molecules, which are either linear or cyclic. There exists a very large number of terpenes and many are major components of essential plant oils. See also GTPases.

**Isotope** Refers to one of the several "varieties" of atoms that exist, of the same element, that differ from each other in the number of neutrons in the atom's nucleus. For example, the element chlorine exists primarily in two forms (isotopes) in nature, with 18 neutrons (76% of the time) and with 20 neutrons (24% of the time). The chemical properties of isotopes of a given element are virtually identical. See also ATOMIC WEIGHT.

**Isozymes** (isoenzymes) Multiple forms of an enzyme that differ from each other in their substrate (substance acted upon) affinity, in their maximum activity, or in their regulatory properties. See also ENZYME, SUBSTRATE (CHEMICAL), RIBOZYMES.

**ISPM** Acronym for International Standards for Pest Management. See also INTERNATIONAL PLANT PROTECTION CONVENTION (IPPC).

# J

**Japan Bio-Industry Association** An association of the largest Japanese companies that are engaged in at least some form of genetic engineering research or production. Similar to America's Biotechnology Industry Organization (BIO), it is headquartered in Tokyo. See also BIOTECHNOLOGY INDUSTRY ORGANIZATION (BIO), BIOTECHNOLOGY, GENETIC ENGINEERING, RECOMBINANT DNA (rDNA), SENIOR ADVISORY GROUP ON BIOTECHNOLOGY (SAGB), INTERNATIONAL FOOD BIOTECHNOLOGY COUNCIL.

**Jasmonic Acid** Jasmonic Acid is a signaling molecule in Systemic Acquired Resistance (SAR) when SAR is triggered in plants (via spray application of harpin protein to various plants, via chewing of insects on the leaves of certain plants, and/or via the entry-into-plant of certain pathogenic bacteria/fungi, etc.). See also SYSTEMIC ACQUIRED RESISTANCE (SAR), SIGNALING MOLECULE, SOYBEAN PLANT, FUNGUS, PATHOGEN, PROTEIN, PATHOGENESIS RELATED PROTEINS, HARPIN, PHYTOALEXINS.

**Jumping Genes** Genes that move (change positions) within the genome. Genes associated with transposable elements. A segment fragment of deoxyribonucleic acid (DNA) that can move from one position in the genome to another. See also GENE, GENOME, DEOXYRIBONUCLEIC ACID (DNA), GENETIC CODE, TRANSPOSITION, TRANSPOSON, TRANSLOCATION, INTROGRESSION, HOT SPOTS.

***Juncea*** Refers to a group of related plants; often commonly called "wild mustard." See also BRASSICA.

**Junk DNA** A term historically utilized by some, to refer to portions of an organism's DNA that were not obviously genes (i.e., not transcribed into mRNA; thus not part of the DNA "tagged" with ESTs, etc.). However, it has recently been discovered that at least some of what was formerly called "junk DNA" (e.g., introns) helps enable more than one specific protein molecule to be expressed from certain genes. See also DEOXYRIBONUCLEIC ACID (DNA), GENE, INTRON, PROTEIN, EXPRESS, EXPRESSED SEQUENCE TAG (EST), CENTRAL DOGMA (NEW).

# K

**Karnal Bunt** A plant disease that can be caused by the smut fungus *Tilletia indica* in wheat. See also FUNGUS, WHEAT.

**Karyotype** A size-order alignment of an organism's chromosome pairs in the format of a chart. It enables the connecting of chromosomes to symptoms (e.g., of genetic diseases in the organism) and traits. See also CHROMOSOMES, GENE, GENOTYPE, TRAIT, LINKAGE, LINKAGE GROUP, MUSCULAR DYSTROPHY (MD), CHROMATIDS, CHROMATIN.

**Karyotyper** A scientist (or more frequently an automated analytical machine) that

- Takes a video picture of a given cell under a microscope
- Digitizes that picture within a computer
- "Cuts out" the individual chromosomes contained within that cell's genome
- Arranges the cell's chromosomes in pairs by size order into a chart (called a karyotype).

See also CHROMOSOMES, GENOME, KARYOTYPE.

**Kb** An abbreviation for 1,000 (kilo) base pairs of deoxyribonucleic acid (DNA). See also DEOXYRIBONUCLEIC ACID (DNA), KILOBASE PAIRS (Kbp).

**Kd** An abbreviation for kilodalton. See also KILODALTON (Kd).

**Kefauver Rule** A 1962 U.S. law that mandates that the Food and Drug Administration (FDA) requires proof of pharmaceutical efficacy for drugs to be sold in the U.S.. See also FOOD AND DRUG ADMINISTRATION (FDA).

**Kenya Biosafety Council** The country of Kenya's national regulatory body for granting approval to a new genetically engineered plant (e.g., a new genetically engineered crop to be planted). The Kenya Biosafety Council is analogous to Germany's ZKBS (Central Commission on Biological Safety), Australia's GMAC (Genetic Manipulation Advisory Committee), or Brazil's CTNBio (National Biosafety Commission). See also GMAC, RECOMBINANT DNA ADVISORY COMMITTEE (RAC), ZKBS (CENTRAL COMMISSION ON BIOLOGICAL SAFETY), GENETIC ENGINEERING, CTNBio.

**Keratins** Insoluble protective or structural proteins consisting of parallel polypeptide chains arranged in an α-helical or β conformation.

**Ketose** A simple monosaccharide having its carbonyl groups at other than a terminal position. See also MONOSACCHARIDES.

**Killer T Cell** See CYTOTOXIC T CELLS.

**Kilobase Pairs (Kbp)** A unit of DNA equals 1,000 base pairs. See also BASE PAIR (bp), DEOXYRIBONUCLEIC ACID (DNA).

**Kilodalton (Kd)** A unit of mass equal to 1,000 Daltons. See also DALTON.

**Knockout (gene)** See GENE SILENCING, GPA1, NUCLEAR TRANSFER.

**Konzo** A term used in some countries to refer to lathyrism. See also LATHYRISM, GLUCOSINOLATES.

**Koseisho** The Japanese government agency that must approve new pharmaceutical products for sale with Japan. It is the equivalent of the U.S. Food and Drug Administration. See also NDA (TO KOSEISHO), FOOD AND DRUG ADMINISTRATION (FDA), COMMITTEE FOR PROPRIETARY MEDICINAL PRODUCTS (CPMP), COMMITTEE ON SAFETY IN MEDICINES, MEDICINES CONTROL AGENCY (MCA), EUROPEAN MEDICINES EVALUATION AGENCY (EMEA), BUNDESGESUNDHEITSAMT (BGA).

**Krebs Cycle** See CITRIC ACID CYCLE.

**Kunitz Trypsin Inhibitor (TI)** See TRYPSIN INHIBITORS.

# L

**L-Selectin** Also known as the homing receptor. See also SELECTINS, LECTINS, ADHESION MOLECULES.

**Lab-On-A-Chip** See BIOCHIP, NANOTECHNOLOGY, MICROFLUIDICS, GENOSENSORS, GENE EXPRESSION, BIOSENSORS (ELECTRONIC), BIOSENSORS (CHEMICAL), GENE EXPRESSION ANALYSIS.

**Label (radioactive)** A radioactive atom, introduced into molecule(s) in order to:

1. enable observation of that molecule's metabolic transformation (within an organism). For example, if radioactive hydrogen in the form of water (known as deuterium) is supplied to a living cell, a series of "photographs" (e.g., taken via an electron microscope, which has photographic film in it that is sensitive to radiation) will reveal how rapidly that deuterium enters the cell, and into what structures within the cell that water is incorporated.

2. quantify the rate at which cetain (non-) radioactive atoms are being introduced into a polymer (e.g., DNA) that is being polymerized (manufactured) as part of a biological test or testing process (QPCR-Quantitative PCR, RT-PCR-Reverse Transcriptase PCR, etc.).

See also AUTORADIOGRAPHY, CELL, DEOXYRIBONUCLEIC ACID (DNA), GENE EXPRESSION ANALYSIS, QPCR, RT-PCR, RADIOIMMUNOASSAY, RADIOIMMUNOTECHNIQUE.

**Lac Operon** An operon in *Escherichia coli* (*E. coli*) that codes for three enzymes involved in the metabolism of lactose. See also OPERON, CODING SEQUENCE, *ESCHERICHIA COLIFORM* (*E. COLI*).

**Lachrymal Fluid (tears)** A salty solution produced by the tear glands to bathe and lubricate the eye. Possesses antimicrobial properties.

**Lactoferricin** A protein compound that acts to inhibit pathogenic (disease-causing) bacteria and yeasts (e.g., in the human body). See also PROTEIN, PATHOGEN, BACTERIA, YEAST, LACTOFERRIN.

**Lactoferrin** A protein compound that is naturally produced in human breast milk. Also produced in cow's milk. Consumption of lactoferrin by infants (e.g., via nursing) helps strengthen their immune system. Consumption of lactoferrin by older humans helps their immune system to resist infectious diseases. Lactoferrin binds free iron (e.g., in body fluids), thereby denying that iron to pathogenic baceria (which need that iron to grow/infect). Pepsin and some other proteases (enzymes) can convert lactoferrin to lactoferricin. See also PROTEIN, PATHOGEN, BACTERIA, GROWTH (MICROBIAL), LACTOFERRICIN, PEPSIN, PROTEASE, HIGH-LACTOFERRIN RICE, LACTOPEROXIDASE.

**Lactonase** An enzyme that "breaks open" the lactone ring in (molecular structure of) the mycotoxin zearalenone. See also ENZYME, MYCOTOXIN, ZEARALENONE, TOXIN.

**Lactoperoxidase** A protein compound (enzyme) that acts to inhibit pathogenic bacteria (e.g., in human body). See also PROTEIN, ENZYME, PATHOGEN, BACTERIA.

**Lambda Phage** A bacteriophage that infects *Escherichia coli* (*E. coli*). It is commonly used as a vector in recombinant DNA (deoxyribonucleic acid) research. See also PHAGE, *ESCHERICHIA COLIFORM* (*E. COLI*).

**Langerhans Cells** See DENDRITIC LANGERHANS CELLS, ISLETS OF LANGERHANS.

**Lathyrism** See GLUCOSINOLATES.

**Laurate** A medium chain length (i.e., C12) fatty acid that is naturally produced by coconut trees, oil palm trees, and certain species of wild plants. In 1992, some canola varieties were genetically engineered so that they could also produce (desirable) laurate in their seeds. See also FATTY ACID, FATS, CANOLA, GENETIC ENGINEERING, GENETIC CODE, LPAAT PROTEIN, ACP, LAUROYL-ACP THIOESTERASE, HIGH-LAURATE CANOLA.

**Lauric Acid** See LAURATE.

**Lauroyl-ACP Thioesterase** The enzyme that is required for the synthesis (manufacturing) of laurate in plants. For example, the presence of this enzyme in the California bay tree (*Umbellularia californica*) causes its seed oil to contain as much as 45% laurate. See also LAURATE, ENZYME, LPAAT PROTEIN, HIGH-LAURATE CANOLA.

**Lazaroids** A class of drugs being developed to "bring back from the dead" tissues that have been (almost) killed due to a lack of oxygen (e.g., Krebs Cycle L caused by a clot blocking a vital artery). See also HUMAN SUPEROXIDE DISMUTASE (hSOD), FIBRIN, REPERFUSION.

**LDL** See LOW-DENSITY LIPOPROTEINS (LDLP).

**LDLP** See LOW-DENSITY LIPOPROTEINS.

**LDLP Receptors** See LOW-DENSITY LIPOPROTEINS (LDLP).

**Leader** See LEADER SEQUENCE.

**Leader Sequence** The nontranslated sequence at the 5′ end of mRNA that precedes the initiation codon. See also MESSENGER RNA (mRNA), CODON.

**Leaky Mutants** A mutant in which the mutated gene product, such as an enzyme, still possesses a fraction of its normal biological activity. See also MUTATION, GENE, PROTEIN, BIOLOGICAL ACTIVITY, ENZYME.

**Lear** See CANOLA.

**Lecithin** See LECITHIN (crude, mixture), LECITHIN (refined, specific).

**Lecithin (crude, mixture)** A mixture of phospholipids (i.e., lecithin-phosphatidylcholine, cephalin, inositol phosphatides, glycerides, tocopherols, glucosides, and certain pigments). Historically, crude (mixture) lecithin has often been utilized commercially in food processing as an emulsifier, instantizing agent, and lubricating agent. Because lecithin-phosphatidylcholine naturally contains a high content of linoleic acid, consumption by humans of lecithin-phosphatidylcholine results in similar impact (e.g., lowered cholesterol levels in blood) as consumption of linoleic acid. Because dietary fats are generally not absorbed directly through the intestinal wall (when eaten), they must first be emulsified, to form micelles that can pass through the intestinal wall and thus be absorbed by the body. That emulsification/micelle-formation is aided by lecithin, since it is an emulsifier. See also LECITHIN (refined, specific), LIPOPROTEIN, LIPIDS, CONJUGATED PROTEIN, HIGH-DENSITY LIPOPROTEINS (HDLP), LOW-DENSITY LIPOPROTEINS (LDLP), SOYBEAN PLANT, SOYBEAN OIL, CHOLINE, SIGNAL TRANSDUCTION, LINOLEIC ACID, ACETYLCHOLINE, FATS, MICELLE, DIGESTION (WITHIN ORGANISMS), BILE ACIDS.

**Lecithin (refined, specific)** A by-product of the refining process for soybean oil (deoiled lecithin from processed soybeans is composed of approximately 20–25% phosphatidyl choline by weight). The lecithin molecule (i.e., phosphatidyl choline) naturally contains a high content of linoleic acid, so consumption of lecithin by humans results in similar impact (e.g., lowered cholesterol levels in blood) as consumption of linoleic acid. Because dietary fats are generally not absorbed directly through the intestinal wall (when eaten), they must first be emulsified to form micelles that can pass through the intestinal wall and be absorbed by the body. That emulsification/micelle-formation is aided by lecithin, since it is an emulsifier.

Lecithin (also known as phosphatidylcholine) is a source of choline when digested, and is a critical component of the lipoproteins that transport fat and cholesterol molecules in the bloodstream (e.g., from the digestive system, to body cells, to the liver, etc.). Lecithin (phosphatidylcholine) promotes synthesis of high-density lipoproteins (HDLP, also known as "good" cholesterol) by the liver, when it is consumed by humans. Phosphatidyl choline (PC) is involved in cell signal transduction (e.g., via which a cell reacts to an external chemical "signal"). Some other common dietary sources of lecithin include eggs, red meats, spinach, and nuts. See also LIPOPROTEIN, LIPIDS, CONJUGATED PROTEIN, HIGH-DENSITY LIPOPROTEINS (HDLP), LOW-DENSITY LIPOPROTEINS (LDLP), SOYBEAN PLANT, SOYBEAN OIL, CHOLINE, SIGNAL TRANSDUCTION, LINOLEIC ACID, ACETYLCHOLINE, LECITHIN (crude, mixture), FATS, MICELLE, DIGESTION (WITHIN ORGANISMS).

**Lectins** A class of proteins that have the capability to rapidly (and reversibly) combine with specific sugar molecules (e.g., those sugar molecules or glycoproteins on the surface of adjacent cells, within an organism). Lectins are a common component of the surface (membranes) of plant and animal cells, and are so specific (regarding sugar molecules that they will or won't combine with/attach to) that they discriminate between different monosaccharides and different oligosaccharides (i.e., on the surfaces of adjacent cells within an organism). This capability to reversibly combine with sugar (i.e., carbohydrate) molecules (on the surface of adjacent cells) is utilized by:

- Bacteria and other microorganisms, to adhere to (sugar molecules on surface of ) host cells, as the first step in the process of infecting those host cells
- White blood cells (e.g., lymphocytes), to adhere to the walls of blood vessels (endothelium), as the first step to leaving the bloodstream to go fight infection (pathogens, trauma) in tissue adjacent to that blood vessel. The lectin (glycoprotein) that adheres to the (endothelial sugar molecule on) blood vessel wall is called L-selectin, or the homing receptor. The two sugar molecules (glycoproteins) on the blood vessel wall (endothelium) are called P-selectin and E-selectin (also known as ELAM-1)
- Cancerous tumor cells, to adhere to the walls of blood vessels (endothelium) as part of the tumor-proliferation process known as metastasis (i.e., new tumors are "seeded" throughout the body via this process).

Separate and apart from the above impacts, some plant lectins (e.g., in the seeds of certain plants) are toxic to some of the animals that consume those seeds. See also PROTEIN, SUGAR MOLECULES, GLYCOPROTEIN, LEUKOCYTES, SELECTINS, LYMPHOCYTES, MONOCYTES, NEUTROPHILS, ENDOTHELIAL CELLS, ENDOTHELIUM, CANCER, METASTASIS, SIGNAL TRANSDUCTION.

**Leptin** A protein hormone that is produced by fat cells (adipose tissue) in the body. When leptin is produced and travels to cells whose surface bears leptin receptors (e.g., in the brain), those (brain) cells receive signal (transduction) indicating fullness/satiety. Leptin has been found to be present in the bloodstream of obese humans at a concentration of approximately four times the concentration found in bloodstreams of lean humans. High levels of leptin present in the bloodstream disrupt some of the activities of insulin (hormone which regulates blood sugar levels), and may possibly lead to diabetes. See also HORMONE, PROTEIN, BIOLOGICAL ACTIVITY, INSULIN, ADIPOSE.

**Leptin Receptors** Cellular receptors which are specific to leptin. In 1996, H. Ralph Snodgrass discovered that leptin receptors are involved in the "sorting" of immature blood cells (from bone marrow) to create subpopulations. See also LEPTIN, RECEPTORS.

**Lethal Mutation** Mutation of a gene to yield no, or a totally defective, gene product (protein), thereby making it unable to function, and hence unable to sustain the life of the organism.

**Leucine (leu)** A monocarboxylic essential amino acid. See also AMINO ACID, ESSENTIAL AMINO ACIDS, ALS GENE.

**Leukocytes** (white blood cells) A diverse family of nucleated cells that has many immunological functions. See also NEUTROPHILS, EOSINOPHILS, BASOPHILS, LYMPHOCYTE, B LYMPHOCYTES, MONOCYTES, GRANULOCYTES.

**Leukotrienes** Lipid mediator molecules (synthesized from arachidonic acid) released by certain cells (T cells), which "signal" leukocytes (white blood cells) during the initial stages of an infection or an allergic reaction. When thus activated, the leukocytes migrate to the site of infection to combat the pathogens (or allergens), and mediate the inflammation. See also LIPIDS, LEUKOCYTES, MAST CELLS, SIGNALING, SIGNAL TRANSDUCTION,

**L**

T CELLS, PATHOGEN, ARACHIDONIC ACID, ALLER-
GIES, SIGNALING MOLECULE.

**Levorotary (L) Isomer** An isomer of an optically active compound; rotates (when illuminated) the plane of plane-polarized light
to the left. See also STEREOISOMERS, DEX-
TROROTARY (D) ISOMER.

**LH** See LUTEINIZING HORMONE.

**Library** A set of cloned DNA fragments
together representing the entire genome. See
also DEOXYRIBONUCLEIC ACID (DNA), GENOME.

**Ligand (in biochemistry)** In general, a molecule or ion that can bind to (interact with) a
protein molecule. For example, a pharmaceutical that binds to a receptor protein molecule on the surface of a cell may be called
a ligand. See also PROTEIN, RECEPTORS, T CELL
RECEPTORS, ENDOCYTOSIS, CD4 PROTEIN, INVA-
SIN, LIGAND (IN CHROMATOGRAPHY), CHELATION.

**Ligand (in chromatography)** A term used to
describe a substance (the ligand) that has the
capacity for specific and noncovalent
(reversible) binding to some protein. A
ligand may be a coenzyme for a specific
enzyme. The ligand can be covalently
attached (immobilized) by means of the
appropriate chemical reaction to the surface
of certain porous column material. When a
mixture of proteins containing the enzyme
to be isolated is passed through the column,
the enzyme, which is capable of tightly binding to the ligand, does so, and is in this
manner held to the column. The other proteins present, which have no specific affinity
for the ligand, pass on through the column.
The protein/ligand complex is then dissociated and the enzyme eluted from the column,
which may be accomplished by passing
more free (unbound) coenzymes through the
column. The ligand may be hormones (i.e.,
used to isolate receptor molecules) or any
other type of molecule that is capable of
binding specifically and reversibly to the
desired protein or protein complex. See also
AFFINITY CHROMATOGRAPHY, SUBSTRATE (IN
CHROMATOGRAPHY), CHROMATOGRAPHY, PROTEIN,
PEPTIDE, ANTIBODY, MONOCLONAL ANTIBODIES
(MAb).

**Ligase** An enzyme used to catalyze the joining
of single-stranded DNA segments. See also
DEOXYRIBONUCLEIC ACID (DNA).

**Ligation** The formation of a phosphodiester
bond to link two adjacent bases separated by
a nick in one strand of a double helix of DNA
(deoxyribonucleic acid). The term can also be
applied to blunt-end ligation and to the joining of RNA (ribonucleic acid) strands. See
also DEOXYRIBONUCLEIC ACID (DNA), LIGASE.

**Light-Chain Variable (VL) Domains** The
regions (domains) of the antibody (molecule's) "light chain" that vary in their amino
acid sequence. The "chains" (of atoms) comprising the antibody (immunoglobulin) molecule consist of a region of variable (V)
amino acid sequence and a region in which
the amino acid sequence remains constant
(C). An antibody molecule possesses two
antigen binding sites, and it is the variable
domains of the light (VL) and heavy (VH)
chains which contribute to this (antigen
binding ability). See also ANTIBODY, IMMUNO-
GLOBULIN, PROTEIN, SEQUENCE (OF A PROTEIN
MOLECULE), ANTIGEN, AMINO ACID, COMBINING
SITE, DOMAIN (OF A PROTEIN), HEAVY-CHAIN
VARIABLE (VH) DOMAINS.

**Lignans** A category of phytochemicals that
play defensive roles (e.g., against infections
by bacteria, fungi, etc.) within land plants
(e.g., those grown by man for crops). Lignans are also sometimes referred to by some
people as "phytoestrogens," and are typically beneficial to the health of humans that
consume them. Lignans are found in virtually all fruits, vegetables, and cereals
(grains); generally within the seed coats,
stems, leaves, or flowers. One of the beneficial lignans commonly consumed by
humans is sesamin, found in seeds of the
sesame plant (*Sesamum indicum*); which
acts as an antioxidant. See also PHYTOCHEM-
ICALS, PHYTOESTROGENS, ISOFLAVONES, ANTI-
OXIDANTS, OXIDATIVE STRESS.

**Lignins** A category of phenolic ("ring-shaped"
molecules) polymeric (i.e., composed of
more than one molecular unit) compounds
produced by land plants within the cell walls
(i.e., exterior of cell's plasma membrane) of
those plants, to reinforce/strengthen those
cell walls. See also CELL, POLYMER, PLASMA
MEMBRANE.

**Lignocellulose** A complex biopolymer comprising the bulk of woody plants. It consists

of polysaccharides and polymer phenols. See also POLYSACCHARIDES, LIGNINS.

**Limonene** See PHYTOCHEMICALS.

**Linkage** A phenomenon discovered by Thomas Hunt Morgan in the early 1900s via his experiments with fruit flies. This term describes the tendency of genes to be inherited together as a result of their locations being physically close to each other on the same chromosome; measured by percent recombination between loci. Because the locus (location of gene on the chromosome) determines the likelihood that two genes will go together into offspring, "marker genes" that are linked to a gene (e.g., for a given trait or disease) of interest can be utilized to predict the presence of that (trait or disease-causing) gene. See also GENE, LOCUS, CHROMOSOMES, LINKAGE GROUP, MARKER (GENETIC MARKER), MAP DISTANCE, LINKAGE MAP.

**Linkage Group** Includes all loci (in DNA molecule) that can be connected (directly or indirectly) by linkage relationships; equivalent to a chromosome. See also LOCUS, CHROMOSOMES, LINKAGE, CHROMATIDS, CHROMATIN, LINKAGE MAP, DEOXYRIBONUCLEIC ACID (DNA).

**Linkage Map** A depiction of gene loci (on chromosomes) based on the frequency of recombination (of linked genes) in the offspring's genome. See also LINKAGE, LINKAGE GROUP, GENE, LOCUS, MARKER (GENETIC MARKER).

**Linker** A short synthetic duplex oligonucleotide containing the target site for some restriction enzyme. It may be added to the ends of a DNA (deoxyribonucleic acid) fragment prepared by cleavage with some other enzyme reconstructions of recombinant DNA.

**Linking** The process of "attaching" a drug or a toxin to a monoclonal antibody, or another homing molecule of the immune system. Because this attachment must be reversible, so that the homing molecule can release the drug or toxin after delivering that drug or toxin to the desired site in the body (e.g., delivery of a toxin to a tumor, to kill the tumor), linking is a difficult process to reliably achieve. See also IMMUNOTOXIN, CONJUGATE, MONOCLONAL ANTIBODIES (MAb), TOXIN.

**Linoleic Acid** One of the so-called "omega-6" (n-6) polyunsaturated fatty acids (PUFA), it has historically comprised approximately 53% of the total fatty acid content of soybean oil. It is an essential fatty acid for humans. When consumed by humans, linoleic acid causes LDLP cholesterol levels in the blood to decrease, which reduces risk of coronary heart disease (CHD). The human body converts linoleic acid to the n-6 highly unsaturated fatty acid (HUFA) arachidonic acid. See also POLYUNSATURATED FATTY ACIDS (PUFA), N-6 FATTY ACIDS, FATS, UNSATURATED FATTY ACIDS, ESSENTIAL FATTY ACIDS, LOW DENSITY LIPOPROTEINS (LDLP), CHOLESTEROL, LECITHIN, CONJUGATED LINOLEIC ACID (CLA), CORONARY HEART DISEASE (CHD), VOLICITIN, SOYBEAN OIL, ARACHIDONIC ACID, COSUPPRESSION.

**Linolenic Acid** Also known as α-linolenic acid. One of the so-called "omega-3" (n-3) polyunsaturated fatty acids (PUFA), it has historically comprised approximately 8% of the total fatty acid content of soybean oil. It is an essential fatty acid for humans (i.e., required by the human body). The human body converts linolenic acid to the n-3 highly unsaturated fatty acids (HUFA) docosahexanoic acid (DHA) and eicosapentanoic acid (EPA). When consumed by humans, both DHA and EPA confer various health benefits to the human body. See also N-3 FATTY ACIDS, POLYUNSATURATED FATTY ACIDS (PUFA), UNSATURATED FATTY ACIDS, ESSENTIAL FATTY ACIDS, CORONARY HEART DISEASE (CHD), CANCER, HIGHLY UNSATURATED FATTY ACIDS (HUFA), DOCOSAHEXANOIC ACID (DHA), EICOSAPENTANOIC ACID (EPA), FATS.

**Lipase** An enzyme (one of a class of enzymes) that catalyzes the hydrolytic cleavage of lipid molecules (triglycerides) to yield free fatty acids. Lipase was the first enzyme to be produced via genetic engineering and marketed. Lipase also occurs naturally in cow's milk, and in the intestines of many animals (where it aids/assists digestion of fats that the animal consumes). See also ENZYME, HYDROLYTIC CLEAVAGE, TRIGLYCERIDES, FATS, FATTY ACID, FREE FATTY ACIDS, DIGESTION (WITHIN ORGANISMS).

**Lipid Bilayer** A membrane (i.e., thin sheet-type) structure composed of relatively small lipid molecules which possess both a hydrophilic ("water loving") and a hydrophobic ("water hating") moiety. These (membrane) lipids thus spontaneously form closed bimolecular sheets in aqueous (water-containing) media, in which the hydrophobic ends of each lipid molecule are in the center of the bimolecular membrane and the hydrophilic ends of the lipid molecules are on the outside (i.e., touching the water molecules). See also LIPIDS, PLASMA MEMBRANE, MOIETY.

**Lipid Vesicles** See LIPOSOMES.

**Lipids** From the Greek word *lipos,* fat, lipids are water-insoluble (fat) biomolecules that are highly soluble in organic solvents such as chloroform. Lipids serve as fuel molecules, highly concentrated energy stores, "signaling" molecules, and components of cell membranes. Membrane lipids are relatively small molecules that have both a hydrophilic ("water loving") and a hydrophobic ("water hating") moiety. These (membrane) lipids spontaneously form closed bimolecular sheets in aqueous media (water) which are barriers to the free movement (flow) of polar molecules. See also FATS, MOIETY, LIPOPROTEIN, CHOLESTEROL, SIGNALING, SIGNALING MOLECULE, SIGNAL TRANSDUCTION, PLASMA MEMBRANE, ANTIOXIDANTS, OXIDATIVE STRESS, LIPID BILAYER, LEUKOTRIENES, OLEOSOMES.

**Lipolytic Enzymes** See LIPASE.

**Lipophilic** A "fat loving" molecule, or portion of a molecule. Relating to, or having strong affinity for, fats or other lipids. See also LIPIDS, FATS.

**Lipopolysaccharide (LPS)** See ENDOTOXIN.

**Lipoprotein** A conjugated protein containing a lipid or a group of lipids. For example, low-density lipoproteins (also known as "bad" cholesterol) are a "package" of cholesterol (lipid) surrounded by a hydrophilic protein. Low-density lipoproteins (LDLPs) and very low-density lipoproteins (VLDLs) are the specific lipoproteins that are most likely to deposit cholesterol (plaque) on artery walls, which increases risk of coronary heart disease (CHD). See also PROTEIN, LOW-DENSITY LIPOPROTEINS (LDLP), VERY LOW-DENSITY LIPOPROTEINS (VLDL), CONJUGATED PROTEIN, HYDROPHILIC, LIPIDS, CHOLESTEROL, APOLIPOPROTEINS.

**Lipoprotein-Associated Coagulation (Clot) Inhibitor (LACI)** A protein that prevents formation of blood clots. This occurs because LACI inhibits the controlled series of zymogen activations (enzymatic cascade) which causes the formation of fibrinogen (precursor to fibrin), leading subsequently to clot formation. See also FIBRIN, FIBRONECTIN, ZYMOGENS.

**Liposomes** Also called lipid vesicles or vesicle. Aqueous (watery) compartments enclosed by a lipid bilayer. They can be formed by suspending a suitable lipid, such as phosphatidyl choline, in an aqueous medium. This mixture is then sonicated (i.e., agitated by high-frequency sound waves) to give a dispersion of closed vesicles (i.e., compartments) that are quite uniform in size. Alternatively, liposomes can be prepared by rapidly mixing a solution of lipid in ethanol with water, which yields vesicles that are nearly spherical in shape and have a diameter of 500 Å (Angstroms). Larger vesicles (10,000 Å or 1 mm, or 0.00003937 inch in diameter) can be prepared by slowly evaporating the organic solvent from a suspension of phospholipid in a mixed solvent system.

Liposomes can be made to contain certain drugs for protective, controlled release delivery to targeted tissues. For example, pharmaceuticals which tend to be rapidly degraded in the bloodstream could be enclosed within liposomes so that more of the nondegraded pharmaceutical would remain by the time it reached the targeted tissue. The controlled release property enables larger doses (of drugs possessing toxic side effects) to be prescribed, knowing that the drug will be released in the body over an extended period of time. See also LIPIDS, MICRON, ANGSTROM (Å).

**Lipoxidase** See LIPOXYGENASE (LOX).

**Lipoxygenase (LOX)** A "family" of enzymes that is naturally produced within its seeds (soybeans) by the soybean plant (*Glycine max* (L.) Merrill). In the presence of moisture and certain other conditions, lipoxygenase enzymes catalyze a chemical reaction in which objectionable "beany" flavor can be produced from certain components of the soybean. That "beany" flavor decreases the suitability of resultant soybean raw materials for manufacture of human foods in some countries.

Prevention of the reactions that create the "beany" flavor can be accomplished via heat denaturation (of lipoxygenases present in the soybeans) or via creation of soybeans that do not contain any lipoxygenase enzymes (known as "LOX null" soybeans). Lipoxygenase enzymes also catalyze a reaction in which certain volatile chemicals are produced that inhibit growth of any *Aspergillus flavus* fungus. See also ENZYME, SOYBEAN PLANT, LOX NULL SOYBEANS, LOX-1, LOX-2, LOX-3.

**Lipoxygenase Null** See LOX NULL SOYBEANS, LIPOXYGENASE (LOX).

**Listeria monocytogenes** Refers to the "family" (numerous strains) of *Listeria monocytogenes* bacteria, that can grow in many different foodstuffs (e.g., meats) under specific conditions, and can cause food poisoning (Listeriosis) in humans who subsequently consume those foodstuffs. When consumed by humans, certain strains/serotypes of *Listeria monocytogenes* can cause fever, severe headaches, stiffness, nausea, diarrhea, and possibly miscarriages in pregnant women. As of January 19, 2001, all meat processed in the U.S. is required to be tested for the presence of *Listeria monocytogenes*. See also BACTERIA, STRAIN, SEROTYPES, ENTEROTOXIN, BACTERIOCINS, CADHERINS.

**Living Modified Organism (LMO)** See GMO.

**LMO (Living modified organism)** See GMO.

**Loci** The plural of locus. See also LOCUS.

**Locus** The position of a gene on a chromosome. See also GENE, CHROMOSOMES.

**Loop** A single-stranded region at the end of a hairpin in RNA (or single-stranded DNA). It corresponds to the sequence between inverted repeats in duplex DNA. See also RIBONUCLEIC ACID (RNA), DEOXYRIBONUCLEIC ACID (DNA), SEQUENCE (OF A DNA MOLECULE).

**LOSBM** Low-oligosaccharide soybean meal. See also LOW-STACHYOSE SOYBEANS, SOYBEAN PLANT.

**Low-Density Lipoproteins (LDLP)** So-called "bad" cholesterol (i.e., LDL cholesterol), which carries cholesterol molecules from the digestive system (e.g., intestine) to body

cells and can sometimes clog arteries over time (a disease called atherosclerosis, or coronary heart disease). Since cholesterol does not dissolve in water (which constitutes most of the volume of blood), the body makes LDL cholesterol (derived from the digestion of fatty foods) into little "packages" surrounded by a hydrophilic ("water loving") protein. That protein "wrapper" is known as apolipoprotein B-100, or apo B-100, and it enables LDL cholesterol to be transported in the bloodstream because the apolipoprotein B-100 is attracted to water molecules in the blood. Part of the apolipoprotein B-100 molecule also will bind to special LDLP receptor molecules in the liver, which then clears those (bound) cholesterol packages out of the body as part of regular liver functions. See also HIGH-DENSITY LIPOPROTEINS (HDLPs), HYDROPHILIC, RECEPTORS, PROTEIN, SITOSTANOL, ISOFLAVONES, WATER SOLUBLE FIBER, CHOLESTEROL, CORONARY HEART DISEASE (CHD), APOLIPOPROTEINS, VERY LOW-DENSITY LIPOPROTEINS (VLDL).

**Low-Linolenic Oil Soybeans** Soybeans from soybean (*Glycine max*) plant varieties which have been bred specifically to produce soybeans bearing oil that contains less than 4% linolenic acid, instead of the typical 8% linolenic acid content of soybean oil produced from traditional varieties of soybeans. Low-linolenic soybean oil would tend to have greater flavor stability (especially at elevated temperatures utilized in frying foods) than soybean oil from traditional varieties of soybeans. See also SOYBEAN PLANT, SOYBEAN OIL, FATTY ACID, LINOLENIC ACID, POLYUNSATURATED FATTY ACIDS (PUFA).

**Low-lipoxygenase Soybeans** See LOX-NULL SOYBEANS.

**Low-Phytate Corn** Developed in the U.S. during the 1990s, these are corn (maize) hybrids possessing the Lpa1 gene, the Lpa2 gene, or the HAP (highly available phosphorous) gene (which was discovered by Victor Raboy). That gene causes corn (maize) hybrids possessing it to produce much less phytate than the 0.15% typically present in traditional varieties of corn (maize).

Because phytate is not digestible in humans and other monogastric animals (swine, poultry, etc.), substituting low-phytate corn in place of traditional corn varieties in those animals' diets helps lessen adverse environmental impact of animal feeding (e.g., phosphorous emissions in excess of annual cropland requirements). Swine fed a diet in which traditional corn (maize) varieties have been replaced by low-phytate corn (maize) produce up to 30% less phosphorous in their manure, thereby lessening the phosphorous impact of those swine on the environment. Humans consuming a diet based heavily on corn/maize (e.g., tortillas) absorb 50% more iron when traditional corn varieties are replaced by low-phytate corn varieties. That is because the phytate (inositol hexaphosphate) molecule "binds"/chelates iron (and some other metals) within the digestive system and prevents their absorption into the body. See also CORN, PHYTATE, HIGH-PHYTASE CORN, PHYTASE, VALUE-ENHANCED GRAINS, HIGHLY-AVAILABLE PHOSPHOROUS (HAP) GENE, CHELATION, CHELATING AGENT, IRON DEFICIENCY ANEMIA (IDA).

**Low-Phytate Soybeans** Developed in the U.S. during the 1990s, these are soybean varieties possessing less than 0.30% (of total soybean weight) phytate, vs. the typical 0.45% phytate content of soybeans from traditional soybean varieties.

Because phytate is not digestible in humans and other monogastric animals (swine poultry, etc.), substituting low-phytate soybeans in place of traditional soybean varieties in those animals' diets helps to lessen adverse environmental impact of animal feeding (e.g., manure phosphorous emissions in excess of cropland requirements). Swine fed a diet in which traditional soybean varieties have been replaced by low-phytate soybeans produce up to 20% less phosphorous in their manure, thereby lessening the phosphorous impact of those swine on the environment. Due to the fact that the amino acids lysine, methionine, cysteine, arginine, and threonine all become more "bioavailable" (i.e., available for the animal to build its body tissue, or otherwise utilize) in a low-phytate diet, low-phytate diets also help reduce exess nitrogen emissions. See also SOYBEAN PLANT, PHYTATE, LOW-PHYTATE

CORN, HIGH-PHYTASE CORN/SOYBEANS, LYSINE, CYSTEINE, METHIONINE, ARGININE, THREONINE, DEAMINATION.

**Low-Stachyose Soybeans** Those soybean varieties that contain lower than 1% levels of the relatively indigestible stachyose carbohydrate (and thus higher levels of easily digestible other nutrients) than traditional varieties of soybeans (which typically contain 1.4–4.1% stachyose in traditional soybean varieties). Compared to traditional varieties of soybeans, low-stachyose soybeans have approximately 10% more metabolizable (i.e., useable by animals) energy content and a 3% increase in amino acid digestibility. Low-stachyose soybeans are particularly useful for feeding of monogastric animals (swine, poultry, etc.), since their single stomach cannot digest stachyose. Thus, stachyose tends to "ferment" (promote excess bacterial growth) in their intestines, causing them to feel prematurely full. See also STACHYOSE, CARBOHYDRATES (SACCHARIDES), VALUE-ENHANCED GRAINS, SOYBEAN PLANT, HIGH-SUCROSE SOYBEANS, DIGESTION (WITHIN ORGANISMS), METABOLISM.

**Low-Tillage Crop Production** A methodology of crop production in which the farmer utilizes a minimum of mechanical cultivation (i.e., only two to four passes over the field with tillage equipment instead of the conventional five passes per year utilized for traditional crop production). This reduced mechanical tillage leaves more carbon in the (less disturbed) soil, leaves more earthworms (*Eisenia foetida*) per cubic foot or per cubic meter living in the topsoil, and reduces soil compaction (i.e., the reduction in interstitial spaces between individual soil particles); thereby increasing the fertility of "low till" farm fields.

The plant residue remaining on the field's surface helps control weeds and reduce soil erosion; it also provides sites for insects to shelter and reproduce, leading to a need for increased pest insect control via methods such as inserting a *Bacillus thuringiensis* (*B.t.*) gene into certain crop plants. But if a farmer needs to apply synthetic chemical pesticides, the plant residue remaining on the field's surface helps cause breakdown (into substances such as carbon dioxide and water) of those pesticides. That is because that plant residue helps to retain moisture in the field-surface environment, thereby enhancing growth of the types of microorganisms that help break down pesticides. See also NO-TILLAGE CROP PRODUCTION, GLOMALIN, EARTHWORMS, MICROORGANISMS, INTEGRATED PEST MANAGEMENT (IPM), CORN, SOYBEAN PLANT, *BACILLUS THURINGIENSIS* (*B.t.*), GENE, GENETIC ENGINEERING, EUROPEAN CORN BORER (ECB), *HELICOVERPA ZEA* (*H. ZEA*), CORN ROOTWORM, COLD HARDENING.

**LOX Null Soybeans** Refers to soybeans that do not contain any of the three lipoxygenase enzymes (thus, they result in a "null" test reading). See also LIPOXYGENASE (LOX), LOX-1, LOX-2, LOX-3, SOYBEAN PLANT, ENZYME.

**LOX-1** One of the isozymes (enzyme molecule variations) of the lipoxygenase (LOX) enzyme "family." See also LIPOXYGENASE (LOX), ISOZYMES (ISOENZYMES).

**LOX-2** One of the isozymes (enzyme molecule variations) of the lipoxygenase (LOX) enzyme "family." See also LIPOXYGENASE (LOX), ISOZYMES (ISOENZYMES).

**LOX-3** One of the isozymes (enzyme molecule variations) of the lipoxygenase (LOX) enzyme "family." See also LIPOXYGENASE (LOX), ISOZYMES (ISOENZYMES).

**LPAAT Protein** A protein consisting of lysophosphatidic acid acyl transferase (enzyme), which (when present in a plant) causes production of triglycerides (in the seeds) possessing saturated fatty acids in the "middle position" of the triglycerides' molecular (glycerol) "backbone." For example, canola (rapeseed) plants genetically engineered to contain LPAAT protein are able to produce high levels of saturated fatty acids (including laurate) in their oil. See also PROTEIN, LAURATE, ENZYME, TRIGLYCERIDES, SATURATED FATTY ACIDS, MONOUNSATURATED FATS, CANOLA, GENETIC ENGINEERING.

**LPE** See LYSOPHOSPHATIDYLETHANOLAMINE.

**LPS** See ENDOTOXIN.

**Luciferase** Refers to a group of enzymes that can catalyze a chemical reaction that results in the production of light (i.e., bioluminescense) within certain living oganisms. For example, the common firefly is able to emit light from its tail (photophores) via luciferase-catalyzed bioluminescence. The ocean jellyfish known as the sea pansy (*Renilla reniformis*) is able to emit light via similar use of a slightly different luciferase molecule. See also BIOLUMINESCENCE, ENZYME, CATALYST, ORGANISM, NITRIC OXIDE.

**Luciferin** See BIOLUMINESCENCE.

**Lumen** The interior (opening through which blood flows); e.g., within a blood vessel. See also ENDOTHELIUM.

**Luminesce** See BIOLUMINESCENCE.

**Luminescence** See BIOLUMINESCENCE.

**Luminescent Assays** Refers to assays (i.e., tests/test techniques) which detect or measure the presence of a specific substance (e.g., bacteria ATP on surfaces in a slaughterhouse) and the efficacy (i.e., effectiveness) of a specific substance via the enzyme (e.g., luciferase)-catalyzed production of light. For example, one (rapid) luminescent assay utilizes two chemical reagents which first break down bacteria cell membranes, then cause ATP from those broken-open cells to luminesce. Subsequent measurement of that light is the assay's proof (e.g., that bacteria had been present on the tested surface in a slaughterhouse). See also ASSAY, BIOLUMINESCENCE, ENZYME, BACTERIA, PLASMA MEMBRANE, ADENOSINE TRIPHOSPHATE (ATP).

**Lupus** An autoimmune disease of the body, in which anti-DNA antibodies bind to DNA. The resulting complexes (of DNA and antibodies) travel to the kidneys via the bloodstream, and become lodged in the kidneys, where they cause inflammatory reactions (that can lead to kidney failure). Sometimes joints, blood vessels, bone marrow, and the liver are also damaged by this disease. See also ANTIBODY, DEOXYRIBONUCLEIC ACID (DNA), AUTOIMMUNE DISEASE, SUPERANTIGENS.

**Lupus Erythematosus** See LUPUS.

**Lutein** A carotenoid (i.e., "light harvesting" compound utilized in photosynthesis) that is naturally produced in carrots, summer squash, broccoli, dark lettuce, and green peas. Lutein is a phytochemical/nutraceutical conducive to good eye health, and regular consumption of large amounts of lutein has been shown to reduce the risk of the disease age-related macular degeneration, a leading cause of blindness in elderly people. Research

**L**

indicates that consumption of lutein by humans also reduces risk of prostate cancer and breast cancer. See also PHYTOCHEMICALS, NUTRACEUTICALS, CAROTENOIDS, PHOTOSYNTHESIS.

**Luteinizing Hormone (LH)** A reproductive hormone that acts upon the ovaries to stimulate ovulation. It is secreted by the pituitary gland. See also HORMONE, PITUITARY GLAND, ENDOCRINE HORMONES, ESTROGEN.

**Luteolin** See NODULATION.

**Lycopene** An antioxidant carotenoid ("light harvesting" pigment utilized by plants in the photosynthesis process) that is a naturally occurring phytochemical in tomatoes, watermelon, guava, pink grapefruit (and some other fruits). Consumption of significant amounts of lycopene by humans causes an increase in the concentration of lycopene in the blood plasma. Lycopene is a natural constituent of blood plasma and certain tissues in the human body, but it must be consumed in the diet, because the human body does not synthesize (manufacture) lycopene. Consumption of lycopene by humans has been linked to a reduction in atherosclerosis, coronary heart disease, some cancers (e.g., prostate cancer), and inhibition of oxidation of low-density lipoproteins (LDLP).

Lycopene is also converted (in some instances) into alpha-carotene and/or beta-carotene. Because beta-carotene is processed into vitamin A by the human body, consumption of this phytochemical can help prevent human diseases (e.g., in developing countries) that result from deficiency of vitamin A, e.g.: coronary heart disease; certain cancers (cancer of prostate, lung, etc.), childhood blindness, macular degeneration (a leading cause of blindness in older people), and various childhood diseases that can cause death due to a weakened immune system. See also PHYTOCHEMICALS, NUTRACEUTICALS, CANCER, ANTIOXIDANTS, CAROTENOIDS, CORONARY HEART DISEASE (CHD), PLASMA, ATHEROSCLEROSIS, PROSTATE-SPECIFIC ANTIGEN (PSA), TOMATO, BETA CAROTENE, VITAMIN, LUTEIN, PHOTOSYNTHESIS, LOW-DENSITY LIPOPROTEINS (LDLP).

**Lymphocyte** A type of cell found in the blood, spleen, lymph nodes, etc. of higher animals. They are formed very early in fetal life, arising in the liver by the sixth week of human gestation. There exist two subclasses of lymphocytes: B lymphocytes and T lymphocytes. B lymphocytes make antibodies (immunoglobins) of which there are five classes: IgM, IgA, IgG, IgD, and IgE. The antibodies circulate in the bloodstream. T lymphocytes recognize and reject foreign tissue, modulate B cell activity, kill tumor cells, and kill host cells infected with virus. T-lymphocytes are also called T cells. See also B LYMPHOCYTES, T CELLS, ANTIBODY, HELPER T CELLS (T4 CELLS), BLAST CELL, CYTOTOXIC T CELLS, ANTIGEN.

**Lymphokines** Peptides and proteins secreted by (immune system response) stimulated T cells. These hormone-like (peptide and protein) molecules direct the movements and activities of other cells in the immune system. Some examples of lymphokines are interleukin-1, interleukin-2, tumor necrosis factor (TNF), gamma interferon, colony stimulating factors, macrophage chemotactic factor, and lymphocyte growth factor. The suffix "-kine" comes from the Greek word kinesis, meaning movement.

**Lyochrome** See FLAVIN.

**Lyophilization** The process of removing water from a frozen biomaterial (e.g., a microbial culture or an aqueous protein solution) via application of a vacuum. It is a drying method for long-term preservation of proteins in the solid state, and for long-term storage of live microbial cultures. See also CULTURE, PROTEIN.

**Lyse** To rupture a membrane (cell). The act of lysis (rupturing a membrane). See also LYSIS.

**Lysine (lys)** An essential amino acid that can be obtained from many proteins by hydrolysis (i.e., cutting apart the protein molecule). See also ESSENTIAL AMINO ACIDS, PROTEIN, OPAGUE-2, PHOTORHABDUS LUMINESCENS, HYDROLYSIS.

**Lysis** The process of cell disintegration; membrane rupturing; breaking up of the cell wall. See also CYTOLYSIS, CELL, LYSOZYME, MEMBRANE TRANSPORT, BIOCIDE.

**Lysophosphatidylethanolamine** Also known by the abbreviation LPE; also known as phosphatidyl ethanolamine. It is one of the lipids (phospholipids) naturally found in soybean oil. In plants, it functions as a signaling molecule (e.g., speeding the ripening process). See also LIPIDS, SOYBEAN OIL, SIGNALING MOLECULE.

**Lysosome** A membrane-surrounded organelle in the cytoplasm of eucaryotic cells which contains many hydrolytic enzymes. The lysosome internalizes and digests foreign proteins as well as cellular debris. The protein fragments (epitopes) are "presented" to T cells by the major histocompatibility complex (MHC) proteins on the surface of the eucaryotic cell. See also ANTIGEN, MAJOR HISTOCOMPATIBILITY COMPLEX (MHC), T CELLS.

**Lysozyme** An enzyme, naturally produced by some animals, which possesses antibacterial (bacteria-killing) properties. Discovered in 1922 by Alexander Fleming, in his nasal mucus, Mr. Fleming named it from the Greek *lyso* — due to its ability to lyse (cut) bacteria — and *zyme* — due to its being an enzyme.

Lysozyme lyses certain kinds of bacteria, by dissolving the polysaccharide components of the bacteria's cell wall. When that cell wall is weakened, the bacteria cell bursts because osmotic pressure (inside that bacteria cell) is greater than the weakened cell wall can contain. Tears and egg whites both contain significant amounts of lysozyme, as agents to prevent bacterial infections (e.g., against bacteria entering the body via eye openings; against bacteria entering the chicken embryo through the eggshell). See also ENZYME, LYSIS, CELL, CYTOLYSIS, POLYSACCHARIDES, BACTERIA.

**Lytic Infection** A viral infection in which the final act of the infection is to lyse (i.e., burst, or destroy) the cell. This releases the new (progeny) viruses so they can go on to infect other cells. See also LYSE, LYSIS.

L

# M

**MAA Marketing Authorization Application**
It is the European Union (EU) equivalent to a U.S. NDA (New Drug Application). An MAA is an application to the EU's Committee for Proprietary Medicinal Products (CPMP) seeking approval of a new drug that has undergone Phase 2 and Phase 3 clinical trials. See also NDA (TO FDA), CANDA, FOOD AND DRUG ADMINISTRATION (FDA), MAA, NDA (TO KOSEISHO), CPMP, PHASE I CLINICAL TESTING, PHASE II CLINICAL TESTS, PHASE III CLINICAL TESTS.

**MAB** See MARKER ASSISTED BREEDING.

**MAb** See MONOCLONAL ANTIBODIES (MAb).

**Macromolecules** Large molecules with molecular weights ranging from a few thousand to hundreds of millions. See also MOLECULAR WEIGHT.

**Macrophage** A phagocytic cell that is the counterpart of the monocyte. A monocyte that has left the bloodstream and has moved into the tissues. Macrophages have basically the same functions as monocytes, but they carry these out in the tissues. In summary, they engulf and kill microorganisms, present antigen to the lymphocytes, kill certain tumor cells, and their secretions regulate inflammation.

Macrophages utilize nitric oxide (which they synthesize) to kill the microorganisms they engulf (via oxidation), and the nitric oxide also helps to regulate the immune system. In the spleen, macrophages engulf and destroy old red blood cells. When they reside in the bone marrow, they store iron and then transfer it to red blood cells. In the lungs and GI tract, they are scavengers and keep tissues clean. They also serve as a reservoir for the AIDS virus.

They (and other phagocytic cells) are largely responsible for the localization and degradation of foreign materials at inflammatory sites. Macrophages display chemotaxis (i.e., the sensing of, and movement toward or away from a specific chemical). For example, consumption (in food/feed) of mannanoligosaccharides by mammals causes macrophages (within that mammal's bloodstream) to depart from the bloodstream and move toward the gastrointestinal tract (tissues) where those macrophages eliminate some pathogens (i.e., those growing/reproducing in the gastrointestinal tract). See also CELL CELLULAR IMMUNE RESPONSE, CHEMOTAXIS, MONOCYTES, PHAGOCYTE, ADHESION MOLECULE, LYSOSOME, NITRIC OXIDE, NITRIC OXIDE SYNTHASE, MANNANOLIGOSACCHARIDES (MOS), PATHOGEN, LEUKOTRIENES.

**Macrophage Colony Stimulating Factor (M-CSF)** A colony stimulating factor (CSF) that stimulates production of macrophages in the body. See also COLONY STIMULATING FACTORS (CSFs), MACROPHAGE.

**MACS** Acronym for Magnetic Cell Sorting. See also MAGNETIC PARTICLES.

**Magainins** Discovered within frog skin tissues by Michael Zasloff in 1987, magainins are antimicrobial, amphopathic peptides that lyse (burst) certain cells upon contact by "worming" their hydrophobic portion into the cell's membrane, which creates a transmembrane (i.e., through the surface) pore (allowing ions to flow into the cell, causing osmotic bursting). Magainins are selective against bacteria, fungi, and protozoa cells. The word magainin comes from the Hebrew word for "shield." See also AMPHIPHILIC MOLECULES, CELL, PEPTIDE, BACTERIA, FUNGUS, ANTIBIOTICS, PLASMA MEMBRANE.

**"Magic Bullet"** When this term was first coined by Paul Ehrlich in 1905, it initially referred only to antibodies (e.g., because antibodies seek their own target, without damaging other nearby tissues). However, over time, this term has come to be applied to immunotoxins and other immunoconjugates (i.e., toxic or pharmacological molecules which are "attached" to an antibody that "steers/guides" the toxic or pharmacological molecule to the intended "target" in the body such as a tumor). See also ANTIBODY, IMMUNOCONJUGATE, IMMUNOTOXIN, GENISTEIN, RICIN, MONOCLONAL ANTIBODIES (MAb), HER2 GENE.

**Magnetic Antibodies** See MAGNETIC PARTICLES.

**Magnetic Beads** See MAGNETIC PARTICLES.

**Magnetic Cell Sorting** See MAGNETIC PARTICLES.

**Magnetic Labeling** See MAGNETIC PARTICLES.

**Magnetic Particles** Refers to various tiny pieces of naturally magnetic materials, that are bonded (attached) to antibodies (e.g., monoclonal antibodies that are specific to a particular type of cell). These can then be mixed with a large population of many cell types (crude tissue samples, cells grown in a vat/reactor, etc.), where the magnetic antibodies will attach themselves to only the desired cells, then the desired cells are separated out using a magnetic field (and the magnetic particles/antibodies are subsequently removed from those cells). See also ANTIBODY, MONOCLONAL ANTIBODIES (MAb), CELL, IMMUNOCONJUGATE, CELL SORTING.

**Maize** See CORN.

**Major Histocompatibility Complex (MHC)** A chromosomal region (approximately 3,000 Kb) which encodes for three classes of transmembrane (cell) proteins. MHC I proteins (located on the surface of nearly all cells) present foreign epitopes (i.e., fragments of antigens that have been ingested; peptides) to cytotoxic T cells (killer T cells). MHC II proteins (located on the surface of immune system cells and phagocytes) present foreign epitopes to helper T cells, and MHC III proteins are components of the complement cascade. Genes in the MHC must be matched (between an organ donor and organ recipient) to prevent rejection of organ transplants. See also COMPLEMENT CASCADE, GRAFT-VERSUS-HOST DISEASE (GVHD), Kb, MACROPHAGE, PROTEIN, CELL, T CELL RECEPTORS, ANTIGEN, T CELLS, CYTOTOXIC T CELLS, EPITOPE, GENE, TUMOR-ASSOCIATED ANTIGENS, HUMAN LEUKOCYTE ANTIGENS (HLA).

**MAL (Multiple Aleurone Layer) Gene** A gene in corn (maize) that (when present in the DNA of a given plant) causes that plant to produce seed that contains higher-than-normal levels of calcium, magnesium, iron, zinc, and manganese. These higher mineral levels are particularly useful for feeding of swine, since traditional No. 2 yellow (dent) corn does not contain enough for optimal pig growth. See also GENE, DEOXYRIBONUCLEIC ACID (DNA), HIGH-METHIONINE CORN, HIGH-LYSINE CORN, FLOURY-2, OPAGUE-2.

**MALDI-TOF-MS** Acronym for Matrix-Associated Laser Desorption Ionization Time of Flight Mass Spectrometry. A mass spectrometry methodology/technology that can establish, in seconds, the identity, purity, etc. of a sample of proteins, oligonucleotide, or (poly)peptides. Also the identification of gram-positive microorganisms, or characterization of genetic materials (DNA, RNA, etc.) on hybridization surfaces. MALDI-TOF utilizes measurement of the time for particles (e.g., proteins) to transit a specific distance after being "dislodged" from ('adhered') surface by specific amount of energy to precisely determine the molecular weight (of proteins, etc.). See also MASS SPECTROMETER, MICROORGANISM, OLIGONUCLEOTIDE, GRAM-POSITIVE, RIBONUCLEIC ACID (RNA), HYBRIDIZATION SURFACES, DEOXYRIBONUCLEIC ACID (DNA), IN SILICO BIOLOGY, PROTEIN, PEPTIDE.

**Male-sterile** See BARNASE.

**Malonyl CoA** See FATS.

**Mammalian Cell Culture** Technology to artificially cultivate cells, of mammal origin, in a laboratory or production-scale device (i.e., *in vitro*). Can be either a batch or continuous process device. The first mammalian cell culture was performed by a neurobiologist named R. G. Harrison in 1907, when he added chopped-up spinal cord tissue to clotted (blood) plasma in a humidified growth chamber. The nerve cells from this spinal cord tissue successfully grew, divided, and extended long fibers into the clot. Many improvements to cell culture process have been made over the years, including special growth media (fluids that bathe the cultured cells with the right amounts of amino acids, salts, and other minerals). See also CONTINUOUS PERFUSION, DISSOCIATING ENZYMES, HARVESTING ENZYMES, *IN VITRO*, PLASMA, CELL, MEDIUM, AMINO ACID.

**Mannan Oligosaccharides** See MANNANOLIGOSACCHARIDES (MOS).

**Mannanoligosaccharides (MOS)** A family of oligosaccharides that can be produced by man in commercial quantities via certain yeast cells. When consumed (e.g., by humans or monogastric livestock such as swine or poultry), mannose sugars in the MOS stimulate the liver to secrete the mannose-binding protein. Mannose-binding protein enters the digestive system and binds to the (mannose- containing) capsule (surface membrane) of pathogenic bacteria. That binding to pathogens triggers the immune system's complement cascade to combat those pathogenic bacteria. Consumption of mannanoligosaccharides by mammals also causes macrophages to move toward the gastrointestinal tract (in body's tissues), where those macrophages eliminate some pathogens (i.e., growing/reproducing in the gastrointestinal tract). See also OLIGOSACCHARIDES, FRUCTOSE OLIGOSACCHARIDES, SUGAR MOLECULES, YEAST, COMPLEMENT CASCADE, PATHOGENIC, BACTERIA, IMMUNE RESPONSE, COMPLEMENT, CAPSULE, MACROPHAGE, FOSHU, NUTRACEUTICALS.

**Map Distance** A number proportional to the frequency of recombination between two genes. One map unit corresponds to a recombination frequency of 1%. See also GENETICS, GENETIC CODE, GENETIC MAP, GENE, LINKAGE, QUANTITATIVE TRAIT LOCI (QTL).

**Mapping (of genome)** See GENETICS, GENETIC CODE, GENETIC MAP, QUANTITATIVE TRAIT LOCI, POSITION EFFECT.

**Marker (DNA marker)** A DNA fragment of known size used to calibrate an electrophoretic gel. See also ELECTROPHORESIS, TWO-DIMENSIONAL (2D) GEL ELECTROPHORESIS, DEOXYRIBONUCLEIC ACID (DNA).

**Marker (DNA sequence)** A specific sequence of DNA that is virtually always associated with a specified trait, because of "linkage" between that DNA sequence (the "marker") and the gene(s) that cause that particular trait. Such markers have been utilized to aid/speed up the process of plant (e.g., crop) breeding since the mid-1970s, via Marker Assisted Selection. See also DEOXYRIBONUCLEIC ACID (DNA), TRAIT, LINKAGE, LINKAGE GROUP, LINKAGE MAP, GENE, SEQUENCE (OF A DNA MOLECULE), MARKER ASSISTED SELECTION.

**Marker (genetic marker)** A trait that can be observed to occur or not to occur in an organism such as, e.g., bacteria or plant(s). Genetic markers include such traits as: expression of luciferase-catalyzed bioluminescence in leaf cells (causing leaves to glow when illuminated by certain light sources); resistance to specific antibiotics; the nature of the cell wall and capsule characteristics; requirements for a particular growth factor; and carbohydrate utilization, to mention a few. For example, if a culture of dividing (growing) bacteria that is not resistant to a particular antibiotic (i.e., lacks the trait of antibiotic resistance) is exposed to only the DNA isolated from bacteria that are resistant to the antibiotic, then a fraction of the cells exposed will directly incorporate this trait (some DNA) into their genome, hence acquiring the trait. The first genetically engineered plants bearing a marker gene were field tested in 1986. See also ALLELE, GENETIC ENGINEERING, POSITIVE AND NEGATIVE SELECTION (PNS), TRANSFORMATION, TRANSFECTION, NPTII GENE, BIOLUMINESCENCE, MARKER ASSISTED SELECTION, GUS GENE, bla GENE, RECOMBINASE.

**Marker Assisted Breeding** See MARKER ASSISTED SELECTION.

**Marker Assisted Selection** The utilization of DNA sequence "markers" by commercial breeders to select the organisms (crops, livestock, etc.) that possess gene(s) for a particular performance trait (rapid growth, high yield, etc.) desired, for subsequent breeding/propagation. Marker Assisted Selection has been utilized in many plant (e.g., crop) breeding programs since the mid-1970s. See also DEOXYRIBONUCLEIC ACID (DNA), SEQUENCE (OF A DNA MOLECULE), MARKER (DNA SEQUENCE), GENE, TRAIT, GENETIC MAP, LINKAGE, LINKAGE GROUP, MOLECULAR BREEDING, LINKAGE MAP, QUANTITATIVE TRAIT LOCI (QTL).

M

**MAS** See MARKER ASSISTED SELECTION.

**Mass Applied Genomics** See GENOMICS, BIO-CHIPS, MICROARRAYS (TESTING), BIOINFORMATICS.

**Mass Spectrometer** An analytical device that can be used to determine the molecular weights (mass) of proteins and nucleic acids, the sequence of (composition and order of amino acids comprising) protein molecules, the chemical composition of virtually any material, and the rapid identification of intact gram-negative and gram-positive microorganisms (the latter, using matrix-assisted laser desorption ionization time of flight mass spectrometry). See also GRAM-NEGATIVE, GRAM-POSITIVE, MOLECULAR WEIGHT, SEQUENCING (OF DNA MOLECULES), PROTEIN, AMINO ACID, NUCLEIC ACIDS, GENE MACHINE, MALDI-TOF-MS.

**Mast Cells** Fixed (noncirculating) cells that are present in many different kinds of body tissues. When two IgE molecules of the same antibody "dock" at adjacent receptor sites on a mast cell, then (the two IgE molecules) capture an allergen (e.g., a particle of pollen) between them, a chemical-energetic signal is sent to the interior (inside mast cell) portion of receptor molecules, which causes that interior portion of molecule to change (i.e., transduction). That signal transduction causes a protein named "syk" to set off a chemical chain reaction inside the mast cell; thereby causing that mast cell to release leukotrienes, histamine, serotonin, bradykinin, and "slow reacting substance." Release of these chemicals into the body causes the blood vessels to become more permeable (leaky) and causes the nose to run, and itchy and watery eyes. These chemicals also cause smooth muscle contraction, causing sneezing, breath constriction, coughing, wheezing, etc. See also BASOPHILS, ANTIGEN, ANTIBODY, RECEPTORS, SIGNAL TRANSDUCTION, HISTAMINE, ALLERGIES (FOODBORNE), SIGNALING, LEUKOTRIENES.

**Matrix Metalloproteinases (MMP)** A family of enzymes that contain the zinc metal ion ($Zn^{2+}$) at their active sites. Among this family are the collagenases. See also ENZYME, ION, ACTIVE SITE, CATALYTIC SITE, STROMELYSIN (MMP-3), COLLAGENASE.

**Maximum Residue Level (MRL)** Term used for an officially established upper allowable limit of a given compound (e.g., a synthetic hormone) in a particular product, such as meat. For example, in 1994, the Codex Alimentarius Commission in Rome, Italy, decided to establish maximum residue levels for each of five growth promotants commonly utilized by the U.S. beef industry. Because the World Trade Organization (WTO) subsequently stated that it would respect MRLs, a WTO member nation cannot legally refuse to allow import of meat products on growth promotant-content basis if the content of the promotant contained in the meat is less than its maximum residue level. See also GROWTH HORMONE, GROWTH FACTOR, CODEX ALIMENTARIUS COMMISSION, WORLD TRADE ORGANIZATION (WTO).

**MCA** See MEDICINES CONTROL AGENCY (MCA).

**MEA** Acronym for Multilateral Environmental Agreement; an agreement (treaty) between a number of nations intended to protect/benefit the environment. See also CONVENTION ON BIOLOGICAL DIVERSITY (CBD).

**Medicines Control Agency (MCA)** The British Government agency that, in concert with the Committee on Safety in Medicines, regulates the approval and sale of pharmaceutical products in the United Kingdom. See also COMMITTEE ON SAFETY IN MEDICINES, FOOD AND DRUG ADMINISTRATION (FDA), COMMITTEE FOR PROPRIETARY MEDICINAL PRODUCTS (CPMP), KOSEISHO, NDA (TO KOSEISHO), IND, BUNDESGE-SUNDHEITSAMT (BGA).

**Medifoods** See NUTRACEUTICALS, PHYTOCHEMI-CALS.

**Medium** A substance used to provide nutrients for cell growth. It may be liquid (e.g., broth) or solid (e.g., agar). See also CULTURE MEDIUM, AGAR, MAMMALIAN CELL CULTURE.

**Mega-Yeast Artificial Chromosomes (mega YAC)** A large (greater than 500 base pairs in length) piece of DNA that has been cloned (made) inside a living yeast cell. While most bacterial vectors cannot carry DNA pieces that are larger than 50 base pairs, and "standard" YACs typically cannot carry DNA pieces that are larger than 500 base pairs, mega YACs can carry DNA pieces (chromosomes) as large as one million base pairs in length. See also YEAST, CHROMOSOMES, HUMAN ARTIFICIAL CHROMOSOMES (HAC), *ARABIDOPSIS*

*THALIANA*, DEOXYRIBONUCLEIC ACID (DNA), CLONE (A MOLECULE), VECTOR, BASE PAIR (bp), YEAST ARTIFICIAL CHROMOSOMES (YAC).

**Megakaryocyte Stimulating Factor (MSF)** A colony stimulating factor (protein) involved in the regulation of platelet production, white blood cell production, and red blood cell production from stem cells in bone marrow. See also COLONY STIMULATING FACTORS (CSFs), PLATELETS, STEM CELLS.

**Meiosis** Discovered by Edouard Van Beneden in the 1870s, meiosis is the sequence of complex cell nucleus changes resulting in the production of cells (as gametes) with half the number of chromosomes present in the original cell. It typically involves an actual reduction division in which the chromosomes without undergoing prior splitting join in pairs with homologous chromosomes (of maternal and paternal origin) and then separate (i.e., pulled apart by microtubules within the cell), so that one member of each pair enters each product cell nucleus and undergoes a second division not involving reduction. Occurs by two successive divisions (meiosis I and II) that reduce the starting number of 4n chromosomes to 1n in each of four product cells. Product cells may mature to germ cells (sperm or eggs). See also OOCYTES, CELL, CHROMOSOME, NUCLEUS, MICROTUBULES.

**Melting (of DNA)** Melting DNA means to heat-denature it. When this happens, the hydrogen bonds holding the DNA molecule together in the normal way are disrupted, allowing a more random polymer structure to exist. See also DENATURED DNA.

**Melting (of substance other than DNA)** To change from a solid to a nonsolid (e.g., liquid) state by the addition of heat (to the solid substance).

**Melting Temperature (of DNA) (Tm)** The midpoint of the temperature range over which DNA is denatured. See also MELTING (OF DNA).

**Membrane Transport** The facilitated transport of a solute across a membrane, usually by a specific membrane protein (e.g., adhesion molecule). See also ENDOCYTOSIS, EXOCYTOSIS, SIGNAL TRANSDUCTION, G-PROTEINS, VAGINOSIS, RECEPTORS, ADHESION MOLECULE, VESICULAR TRANSPORT, GATED TRANSPORT, CALCIUM CHANNEL-BLOCKERS.

**Membrane Transporter Protein** A class of transmembrane proteins (i.e., protein molecules embedded in a cell's membrane, extending through both sides of the membrane) that function to transport certain molecules through the cell's membrane. Such molecules which are thus "transported" include: sugar molecules (utilized by the cell as "fuel"); inorganic ions (which catalyze certain cellular processes); polypeptides [e.g., "manufactured" in the cell's ribosome(s) and then secreted from the cell to perform some function elsewhere in the body of the organism]; anticancer drugs; antibiotics. See also PROTEIN, CELL, PLASMA MEMBRANE, MEMBRANE TRANSPORT, RIBOSOMES, POLYPEPTIDE (protein), ABC TRANSPORTERS.

**Membranes (of a cell)** Refers to the thin "skin-like" structures that surround the exterior of a cell (i.e., plasma membrane), and also surround various specialized bodies (nucleus, mitochondria, etc.) within the cell itself (e.g., the membrane that surrounds the cell's nucleus is called the "nuclear envelope"). Membranes are lipoidal, i.e., made of fat-like material, in which proteins and protein complexes are embedded. For example, protein molecules known as receptors are embedded in the plasma membrane (i.e., the outermost membrane of the cell) and in the nuclear envelope. See also CELL, CECROPHINS (LYTIC PROTEINS), MAGAININS, PLASMA MEMBRANE, TRANSMEMBRANE PROTEINS, ION CHANNELS, RECEPTORS, NUCLEAR RECEPTORS.

**MEMS (nanotechnology)** Acronym utilized by Americans to refer to "micro-electromechanical systems" (which Europeans tend to refer to as "microsystems technology" — MST). See also NANOTECHNOLOGY, BIOCHIP, GENOSENSORS, BIOSENSORS (ELECTRONIC), BIOSENSORS (CHEMICAL), NANOCRYSTAL MOLECULES MICROFLUIDICS, QUANTUM WIRE, QUANTUM DOT, MOLECULAR MACHINES, BIOMOTORS, BIOMEMS.

**mEPSPS** The "m" variant (of the many forms of) the enzyme 5-enolpyruvyl-shikimate-3-phosphate synthase. mEPSPS is unaffected by glyphosate- or sulfosate-containing herbicides, so introduction of the gene (coding for mEPSPS) into crop plants (e.g.,

M

corn/maize) makes those crop plants essentially impervious to glyphosate- or sulfosate-containing herbicides. See also ENZYME, GENE, GENETIC ENGINEERING, EPSP SYNTHASE, GLYPHOSATE, SULFOSATE, CORN, HERBICIDE-TOLERANT CROP, ARO A.

**Mesenchymal Adult Stem Cells** See MESODERMAL ADULT STEM CELLS.

**Mesodermal Adult Stem Cells** Certain stem cells present within (adult) bodies of organisms, that can be differentiated (via chemical signals) to give rise to bone, muscle, and/or fat cells. See also STEM CELLS, MULTIPOTENT ADULT STEM CELLS, CELL, ORGANISM, SIGNALING.

**Mesophile** An organism that grows best in the temperature range of 25°C (77°F) to 40°C (104°F). See also THERMOPHILE, PSYCHROPHILE.

**"Messenger" Molecule** See SIGNALING MOLECULE, HORMONE, NITRIC OXIDE.

**Messenger RNA (mRNA)** Messenger ribonucleic acid. The intermediary molecule between DNA and ribosomes (in a cell) which synthesize (manufacture) those proteins coded for by the cell's DNA. Upon receiving the "message" encoded in the DNA, the messenger RNA passes through the ribosomes like a reel of punched paper passes through an old player piano (pianola), giving the ribosomes the specifications for making the coded-for proteins. This process is aided by transfer RNA (tRNA) molecules, which forage for amino acids that float around in the cell (outside of the cell's nucleus and ribosomes). The transfer RNA (tRNA) molecules attach to, and escort, individual amino acids to the ribosome, as and when the messenger RNA (mRNA) directs. Each of the 20 different amino acids has at least one of its own purpose-built tRNA molecules, which possess a three-letter code of nucleotides at the stem of the cloverleaf-shaped rRNA molecule.

The ribosome has room for only two tRNA molecules at a time. The messenger RNA (mRNA) molecule (which itself is passing through the ribosome) calls over the first tRNA molecule, which brings with it the specified amino acid. Short sections of the messenger RNA (mRNA) and transfer RNA (tRNA) molecules lock together inside the ribosome (because where these two molecules meet, their three nucleotides are complementary), the whole (locked together) apparatus shifts along by three notches (i.e., nucleotides), and a second tRNA molecule (bearing another amino acid) slips in next to the first tRNA molecule.

Next, the first amino acid (brought in by the first tRNA molecule) jumps over to the second tRNA molecule, joining to the amino acid that was brought in by the second tRNA molecule, thus making the start of a protein (i.e., a poly-amino acid molecule, also known as polypeptide or protein molecule). The empty (first) tRNA molecule falls out of the ribosome, and the whole (locked together) apparatus (i.e., mRNA plus second tRNA molecule) moves three more notches (i.e., nucleotides) along the mRNA molecule to make room for a third tRNA molecule bearing another amino acid, and so on.

This process of creating ever-longer chains of amino acids continues to repeat itself inside the ribosome until the protein (coded for by the DNA, which code was transferred to mRNA, which transferred it to the ribosome) is completed. See also TRANSCRIPTION, COMPLEMENTARY DNA (c-DNA), CENTRAL DOGMA, DEOXYRIBONUCLEIC ACID (DNA), RIBONUCLEIC ACID (RNA), NUCLEIC ACIDS, CODING SEQUENCE, GENETIC CODE, CELL, INFORMATIONAL MOLECULES, CODON, RIBOSOMES, POLYRIBOSOME (POLYSOME), rRNA (RIBOSOMAL RNA), NUCLEOTIDE, POLYMER, TRANSFER RNA (tRNA), PROTEIN, AMINO ACID, POLYPEPTIDE (PROTEIN), ANTISENSE (DNA SEQUENCE).

**Messenger™** See HARPIN.

**Metabolic Engineering** The selective, deliberate alteration of an organism's metabolic pathway(s) via genetic engineering of the genes that define/control the organism's metabolism. Some reasons to do metabolic engineering of an organism include:

• Altering cell "behavior" and organism metabolic patterns to induce production of proteins/polypeptides and/or metabolites that are desired by mankind (e.g., "golden rice").

- Altering cell "behavior" and organism metabolic patterns to induce a given organism to consume or accumulate toxic wastes or valuable materials (e.g., gold) that are present at a site in low concentration or highly dispersed.
- Altering cell "behavior" and organism metabolic patterns to cure disease.

See also METABOLISM, INTERMEDIARY METABOLISM, CELL, PATHWAY, METABOLIC PATHWAY, GENETIC ENGINEERING, ORGANISM, GENE, GENE SPLICING, PROTEIN, PHYTO-MANUFACTURING, POLYPEPTIDES, BIOLEACHING, BIODESULFURIZATION, BIORECOVERY, BIOREMEDIATION, GOLDEN RICE, PHYTOREMEDIATION.

**Metabolic Pathway** Refers to a particular pathway [i.e., series of chemical reactions, each of which is dependent on previous one(s)] within the overall process of metabolism in an organism. For example, when humans consume the herb known as Saint John's Wort (*Hypericum perforatum*), certain components in that herb induce a (new) metabolic pathway — catalyzed by cytochrome P450 enzymes — that (more) rapidly metabolizes (i.e., breaks down) a number of commercial pharmaceuticals (thereby lowering the effectiveness of a given dose of that particular pharmaceutical). See also METABOLISM, PATHWAY, ORGANISM, INTERMEDIARY METABOLISM, CYTOCHROME p450, CYTOCHROME P4503A4, CATALYST, GOLDEN RICE.

**Metabolism** The entire set of enzyme-catalyzed transformations of organic nutrient molecules (to sustain life) in living cells. Conversion of food and water into nutrients that can be used by the body's cells, and the use of those nutrients by those cells (to sustain life, grow, etc.). See also ENZYME, CELL, INTERMEDIARY METABOLISM, METABOLITE, COMBINATORIAL BIOLOGY, CITRIC ACID, AFLATOXIN, *FUSARIUM*, CYTOCHROME P4503A4, PATHWAY, METABOLIC PATHWAY.

**Metabolite** A chemical intermediate in the enzyme-catalyzed chemical reactions of metabolism. See also METABOLISM, ENZYME, CELL, INTERMEDIARY METABOLISM, AFLATOXIN, *FUSARIUM*.

**Metalloenzyme** An enzyme having a metal ion as its prosthetic group. See also ENZYME, PROSTHETIC GROUP, METALLOPROTEINS.

**Metalloproteins** A term that is utilized to refer to any protein molecule that contains within it (i.e., in "peptide chain") a metal atom (zinc, iron, copper, etc.). Approximately one third of all proteins are metalloproteins. Those that contain a zinc atom ($Zn^{2+}$) are generally enzymes (thus called metalloenzymes), because that metal acts as a catalyst. See also PROTEIN, PEPTIDE, ENZYME, CATALYST, METALLOENZYME.

**Metamodel Methods (of Bioinformatics)** These refer to methods utilized to integrate data that has been independently generated/created (and generally stored in separate database models) via independent genomics research projects, combinatorial chemistry projects, high-throughput screening projects (e.g., via biochip use), etc. Metamodel methods sometimes reveal important interrelationships that were not apparent in the individual models (i.e., created solely for the genomics project data, or created solely for the combinatorial chemistry project data, or created solely for the high-throughput screening project data, etc.). See also BIOINFORMATICS, GENOMICS, FUNCTIONAL GENOMICS, STRUCTURAL GENOMICS, COMBINATORIAL CHEMISTRY, HIGH-THROUGHPUT SCREENING, BIOCHIP.

**Metastasis** The process via which a given cancer (e.g., initial tumor) spreads from the site of its initial formation (in body) to other parts of the body. See also CANCER, OLIGOSACCHARIDES, LECTINS, ANGIOGENESIS, GENISTEIN (Gen), ISOFLAVONES.

**Meter** A unit of measurement that was contrived by French scientists during the 1670s. It was initially defined to be one ten-millionth of the distance from the earth's equator to its poles. See also NANOMETERS (NM).

**Methionine (met)** An essential amino acid; furnishes (to organism) both labile methyl groups and sulfur necessary for normal metabolism. See also ESSENTIAL AMINO ACIDS, METABOLISM, CYSTINE, HIGH-METHIONINE CORN.

**Methyl Jasmonate** The volatile chemical compound that results when methyl groups ($CH_3$) are chemically added to a molecule of jasmonic acid. See also JASMONIC ACID.

**Methyl Salicylate** The volatile chemical compound that results when methyl groups

M

(CH$_3$) are added to a molecule of salicylic acid. During 1997, Ilya Raskin showed that methyl salicylate emitted by one tobacco plant (e.g., under 'attack' by insects, fungi, bacteria, or viruses) could cause other nearby tobacco plants to "turn on" their self-defense mechanism (systemic acquired resistance). See also SALICYLIC ACID (SA), BACTERIA, SYSTEMIC ACQUIRED RESISTANCE (SAR), FUNGUS.

**Methylated** Refers to a DNA molecule that is saturated with methyl groups (i.e., methyl submolecule groups, -CH$_3$, have attached themselves to the DNA molecule at all possible locations). Generally, when a DNA molecule is methylated, the genes comprising that DNA molecule are "turned off" (inactivated). See also DNA METHYLATION, DEOXYRIBONUCLEIC ACID (DNA), TRANSCRIPTION, MESSENGER RNA (mRNA), GENE, GENETIC CODE, p53 GENE, TUMOR-SUPPRESSOR GENES.

**MHC** See MAJOR HISTOCOMPATIBILITY COMPLEX (MHC).

**Micelle** The spherical structure formed by the association of a number of amphiphilic molecules dissolved in water. Structurally, the outer surface of the micelle (sphere) is covered with the polar domains (head groups) which are directed toward (stick into) the water while the interior of the micelle contains the nonpolar domains (tails), which self-associate to create an "oil droplet" microenvironment. Micelles may be used to solubilize nonwater (oil) soluble or sparingly water soluble molecules in water. They may be formed by ionic or nonionic surfactants. See also AMPHIPHILIC MOLECULES, SUPERCRITICAL CARBON DIOXIDE, CRITICAL MICELLE CONCENTRATION, REVERSE MICELLE (RM), SURFACTANT, FATS, SELF-ASSEMBLY.

**Micro Sensors** See BIOCHIP, MICROARRAY (TESTING), BIOSENSOR.

**Micro Total Analysis Systems** Abbreviated mTAS. See also GENE EXPRESSION ANALYSIS, BIOCHIP, GENOSENSORS, NANOTECHNOLOGY, BIOSENSORS (ELECTRONIC), BIOSENSORS (CHEMICAL).

**Micro-electromechanical Systems** See MEMS (NANOTECHNOLOGY).

**Microaerophile** An organism that grows best in the presence of a small amount of oxygen.

See also ORGANISM, MICROORGANISM, FACULTATIVE ANAEROBE.

**Microarray (testing)** Refers to a piece of glass, plastic, or silicon onto which has been placed a large number of biosensors. These microarrays (sometimes called "biochips" or "DNA chips") can then be utilized to test a single biological sample for a variety of attributes or effects. For example, by placing protein-detection molecules (e.g., ligands, which change color or cause electronic signal upon contact with specific protein molecules) onto a microarray, a scientist can perform gene expression analysis (i.e., evaluation of the protein expression and expression levels of genes in a biological sample).

Another application would be to place (cellular) receptors, nucleic acids/probes, adhesion molecules, messenger RNA (specific to which gene is "turned on" in a given disease state), cDNA (complementary to mRNA coded for by each gene that is "turned on"), or cells (indicating which cellular pathway is "turned on," etc.) onto a microarray, to utilize that microarray to screen for proteins or other chemical compounds that act against a disease (i.e., therapeutic target); as indicated by (the relevant component from biological sample) adhesion or hybridization to the specific spot on the microarray where a specific (target molecule) was earlier placed/attached. "Quantum dots" could potentially be used on microarrays in place of cellular receptors in the future. See also DNA CHIP, BIOCHIP, GENE, CODING SEQUENCE, GENE EXPRESSION, GENE EXPRESSION ANALYSIS, GENOSENSORS, NANOTECHNOLOGY, GENOMICS, FUNCTIONAL GENOMICS, BIOSENSORS (ELECTRONIC), BIOSENSORS (CHEMICAL), HIGH-THROUGHPUT SCREENING (HTS), TARGET-LIGAND INTERACTION SCREENING, RECEPTORS, BIORECEPTORS, COMBINATORIAL CHEMISTRY, TARGET (OF A THERAPEUTIC AGENT), TARGET (OF A HERBICIDE OR INSECTICIDE), ADHESION MOLECULE, MICROFLUIDICS, BIOELECTRONICS, ASSAY, BIOASSAY, MESSENGER RNA (mRNA), CHARACTERIZATION ASSAY, PROBE, HYBRIDIZATION (MOLECULAR BIOLOGY), BIOINFORMATICS, HYBRIDIZATION SURFACES, PATHWAY, DEOXYRIBONUCLEIC ACID (DNA), QUANTUM DOT, PROTEOME CHIP.

**Microbe** A microscopic organism; applied particularly to bacteria. The word "microbe" was coined by Monsieur Sedillot, a colleague of Louis Pasteur. See also BACTERIA, GENETICALLY ENGINEERED MICROBIAL PESTICIDES (GEMP), PHYTOALEXINS.

**Microbial Physiology** The cell structure, growth factors, metabolism, and genetics of microorganisms. See also MICROORGANISM, CELL, METABOLISM, GENETICS, MICROBIOLOGY.

**Microbial Source Tracking (MST)** The process of systematically determining the original source (in a specific environment) of a microbe (e.g., the one that has caused a given disease outbreak). Some of the technologies utilized in MST include genetic fingerprinting, polymerase chain reaction (PCR), serotyping, etc. See also MICROBE, PATHOGEN, POLYMERASE CHAIN REACTION (PCR) TECHNIQUE, SEROTYPES.

**Microbicide** Any chemical that will kill microorganisms. Used synonymously with the terms biocide and bactericide. See also MICROORGANISM, BIOCIDE.

**Microbiology** The science dealing with the structure, classification, physiology, and distribution of microorganisms, and with their technical and medical significance. The term microorganism is applied to the simple unicellular and structurally similar representatives of the plant and animal kingdoms. With few exceptions, the unicellular organisms are invisible to the naked eye and generally have dimensions of between a fraction of a micron and 200 microns. See also MICRON.

**Microchannel Fluidic Devices** See MICROFLUIDICS.

**Microfilaments** Very thin filaments found in the cytoplasm of cells. See also CELL, CYTOPLASM, MICROTUBULES.

**Microfluidic Chips** See BIOCHIP, MICROFLUIDICS, NANOTECHNOLOGY.

**Microfluidics** Refers to the science and properties of fluids when flowing through very small passages (e.g., micron or nanometer dimensions) and/or in very small amounts (e.g., femtogram quantities). For example, to move fluid (samples), microfluidic chips utilize either capillary action or else they "pump" fluid (through microchannels in those chips) electrokinetically (i.e., cause the flow to occur by applying a controlled electrical field, so liquid is attracted to electrical charge, and thereby flows). Such "pumping" could also be utilized to deliver certain medicines in very small, precisely timed and metered doses (e.g., if the microfluidic chip is embedded into diseased tissue within the body). Another potential application of such "pumping" could be to perform multiple chemical analyses (e.g., of body fluids within diseased tissues), in which case such microfluidic chips are known as "lab-on-a-chip"/laboratory-on-a-chip analytical devices. See also BIOCHIP, NANOTECHNOLOGY, MICROARRAY (TESTING), NANOSCIENCE, MICRON.

**Microgram** $10^{-6}$ gram, or $2.527 \times 10^{-8}$ ounce (avoirdupoir).

**Micromachining** Refers to the technology and tools or methods utilized to create the very small parts, grooves (in chips/arrays), etc. in NEMS (nanoelectromechanical systems), biochips, microarrays, and other devices of the field of nanotechnology. See also NANOTECHNOLOGY, NANOELECTROMECHANICAL SYSTEMS (NEMS), BIOCHIP, MICROARRAY (TESTING).

**Micron** Also called micrometer. A unit of length convenient for describing cellular dimensions; the Greek letter μ is used as its symbol. A micron is equal to $10^{-3}$ mm (millimeter) or 104 Å (Angstroms) or 0.00003937 inch. See also MICROBIOLOGY, CELL, MICROFLUIDICS.

**Microorganism** Any organism of microscopic size (i.e., requires a microscope to be seen by man). First viewed by Antoni van Leeuenhoek in 1676. Some microorganisms are pathogenic (disease-causing) and some are not. See also MICROBIOLOGY, BACTERIA, PATHOGENIC, NEMATODES, CAPSULE.

**Microparticles** Refers to the metal particles (R gene gun). See also BIOLISTIC R GENE GUN, VECTORS, MICRON, GENE.

**Microphage** See POLYMORPHONUCLEAR LEUKOCYTES.

**Micropropagation** A technique used by man to replicate (mass-produce) a given (e.g., valuable) plant by making genetic clones ("copies") of that original plant. See also CLONE (AN ORGANISM), GENETICS.

**Microsatellite DNA** Pieces of the same small segment (i.e., a DNA sequence) which are "repeated" (appear repeatedly in sequence within the DNA molecule) adjacent to a specific gene within the DNA molecule. Thus, these "microsatellites" are linked to that specific gene. See also DEOXYRIBONUCLEIC ACID (DNA), LINKAGE, SEQUENCE (OF A DNA MOLECULE), SATELLITE DNA, GENE, LINKAGE GROUP.

**Microsystems Technology** See MST (NANOTECHNOLOGY).

**Microtubules** Tiny hollow filaments (i.e., string-like structures) within eucaryotic cells, that are made of tubulin ($\alpha$ and $\beta$ proteins). Some microtubules give the cell its shape (e.g., act as structural components of cell). Other microtubules are the "tow ropes" utilized to move proteins within cells via vesicular transport (vesicles are small hollow structures that contain those protein molecules).

Microtubules also "tow" apart the paired chromosomes within cells undergoing meiosis. Within neurons (cells of the mammal nervous system), microtubules transport messenger RNAs (mRNA) from the nucleus (where they are manufactured) to the ribosomes in the dendrites (long extensions of the neuron cell), where the mRNAs are "translated" into protein molecules (i.e, proteins are manufactured by ribosome). See also CELL, MEIOSIS, MESSENGER RNA (mRNA), DENDRITES, PROTEIN, VESICULAR TRANSPORT (OF A PROTEIN), EUCARYOTE.

**Mid-Oleic Sunflowers** Refers to sunflower (crop) plant varieties which have been bred so their seeds contain 50–75% oleic acid within the oil in those seeds; vs. historical average of 20% oleic acid in the oil of traditional sunflower (crop) plant varieties. See also FATTY ACID, OLEIC ACID, HIGH-OLEIC OIL SOYBEANS.

**Mid-Oleic Vegetable Oils** Refers to any vegetable oils (other than sunflower oil) that contain 50–70% oleic acid. The range of oleic acid content is slightly different for mid-oleic sunflower oil definition. See also MID-OLEIC SUNFLOWERS, FATTY ACID, OLEIC ACID.

**Mimetics** See BIOMIMETIC MATERIALS.

**Minimized Domains** See MINIMIZED PROTEINS.

**Minimized Proteins** The domain/active site of a (former) native protein after all or most of its extraneous (unneeded) portions (peptides) have been removed. In 1995, Brian Cunningham and James A. Wells reduced the 28-residue (peptide) protein (hormone) Atrial Natriuretic Factor to 15-residues (peptides) size without reducing its potency (biological activity). Minimized proteins that retain their potency hold the potential for medicines possessing a greater serum lifetime (when injected into a patient's body), and as "models" for the creation of organic-chemical-synthesized mimetic drugs possessing the same therapeutic effect as the native protein did. See also PROTEIN, PEPTIDE, ACTIVE SITE, ENZYME, CATALYTIC SITE, DOMAIN (OF A PROTEIN), HORMONE, ATRIAL NATRIURETIC FACTOR, BIOMIMETIC MATERIALS, SERUM LIFETIME, BIOLOGICAL ACTIVITY.

**Minimum Tillage** See LOW-TILLAGE CROP PRODUCTION, NO-TILLAGE CROP PRODUCTION.

**"Miniprotein Domains"** See MINIMIZED PROTEINS.

**"Miniproteins"** See MINIMIZED PROTEINS.

**Mitochondria** Granular or rod-shaped bodies (organelles) in a cell's cytoplasm, that contain the zyme systems required in the citric acid cycle, electron transport, beta oxidation of fatty acids, and synthesis of ATP via oxidative phosphorylation. See also ZYME SYSTEMS, CELL, MITOCHONDRIAL DNA, CARNITINE, ADENOSINE TRIPHOSPHATE, FATTY ACIDS, FATS, PHOSPHOLIPIDS, CYTOCHROME, CYTOPLASM, CITRIC ACID CYCLE, ATP, Ac-CoA.

**Mitochondrial DNA** The DNA within an organism's (e.g., human) cells that is located inside the mitchondria (organelles); not inside the cell nucleus. Mitochondrial DNA is only passed down from mother to offspring; not from father to offspring, as nuclear DNA is. See also DEOXYRIBONUCLEIC ACID (DNA), CELL, MITOCHONDRIA, NUCLEUS, CYTOPLASMIC DNA.

**Mitogen** A substance (growth factor, hormone, etc.) that initiates cell division within the body. For example, most Angiogenic Growth Factors (e.g., fibroblast growth factor) stimulate cell division of the endothelial cells which line blood vessel walls. See also MITOSIS, GROWTH FACTOR, HORMONE, ANGIOGENIC GROWTH FACTORS, ENDOTHELIAL CELLS.

M

**Mitosis** A process of cell duplication, or reproduction, during which one cell gives rise to two identical daughter cells. See also MITOGEN, TUBULIN.

**Mixed-Function Oxygenases** Enzymes catalyzing simultaneous oxidation of two substances by oxygen, one of which is usually NADPH or NADH. See also NADPH, NADH, OXIDATION, ENZYME.

**Model Organism** Refers to an organism that is utilized (e.g., in scientific experiments) to conduct tests, etc. in an attempt to infer results applicable to larger, more complex organisms. For example, the use of the microscopic roundworm *C. elegans* in high-throughput screening to attempt to find pharmaceuticals that will be useful for humans. See also ORGANISM, DROSOPHILA, *CAENORHABDITIS ELEGANS* (*C. ELEGANS*), HIGH-THROUGHPUT SCREENING (HTS), *ARABIDOPSIS THALIANA*.

**Moiety** Referring to a part or portion of a molecule, generally complex, having a characteristic chemical or pharmacological property. See also ANALOGUE, PHARMACOPHORE.

**Mold** See FUNGUS.

**Mole** An Avogadro's number ($6.023 \times 10^{23}$) of whatever units are being considered. One gram molecular weight of an element or a compound (i.e., same number of grams of an element or a compound as that substance's molecular weight, equal to $6.023 \times 10^{23}$ molecules). See also MOLECULAR WEIGHT.

**Molecular Beacon** Term that is used to refer to specific oligonucleotides possessing a "hairpin loop" and bearing a fluorescent dye. A "quencher dye" located on a nearby portion of the hairpin loop prevents fluorescence until the hairpin loop is opened up. Molecular beacons (sometimes called fluorogenic probes) are utilized (e.g., in high-throughput screening or high-throughput identification) to detect the presence of a desired "target" molecule. When the "target" (i.e., a molecule possessing the desired functional group or desired property) is present within a given sample being evaluated, the "hairpin loop" opens up because a portion of it forms a stronger bond to the "target," than to the rest of the loop thereby allowing the fluorescent dye to emit light. See also OLIGONUCLEOTIDE, HAIRPIN LOOP, FLUORESCENCE, TARGET (OF A THERAPEUTIC AGENT), TARGET (OF A HERBICIDE OR INSECTICIDE), HIGH-THROUGHPUT IDENTIFICATION, HIGH-THROUGHPUT SCREENING (HTS).

**Molecular Biology** A term coined by Vannevar Bush during the 1940s that eventually came to mean the study and manipulation of molecules that constitute, or interact with, cells. Molecular biology as a distinct scientific discipline originated largely as a result of a decision to provide "support for the application of new physical and chemical techniques to biology" during the 1930s by Warren Weaver, director of the biology (funding) program at America's Rockefeller Foundation (a philanthropic organization). See also MOLECULAR GENETICS, GENETICS, GENETIC ENGINEERING, BIOLOGICAL ACTIVITY, BIOPOLYMER, BIOGENESIS, BIOCHEMISTRY, DEOXYRIBONUCLEIC ACID (DNA), MITOSIS, MEIOSIS.

**Molecular Breeding™** A trademarked term that refers to certain "molecular evolution" technologies developed by Maxygen Company. This term is also sometimes used to refer to the utilization of molecular genetics and/or marker assisted selection in a breeding program (e.g., within a seed company or within a university) to select the organisms (e.g., crop varieties) that possess gene(s) for a particular trait (higher yield, disease resistance, etc.). See also MARKER ASSISTED SELECTION, MOLECULAR EVOLUTION, GENE, TRAIT, MARKER (DNA SEQUENCE), QUANTITATIVE TRAIT LOCI (QTL).

**Molecular Chaperones** See CHAPERONES, PROTEIN FOLDING.

**Molecular Diversity** Sometimes referred to as "irrational drug design," this refers to the drug design technique of generating large numbers of diverse candidate molecules (e.g., pieces of DNA, RNA, proteins, or other organic moieties) at random (via a variety of methods). These diverse candidate molecules are then tested to see which is best at working against a disease/condition (e.g., fitting a cell receptor, or category of receptors relevant to the disease in question). Molecular candidates that show promise (e.g., via a "pretty good fit" to receptor) are then produced in larger quantities (e.g., via Polymerase Chain Reaction techniques) along with additional molecules that are similar though slightly different in structure (e.g., via site-directed mutagenesis) in an attempt to create a molecule that is a "perfect fit" (e.g., to receptor). See also RATIONAL DRUG DESIGN, DEOXYRIBONUCLEIC ACID (DNA), RIBONUCLEIC ACID (RNA), RECEPTORS, RECEPTOR FITTING (RF), RECEPTOR MAPPING (RM), MOIETY, POLYMERASE CHAIN REACTION (PCR), SITE-DIRECTED MUTAGENESIS, DIVERSITY BIOTECHNOLOGY CONSORTIUM, COMBINATORIAL CHEMISTRY, COMBINATORIAL BIOLOGY.

**Molecular Evolution** See COMBINATORIAL CHEMISTRY.

**Molecular Fingerprinting** See COMBINATORIAL CHEMISTRY.

**Molecular Genetics** The science dealing with the study of the nature and biochemistry of the genetic material. Includes the technologies of genetic engineering. See also GENETICS, GENETIC ENGINEERING, MOLECULAR BIOLOGY, BIOLOGICAL ACTIVITY, BIOPOLYMER, BIOGENESIS, BIOCHEMISTRY, DEOXYRIBONUCLEIC ACID (DNA), MITOSIS, MEIOSIS, MOLECULAR DIVERSITY, CENTRAL DOGMA.

**Molecular Machines** Refers to nanometer-dimension "machines" capable of doing various tasks. See also NANOTECHNOLOGY, NANOMETERS (NM), BIOMOTORS, NANOBOTS, NANOELECTROMECHANICAL SYSTEM (NEMS), NANOSCIENCE.

**Molecular Pharming™** A trademark of the Groupe Limagrain company, it refers to the production of pharmaceuticals and certain other chemicals (e.g., intermediates utilized to manufacture pharmaceuticals) in agronomic plants (which have been genetically engineered). See also ANTIBIOTIC, GENETIC ENGINEERING, PHYTOCHEMICALS, "EDIBLE VACCINES", CORN, PLANTIBODIES™.

**Molecular Weight** The sum of the atomic weights of the constituent atoms in a molecule. See also ATOMIC WEIGHT.

**Monarch Butterfly** Refers to the insect (*Lepidoptera*: *Danaidae* or *Danaus plexippus*) whose pupae (caterpillars) feed exclusively on tissue of the plant known as common milkweed (*Asclepias syriaca*), and whose territory extends from northern Mexico to approximately Canada's southern border. See also BACILLUS THURINGIENSIS (*B.t.*), *B.t.* KURSTAKI, *B.t.* TOLWORTHI, CRY1A (b) PROTEIN.

**Monoclonal Antibodies (MAb)** Discovered and developed in the 1970s by Cesar Milstein and Georges Kohler, monoclonal antibodies are the name for antibodies derived from a single source or clone of cells that recognize only one kind of antigen. Made by fusing myeloma cancer cells (which multiply very fast) with antibody-producing cells, then spreading the resulting conjugate colony so thin that each cell can be grown into a whole, separate colony (i.e., cloning). In this way, one gets whole batches of the same (monoclonal) antibody, which are all specific to the same antigen.

Monoclonal antibodies have found markets in diagnostic kits and show potential for use in drugs (e.g., to shrink tumors), imaging agents, and in purification processes. One example of a diagnostic use is the invention in 1997 by Bruno Oesch of a monoclonal antibody-based rapid test to detect the prion (PrP 5c) that causes bovine spongiform encephalopathy (BSE) in cattle. See also ASCITES, MYELOMA, TUMOR, CORN, IMMUNOTOXIN, BLAST CELL, ANTIGEN, ANTIBODY, SINGLE-DOMAIN ANTIBODIES (dAbs), MURINE, CATALYTIC ANTIBODY, SEMISYNTHETIC CATALYTIC ANTIBODY, BSE, PRION, HER-2 GENE.

**Monocytes** Also called monocyte macrophages. The round-nucleated cells that circulate in the blood. In summary they engulf and kill microorganisms, present antigen to the lymphocytes, kill certain tumor cells, and are involved in the regulation of inflammation.

These cells are often the first to encounter a foreign substance or pathogen or normal cell debris in the body. When they do, the material is taken up (engulfed) and degraded by means of oxidative and hydrolytic enzymatic attack. Peptides that result from the degradation of foreign protein are then bound to a monocyte protein called class II MHC (major histocompatibility complex) and this self-foreign complex then migrates to the surface of the cell where it is embedded into the cell membrane in such a way as to present the peptide to the outside of the cell. This positioning allows T lymphocytes to recognize (inspect) the peptide. Whereas self-peptides derived from normal cellular debris are ignored, foreign peptides activate precursors of helper T cells to further mature into active, lymphokine-secreting helper T lymphocytes, also known as TH cells. When monocytes move out of the bloodstream and into the tissues they are then called macrophages. See also MACROPHAGE, CELLULAR IMMUNE RESPONSE, PATHOGEN, MHC.

**Monoecious** A category of plants (e.g., the soybean plant) that possess both male and female reproductive structures on the same plant. Thus, such plants are capable of self-pollination. For example, 95% of the pollen from a soybean plant (*Glycine max*) does not leave the flower in which it was produced. Virtually none of a given soybean plant's pollen leaves the plant in which it was produced. See also SOYBEAN PLANT, BARNASE.

**Monomer** The basic molecular subunit from which, by repetition of a single reaction, polymers are made. For example, amino acids (monomers) link together via condensation reactions to yield polypeptides or proteins (polymers). A monomer is analogous to a link (monomer) in a metal chain (polymer). See also POLYMER.

**Monosaccharides** The chemical building blocks of carbohydrates, hence known as "simple sugars." They are classified by the number of carbon atoms in the (monosaccharide) molecule. For example, pentoses have five and hexoses have six carbon atoms. They normally form ring structures. The empirical formula for monosaccharides is $(CH_2O)n$. See also OLIGOSACCHARIDES, CARBOHYDRATES, SUGAR MOLECULES.

**Monounsaturated Fats** Fat molecules possessing one less than the maximum possible number of hydrogen atoms (on that given fat molecule). Diets that are high in monounsaturated fat content have been shown to reduce low-density lipoproteins ("bad" cholesterol) blood content, while leaving blood levels of high-density lipoproteins ("good" cholesterol) essentially unchanged. See also FATTY ACID, SATURATED FATTY ACIDS, DEHYDROGENATION, UNSATURATED FATTY ACID, LOW-DENSITY LIPOPROTEINS (LDLP), HIGH-DENSITY LIPOPROTEINS (HDLPs), OLEIC ACID, FATS.

**Monounsaturated Fatty Acids (MUFA)** Refers to the category of those fatty acids (e.g., oleic acid) that possess one less than

the maximum possible number of hydrogen atoms (e.g., possible to be attached to the molecular structure of oleic acid). Enzymes (e.g., Δ12 desaturase) present in some oilseed plants (soybean, corn/maize, canola, etc.) convert some MUFAs to polyunsaturated fatty acids (PUFAs) within their developing seeds. Diets that are high in monounsaturated fatty acid content have been shown to reduce low-density lipoproteins ("bad" cholesterol) blood content while simultaneously leaving blood levels of high-density lipoproteins ("good" cholesterol) essentially unchanged. Soybean oil has historically averaged approximately 24.5% monounsaurated fatty acid content by weight. See also MONOUNSATURATED FATS, FATTY ACID, UNSATURATED FATTY ACID, SOYBEAN OIL, OLEIC ACID, LOW-DENSITY LIPOPROTEINS (LDLP), DELTA 12 DESATURASE, POLYUNSATURATED FATTY ACIDS (PUFA).

**Morphogenetic** An adjective referring to formation and differentiation of tissues and organs in an organism. See also MORPHOLOGY, STEM CELLS, TOTIPOTENT STEM CELLS.

**Morphology** First used in print by the poet Johann Wolfgang von Goethe, this word is utilized to refer to the form/structure of an organism or any of its parts. See also TRAIT, PHENOTYPE.

**MOS** See MANNANOLIGOSACCHARIDES.

**MRA** See MUTUAL RECOGNITION AGREEMENTS, MUTUAL RECOGNITION ARRANGEMENTS.

**MRL** See MAXIMUM RESIDUE LEVEL.

**mRNA** See MESSENGER RNA.

**MSF** See MEGAKARYOCYTE STIMULATING FACTOR.

**MST (microbes)** See MICROBIAL SOURCE TRACKING.

**MST (nanotechnology)** Acronym utilized by Europeans to refer to "microsystems technology" (i.e., their common term for "microelectromechanical systems" — MEMS). See also NANOTECHNOLOGY, BIOCHIP, GENOSENSORS, BIOSENSORS (ELECTRONIC), BIOSENSORS (CHEMICAL), QUANTUM WIRE, QUANTUM DOT, NANOCRYSTAL MOLECULES, MICROFLUIDICS, BIOMOTORS, MOLECULAR MACHINES.

**MTAS** See MICRO TOTAL ANALYSIS SYSTEMS.

**MUFA** See MONOUNSATURATED FATTY ACIDS (MUFA).

.M

**Multi-Copy Plasmids** Plasmids present inside bacteria in quantities greater than one plasmid per (host) cell. See also PLASMID, VECTOR, COPY NUMBER.

**Multienzyme System** A sequence of related enzymes participating in a given metabolic (chemical reaction) pathway.

**Multiple Sclerosis** A disease in which the human body's immune cells attack myelin (the "insulation" that surrounds nerve fibers in the spinal cord and brain) and the body's acetyl choline receptors. That leads to recurrent muscle weakness, loss of muscle control, and (potentially) eventual paralysis. See also AUTOIMMUNE DISEASE, THYMUS, ACETYL-CHOLINE, RECEPTORS, IMMUNE RESPONSE, NEUROTRANSMITTER, EXCITATORY AMINO ACIDS (EAAs).

**Multipotent Adult Stem Cell** Certain stem cells present within (adult) bodies of organisms, that can be differentiated (via chemical signals) to give rise to a variety of different cell/tissue types (bone, cartilage, fat, muscle, red blood cells, B cells, T cells, etc.). See also STEM CELLS, CELL, ORGANISM, SIGNALING, RED BLOOD CELLS, B CELLS, T CELLS, MESODERMAL ADULT STEM CELLS.

**Murine** Of, or pertaining to, mice. For example, the first monoclonal antibodies were produced using cells from mice. This frequently caused adverse immune responses to monoclonal antibodies when they were injected into the human body (e.g., thus limiting their use in therapeutic purposes). However, researchers have recently discovered how to make monoclonal antibodies in human cells. See also MONOCLONAL ANTIBODIES (MAb).

**Muscular Dystrophy (MD)** A genetic disease caused by a defect in the X chromosome (resulting in nonexpression of the Duchenne Muscular Dystrophy gene); first recognized by G. A. B. Duchenne in 1858. The disease afflicts males almost exclusively because males have only one X chromosome, whereas females inherit two copies of the X chromosome and have a "backup" in case one X chromosome is damaged (as is the case for MD victims). In 1981, Kay E. Davies used DNA probes (genetic probes) to discover that the Duchenne Muscular Dystrophy (DMD) gene must lie somewhere between two unique (to MD victims) segments on the upper, shorter arm of the X chromosome. See also DNA PROBE, CHROMOSOMES, KARYOTYPE, CHROMATIDS, CHROMATIN, SINGLE-NUCLEOTIDE POLYMORPHISMS (SNPs).

**Mutagen** A chemical substance capable of producing a genetic mutation (change), by causing changes in the DNA of living organisms. For example, Dr. Gary Shaw discovered in 1996 that women who smoke cigarettes during their pregnancies are twice as likely to have babies with the genetic deformity known as cleft lip and palate. If those women have a particularly susceptible (to smoke) gene variant (allele) within their DNA, they are as much as eight times as likely to have babies with cleft lip and palate. According to the World Health Organization (WHO), 60–80% of all known mutagens are also carcinogens (cancer-causing). See also MUTATION, GENE, GENETICS, HEREDITY, GENETIC CODE, CANCER, CARCINOGEN, ALLELE, DEOXYRIBONUCLEIC ACID (DNA), ONCOGENES, MUTANT, ANTIOXIDANTS.

**Mutant** An altered cell or organism resulting from mutation (an alteration) of the original wild (normal) type. A change from the normal to the unique or abnormal. See also MUTAGEN, HEREDITY, WILD TYPE.

**Mutase** An enzyme catalyzing transposition of a functional group in the substrate (substance acted upon by the enzyme). Intramolecular transfer of a chemical group from one position (i.e., carbon atom) to another within the same molecule. An example of a mutase is phosphoglucomutase. It has a molecular weight of about 60,000 Daltons with about 600 amino acid residues (monomers). The mutase can interchange (move) a phosphate unit between the 1 and 6 position. The 1 refers to a carbon atom designated as "#1" and the 6 refers to a different carbon atom designated as "#6."

**Mutation** From the Latin term *mutare*, meaning to change. Any change that alters the sequence of the nucleotide bases in the genetic material (DNA) of an organism or cell; with alteration occurring either by displacement, addition, deletion, cross-linking, or other destruction. The mutation alteration to the DNA sequence would alter its meaning, i.e., its ability to produce the normal amount or normal kind of protein, so the organism or cell is itself altered. Such an altered organism is called a mutant. See also MUTANT, INFORMATIONAL MOLECULES, HEREDITY, GENETIC CODE, GENETIC MAP, PROTEIN, DEOXYRIBONUCLEIC ACID (DNA).

**Mutation Breeding** Refers to several techniques, involving induced mutations, that were utilized by some crop plant breeders (primarily in the 1960s and 1970s) to introduce desirable genes into the plants with which they were working. For example, gene(s) to confer resistance to plant diseases, increased yield per acre/hectare or improvements in composition that were not present within the historic/natural germplasm of that plant species. These new-to-that-species genes were "created" via soaking its seeds or pollen in mutation-causing chemicals (i.e., mutagens), or via bombardment of seeds with ionizing radiation; followed by grow-out of the resultant plants and selection of the particular mutation (i.e., beneficial trait) desired by the plant breeder. That plant was then propagated via straightforward breeding to yield seeds that are still sown today. See also TRADITIONAL BREEDING METHODS, MUTATION, MUTAGEN, GENE, TRAIT, WHEAT, BARLEY, POINT MUTATION.

**Mutual Recognition Agreements (MRAs)** Legal agreements (treaties) between two or more nations, to recognize and respect each other's approval process (e.g., for new crops derived via biotechnology). See also GMO, COMMITTEE FOR VETERINARY MEDICINAL PRODUCTS (CVMP), ORGANIZATION FOR ECONOMIC COOPERATION AND DEVELOPMENT (OECD), EVENT, EUROPEAN MEDICINES EVALUATION AGENCY (EMEA), COMMITTEE FOR PROPRIETARY MEDICINAL PRODUCTS (CPMP), UNION FOR PROTECTION OF NEW VARIETIES OF PLANTS (UPOV).

**Mutual Recognition Arrangements** See MUTUAL RECOGNITION AGREEMENTS (MRAs).

*Mycobacterium tuberculosis* The pathogen that causes tuberculosis, a human disease in which the lungs are destroyed as this bacteria grows (within lung tissue). In 1998, scientists completed sequencing of the genome of *Mycobacterium tuberculosis*. Recently, a new strain of *M. tuberculosis*, that is resistant to virtually all commercial antibiotics, has begun to infect some people. See also BACTERIA, PATHOGEN, SEQUENCING (OF DNA MOLECULES), ANTIBIOTIC, ANTIBIOTIC RESISTANCE, GENOME, STRAIN.

**Mycotoxins** Toxins produced by fungi. More than 350 different mycotoxins are known to man, but the first ones to be isolated and scientifically characterized (i.e., described) were the aflatoxins, in 1961. The second group of mycotoxins to be isolated and characterized were the ochratoxins, in 1965.

Almost all mycotoxins possess the capacity to harmfully alter the immune systems of animals. Consumption by animals (including humans) of certain mycotoxins (via eating infected corn/maize, wheat, certain tree nuts, peanuts, cottonseed products, etc.) can result in liver toxicity, gastrointestinal lesions, cancer, muscle necrosis, etc. See also TOXIN, FUNGUS, *FUSARIUM*, AFLATOXIN, VOMITOXIN, *FUSARIUM MONILIFORME*, FUMONISINS, ZEARALENONE, OCHRATOXINS, ERGOTAMINE.

**Myeloma** A tumor cell line derived from a lymphocyte. It usually produces a single type of immunoglobulin. See also HYBRIDOMA, LYMPHOCYTE, AGING.

**Myoelectric Signals** The nerve signals that are sent by the body in order to control muscle movement.

**Myristoylation** Transformation of proteins in cells in such a manner that these cells then cause cancer. See also CANCER.

M

# N

**N Glycosylation** See GLYCOSYLATION.

**n-3 Fatty Acids** Also known as "omega-3" fatty acids. Research indicates there are human health benefits (e.g., antithrombotic, reduce/avoid coronary heart disease) if the ratio of n-6 to n-3 fatty acids contained in the diet is higher than 3, but less than 10. Soybean oil has an n-6/n-3 ratio of approximately 7:1. Examples of n-3 fatty acids include linolenic acid (C18:3n-3). Research indicates that human consumption of n-3 fatty acid(s) imparts anti-thrombotic and anti-inflammatory health benefits; plus it lowers levels of triglycerides content in the bloodstream. During 2000, research was published that indicated a 66% reduction in probability for children to develop juvenile (Type I) diabetes, if their mothers consumed significant quantities of n-3 fatty acids during pregnancy. See also POLYUNSATURATED FATTY ACIDS (PUFA), DOCOSAHEXANOIC ACID (DHA), EICOSAPENTANOIC ACID (EPA), LINOLENIC ACID, SOYBEAN OIL, THROMBOSIS, TRIGLYCERIDES, CORONARY HEART DISEASE (CHD), DIABETES, INSULIN.

**n-6 Fatty Acids** Also known as "omega-6" fatty acids. Research indicates there are human health benefits (e.g., antithrombotic, reduce/avoid coronary heart disease) if the ratio of n-6 to n-3 fatty acids contained in the diet is higher than 3 but less than 10. Soybean oil has an n-6/n-3 ratio of approximately 7:1. Examples of n-6 fatty acids include linoleic acid (C18:2n-6). Research indicates that consumption of n-6 fatty acids has been related to decreased cholesterol levels in the bloodstream, and decreased incidence of coronary heart disease (CHD). See also POLYUNSATURATED FATTY ACIDS (PUFA), ARACHIDONIC ACID, LINOLEIC ACID, SOYBEAN OIL, THROMBOSIS, CORONARY HEART DISEASE (CHD), CHOLESTEROL.

**NAD (NADH, NADP, NADPH)** Nicotinamide-adenine dinucleotide, also known as diphosphopyridine nucleotide, codehydrogenase 1, coenzyme 1, and coenzymase by its discoverers, Harden and Young. C21H27O14N7 P2. An organic coenzyme (molecule) that functions as a distinct yet integral part of certain enzymes. NAD plays a role in certain enzymes concerned with oxidation/reduction reactions. Meanings: NADH, nicotinamide-adenine dinucleotide, reduced; NADP, nicotinamide-adenine dinucleotide phosphate; and NADPH, nicotinamide-adenine dinucleotide phosphate, reduced. See also ENZYME, COENZYME, OXIDATION-REDUCTION REACTION, NITRIC OXIDE SYNTHASE.

**NADA (New Animal Drug Application)** An application to the U.S. Food and Drug Administration (FDA) to begin testing/studies of a new drug for animals (e.g., livestock), that might (eventually) lead to its FDA approval. See also IND.

**NADH** Nicotine-adenine dinucleotide, reduced. See also NAD.

**NADP** Nicotine-adenine dinucleotide phosphate. See also NAD, NITRIC OXIDE SYNTHASE.

**NADPH** Nicotinamide-adenine dinucleotide phosphate, reduced. See also NAD.

**Naked DNA** See NAKED GENE.

**Naked Gene** A bare gene (strand of DNA that codes for a protein) that has been extracted from an organism, or otherwise derived (e.g., synthesized from sequence data). During the 1990s, it was discovered that:

- Injecting the Duchenne Muscular Dystrophy "naked gene" into muscle tissue in the bodies of people suffering from

Muscular Dystrophy (MD) resulted in temporary production of the relevant protein in that muscle tissue (i.e., temporary MD symptom reduction).

- Injecting the VEGF "naked gene" into relevant tissue in the bodies of people suffering from inadequate local blood supply (the shortage of blood flow to heart known as myocardial ischemia, lack of blood flow in legs or other extremities, etc.) resulted in (new) growth of blood vessels/endothelium, and reduction in symptoms of those inadequate blood-supply conditions.

- Injecting the "naked gene" for the relevant antigen of certain pathogens into some tissues in the (usual disease host) organism sometimes resulted in those (host organism) tissues taking up the "naked gene" and expressing some of the (pathogen's) antigen(s), such that the (putative host organism's) immune system initiates an immune response (thereby resulting in vaccination against the disease conferred by the pathogen). When that happens, such "naked genes" are referred to as "DNA vaccines."

See also GENE, DEOXYRIBONUCLEIC ACID (DNA), PROTEIN, ORGANISM, SYNTHESIZING (OF DNA MOLECULES), SEQUENCING (OF DNA MOLECULES), DUCHENNE MUSCULAR DYSTROPHY GENE, MUSCULAR DYSTROPHY (MD), VASCULAR ENDOTHELIAL GROWTH FACTOR (VEGF), PATHOGEN, EXPRESS, DNA VACCINES, IMMUNE RESPONSE, CELLULAR IMMUNE RESPONSE, HUMORAL IMMUNITY, ANTIBODY, DNA VECTOR.

**Nanobiology** See NANOTECHNOLOGY, NANOCOMPOSITES, BIOINORGANIC, NANOCRYSTALS, NANOELECTROMECHANICAL SYSTEM (NEMS).

**Nanobots** Refers to very small "robots" whose dimensions could be measured in terms of nanometers (nm), and could perform specific tasks. See also NANOELECTROMECHANICAL SYSTEM (NEMS), NANOSCIENCE, MEMS (NANOTECHNOLOGY), BIOMEMS, NANOMETERS (NM), NANOTECHNOLOGY.

**Nanocomposites** Nanometer-scale composite structures composed of organic molecules intimately incorporated with inorganic molecules. For example, abalone shellfish make mother-of-pearl shells via an intimate combination of protein and calcium carbonate. Researchers are working on making semiconductor devices (chips) containing peptides and other organic molecules attached to silicon or gallium arsenide. They are also working on nanoelectromechanical systems (NEMS) that would have tiny "moving parts" to be able to do "work" at nanometer scale. See also NANOMETERS (NM), NANOTECHNOLOGY; PROTEIN, BIOCHIP, PEPTIDE, BIOSENSORS (ELECTRONIC), BIOINORGANIC, NANOELECTROMECHANICAL SYSTEM (NEMS).

**Nanocrystal Molecules** Coined by researchers A. Paul Alivisatos and Peter G. Schultz, it is a term used to describe double-stranded DNA molecules that have several multiatom clusters of gold attached to them. As of 1996, these researchers were working to try to create nanometer-scale electrical circuits, semiconductors, etc. A separate methodology, researched by Chad A. Mirkin et al., utilizes strands of DNA to reversibly assemble gold nanoparticles (nanometer-scale multi-atom particles) into supramolecular (many molecule) agglomerations, in which the gold particles are separated from each other by a distance of approximately 60 Angstroms. The aggregation of these DNA-metal nanoparticles causes a visible color change.

As of 1996, these researchers were working to try to create simple and rapid tests that would indicate the presence of a virus (e.g., HIV-1 or HIV-2) via a visible color change. Such a test would use two noncomplementary DNA sequences, each of which has attached to it a gold nanoparticle (via a thiol group). The two sequences would be selected for their ability to latch onto a target sequence in the desired virus, but they would be unable to combine with each other, since they are noncomplementary. When double-stranded DNA molecules possessing two "sticky ends" (that are complementary to the sequences attached to virus) are added, the resultant color change indicates virus presence. See also DOUBLE HELIX, DEOXYRIBONUCLEIC ACID (DNA), ANGSTROM (Å), NANOMETERS (nm), HYBRIDIZATION SURFACES, BASE PAIR (bp), SELF-ASSEMBLY, NANOTECHNOLOGY, STICKY ENDS,

HYBRIDIZATION (MOLECULAR GENETICS), SEQUENCE (OF A DNA MOLECULE), VIRUS, BIO-SENSORS (CHEMICAL), BIOCHIP, MICROFLUIDICS, NANOCRYSTALS.

**Nanocrystals** A term used to refer to any crystalline structure possessing dimensions (e.g., overall width) measured in terms of nanometers. See also NANOMETERS (NM), QUANTUM DOT, NANOSCIENCE, NANOTECHNOLOGY, NANOCRYSTAL MOLECULES, NANOCOMPOSITES.

**Nanoelectromechanical System (NEMS)** Refers to working (i.e., those with moving "mechanical parts") systems of a scale whose relevant dimensions are measured in terms of nanometers (nm). For example, in 2000, Carlo Montemagno and colleagues assembled a NEMS in which a tiny metal "propeller" was caused to spin within the domain of the enzyme ATP Synthase. The metal propeller was attached (via a biotin-streptavidin "molecular linkage") to the one subunit (designated alpha) of ATP Synthase that rotates within the other (hollow) part of ATP Synthase molecule when ATP is "fed" to a free standing (i.e., not in cell) molecule of ATP Synthase. See also NANOMETERS (NM), ATP SYNTHASE, ENZYME, ADENOSINE TRIPHOSPHATE (ATP), BIOTIN, AVIDIN, NANOCOMPOSITES, NANOSCIENCE, MICROMACHINING.

**Nanofluidics** See MICROFLUIDICS.

**Nanogram (ng)** $10^{-9}$ gram, or $3.527 \times 10^{-11}$ ounce (avoirdupoir).

**Nanometers (nm)** $10^{-9}$ meter. Often used to express wavelengths of light (e.g., in a spectrophotometer), or to express dimensions of nanocomposites, devices (e.g., of miniature "machines" called nanoelectromechanical systems), etc. in the field of nanotechnology. See also SPECTROPHOTOMETER, NANOTECHNOLOGY, NANOCOMPOSITES, NANOELECTROMECHANICAL SYSTEM (NEMS), MICROFLUIDICS, METER.

**Nanoparticles** See NANOCRYSTALS, NANOCRYSTAL MOLECULES, NANOTECHNOLOGY.

**Nanopore** A device that can distinguish between DNA strands (molecules) that differ from each other by a single nucleotide (in the makeup of those molecular strands). Developed by David Deamer and Mark Akeson in 2001, it consists of an artificial membrane (lipid bilayer) with a "hole" (nanopore) punctured in that membrane by the protein alpha-hemolysin. Because a DNA molecule moving through such a nanopore temporarily blocks the nanopore (until it dissociates into a single DNA strand and "slides" through), an electrical current/voltage applied to that nanopore varies (in amplitude, modulation, duration, etc.) as the DNA strand "slides through," in a way that provides information (e.g., to scientist) about the nucleotides that makeup that DNA strand. It is expected that nanopores will also be used for DNA sequencing. See also NANOSCIENCE, NANOMETERS (NM), NANOTECHNOLOGY, PLASMA MEMBRANE, MICELLE, DEOXYRIBONUCLEIC ACID (DNA), NUCLEOTIDE, SINGLE-NUCLEOTIDE POLYMORPHISM (SNP), ION CHANNELS, SEQUENCING (OF DNA MOLECULES).

**Nanoscience** A term utilized to refer to the science underlying nanotechnology, nanocrystals, nanocrystal molecules, nanocomposites, "quantum dots," nanoelectromechanical systems (NEMS), etc. "Nanoscale" materials (i.e., those whose dimensions are approximately 1 to 100 nanometers) generally possess different chemical and physical properties than "bulk" materials. For example, when bulk gold metal is formed into nanoscale rods, the intensity of its fluorescence increases by a factor of approximately 10 million. Another example is that silicon nanocrystals (i.e., quantum dots) dispersed in a silicon dioxide matrix, emit larger-than-typical-for-silicon amounts of light, when stimulated (i.e., bombarded) with pulses of ultraviolet light. See also NANOTECHNOLOGY, NANOCRYSTALS, QUANTUM DOT, NANOCRYSTAL MOLECULES, NANOCOMPOSITES, NANOELECTROMECHANICAL SYSTEM (NEMS), SELF-ASSEMBLY (OF A LARGE MOLECULAR STRUCTURE), NANOPORE, MICROFLUIDICS.

**Nanotechnology** From the Latin *nanus*, dwarf, so it literally means "dwarf technology." The word was originally coined by Norio Taniguchi in 1974, to refer to high precision machining. However, Richard Feynman and K. Eric Drexler later popularized the concept of nanotechnology as a new and developing technology in which man manipulates objects whose dimensions are approximately 1 to 100 nanometers. Theoretically, it is possible that in the future a variety of man-made

N

"nano-assemblers" [tiny (molecular) machines smaller than a grain of sand] would manufacture those things that are produced today in factories. For example, enzyme molecules function essentially as jigs and machine tools to shape large molecules as they are formed in biochemical reactions. The technology also encompasses biochips, biosensors, and manipulating atoms and molecules in order to form (build) bigger, but still vanishingly small functional structures and machines. See also ENZYME, GENOSENSORS, NANOMETERS (NM), BIOSENSORS (ELECTRONIC), BIOCHIP, MICROFLUIDICS, NANO-CRYSTALS, NANOCRYSTAL MOLECULES, BIOSENSORS (CHEMICAL), QUANTUM DOT, NANOCOMPOSITES, NANOELECTROMECHANICAL SYSTEM (NEMS), SELF-ASSEMBLY (OF A LARGE MOLECULAR STRUCTURE), NANOPORE, BIOMEMS.

**Nanotube** See NANOSCIENCE, NANOTECHNOLOGY, SELF-ASSEMBLY (OF A LARGE MOLECULAR STRUCTURE).

**Napole Gene** See REDEMENT NAPOLE (RN) GENE.

**NAS** See NATIONAL ACADEMY OF SCIENCES.

**National Academy of Sciences (NAS)** A private, self-perpetuating society of distinguished scholars in scientific and engineering research, dedicated to the advancement of science and technology and their use for the general welfare. Under the authority of its congressional charter of 1863, the NAS has a working mandate that calls upon it to advise the U.S. Federal Government on scientific and technical matters. See also VITAMIN E.

**National Cancer Institute (NCI)** One of the National Institutes of Health. See also NATIONAL INSTITUTE OF HEALTH (NIH).

**National Heart, Lung, and Blood Institute (NHLBI)** One of the National Institutes of Health. See also NATIONAL INSTITUTES OF HEALTH (NIH).

**National Institute of Allergy and Infectious Diseases (NIAID)** The main agency of the National Institutes of Health. See also NATIONAL INSTITUTES OF HEALTH (NIH).

**National Institute of General Medical Sciences (NIGMS)** One of the National Institutes of Health. See also NATIONAL INSTITUTES OF HEALTH (NIH).

**National Institutes of Health (NIH)** The major U.S. Government sponsor of biotechnology research. It is composed of a group of government institutes that each focus on specific medical areas. See also RECOMBINANT DNA ADVISORY COMMITTEE (RAC).

**Native Conformation** The normal, biologically active conformation (i.e., the three-dimensional arrangement of its atoms) of a protein molecule. See also CONFORMATION.

**Naturaceuticals** See NUTRACEUTICALS.

**Natural Killer Cells** These cells are involved in tumor surveillance. They also kill virus-laden cells.

**NCI** See NATIONAL CANCER INSTITUTE (NCI).

**NDA (to FDA)** New Drug Application (to the U.S. Food and Drug Administration). A (paper) application to the U.S. Food and Drug Administration (FDA) seeking approval of a new drug that has undergone Phase 2 and Phase 3 clinical trials. An NDA is submitted in the form of (thousands of) pages of (clinical and other) data, along with various analyses (e.g., statistical) of that data for efficacy, safety, etc. See also CANDA, FOOD AND DRUG ADMINISTRATION (FDA), MAA, NDA (TO KOSEISHO), PHASE I CLINICAL TESTING.

**NDA (to Koseisho)** New drug application. It is the Japanese equivalent to a U.S. IND (investigational new drug) application; to the Koseisho, the Japanese equivalent of the U.S. Food and Drug Administration (FDA). See also IND, KOSEISHO, FOOD AND DRUG ADMINISTRATION (FDA).

**Near-Infrared Spectroscopy (NIR)** Refers to analytical instruments which shine light (possessing wavelengths between that of visible light and infrared light spectrum) onto samples (e.g., kernels of grain) and measure the reflected or transmitted (near-infrared) light in order to quickly determine the amounts of protein, fat, moisture, lignans, etc. present in the sample. In certain samples, the near-infrared light causes cells (or specific molecules) to fluoresce (i.e., as light of very defined wavelength), which can subsequently be utilized for measurement/identification of compounds within the sample. NIR is also being developed for use in quantifying (e.g., amounts that are present within the sample) of immunoassays and detection of specific molecules (e.g., in DNA sequencing process). See also PROTEIN, FATS, LIGNANS, IMMUNOASSAY, FLUORESCENCE, SEQUENCING (OF DNA MOLECULES).

**Near-Infrared Transmission (NIT)** Refers to certain analytical instruments which shine light (possessing wavelengths between that of visible light and infrared spectrum) through samples (e.g., kernels of grain) in order to quickly determine the amounts of protein, fat, moisture, lignans, etc. present in the sample. See also PROTEIN, FATS, LIGNANS, NEAR-INFRARED SPECTROSCOPY (NIR).

**Necrosis** Refers to cell death caused by physical injury to the cell (e.g., exposure to toxin, exposure to ultraviolet light, lack of oxygen, etc.). See also CELL, TOXIN, RESPIRATION, TUMOR NECROSIS FACTOR.

**Neem Tree** A tropical tree (*Azadirachta indica*) found in India, Somalia, Mauritania, Australia, and other tropical countries; that resists insect (e.g., whiteflies, mealybugs, aphids, mites) depradations and certain fungal diseases (rusts, powdery mildew, etc.) via secretions of liquids that contain Azadirachtin (an insect-repelling chemical). See also AZADIRACHTIN, FUNGUS.

**Negative Supercoiling** Comprises the twisting of a duplex of DNA (deoxyribonucleic acid) in space in the opposite sense to the turns of the strands in the double helix. See also DOUBLE HELIX.

**Nematodes** Microscopic roundworms, which are the most abundant multi-celled creatures on earth. They are primarily found living in soil. One nematode named *Caenorhabditis elegans* (*C. elegans*) is commonly used by scientists in genetics experiments, so a large base of knowledge about its genetics has been accumulated by the world's scientific community. For example, of the nearly 300 "disease-causing" genes in the human genome, more than half of them have an analogous gene within the genome of *C. elegans*. One Antarctic nematode (*Panagrolaimus davidi*) is able to survive Antarctic winters by drying out and achieving a state of "suspended animation" (anhydrobiosis) for as long as thirty-nine years. See also CELL, *CAENORHABDITIS ELEGANS* (*C. ELEGANS*), GENETICS, GENE, GENOME, GENETIC MAP, MODEL

ORGANISM, SOYBEAN CYST NEMATODES (SCN), CYSTX.

**NEMS** See NANOELECTROMECHANICAL SYSTEM (NEMS).

**Neoplasia** New growth. See also NEOPLASTIC GROWTH.

**Neoplastic Growth** A new growth of animal or plant tissue resembling (more or less) the tissue from which it arises but having distinct biochemical differences from the parent cell. The neoplastic tissue is a mutant version of the original and appears to serve no physiologic function in the same sense as did the original tissue. It may be benign or malignant (i.e., a cancerous tumor). See also TUMOR, CANCER, SELECTIVE APOPTOTIC ANTI-NEOPLASTIC DRUG (SAAND), METASTASIS.

**Nerve Growth Factor (NGF)** A protein produced by the salivary glands (and also in tumors) that greatly increases growth/reproduction of nerve cells and guides the formation of neural networks. In the brain, NGF is thought to increase the production of the messenger chemical, acetylcholine, by protecting and stimulating those neurons that produce acetylcholine. Because those (acetylcholine-producing) neurons are typically the first to be destroyed in an Alzheimer's disease victim, NGF holds potential to be used to counteract (some of) the effects of the disease.

NGF is also necessary for normal development of the hypothalamus, a brain structure that regulates a number of hormones. Human T cells appear to have receptors for NGF, which could explain the "mind–body connection" between a person's emotional well-being and physical health (i.e., NGF may be a go-between for the brain and the immune system). NGF was discovered by Rita Levi-Montalcini in 1954. See also GROWTH FACTOR, EPIDERMAL GROWTH FACTOR (EGF), HYPOTHALAMUS, HORMONE, PROTEIN.

**Nested PCR** Refers to a specific PCR (polymerase chain reaction) technique of two consecutive-run PCRs, in which the second PCR amplifies (i.e., makes multiple copies of) a DNA sequence within the product (amplicon) of the first PCR. See also POLYMERASE CHAIN REACTION (PCR), POLYMERASE CHAIN REACTION (PCR) TECHNIQUE, SEQUENCE (OF A

DNA MOLECULE), DEOXYRIBONUCLEIC ACID (DNA), AMPLICON.

**Neuraminidase (NA)** A transmembrane (i.e., through the membrane) glycoprotein enzyme that appears in the (external) membrane of the influenza virus. See also ENZYME, GLYCOPROTEIN, VIRUS.

**Neuron** Cells of the body's nervous system, which transmit nerve impulses (electrical signals conducted by the flow of ions across the plasma membrane of neuron cells). Neurons are involved in controlling movement (known as motor control), emotions, and memory. There are approximately 100 billion neurons in the typical human brain. The nerve impulses within them move at a speed of approximately 400 kilometers per hour (300 miles per hour). See also NEUROTRANSMITTER, ACETYLCHOLINE, SEROTONIN, CELL, PARKINSON'S DISEASE, PLASMA MEMBRANE, ION, DENDRITES.

**Neurotransmitter** An organic, low molecular weight compound that is secreted from the (axon) terminal end of a neuron (in response to the arrival of an electrical impulse) into a liquid-filled gap that exists between neurons. The transmitter molecule then diffuses across the small gap and attaches to the next neuron. This attachment causes structural changes in the membrane of the neuron and initiates the conductance of an electrical impulse. In this way, an electrical impulse is transmitted (via this "cascade") along a neuron network of which the neurons themselves do not physically touch.

A neurotransmitter serves to transmit a nerve impulse between different neurons. Examples of neurotransmitters include dopamine and norepinephrine. A shortage of dopamine in the brain causes the disease known as Parkinson's disease. See also MOLECULAR WEIGHT, NEURON, SEROTONIN, ACETYLCHOLINE, PARKINSON'S DISEASE, CASCADE, DENDRITES.

**Neutraceuticals** See NUTRACEUTICALS.

**Neutriceuticals** See NUTRACEUTICALS.

**Neutrophils** Phagocytic (ingesting, scavenging) white blood cells produced in the bone marrow. They ingest and destroy invading microorganisms and facilitate post-infection tissue repair. They can secrete collagenase and plasminogen activator. They are the immune system's "first line" of defense against invading pathogens, and large reserves are called forth within hours of the start of a "pathogen invasion." See also PATHOGEN, COLLAGENASE, MICROORGANISM.

**New Drug Application** See NDA (TO KOSEISHO), NDA (TO FDA), MAA, IND, CANDA.

**NIAID** See NATIONAL INSTITUTE OF ALLERGY AND INFECTIOUS DISEASES.

**Nick** A break in one strand of a double-stranded DNA molecule. One of the phosphodiester bonds between two adjacent nucleotides is ruptured. No bases are removed from the strand, it is just opened at that point. See also DEOXYRIBONUCLEIC ACID (DNA).

**Nicotine-Adenine Dinucleotide (NAD)** See NAD.

**Nicotine-Adenine Dinucleotide Phosphate (NADP)** See NAD.

**Nicotine-Adenine Dinucleotide Phosphate, reduced (NADPH)** See NAD.

**Nicotine-Adenine Dinucleotide, reduced (NADH)** See NAD.

**NIH** See NATIONAL INSTITUTES OF HEALTH (NIH).

**NIHRAC** See RECOMBINANT DNA ADVISORY COMMITTEE (RAC).

**Ninhydrin Reaction** A color reaction given by amino acids and peptides on heating with the chemical ninhydrin. The technique is widely used for the detection and quantitation (measurement) of amino acids and peptides. The concentration of amino acid in a solution (of hydrochloric acid) is proportional to the optical absorbance of the solution after heating it with ninhydrin. α-Amino acids give an intense blue color, and amino acids (such as proline) give a yellow color. One is able to determine concentration of a protein or peptide and also obtain an idea of the type of protein or peptide that is present. See also ABSORBANCE (A), AMINO ACID, PEPTIDE.

**Nitrate Bacteria** See NITRATES, NITRITES, BACTERIA.

**Nitrate Reduction** The reduction of nitrate to nitrite or ammonia by an organism. See also NITRATES, REDUCTION (IN A CHEMICAL REACTION), NITRITES.

**Nitrates** Refers to nitrogen compounds that exist in a chemical form which plant roots are able to take in (i.e., utilized by the plant to make nitrogen-containing molecules such as proteins). Nitrates are produced from nitrogen:

- Taken out of the atmosphere by nitrogen-fixing bacteria (living among the roots of legume plants such as the soybean, etc.)
- Taken out of nitrites (in soil) by nitrate bacteria
- Taken out of the atmosphere by blue-green algae

See also PROTEIN, NITROGEN FIXATION, SOYBEAN PLANT, NITRITES.

**Nitric Oxide** Abbreviated NO, it is a molecule produced in the body of an organism, which can act as:

- A signaling molecule (e.g., to cause a firefly's tail to begin the chemical reaction of luciferin with luciferase that results in the light emission known as bioluminescence)
- An oxidant utilized against pathogens by the immune system
- An instigator of (destructive) free radicals
- An inducer of genes (e.g., in soybean plants) that cause production of certain chemical compounds which protect the organism (e.g., soybean plant) from bacterial diseases

As a signaling molecule, or "messenger molecule," nitric oxide is utilized by the human body for control of blood pressure (i.e., when the endothelial cells that line blood vessels produce NO that causes neighboring smooth-muscle cells to relax so entire blood vessel dilates; thereby lowering blood pressure). Nitric oxide is also utilized by the human body for immune system regulation, and its synthesis in macrophages is required for macrophages to kill pathogens and tumor cells (by oxidizing them after the macrophage has engulfed them).

During the 1980s, John Garthwaite and Solomon H. Snyder showed that nitric oxide is an important messenger molecule utilized in neural signaling (i.e., NO is an important signaling molecule in the human brain).

Nitric oxide increases the effectiveness of reactive free radicals (e.g., superoxide $O_2$) in killing off any infected cells within a soybean plant. Nitric oxide also induces certain genes to code for the production of certain chemical compounds that protect the soybean plant from bacterial plant diseases. See also SIGNALING MOLECULE, SIGNALING, OXIDIZING AGENT, PATHOGEN, IMMUNE RESPONSE, HUMAN SUPEROXIDE DISMUTASE (hSOD), SIGNAL TRANSDUCTION, NITRIC OXIDE SYNTHASE, SOYBEAN PLANT, PROTEIN, INDUCERS, GENE, CODING SEQUENCE, FREE RADICAL, ENDOTHELIAL CELLS, ENDOTHELIUM, MACROPHAGE, PATHOGEN, BACTERIA, TUMOR, NEUROTRANSMITTER, BIOLUMINESCENCE.

**Nitric Oxide Synthase** An enzyme that catalyzes the reaction which the body (of animals or plants) utilizes to make nitric oxide from L-arginine (via cleavage, off that molecule). The cofactor for that reaction is nicotine-adenine dinucleotide phosphate (NADP). See also ENZYME, NITRIC OXIDE, COFACTOR, NAD (NADH, NADP, NADPH), ARGININE (arg), LEVOROTARY (L) ISOMER, HYDROLYTIC CLEAVAGE, ENDOTHELIAL CELLS, ENDOTHELIUM, MACROPHAGE.

**Nitrification** The oxidation of ammonia (e.g., from ammonia-containing substances such as liquid wastes excreted by animals, decomposed animals and plants, etc.) to nitrates by a microorganism. See also NITRATES, NITRITES, OXIDATION (chemical reaction).

**Nitrifying Bacteria** See NITRITES.

**Nitrilase** An enzyme that catalyzes the degradation (breaking down) of bromoxynil (an active ingredient in some herbicides). Nitrilase is naturally produced in the soil bacteria *Klebsiella pneumoniae* subs. *Ozaenae*. If a gene (called BXN) that codes for the production of nitrilase is inserted via genetic engineering into crop plants, the resultant plant production of nitrilase would enable such plants to survive post-emergence applications of bromoxynil-containing herbicides. See also ENZYME, BACTERIA, BROMOXYNYL, GENE, CODING SEQUENCE, GENETIC ENGINEERING.

**Nitrites** Refers to nitrogen compounds that exist in a chemical form which plant roots are unable to take in. After conversion to nitrates via internal respiration by nitrate bacteria (in soil), the nitrates can be taken in

**N**

by plant roots (i.e., utilized by the plant to make nitrogen-containing molecules such as proteins). Nitrites are made (via internal respiration) by nitrifying bacteria (e.g., in soil) from ammonia-containing substances (e.g., liquid wastes excreted by animals, decomposed animals and plants, etc.). See also NITRATES, PROTEIN, RESPIRATION.

**Nitrogen Cycle** The cycling of various forms of biologically available nitrogen through the plant, animal, and microbial worlds (kingdoms), as well as the atmosphere and geosphere. See also NITRATES, NITRITES, NITRIFICATION, DENITRIFICATION, NITROGEN FIXATION.

**Nitrogen Fixation** Conversion of atmospheric nitrogen ($N_2$) into ammonia; a soluble, biologically available form (nitrates). The conversion is carried out by nitrogen-fixing organisms (e.g., *Rhizobium* bacteria) which live symbiotically in the roots of legume plants, e.g., alfalfa or soybeans. This is one of nature's ways of fertilizing [e.g., traditional varieties of soybeans typically leave approximately 40 pounds of residual nitrogen per acre (44kg) in fields at the end of the growing season]. When not enough nitrogen fixation occurs (when only nonlegume plants are grown), soil is not able to produce maximum crop yields and farmers may need to spread fixed nitrogen onto the field in the form of the fertilizer anhydrous ammonia, ammonium nitrate, or sodium nitrate. See also NITRATES, SYMBIOTIC, GENISTEIN (Gen), BACTERIA, SOYBEAN PLANT, NITROGENASE SYSTEM, NITROGEN CYCLE, ISOFLAVONES, HEMAGGLUTININ (HA), NODULATION.

**Nitrogen Metabolism** See GLUTAMATE DEHYDROGENASE.

**Nitrogenase System** A system of enzymes capable of reducing atmospheric nitrogen to ammonia in the presence of ATP. See also REDUCTION (IN A CHEMICAL REACTION), ENZYME, NITROGEN FIXATION.

**NO** See NITRIC OXIDE.

**"No-Till" Crop Production** See NO-TILLAGE CROP PRODUCTION.

**No-Tillage Crop Production** A methodology of crop production in which the farmer utilizes virtually no mechanical cultivation (i.e., only one pass over the field, with a planter; instead of the conventional four passes per year with mechanical cultivator equipment plus one pass with planter, used for traditional crop production). This reduction in field soil disturbance leaves more carbon in the soil (thereby reducing "greenhouse gases" in the atmosphere), leaves more earthworms (*Eisenia foetida*) per cubic foot or per cubic meter living in the topsoil, and reduces soil compaction (i.e., the reduction of interstitial spaces between individual soil particles); thereby increasing the fertility of such "no till" farm fields.

The plant residue remaining on the field's surface helps to control weeds and reduce soil erosion (by 90–95% vs. traditional mechanical tillage), and it also provides sites for insects to shelter and reproduce, leading to a need for increased insect control via methods such as inserting a *Bacillus thuringiensis* (*B.t.*) gene into certain crop plants or utilizing integrated pest management (IPM). But, if a farmer needs to apply synthetic chemical pesticides, the plant residue remaining on the field's surface helps to cause breakdown (into substances such as carbon dioxide and water) of pesticides. That is because that plant residue helps to retain moisture in the field-surface environment, thereby enhancing growth of the microorganisms that help break down pesticides. Use of No-Tillage Crop Production (methodology) helps farmers to reduce the incidence of certain plant diseases such as white mold disease. See also INTEGRATED PEST MANAGEMENT (IPM), CORN, GLOMALIN, SOYBEAN PLANT, *BACILLUS THURINGIENSIS* (*B.t.*), GENE, GENETIC ENGINEERING, EUROPEAN CORN BORER (ECB), *HELICOVERPA ZEA* (*H. ZEA*), CORN ROOTWORM, COLD HARDENING, MICROORGANISM, LOW-TILLAGE CROP PRODUCTION, EARTHWORMS, WHITE MOLD DISEASE.

**Nod Gene** See NODULATION.

**Nodulation** The process in which certain strains of soil-dwelling *Rhizobium* bacteria colonize the roots of specific plants (i.e., the legumes) such as soybean (*Glycine max* L.) or alfalfa. As part of that process:

- The *Rhizobium* bacteria are attracted to the vicinity of the plant's roots. For the soybean plant (*Glycine max* L.), that is accomplished by the plant synthesizing the signaling molecules known as isoflavones, which attract *Rhizobium japonicum* bacteria. For the alfalfa plant, that is accomplished by the plant synthesizing luteolin molecules, which attract *Sinorhizobium meliloti* bacteria.

- Certain genes (called nod) within the relevant *Rhizobium* bacteria are expressed (resulting in the synthesis of specific chemical compounds).
- When the plant roots detect those chemical compounds, certain genes within those roots are expressed (resulting in the formation of nodules on those roots).
- The relevant *Rhizobium* bacteria move in and inhabit those plant root nodules, where the bacteria then "fix" nitrogen from the atmosphere; which converts that nitrogen into a chemical form (i.e., nitrates) that is available for use by plants (as fertilizer/plant food). See also RHIZOBIUM (bacteria), CHEMOTAXIS, SOYBEAN PLANT, ISOFLAVONES, GENISTEIN (Gen), TRANSCRIPTION FACTORS, GENE, GENE EXPRESSION, SIGNALING MOLECULE, NITROGEN FIXATION, SYMBIOTIC, HEMAGGLUTININ (HA).

**Non-Starch Polysaccharides** Term — abbreviated NSP — that refers to polysaccharide molecules (in plant seeds) other than starch. These include arabinoxylans, pectins, beta glucans, and alpha galactosides (e.g., raffinose, stachyose, verbascose). See also POLYSACCHARIDES, STACHYOSE.

**Nonessential Amino Acids** Amino acids of proteins that can be made (biochemically synthesized within the body) by humans and certain other vertebrate animals from simple chemical precursors (in contrast to the essential amino acids). These amino acids are thus not required in the diet (of humans and those other vertebrates). See also ESSENTIAL AMINO ACIDS, AMINO ACID, PROTEIN.

**Nonheme-Iron Proteins** Proteins containing iron but no porphyrin groups (within which iron atoms are held) in their structure. See also HEME.

**Nonpolar Group** A hydrophobic ("water hating") group on a molecule; usually hydrocarbon (composed of hydrogen and carbon atoms) in nature. These groups are more at home in a nonpolar (oil-like) environment. See also POLAR GROUP, AMPHIPATHIC MOLECULES, AMPHOTERIC COMPOUND.

**Nonsense Codon** A triplet of nucleotides that does not code for an amino acid. Any one of three triplets (U-A-G, U-A-A, U-G-A) that cause termination of protein synthesis. U-A-G is known as amber and U-A-A is known as ochre. See also GENETIC CODE, CODON, TERMINATION CODON (SEQUENCE).

**Nonsense Mutation** A mutation that converts a codon that specifies an amino acid into one that does not specify any amino acid. A change in the nucleotide sequence of a codon that may result in the termination of a polypeptide chain. See also NONSENSE CODON, GENETIC CODE, CODON.

**Nontranscribed Spacer** A region between transcription units in a tandem gene cluster. See also TRANSCRIPTION, MESSENGER RNA (mRNA), GENETIC CODE, GENE SPLICING, GENE.

**North American Plant Protection Organization (NAPPO)** One of the international SPS standard-setting organizations that develops plant health standards, guidelines, and recommendations (e.g., to prevent transfer of a disease from one country to another). Subsidiary to the International Plant Protection Convention (IPPC), it covers the countries of North America. Its secretariat is located in Nepean, Canada. See also INTERNATIONAL PLANT PROTECTION CONVENTION (IPPC), EUROPEAN PLANT PROTECTION ORGANIZATION (EPPO), SPS.

**Northern Blotting** A research test/methodology used to transfer RNA fragments from an agarose gel (e.g., following gel electrophoresis) to a filter paper without changing the relative positions of the RNA fragments (e.g., re electrophoresis separation grid). See also RIBONUCLEIC ACID (RNA), GEL ELECTROPHORESIS, AGAROSE, CHROMATOGRAPHY, FIELD INVERSION GEL ELECTROPHORESIS.

**Northern Corn Rootworm** Latin name *Diabrotica barberi*. See also CORN ROOTWORM.

**NOS Terminator** A termination codon (sequence of DNA) frequently utilized in genetic engineering of plants to "terminate" expression of the inserted gene (i.e., to halt synthesis of desired protein in the plant, after the desired protein synthesis has occurred).

The NOS terminator was originally extracted from the bacteria species *Agrobacterium tumefaciens*. See also TERMINATION CODON (terminator sequence), SEQUENCE (OF A DNA MOLECULE), DEOXYRIBONUCLEIC ACID (DNA), GENETIC ENGINEERING, EXPRESS, GENE, PROTEIN, SYNTHESIZING (OF PROTEIN MOLECULE), *AGROBACTERIUM TUMEFACIENS*, BACTERIA, CONTROL SEQUENCES.

**NPTII** See NPTII GENE.

**NPTII Gene** A marker gene that codes for (i.e., "causes manufacture of") the enzyme neomycin phosphotransferase II, which can inactivate the antibiotic kanamycin. The NPTII gene is commonly utilized as a "marker gene" for genetically engineered plants. Neomycin phosphotransferase confers kanamycin resistance to cells expressing it (i.e., cells that contain the NPTII gene in addition to the other gene(s) inserted along with it), so those (engineered) cells will live in a laboratory vessel containing kanamycin. See also GENE, MARKER (GENETIC MARKER), CODING SEQUENCE, ENZYME, CELL, GENETIC ENGINEERING.

**NSP** See NON-STARCH POLYSACCHARIDES.

**NT** An acronym for Nuclear Transfer. See also NUCLEAR TRANSFER.

**Nuclear DNA** The DNA contained within the nucleus of a cell. See also DEOXYRIBONUCLEIC ACID (DNA), CELL, GENOME, NUCLEUS, NUCLEAR TRANSFER.

**Nuclear Envelope** See MEMBRANE (of a cell).

**Nuclear Matrix Proteins** Protein molecules present in cancerous cells, but not in normal (nonmutated) cells. See also PROTEIN, CELL, MUTATION, MUTANT, MYRISTOYLATION, NEOPLASTIC GROWTH, "PARP".

**Nuclear Receptors** Receptors in a cell's outer membrane that serve to convey a "signal" from outside the cell all the way into the cell's nucleus. See also RECEPTORS, SIGNALING, SIGNAL TRANSDUCTION, NUCLEUS, G-PROTEINS, ENDOCYTOSIS, VAGINOSIS, CD4 PROTEIN, PROTEIN, CELL, GENE, EXPRESS, GENE EXPRESSION, TRANSCRIPTION, TRANSCRIPTION FACTORS, POLYUNSATURATED FATTY ACIDS (PUFA), MEMBRANES (of a cell), PLASMA MEMBRANE.

**Nuclear Transfer** A method of cloning a living organism, in which that organism's entire genetic information is conveyed via transfer of an (adult) cell nucleus into an unfertilized egg (from another animal of the same species) whose nucleus had previously been removed. This was the method utilized to produce "Dolly," the first cloned sheep, in 1996. It is possible to also delete or substitute genes (e.g., brought in from another species) as part of the nuclear transfer process, so nuclear transfer can be utilized to produce transgenic organisms or "knock out" organisms. See also CLONE (AN ORGANISM), CELL, NUCLEUS, GENOME, NUCLEAR DNA, DEOXYRIBONUCLEIC ACID (DNA), GENE, SPECIES, TRANSGENIC (organism), KNOCK OUT (GENE), GENETIC ENGINEERING.

**Nuclease** An enzyme capable of hydrolyzing the internucleotide linkages of a nucleic acid (e.g., DNA or RNA). Nucleases present in cells tend to degrade (i.e., hydrolyze, cleave) artificially inserted DNA strands, making genetic targeting more difficult. See also GENETIC TARGETING, HYDROLYSIS, DEOXYRIBONUCLEIC ACID (DNA), RIBONUCLEIC ACID (RNA), ANTISENSE (DNA SEQUENCE).

**Nucleic Acid Probes** See DNA PROBE, NUCLEIC ACIDS, POLYMERASE CHAIN REACTION (PCR), RAPID MICROBIAL DETECTION (RMD).

**Nucleic Acids** A nucleotide polymer. A large, chain-like molecule containing phosphate groups, sugar groups, and purine and pyrimidine bases; two types are ribonucleic acid (RNA) and deoxyribonucleic acid (DNA). The bases involved are adenine, guanine, cytosine, and thymine (uracil in RNA). Nucleic acids are either the specific (genetic) informational molecule (DNA), or act as agent (RNA) in causing that information to be expressed (e.g., as a protein). See also NUCLEOTIDE, POLYMER, INFORMATIONAL MOLECULES, GENE, GENETIC CODE, DEOXYRIBONUCLEIC ACID (DNA), RIBONUCLEIC ACID (RNA), EXPRESS, EXTENSION (IN NUCLEIC ACIDS).

**Nucleoid** The compact body that contains the genome in a bacterium. See also GENOME.

**Nucleolus** A round, granular structure situated in the nucleus of eucaryotic cells. It is involved in rRNA (ribosomal RNA) synthesis and ribosome formation. See also RIBOSOMES, NUCLEUS.

**Nucleophilic Group** An electron-rich group with a strong tendency to donate electrons to an electron-deficient nucleus. See also POLAR GROUP, NONPOLAR GROUP.

**Nucleoproteins** Complexes made up of nucleic acid and protein. These two substances are apparently not linked by strong chemical bonds, but are held together by salt linkages and other weak bonds. Most viruses consist entirely of nucleoproteins, although some viruses also contain fatty substances. Nucleoproteins also occur in animal and plant cells and in bacteria. See also PROTEIN, VIRUS.

**Nucleoside** A hybrid molecule consisting of a purine (adenine, guanine) or pyrimidine (thymine, uracil, or cytosine) base covalently linked to a five-membered sugar ring (ribose in the case of RNA and deoxyribose in the case of DNA). See also NUCLEOTIDE.

**Nucleoside Diphosphate Sugar** A coenzyme-like carrier of a sugar molecule functioning in the enzymatic synthesis of polysaccharides and sugar derivatives. See also POLYSACCHARIDES.

**Nucleosome** Spherical particles composed of a special class of basic proteins (histone) in combination with DNA (146 bp of DNA are wrapped 1.75 times around a "core" of histone proteins). The particles are approximately 12.5 nm in diameter and are connected to each other by DNA filaments. Under an electron microscope they appear somewhat like a string of pearls. See also CHROMATIN, HISTONES, PROTEIN, DEOXYRIBONUCLEIC ACID (DNA), BASE PAIR (bp), NANOMETERS (nm).

**Nucleotide** An ester of a nucleoside and phosphoric acid. Nucleotides are nucleosides that have a phosphate group attached to one or more of the hydroxyl groups of the sugar (ribose or deoxyribose). In short, a nucleotide is a hybrid molecule consisting of a purine or pyrimidine base covalently linked to a five-membered sugar ring which is covalently linked to a phosphate group. While (polymerized) nucleotides are the structural units of a nucleic acid, free nucleotides that are not an integral part of nucleic acids are also found in tissues and play important roles in the cell, e.g., ATP and cyclic AMP. See also ATP, CYCLIC AMP, BASE (NUCLEOTIDE), NUCLEOSIDE, NUCLEIC ACIDS, MESSENGER RNA (mRNA), RIBONUCLEIC ACID (RNA), DEOXYRIBONUCLEIC ACID (DNA), TRANSVERSION.

**Nucleus** The usually spherical body with each living cell that contains its hereditary biological material (DNA, genes, chromosomes, etc.) and controls the cell's life functions (e.g., metabolism, growth, and reproduction). The nucleus is a highly differentiated, relatively large organelle lying in the cytoplasm of the cell. The nucleus is surrounded by a (nuclear) membrane which is quite similar to the plasma (cell) membrane, except the nuclear membrane contains holes or pores. It is characterized by its high content of chromatin, which contains most of the cell's DNA. That chromatin is normally (when cell is not in process of dividing) distributed throughout the nucleus in a diffuse manner. See also GENOME, CELL, GENE, GENETIC CODE, RNA, HEREDITY, DEOXYRIBONUCLEIC ACID (DNA), CHROMOSOMES, MEIOSIS, NUCLEAR TRANSFER, METABOLISM, CHROMATIDS, CHROMATIN, PLASMA MEMBRANE, ORGANELLES, NUCLEAR RECEPTORS.

**Nutraceuticals** Coined in 1989 by Stephen DeFelice, this term is used to refer to either a food or portion of food (a vitamin, essential amino acid, etc.) that possesses medical or health benefits (to the organism that consumes that nutraceutical). For example, saponins (present in beans, spinach, tomatoes, potatoes, alfalfa, clover, etc.) possess some cancer-prevention properties. Also sometimes called pharmafoods, functional foods, or designer foods, these are food products that have been designed to contain specific concentrations and/or proportions of certain nutrients (vitamins, amino acids, etc.) that are critical for good health. See also ESSENTIAL AMINO ACIDS, AMINO ACID, VITAMIN, FOOD GOOD MANUFACTURING PRACTICE (FGMP), SAPONINS, ESSENTIAL NUTRIENTS, PHYTOCHEMICALS, ANTIOXIDANTS, ISOFLAVONES, GENISTEIN (Gen), RESVERATROL, PHYTOSTEROLS, BETA CAROTENE, LYCOPENE, CAROTENOIDS, LUTEIN, ANTHOCYANINS, VITAMIN E, XANTHOPHYLLS, STEROLS, SITOSTEROLS, SITOSTANOLS, ELLAGIC ACID, ALICIN, PROANTHOCYANIDINS, POLYPHENOLS, ZEAXANTHIN, PHYTO-MANUFACTURING.

**Nutriceuticals** See NUTRACEUTICALS.

**Nutricines** See NUTRACEUTICALS.

**Nutrient Enhanced™** A phrase that is now a trademark of Garst Seed Company; it refers to plants that have been modified to possess novel traits that make those plants more economically valuable for nutritional uses (e.g., higher-than-normal protein content in certain feedgrains). See also VALUE-ENHANCED GRAINS, HIGH-OIL CORN, PROTEIN, GENETIC ENGINEERING, HIGH-LYSINE CORN, HIGH-METHIONINE CORN, PLANT'S NOVEL TRAIT (PNT), HIGH-PHYTASE CORN AND SOYBEANS.

**Nutrigenomics** See PHARMACOGENOMICS.

# O

**O Glycosylation** See GLYCOSYLATION (TO GLYCOSYLATE).

**OAB (Office of Agricultural Biotechnology)** A unit of the U.S. Department of Agriculture in charge of a part of the federal regulatory process for biotechnology (e.g., field tests of transgenic plants). See also TOXIC SUBSTANCES CONTROL ACT (TSCA), RECOMBINANT DNA ADVISORY COMMITTEE (RAC), FOOD AND DRUG ADMINISTRATION (FDA), TRANSGENIC.

**Ochratoxins** A term that refers to a group of related mycotoxins (i.e., toxic metabolites produced by fungi) that are produced by some *Aspergillus* species and some *Penicillium* species of fungi (e.g., *Penicillium viridicatum*). These particular fungi tend to produce ochratoxins when they grow in damaged grain (e.g., during grain storage), especially when grain temperature is above 4°C (40°F) and grain moisture content is above 18%. Ochratoxin A is a very carcinogenic (cancer-causing) toxin when consumed by humans. When dairy cattle consume ochratoxin A-containing grain, the ochratoxin A soon appears in the milk produced by those cows. See also MYCOTOXINS, TOXIN, FUNGUS, PENICILLIUM, CARCINOGEN.

**Octadecanoid/Jasmonate Signal Complex** A chemical signal created and emitted by certain plants in response to those plants being wounded (e.g., via chewing) by insects. The octadecanoid/jasmonate signal complex then causes the production and also emission of volatile chemicals such as volicitin, which attract certain types of wasps that are natural enemies of those insects which initially wounded the plants. Thus, the octadecanoid/jasmonate signal complex is a crucial part of an (indirect) defense mechanism of such plants. See also SIGNALING MOLECULE, SIGNALING, EUROPEAN CORN BORER, INTEGRATED PEST MANAGEMENT (IPM), VOLICITIN.

**OD** See OPTICAL DENSITY.

**Odorant Binding Protein** A protein that enhances people's ability to smell odorants in trace quantities much lower than those needed to activate olfactory (i.e., smelling) nerves. The protein accomplishes this by latching onto (odorant) molecules and enhancing their aroma. Hence, it acts as a kind of "helper" entity in bringing about the ability to smell certain odorants present in low concentration. See also PROTEIN.

**OECD** See ORGANIZATION FOR ECONOMIC COOPERATION AND DEVELOPMENT.

**Office International des Epizootics** See INTERNATIONAL OFFICE OF EPIZOOTICS (OIE).

**OGM** See GMO.

**OH43** Gene in plants (e.g., corn/maize) that causes production of a seed coat more resistant to tearing. Greater tear-resistance results in a lower incidence of fungi infestation in seed, which results in less mycotoxin production in seed. See also GENE, FUNGUS, AFLATOXIN, MYCOTOXINS.

**OIE** Office International des Epizootics. See also INTERNATIONAL OFFICE OF EPIZOOTICS (OIE).

**OIF** See OSTEOINDUCTIVE FACTOR.

**Oils** See FATTY ACID.

**Oleic Acid** A fatty acid naturally present in the fat of animals and also in oils extracted from oilseed plants (soybean, canola, etc.). For example, the soybean oil produced from traditional varieties of soybeans tends to contain 24% oleic acid. See also MONOUNSATURATED FATS, FATTY ACID, FATS, CANOLA, SOYBEAN PLANT, SOYBEAN OIL, HIGH-OLEIC OIL SOYBEANS, COSUPPRESSION.

**Oleosomes** The storage bodies for lipids (fats) in the seeds of certain plants. See also LIPIDS, FATS, FATTY ACID.

**Oligionucleotide** See OLIGONUCLEOTIDE.

**Oligofructans** See FRUCTAN, FRUCTOSE OLIGO-SACCHARIDES.

**Oligofructose** See FRUCTOSE OLIGOSACCHARIDES.

**Oligomer** A relatively short (the prefix oligo-means few, slight) chain molecule (polymer) that is made up of repeating units (e.g., XAXAXAXA or XXAAXXAAXXAA, etc.). Short polymers consisting of only two repeating units are called dimers, those of three repeating units are called trimers. Longer units are called polymers (i.e., many units). As a rule of thumb, oligomers consisting of 11 or more repeating units are called polymers. See also POLYMER.

**Oligonucleotide** Synonymous with oligode-oxyribonucleotide, they are short chains of nucleotides (i.e., single-stranded DNA or RNA) that have been synthesized (made) by chemically linking together a number of specific nucleotides. Oligonucleotides (also called, simply "oligos") are used as synthetic (man-made) genes, DNA probes, and in site-directed mutagenesis. See also NUCLEOTIDE, GENE, DNA PROBE, OLIGOMER, SITE-DIRECTED MUTAGENESIS, GENE MACHINE, DEOXYRIBO-NUCLEIC ACID (DNA), RIBONUCLEIC ACID, SYN-THESIZING (OF DNA MOLECULES).

**Oligonucleotide Probes** Short chain fragments of DNA that are used in various gene analysis tests (e.g., the single base change in DNA that causes sickle-cell anemia). See also OLIGONUCLEOTIDE, DEOXYRIBONUCLEIC ACID (DNA), DNA PROBE, GENE MACHINE.

**Oligopeptide** A relatively short chain molecule made up of amino acids linked by peptide bonds. See also PEPTIDE, POLYPEPTIDE (PROTEIN), OLIGOMER, AMINO ACID.

**"Oligos"** See OLIGONUCLEOTIDE.

**Oligosaccharides** Relatively short molecular chains made of up to 10–100 simple sugar (saccharide) units. These sugar (i.e., carbohydrate) chains are frequently attached to protein molecules. When this happens, the resulting molecule is known as a glycoprotein, i.e., a hybrid molecule that is part protein and part sugar. The oligosaccharide portion affects a protein's conformation(s) and biological activity. The oligosaccharide (carbohydrate) portion of a glycoprotein functions as a mediator of cellular uptake of that glycoprotein. Glycosylation thus affects the length of time the molecule resides in the bloodstream before it is taken out of circulation (serum lifetime).

It is thought that blood group (A, B, O, etc.) is based upon an oligosaccharide concept. For example, different oligosaccharide "branches" on a given glycoprotein (e.g., tissue plasminogen activator) could cause that glycoprotein to be perceived by the body's immune system to be another (incorrect) blood type, thus provoking an immune response against it. Oligosaccharides play a critical role in numerous disease processes (bacterial and viral infection processes, cancer metastasis processes, inflammation processes, etc.). For example, oligosaccharide "chains" extending from the exterior membrane plasma membrane of cells are utilized by bacteria (and inflammation-triggering immune system cells) to latch onto cells and facilitate entry into cells. See also POLYSAC-CHARIDES, CELL, CONFORMATION, MONOSACCHA-RIDES, FURANOSE, PENTOSE, PYRANOSE, GLYCOGEN, GLYCOFORM, FRUCTOSE OLIGOSAC-CHARIDES, GLYCOPROTEIN, MANNANOLIGOSAC-CHARIDES, TISSUE PLASMINOGEN ACTIVATOR (tPA), OLIGOMER, SEROLOGY, HUMORAL IMMU-NITY, CELLULAR IMMUNE RESPONSE, METASTASIS, ADHESION MOLECULES, HEMAGGLUTININ (HA), TRANSGALACTO-OLIGOSACCHARIDES.

**Omega-3 Fatty Acids** More properly called "n-3 fatty acids." See also N-3 FATTY ACIDS.

**Omega-6 Fatty Acids** More properly called "n-6 fatty acids." See also N-6 FATTY ACIDS.

**Oncogenes** Genes within a cell's DNA that code for receptors (proteins on outer surface of cell membrane) for a cellular growth factor (e.g., epidermal growth factor). Via that coding-for of applicable receptors (or other protein molecules that are part of the signal transduction process of a cell), oncogenes "turn on" the process of cell division (replication) at appropriate time(s) during the life of each cell in an organism. When oncogenes are mutated (via exposure to cigarette smoke or ultraviolet light, etc.), those oncogenes can become cancer-causing genes, some of which (e.g., erythroblastosis virus gene) are

O

almost identical to the gene for epidermal growth factor (EGF) receptor (i.e., oncogene is a "deformed copy" of that gene). Such mutated oncogenes code for (i.e., cause to be made) proteins (protein kinases, protein phosphorylating enzymes, etc.) that trigger uncontrolled cell growth. They sometimes may consist of a human chromosome that has viral nucleic acid material incorporated into it and is a permanent part of that chromosome. See also GENE, CELL, DEOXYRIBONUCLEIC ACID (DNA), BRCA GENES, HER-2 GENE, ras GENE, MEIOSIS, CARCINOGEN, RIBOSOMES, PROTEIN, TYROSINE KINASE, ENZYME, CHROMOSOME, PLASMA MEMBRANE, EPIDERMAL GROWTH FACTOR (EGF), SIGNAL TRANSDUCTION, CODING SEQUENCE, TUMOR, CANCER, PROTO-ONCOGENES, GENETIC CODE, RECEPTORS, MUTAGEN.

**Open Reading Frame (ORF)** Region of a gene (DNA) that contains a series of triplet (bases) coding for amino acids without any termination codons. The ORF sequence is potentially translatable into a protein, but the presence of an open reading frame (sequence) does not guarantee that a protein molecule will be produced (by cell ribosome). See also GENE, DEOXYRIBONUCLEIC ACID (DNA), AMINO ACID, PROTEIN, CODING SEQUENCE, GENETIC CODE, TRANSLATION, CELL, RIBOSOMES.

**Operator** Also known as the "o locus." The site on the DNA to which a repressor molecule binds to prevent the initiation of transcription. The operator locus is a distinct entity and exists independently of the structural genes and the regulatory gene. It is the structural/biochemical "switch" with which the operon is turned on or off, and it controls the transcription of an entire group of coordinately induced genes. One type of mutation of the operator locus is called operator constitutive mutants. Constitutive mutants continually churn out the protein characteristic for that operon because the operon unit cannot be turned off by the repressor molecule. See also OPERON, PROMOTER, REGULATORY GENES, REPRESSION (OF GENE TRANSCRIPTION/TRANSLATION), REPRESSOR (PROTEIN), STRUCTURAL GENE, STRUCTURAL GENOMICS.

**Operon** A gene unit consisting of one or more genes that specify a polypeptide and an operator unit that regulates the structural gene,

i.e., the production of messenger RNA (mRNA) and hence, ultimately, of a number of proteins. Generally an operon is defined as a group of functionally related structural genes mapping (being) close to each other in the chromosome and being controlled by the same (one) operator. If the operator is "turned on," then the DNA of the genes comprising the operon will be transcribed into mRNA, and down the line specific proteins are produced. If, on the other hand, the operator is "turned off," then transcription of the genes does not occur and the production of the operon-specific proteins does not occur. See also OPERATOR, TRANSCRIPTION.

**Optical Activity** The capacity of a substance to rotate the plane of polarization of plane-polarized light when examined in an instrument known as a polarimeter. All compounds that are capable of existing in two forms that are nonsuperimposable mirror images of each other exhibit optical activity. Such compounds are called stereoisomers (or enantiomers or chiral molecules) and the two forms arise because compounds having asymmetric carbon atoms to which other atoms are connected may arrange themselves in two different ways. See also STEREOISOMERS, ENANTIOMERS, CHIRAL COMPOUND.

**Optical Density (OD)** The absorbance of light of a specific wavelength by molecules normally dissolved in a solution. Light absorption depends upon the concentration of the absorbing compound (chemical entity) in the solution, the thickness of the sample being illuminated, and the chemical nature of the absorbing compound. An analytical instrument known as a spectrophotometer is used to (quantitatively) express the amount of a substance (dissolved) in a solution. Mathematically, this is accomplished using the Beer-Lambert Law. See also SPECTROPHOTOMETER, ABSORBANCE (A).

**Optimum Foods** See NUTRACEUTICALS, PHYTOCHEMICALS.

**Optimum pH** The pH (level of acidity) at which maximum growth occurs or maximal enzymatic activity occurs, or at which any reaction occurs maximally. See also ENZYME.

**Optimum Temperature** The temperature at which the maximum growth occurs or maximal enzymatic activity occurs, or at which any reaction occurs maximally. See also ENZYME, ENSILING.

**Optrode** A fiberoptic sensor made by coating the tip of a (glass) optic fiber with an antibody that fluoresces when the antibody comes in contact with its corresponding antigen. Alternatively, the fiber tip is sometimes coated with a dye that fluoresces when the dye comes in contact with specific chemicals (oxygen, glucose, etc.).

Functionally, a beam of light is sent down the fiber and strikes ("pumps") the fluorescent complex, which then fluoresces (releases light of a specific wavelength). The light produced by fluorescence travels back up the same optic fiber and is detected by a spectrophotometer upon its return. By application of the Beer-Lambert Law, quantitative detection/measurement of the antigen or chemical *in vivo* in, e.g., a patient's bloodstream is possible. See also ANTIGEN, *IN VIVO*, ANTIBODY, GLUCOSE (GLc), SPECTROPHOTOMETER.

**Oral Cancer** Also sometimes known as "cancer of the mouth," this is a cancer involving the tissues lining the human mouth. Causes include consumption by humans of carcinogens (tobacco products, certain mycotoxins, etc.). Oral cancerous cells arise from precancerous mouth lesions known as oral leukoplakia. During 2000, research by Frank Meyskins and William Armstrong indicated that consumption of Bowman-Birk trypsin inhibitor (BB T.I.) derived from soybeans, in a manner that 'bathes' mouth tissues in BB T.I. (for an extended period of time) inhibits the development of oral leukoplakia. See also CANCER, TUMOR, MUTAGEN, MYCOTOXINS, TRYPSIN INHIBITORS.

**Oral Leukoplakia** See ORAL CANCER.

**ORF** See OPEN READING FRAME (ORF).

**Organelles** Membrane-surrounded structures found in eucaryotic cells; they contain enzymes and other components required for specialized cell function (e.g., ribosomes for protein synthesis, or lysosomes for enzymatic hydrolysis). Some organelles such as mitochondria and chloroplasts contain DNA and can replicate autonomously (from the rest of the cell). See also NUCLEUS, EUCARYOTE, ENZYME, RIBOSOMES, LYSOSOME.

**Organism** Refers to any living plant, animal, bacteria, fungus, virus, etc. Also (e.g., in certain international treaties such as the Convention on Biological Diversity), this term includes things (e.g., seeds, spores, eggs) possessing the potential to become plants, animals, fungi, etc. See also BIOLOGY, BACTERIA, FUNGUS, VIRUS, CONVENTION ON BIOLOGICAL DIVERSITY (CBD).

**Organismos Geneticamente Modificados** See GMO.

**Organization for Economic Cooperation and Development (OECD)** An international organization comprised of the world's wealthiest (most developed) nations, originally established in 1960 to study trade and related matters. In 1991, the OECD's Group of National Experts on Safety in Biotechnology (GNE) completed a document entitled Report on the Concepts and Principles Underpinning Safety Evaluations of Food Derived from Modern Biotechnology. The "aim of that document was to elaborate the scientific principles to be considered (i.e., by OECD member nations' regulatory agencies) in evaluating the safety of new foods and food components" (e.g., genetically modified soybeans, corn/maize, potatoes, etc.). See also BIOTECHNOLOGY, SOYBEAN PLANT, GNE, CANOLA, MUTUAL RECOGNITION AGREEMENTS (MRA).

**Organogenesis** The production of entire organs, usually from basic cells, such as fibroblasts, and structural material such as collagen. See also COLLAGEN, FIBROBLASTS.

**Origin** Point or region where DNA (deoxyribonucleic acid) replication is begun. Often abbreviated "Ori." See also REPLICATION (OF VIRUS), REPLICATION FORK.

**Orphan Drug** The name of the legal status granted by the Food and Drug Administration's Office of Orphan Products Development (to certain pharmaceuticals). This classification provides the sponsors of those pharmaceuticals with special tax and other financial incentives (e.g., market monopoly for a limited time). If companies feel that they possess a cure (drug) for a certain disease, but the number of potential patients is below a certain number and there is potential competition from rival companies, then the high cost of developing and shepherding the drug through the FDA would be such that the company would not be able to regain its development costs and make a profit. Hence, orphan drug status was designed to encourage drug development efforts for otherwise noneconomic pharmaceuticals with less than 200,000 patients a year.

**Orphan Genes** Genes within an organism's genome/DNA, that have no apparent function. See also GENE, ORGANISM, GENOME, DEOXYRIBONUCLEIC ACID (DNA), FUNCTIONAL GENOMICS.

**Orphan Receptors** Refers to cellular receptors (i.e., embedded in surface of cell membrane) that are not coupled to G-protein (cell) system complexes. See also BIORECEPTORS, RECEPTORS, CELL, PLASMA MEMBRANE, G-PROTEINS, ADHESION MOLECULE, MICROARRAY (TESTING), BIOCHIPS, HIGH-THROUGHPUT SCREENING (HTS), TARGET-LIGAND INTERACTION SCREENING, LIGAND (IN BIOCHEMISTRY), BIOASSAY, GENE EXPRESSION ANALYSIS, TARGET (OF A THERAPEUTIC AGENT).

**Orthophosphate Cleavage** Enzymatic cleavage of one of the phosphate ester bonds of ATP to yield ADP and a single phosphate molecule known as orthophosphate (designated as Pi). The cleavage of the phosphate bond is energy-yielding and is (except in the case of a futile cycle) coupled enzymatically to reactions that utilize the energy to run the cell. An orthophosphate cleavage reaction releases relatively less energy than does a corresponding pyrophosphate cleavage reaction. See also ADENOSINE DIPHOSPHATE (ADP), ADENOSINE TRIPHOSPHATE (ATP), FUTILE CYCLE, PYROPHOSPHATE CLEAVAGE.

**Osmosis** Bulk flow of water through a semipermeable (or more accurately, differentially permeable) membrane into another (aqueous) phase containing more of a solute (dissolved compound). As an example, let us set up an osmotically active system. There are two solutions, A and B. Solution A has less salt dissolved in it than solution B and, furthermore, the two solutions are separated by a differentially permeable membrane (this looks like a plastic film). Water molecules (and only water molecules) will flow from solution A through the membrane and into

solution B. The reason for this is that the membrane allows free passage only to water molecules. The bulk flow of water has the effect of diluting solution B, while concentrating solution A. Water will flow from region A to region B until the salt concentrations of both solutions are equal. Osmosis is therefore a process in which water passes from regions of low salt concentration to regions of high salt concentration. The process can be viewed as equalizing the number of water and solute molecules on both sides of the membrane. See also OSMOTIC PRESSURE.

**Osmotic Pressure** May be defined as the hydrostatic pressure which must be applied to a solution on one side of a semipermeable membrane (solution B in the example for osmosis) in order to offset the flow of solvent (water) from the other side (solution A in the example for osmosis). It is a measure of the tendency or "strength" of water to flow from a region of low salt concentration (and conversely high water concentration) to regions of high salt concentration (and conversely low water concentration). See also OSMOSIS.

**Osmotins** A category of proteins, which are produced by some organisms as a natural defense against pathogenic fungi. See also CECROPHINS, MAGAININS, ORGANISM, FUNGUS, PATHOGENIC.

**Osteoarthritis** A disease that affects primarily women older than 45, in which cartilage within the body's joint breaks down. Osteoarthritis encompasses approximately half of all cases of arthritis.

**Osteoinductive Factor (OIF)** A protein that induces the growth of both cartilage-forming cells and bone-forming cells (e.g., after a bone has been broken). When applied in the presence of transforming growth factor-beta, type 2 (another protein), osteoinductive factor first causes connective tissue cells to grow together to form a matrix of cartilage (e.g., across the bone break), then bone cells slowly replace that cartilage. Osteoinductive factor also seems to thwart a type of cell that tears down bone formation, so OIF may someday be used to combat osteoporosis. See also GROWTH FACTOR, TRANSFORMING GROWTH FACTOR-BETA (TGF-BETA), FIBROBLASTS, FIBROBLAST GROWTH FACTOR (FGF), OSTEOPOROSIS.

O

**Osteoporosis** A disease of humans in which the bones gradually weaken and become brittle. A diet containing a large amount of soy isoflavones (i.e., genistein) has been shown to increase bone density; thereby lowering the risk of osteoporosis. Groups that are especially at risk for osteoporosis include postmenopausal women (particularly of Caucasian or Asian ethnicity), those who have undergone early menopause (i.e., prior to age 45), those who smoked, those who consumed excessive amounts of alcohol, and those who consumed excessive amounts of certain pharmaceuticals (e.g., steroids such as prednisone, thyroid hormone, etc.). See also OSTEOINDUCTIVE FACTOR (OIF), GENISTEIN (Gen), SOY PROTEIN, ISOFLAVONES, STEROID, SOYBEAN PLANT, HIGH-ISOFLAVONE SOYBEANS, HAPLOTYPE.

**Outcrossing** The transfer of a given gene or genes (e.g., one synthesized by man and inserted into a plant via genetic engineering) from a domesticated organism (e.g., crop plant) to a wild type (relative of plant). See also GENE, INTROGRESSION, SYNTHESIZING (OF DNA MOLECULES), GENETIC ENGINEERING, WILD TYPE.

**Overwinding** Positive supercoiling. Winding which applies further tension in the direction of the winding of the two strands about each other in the duplex. See also DEOXYRIBONUCLEIC ACID (DNA), SUPERCOILING, DOUBLE HELIX, DUPLEX.

**Oxalate** A salt or ester of oxalic acid. See also CALCIUM OXALATE.

**Oxidant** See OXIDIZING AGENT.

**Oxidation (chemical reaction)** Loss of electrons from a compound (or element) in a chemical reaction. When one compound is oxidized, another compound is reduced. That is, the other compound must "pick up" the electrons which the first has lost. See also OXIDATION-REDUCTION REACTION, HYDROGENATION, OXIDATION (of fats/oils/lipids).

**Oxidation (of fats/oils/lipids)** A chemical transformation of fat/lipid molecules, in which oxygen (e.g., from air) is combined with those molecules. As a result of that (oxidation chemical reaction), various chemical entities are created (peroxides, aldehydes, etc.) which possess objectionable flavors/odors, and are harmful to animals that consume such (rancid) fats/oils. See also FATS, FATTY ACID, LIPIDS, PLASMA MEMBRANE, OXIDATION (chemical reaction), OXIDATIVE STRESS, HYDROLYSIS.

**Oxidation (of fatty acids)** See CARNITINE.

**Oxidation-Reduction Reaction** A chemical reaction in which electrons are transferred from a donor to an acceptor molecule or atom. See also OXIDATION (chemical reaction), OXIDIZING AGENT, REDUCTION (IN A CHEMICAL REACTION).

**Oxidative Phosphorylation** The enzymatic phosphorylation of ADP to ATP coupled to electron transport from a substrate to molecular oxygen. The synthesis (production) of ATP from the starting materials of ADP and inorganic phosphate (orthophosphate). See also ADENOSINE DIPHOSPHATE (ADP), ADENOSINE TRIPHOSPHATE (ATP), ORTHOPHOSPHATE CLEAVAGE.

**Oxidative Stress** The physiological stress/damage that results from the (chemical reaction) breakdown of all or part of an organism, via oxidation reaction(s). For example, oxidative stress appears to be present in the brains of all victims of neurodegenerative diseases (Alzheimer's disease, Parkinson's disease, etc.). One common result of such oxidation reactions is the generation (within organism's body) of reactive oxygen species ("free radicals") that can adversely affect:

- Endothelial function (i.e., the inner lining of blood vessels)
- Platelet aggregation (e.g., inappropriate blood clotting/clumping)
- Atherosclerosis (i.e., buildup of oxidized fatty deposits known as plaque on internal walls of arteries)
- Myocardial function (e.g., heart failure)
- Eye and kidney tissue (especially in diabetics)

A key indicator of oxidative stress is the peroxidation of membrane lipids to form mono- and bifunctional aldehydes (e.g., 4-hydroxy-2-nonenal, also known as HNE). See also ORGANISM, OXIDATION (chemical reaction), ALZHEIMER'S DISEASE, PARKINSON'S DISEASE, CELL, ANTIOXIDANTS, PLASMA MEMBRANE, LIPIDS, GLUTATHIONE, CAROTENOIDS, ENDOTHELIAL CELLS, PLATELETS, ATHEROSCLEROSIS, INSULIN, CORONARY HEART DISEASE (CHD), HAPTOGLOBIN.

**Oxidizing Agent** (oxidant) The acceptor of electrons in an oxidation-reduction reaction. The oxidant is reduced by the end of the chemical reaction. That is, the oxidizing agent is the entity that seeks and accepts electrons. Electron acceptance is, by definition, reduction. See also OXIDATION-REDUCTION REACTION, PEROXIDASE.

**Oxygen Free Radical** See FREE RADICAL.

**Oxygenase** An enzyme catalyzing a reaction in which oxygen is introduced into an acceptor molecule.

O

# P

**P Element** A transposon, whose genes (within this transposon) resist rearrangement during the process (i.e., transposition) of the P element being incorporated into a new location within an organism's genome (i.e., its deoxyribonucleic acid or DNA). In addition to "carrying" genes to a new location(s) in the genome, the P element itself codes for transposase (an enzyme that makes transposition possible). See also TRANSPOSON, GENE, ENZYME, TRANSPOSITION, TRANSPOSASE, DEOXYRIBONUCLEIC ACID (DNA), GENOME.

***P. gossypiella*** See PECTINOPHORA GOSSYPIELLA.

**P-Selectin** Formerly known as GMP-140 and PADGEM, it is a selectin molecule that is synthesized by endothelial cells before (adjacent) tissues are infected. Thus "stored in advance," the endothelial cells can present P-selectin molecules on the internal surface of the endothelium within minutes after an infection (of adjacent tissue) begins. This presentation of P-selectin molecules attracts leukocytes to the site of the infection, and draws them out of the bloodstream (the leukocytes "squeeze" between adjacent endothelial cells). See also SELECTINS, LECTINS, ELAM-1, ADHESION MOLECULE, LEUKOCYTES, ENDOTHELIUM.

**p53 Gene** A tumor-suppressor gene which controls passage (of a given cell) from the "GI" phase to the "s" (i.e., DNA synthesis) phase. The p53 protein that is coded for by the p53 gene is a transcription factor (i.e., it "reads" DNA to determine if damaged, then acts to control cell division, while the p53 gene codes for more production of additional p53 protein).

Discovered in 1993 by Arnold J. Levine and colleagues, it is believed to be responsible for up to 50% of all human cancer tumors (when the p53 gene is damaged or mutated). Normally, the p53 gene codes for (i.e., causes to be manufactured in cell) the p53 protein, which acts to prevent cells from dividing uncontrollably when the cell's DNA has been damaged (e.g., via exposure to cigarette smoke or ultraviolet light). If, in spite of the presence of p53 protein, a cell begins to divide uncontrollably following damage to its DNA, the p53 gene can cause apoptosis, which is also known as "programmed cell death" (to prevent tumors). See also GENE, TUMOR-SUPPRESSOR GENES, ras GENE, GENETIC CODE, MEIOSIS, DEOXYRIBONUCLEIC ACID (DNA), CARCINOGEN, RIBOSOMES, ONCOGENES, TRANSCRIPTION FACTORS, CANCER, TUMOR, p53 PROTEIN, PROTO-ONCOGENES, PROTEIN, APOPTOSIS.

**p53 Protein** A tumor-suppressor protein, sometimes called the master transcription factor, or the "guardian of the genome;" but whose amino acid sequence alterations (resulting from damage or mutation to the p53 gene) are believed to be responsible for up to 50% of all human cancer tumors. The p53 protein has four domains, one of which (i.e., the core domain) binds to a specific sequence(s) of the cell's DNA, in order to prevent the cell from dividing uncontrollably when the cell's DNA has been damaged (e.g., via exposure to cigarette smoke, ultraviolet light, or other carcinogen), until the damage to that DNA can be repaired. As the amount of DNA within a given (damaged) cell increases, the concentration of p53 protein also increases. Because p53 protein is a transcription factor (i.e., "reads" DNA to determine if damaged, then acts to control cell division, while p53 gene codes for production of more p53), p53 is very efficient

P

at preventing/inhibiting tumors. However, if the cell's DNA cannot be repaired, the p53 protein can then cause apoptosis ("programmed cell death") to prevent development of (cancerous) tumors. See also GENE, p53 GENE, TUMOR-SUPPRESSOR GENES, ras GENE, ras PROTEIN, GENETIC CODE, MEIOSIS, CARCINOGEN, DEOXYRIBONUCLEIC ACID (DNA), AFLATOXIN, RIBOSOMES, ONCOGENES, CANCER, TUMOR, PROTO-ONCOGENES, PROTEIN, TRANSCRIPTION FACTORS, DOMAIN (OF A PROTEIN), APOPTOSIS.

**Paclitaxel** An anticancer compound (pharmaceutical) that was originally isolated from the Pacific yew tree (*Taxus brevifolia*), although it is made synthetically today. In 1966, Maurice Wall first identified antitumor effects in an extract from *Taxus brevifolia*. In 1992, the U.S. Food and Drug Administration approved paclitaxel for use to treat recurrent ovarian cancer. Other anticancer uses were later approved. When injected into the human body, paclitaxel also inhibits growth of the parasitic microorganism *Toxoplasma gondii* (which can cause loss of sight and neurological disease in humans, if not controlled). See also CANCER, TAXOL™, FOOD AND DRUG ADMINISTRATION (FDA), CHEMOTHERAPY, TUBULIN, MICROORGANISM, GROWTH (MICROBIAL).

**PAF** Acronym for Platelet Activating Factor. See also CHOLINE.

**PAGE** See POLYACRYLAMIDE GEL ELECTROPHORESIS (PAGE).

**Palindrome** A DNA molecule sequence that is the same when one strand of the molecule is read left to right and the other strand is read right to left. See also DEOXYRIBONUCLEIC ACID (DNA), READING FRAME.

**Palmitate** See PALMITIC ACID.

**Palmitic Acid** A saturated fatty acid containing sixteen carbon atoms in its molecular "backbone"; which tends to increase cholesterol levels in the bloodstream when consumed by humans. It has been shown that feeding of extruded (whole) high-oleic oil soybeans to dairy cattle did decrease the content of palmitic acid in their milk. See also FATTY ACID, SATURATED FATTY ACIDS (SAFA), CHOLESTEROL, HIGH-OLEIC OIL SOYBEANS.

**Pancreas** An organ (gland) located near the stomach that secretes insulin and glucagon into the bloodstream, and digestive fluids into the intestines. See also DNASE, INSULIN, GLUCAGON, BETA CELLS, TYPE I DIABETES, TYPE II DIABETES, DIABETES.

**Papovavirus** A class of animal viruses, e.g., SV40 and polyoma. See also VIRUS.

**Parkinson's Disease** A disease of the human brain, in which those nerve cells (neurons) associated with emotions and those neurons that are involved in controlling movement (motor control) die. Discovered in 1919 by doctors treating an epidemic of encephalitis lethargica (onset of Parkinson's disease commonly follows encephalitis, but it can also be induced by certain drugs, etc.). The (natural) cause of Parkinson's disease (i.e., causing a dwindling supply of dopamine in the brain) is unknown although it can be induced by drug misuse. When a human brain is functioning normally, cells within a region of the brain called the substantia nigra initiate motor (i.e., muscle) activity by releasing the chemical "messenger" known as dopamine. In the brain of a person suffering from Parkinson's disease, those dopamine-producing cells die off, causing a progressive loss of motor control for that person. See also NEUROTRANSMITTER, CILIARY NEUROTROPHIC FACTOR (CNTF), SIGNALING, GLIAL DERIVED NEUROTROPHIC FACTOR (GDNF), OXIDATIVE STRESS, NEURON.

**PARP** Acronym for Poly ADP-ribose Polymerase (an enzyme naturally present in human cells that is involved in control of apoptosis, among other cellular processes). This enzyme can be commercially produced (e.g., to manufacture tests) by genetically engineered hamster cells grown in cell culture. This enzyme can be utilized by man in order to determine/test if a given substance (e.g., industrial chemical) is carcinogenic to humans. See also ENZYME, ADENOSINE DIPHOSPHATE (ADP), RIBOSE, POLYMERASE, CELL, APOPTOSIS, CELL CULTURE, MAMMALIAN CELL CULTURE, CARCINOGEN, CANCER, NUCLEAR MATRIX PROTEINS, GENETIC ENGINEERING, AMES TEST.

**Particle Cannon** See BIOLISTIC R GENE GUN, MICROPARTICLES.

**Particle Gun** See BIOLISTIC® GENE GUN, MICRO-PARTICLES, "SHOTGUN" METHOD.

**Partition Coefficient** A constant (number) that expresses the ratio in which a given solute will be partitioned (distributed) between two given immiscible liquids (e.g., oil and water) at equilibrium.

**Partitioning Agent** Any one of a number of chemical compounds (e.g., certain hormones, conjugated fatty acids, etc.) which cause a given animal's metabolism to deposit significantly more lean muscle tissue and significantly less fat tissue within that (growing) animal's body. See also BOVINE SOMATOTROPIN (BST), PORCINE SOMATOTROPIN (PST), CONJUGATED LINOLEIC ACID (CLA), CARNITINE, METABOLISM, FATS.

**Passive Immunity** An immune response (to a pathogen) that results from injecting another organism's antibodies into the organism that is being challenged by the pathogen. See also POLYCLONAL ANTIBODIES, HUMORAL IMMUNITY, ANTIBODY, COMPLEMENT, COMPLEMENT CASCADE, IMMUNOGLOBULIN, PATHOGEN, ANTIGEN, MONOCLONAL ANTIBODIES (MAb).

**PAT Gene** A dominant gene isolated from the *Streptomyces viridochromogenes* bacterium which codes for (causes production of) the enzyme phosphinothricin acetyl transferase (PAT). When the PAT gene is inserted into a plant's genome, it imparts resistance to glufosinate-ammonium containing herbicides. Because the glufosinate-ammonium herbicides act via inhibition of glutamine synthetase (an enzyme that catalyzes the synthesis of glutamine), this inhibition of enzyme kills plants (e.g., weeds). That is because glutamine is crucial for plants to synthesize critically needed amino acids. The PAT gene is also often used by genetic engineers as a marker gene. See also GENE, GENOME, GENETIC ENGINEERING, MARKER (GENETIC MARKER), BAR GENE, DOMINANT ALLELE, ESSENTIAL AMINO ACIDS, HERBICIDE-TOLERANT CROP, GTS, SOYBEAN PLANT, CANOLA, CORN, GLUTAMINE, GLUTAMINE SYNTHETASE, PHOSPHINOTHRICIN, PHOSPHINOTHRICIN ACETYLTRANSFERASE (PAT).

**Pathogen** Refers to a virus, bacterium, parasitic protozoan, or other microorganism that causes infectious disease by invading the body of an organism (animal, plant, etc.) known as the host. It should be noted that infection is not synonymous with disease because infection does not always lead to injury of the host. See also VIRUS, BACTERIA, PROTOZOA, MICROORGANISM, STRESS PROTEINS, ANTIGEN, IMMUNE RESPONSE, PHYTOALEXINS, PATHOGENESIS RELATED PROTEINS.

**Pathogenesis Related Proteins** Protective (i.e., disease-fighting) proteins that are produced within certain plants in response to the entry-into-plant of plant pathogens (bacteria, fungi, etc. that infect and cause disease in plants). One pathogenesis-related protein is chitinase, a protein enzyme that degrades (breaks down) the chitin within cell walls of pathogenic fungi. Production of pathogenesis-related proteins is often initiated by signaling molecules (e.g., harpin) produced by the pathogens. See also PROTEIN, PATHOGEN, BACTERIA, FUNGUS, CHITINASE, CHITIN, CELL, ENZYME, SIGNALING, SIGNALING MOLECULE, HARPIN, HYPERSENSITIVE RESPONSE.

**Pathogenic** Disease-causing. See also PATHOGEN.

**Pathway** A sequential series of chemical reactions, each of which is dependent on previous ones in the pathway (e.g., the third reaction requires chemical product produced by first/second chemical reactions), that — overall — yields a beneficial impact. For example, metabolism (i.e., the entire set of enzyme-catalyzed chemical reactions which converts food into nutrients that can be used by the body's cells and the use of these nutrients by the body's cells to sustain life, grow, etc.) occurs via a very specific METABOLIC PATHWAY. See also METABOLISM, ACC SYNTHASE, R GENES, PATHWAY FEEDBACK MECHANISMS.

**Pathway Feedback Mechanisms** Chemically based mechanisms (e.g., series of chemical reactions) that hinder (or increase rate of) a given pathway. For example, when the body of bacteria need catabolism (i.e., energy production) to be slowed down, it uses the mechanism of catabolite repression (to slow down catabolism via chemical/reaction means). See also PATHWAY, METABOLISM, CATABOLISM, CATABOLITE REPRESSION.

**PBR** The intellectual property rights that are legally accorded to plant breeders by laws, treaties, etc. Similar to patent law for inventors. See also PLANT BREEDER'S RIGHTS (PBR),

PLANT'S NOVEL TRAIT (PNT), PLANT VARIETY PROTECTION ACT (PVP), EUROPEAN PATENT CONVENTION, EUROPEAN PATENT OFFICE (EPO), U.S. PATENT AND TRADEMARK OFFICE (USPTO).

**pBR322** An *Escherichia coli* (*E. coli*) plasmid cloning vector that contains the ampicillin resistance and tetracycline resistance genes. It consists of a circle of double-stranded DNA. See also ESCHERICHIA COLIFORM (*E. COLI*), PLASMID, VECTOR, DEOXYRIBONUCLEIC ACID (DNA).

**PC** Phosphatidyl choline. See also LECITHIN (refined, specific), LECITHIN (crude, mixture).

**PCR** See POLYMERASE CHAIN REACTION (PCR).

**PDCAAS** See PROTEIN DIGESTIBILITY-CORRECTED AMINO ACID SCORING (PDCAAS).

**PDE** See PHOSPHODIESTERASES.

**PDGF** See PLATELET-DERIVED GROWTH FACTOR (PDGF).

**PDWGF** See PLATELET-DERIVED WOUND GROWTH FACTOR (PDWGF).

***Pectinophora gossypiella*** Also known as the pink bollworm, this is one of three insect species that are called "bollworms" (when they are on cotton plants). The holes that they chew in cotton plants' bolls have been shown to enable the *Aspergillas flavus* fungus to infect those (chewed) cotton plants. See also *B.t.* KURSTAKI, HELICOVERPA ZEA (*H. ZEA*), *HELIOTHIS VIRESCENS*, BRIGHT GREENISH-YELLOW FLUORESCENCE (BGYF).

**PEG-SOD (polyethylene glycol superoxide dismutase)** A modified version of the enzyme human superoxide dismutase (hSOD), in which polyethylene glycol (a polymer made up of ethylene glycol monomers) is combined with the hSOD molecule. The PEG seems to wrap around or about the enzyme in such a way that the whole complex is able to exist in the blood for longer periods of time than the unmodified hSOD enzyme. This is because the PEG effectively camouflages the hSOD molecule and hence protects it from being inactivated by the body's own defense mechanisms in the bloodstream. This technology is important in that hSOD is used to fight certain diseases by injecting it into the body. However, the SOD must be present in the body for extended periods of time in order to effectively work, and since the injected SOD is a foreign molecule, the body tries to destroy it (and hence its function) as quickly as possible. See also HUMAN SUPEROXIDE DISMUTASE (hSOD), CATALASE, ENZYME.

**Penicillin G (benzylpenicillin)** The original penicillin (antibiotic) molecule, discovered by Alexander Fleming in 1928, in a petri dish (experiment) 'spoiled' by accidental introduction of a mold. Fleming named the antibiotic after the particular mold (*Penicillium notatum*) that had produced it. During the 1940s, scientists at the U.S. Department of Agriculture in Peoria, Illinois (in the U.S.), discovered how to produce commercial quantities of Penicillin G by utilizing the fungus *Penicillium chrysogenum*, which they found growing on a canteloupe in Peoria. Penicillin kills bacteria by blocking an enzyme which is crucial to growth and repair of the bacteria's cell wall (peptidoglycan layer), but penicillin does not harm other species, so it is species-specific to certain pathogenic bacteria (e.g., streptococcus, meningococcus, and diphtheria bacillus). See also ANTIBIOTIC, FUNGUS, BACTERIA, ENZYME, SPECIES SPECIFIC, *PENICILLIUM*, BETA-LACTAM ANTIBIOTICS, BACILLUS.

**Penicillinases (E.C. 3.5.2.6)** Also known as β-lactamases, these are enzymes that hydrolyze (break down) the β-lactam ring (portion) of the penicillin molecule's structure. Some microorganisms (e.g., pathogenic bacteria) have become able to produce these enzymes as a defense to penicillin and cephalosporin antibiotics (drugs). See also ENZYME, HYDROLYZE, PENICILLIN G (BENZYLPENICILLIN), PATHOGENIC, BACTERIA, ANTIBIOTIC, ANTIBIOTIC RESISTANCE.

***Penicillium*** Refers to the genus of fungi (mold) that belongs to the category *Deutromycotina* and often causes (food) spoilage. Some of the genus have been utilized commercially to produce antiboiotics. See also GENUS, FUNGUS, OCHRATOXINS, ANTIBIOTIC, PENICILLIN G (benzylpenicillin).

**Pentose** A simple sugar (monosaccharide molecule) whose backbone structure contains five carbon atoms. There exists many different pentoses. Some examples of pentoses are ribose, arabinose, and xylose. See also MONOSACCHARIDES.

**Pepsin** A crystallizable proteinase (enzyme) that in an acidic medium digests (breaks down) most proteins to polypeptides. It is secreted by glands in the mucous membrane of the stomach of higher animals. In combination with dilute hydrochloric acid, it is the chief active principle (component) of gastric juice. Also used in manufacturing peptones and in digesting gelatin for the recovery (i.e., recycling) of silver from photographic film. See also DIGESTION (WITHIN ORGANISMS), PROTEIN, PEPTIDE, LACTOFERRIN, PEPTONE.

**Peptidase** An enzyme that hydrolyzes (cleaves) a peptide bond. See also PEPTIDE BOND, PEPSIN, PEPTONE, PEPTIDE MAPPING ("FINGERPRINTING").

**Peptide** Two or more amino acids covalently joined by peptide bonds. An oligomer component of a polypeptide. A dipeptide, for example, consists of two (di) amino acids joined together by a peptide bond or linkage. By analogy, this structure would correspond to two joined links of a chain. See also POLYPEPTIDE (PROTEIN), OLIGOMER, AMINO ACID.

**Peptide Bond** A covalent bond (linkage) between the α-amino group of one amino acid and the α-carboxyl group of another amino acid. This is the linkage or bond which holds the amino acids (chain links) together in a polypeptide chain. It is the all-important bond which holds the amino acid monomers together to form the polymer known as a polypeptide. See also PEPTIDE, POLYPEPTIDE (PROTEIN), OLIGOMER.

**Peptide Mapping (fingerprinting)** Refers to the characteristic pattern of peptides (i.e., pieces that make up a protein molecule) resulting from partial hydrolysis (cleavage, digestion) of a protein. The pattern (fingerprint) is obtained by separating the peptides via two-dimensional chromatography, in which the peptides are first subjected to chromatography using one solution which separates many, but not all, peptides. The chromatogram is then turned 90°, and is again chromatographed using a second solution, which then separates all of the peptides; thereby producing the final "fingerprint" of the protein. See also CHROMATOGRAPHY, PEPTIDE, PROTEIN, HYDROLYSIS.

**Peptide Nanotube** See SELF-ASSEMBLY (OF A LARGE MOLECULAR STRUCTURE).

**Peptido-Mimetic** See BIOMIMETIC MATERIALS, PEPTIDE.

**Peptone** A protein that has been partially hydrolyzed (cleaved) by the peptidase pepsin. See also PROTEIN, HYDROLYTIC CLEAVAGE, PEPTIDASE, PEPSIN, PEPTIDE MAPPING ("FINGERPRINTING").

**Perforin** A 70 Kd (kilodalton) protein that is instrumental in the lysis of infected cells. A series of reactions occurs on the surface of a cell which results in the polymerization of certain monomers to form transmembrane (through the membrane) pores 100 Å (Angstroms) wide, which allows ions to rush into the cell (due to osmotic pressure) and thus burst (lyse) that cell, so the (formerly) internal pathogens can be attacked by the body's immune system. Perforin is a protein that is akin to the C9 component of the complement. See also OSMOTIC PRESSURE, COMPLEMENT, COMPLEMENT CASCADE, Kd, CYTOTOXIC T CELLS, CECROPHINS, MAGAININS, OSMOTINS.

**Periodicity** The number of base pairs per turn of the DNA double helix. See also DEOXYRIBONUCLEIC ACID (DNA).

**Periodontium** Tissue that anchors teeth in the jaw. Regrowth of periodontal tissue can be stimulated by a combination of platelet-derived growth factor and insulin-like growth factor-1. See also PLATELET-DERIVED GROWTH FACTOR (PDGF), INSULIN-LIKE GROWTH FACTOR-1 (IGF-1).

**Peritoneal Cavity/Membrane** The smooth, transparent, serous membrane that lines the cavity of the abdomen of a mammal.

**Peroxidase** An enzyme that catalyzes the oxidation of a substrate with hydrogen peroxide (as the electron acceptor, so the hydogen peroxide is reduced). Peroxidase is naturally produced in soybeans by approximately half of all commercial soybean varieties. Peroxidase very effectively inhibits (stops) growth of any *Aspergillus flavus* fungi that might be present (e.g., in the soil). Peroxidase can be used to replace more toxic and environmentally problematic chemicals in certain industrial processes. Among other applications, peroxidase can replace formaldehyde use in paints, varnishes, glues, and computer chip manufacturing. See also ENZYME, OXIDIZING

AGENT (OXIDANT), OXIDATION, SUBSTRATE (CHEMICAL), OXIDATION-REDUCTION REACTION.

**Persistence** The tendency of a compound (e.g., an insecticide) to resist degradation by biological means (e.g., metabolism by microorganisms) after that compound has been introduced into the environment (e.g., sprayed onto a field) or by physical means (degradation caused by exposure to sunlight, moisture, etc.). See also METABOLISM, MICROORGANISM, BIODEGRADABLE.

***Pfiesteria piscicida*** A single-celled microscopic algae which has a predator/prey relationship with fish in its ecosystem. During a large portion of its life cycle, *Pfiesteria piscicida* exists in a nontoxic cyst form at the bottom of a river. When those (cysts) detect certain substances (e.g., excreta) emitted by live fish, the *Pfiesteria piscicida* transform into an amoeboid or dinoflagellate form, which secretes a water-soluble neurotoxin into the water (which incapacitates nearby fish). The *Pfiesteria piscicida* next attach themselves to those fish, and excrete a lipid-soluble toxin which destroys the epidermal layer of the fish's skin, allowing the *Pfiesteria piscicida* to begin "eating" the fish's tissue. Human exposure to the neurotoxin apparently causes short-term memory loss. See also ECOLOGY, CELL, TOXIN, LIPIDS.

**PHA** See POLYHYDROXYALKANOIC ACID (PHA).

**Phage** Abbreviation for bacteriophage. Another name for a specific type of virus. A virus that attacks bacteria is known as a bacteriophage. Bacteriophages are frequently used as vectors for carrying (foreign) DNA into cells by genetic engineers. See also BACTERIOPHAGE, VECTOR, GENETIC ENGINEERING, TRANSFECTION, DEOXYRIBONUCLEIC ACID (DNA).

**Phagocyte** A cell such as a leukocyte that engulfs and digests cells, cell debris, microorganisms, and other foreign bodies in the bloodstream and tissues (phagocytosis). The ingested material is then degraded via enzymes. A whole class of cells is known to be phagocytic. See also MACROPHAGE, MICROPHAGE, MONOCYTES, T CELLS, POLYMORPHONUCLEAR LEUKOCYTES (PMN), CELLULAR IMMUNE RESPONSE, POLYMORPHONUCLEAR GRANULOCYTES, LYSOSOME.

**Pharmacoenvirogenetics** A word coined during 2000 by Tim Studt to describe the fact that environmental factors interact with a given individual's (human/animal/plant) genetic makeup (i.e., genome) to determine those individual's (body's) response to a given pharmaceutical (and/or progression of a disease). Those environmental factors include:

- Foods eaten
- The stress the individual is exposed to
- Air and water pollution the individual is exposed to
- Temperature and humidity the individual is exposed to
- Geographical elevation the individual is exposed to
- Bacteria the individual is exposed to

For example, when *Rhizobium japonicum* bacteria grow in the soil near the roots of a soybean plant (*Glycine max* L.), that causes certain specific genes in the soybean plant to be expressed (i.e., "turned on") so that soybean plant's roots become more hospitable as a "home" for those *Rhizobium japonicum* bacteria to live symbiotically (in nodules on the roots) with the soybean plant. See also PHARMACOKINETICS, GENETICS, PHARMACOLOGY, PHARMACOGENETICS, ABSORPTION, METABOLISM, SNP, ALLELE, HAPLOTYPE, HAPTOGLOBIN, RHIZOBIUM (BACTERIA), NODULATION, SYMBIOTIC, CENTRAL DOGMA (NEW), ACCLIMATIZATION.

**Pharmacogenetics** A branch of pharmacokinetics that deals with the reactions between drugs, or free radicals, or synthetic food ingredients, and specific individuals due to the genetics of those individuals. The subgroup of all those individuals whose DNA causes their bodies to respond in a specific way to a given drug or synthetic food ingredient, is known as a HAPLOTYPE. For example, one haplotype (subgroup) of pediatric leukemia patients suffers severe and life-threatening reactions to some commonly used leukemia treatment drugs, due to the variation (i.e., SNP) in the thiopurine S-methyl transferase gene (allele) in their genome. Another example is that consumption of sodium-containing food ingredients tends to cause a dangerous increase in blood pressure (hypertension) among the African-American people living in the U.S., more often than among other ethnic groups living in the U.S. See also PHARMACOKINETICS, PHARMACOGENOM-ICS, GENETICS, PHARMACOLOGY, ABSORPTION, METABOLISM, HAPLOTYPE, DEOXYRIBONUCLEIC ACID (DNA), MUTATION, SNP, ALLELE, CANCER, HAPTOGLOBIN, TRANSVERSION.

**Pharmacogenomics** A branch of pharmacokinetics that deals with the biological impacts of pharmaceuticals or synthetic food ingredients, and the specific differences in response/reaction of living structures (tissues, organs, etc.) due to different genomes (DNA) of those individual organisms that consume those pharmaceuticals or food ingredients. The subgroup consisting of all those individuals whose genome (DNA) causes their body to respond in a specific way to a given pharmaceutical, free radical, or synthetic food ingredient, is known as a HAPLOTYPE. A haplotype could (theoretically) be as small as one individual (e.g., one woman, possessing an as yet unknown genome), because that woman's particular specific response to a pharmaceutical could result from one single-nucleotide polymorphism (SNP) that only her genome possesses. Thus, pharmacogenomics is the pharmacokinetics (of a given pharmaceutical or food ingredient) within a specific haplotype.

For example, some ethnic minorities, genders, and individuals have far different biological reactions/responses to certain pharmaceuticals (e.g., the painkiller morphine works better in women, aspirin "thins" men's blood better than women's blood, the painkiller ibuprofen works better in men, the diuretic drug thiazide works to control hypertension in 60% of U.S. African Americans but only 8% of U.S. Caucasian people, etc.), and food ingredients (e.g., monosodium glutamate, lactase, ethanol, etc.) impact some members of some ethnic minorities more than they do the majority of humans. That is due to the fact that different gene(s) within their genomes (DNA) cause synthesis of certain different proteins (generally enzymes),

which thereby cause the tissues/bodies of those individuals/ethnic minorities to react differently to specific pharmaceuticals or food ingredients in terms of:

- Absorption — transport of the drug (pharmaceutical) or food ingredient into the bloodstream (e.g., from the intestinal tract, in the case of food ingredients or orally administered drugs).
- Distribution — initial physical disposition/behavior of the substance in the body after the substance enters the body tissues. For example, does the substance preferentially concentrate in the fat cells (adipose tissue) of the body, or in other specific tissues?
- Metabolism — breakdown of the substance (if breakdown does occur) into other chemical compounds, and the ultimate disposition in the body of those compounds (or the original substance, if breakdown does not occur).
- Elimination — the speed and thoroughness with which the substance is excreted or is otherwise removed from the body.

See also PHARMACOKINETICS, GENOMICS, PHARMACOLOGY, ADME TESTS, ABSORPTION, METABOLISM, GENOME, DEOXYRIBONUCLEIC ACID (DNA), DIGESTION (WITHIN ORGANISMS), PHASE I CLINICAL TESTING, HAPLOTYPE, CONSENSUS SEQUENCE, PHARMACOGENETICS, GENE, ALLELE, SINGLE-NUCLEOTIDE POLYMORPHISMS (SNPs), PROTEIN, ENZYME, HAPTOGLOBIN, ADIPOSE.

**Pharmacokinetics** (pharmacodynamics) A branch of pharmacology dealing with the reactions between drugs or synthetic food ingredients and living structures (e.g., tissues, organs). The study of the:

- Absorption — transport of the drug (pharmaceutical) or food ingredient into the bloodstream (e.g., from the intestinal tract, in the case of food ingredients).
- Distribution — initial physical disposition/behavior of the substance in the body after the substance enters the

body. For example, does the substance preferentially concentrate in the fat cells of the body?

- Metabolism — breakdown of the substance (if breakdown does occur) into other compounds, and ultimate disposition of those compounds (or the original substance, if breakdown does not occur). For example, some pharmaceuticals break down into smaller compound(s); one of which then acts upon the relevant body cells (to relieve pain, lower blood pressure, etc.).
- Elimination — the speed and thoroughness with which the substance is excreted or otherwise removed from the body.

In short, pharmacokinetics deals with what happens to a substance that is introduced into a living system. For example, how quickly it is broken down, to what intermediates and metabolites it is broken down, and what the pathway of this breakdown is. See also PHARMACOLOGY, ADME TESTS, ABSORPTION, METABOLISM, INTERMEDIARY METABOLISM, DIGESTION (WITHIN ORGANISMS), PHASE I CLINICAL TESTING, PHARMACOGENOMICS, PHARMACOGENETICS, PHARMACOENVIROGENETICS, PATHWAY.

**Pharmacology** The study of chemicals (e.g., pharmaceuticals) and their effects on living organisms. See also PHARMACOKINETICS, PHARMACOGENOMICS, PHARMACOGENETICS.

**Pharmacophore** The portion of a molecule (e.g., a pharmaceutical) that is responsible for its biological activity (i.e., therapeutic action on recipient's tissue, etc.). See also BIOLOGICAL ACTIVITY, ACTIVE SITE, CATALYTIC SITE, MINIPROTEINS.

**Phase I Clinical Testing** The first in a series of human tests of new pharmaceuticals, mandated by the U.S. Food and Drug Administration (FDA). The primary purpose of the Phase I clinical test is to detect if the new pharmaceutical is toxic or otherwise harmful to normal, healthy humans. The conclusion of Phase I testing leads to Phase II and Phase III testing.

During the 1990s, the FDA began to require the inclusion of ethnic minorities and women (in addition to men) as subjects in these tests, to enable pharmacogenomics (i.e., the testing to determine if a given pharmaceutical causes nontypical response in the bodies of members of these subgroups). See also FOOD AND DRUG ADMINISTRATION (FDA), KEFAUVER RULE, KOSEISHO, BUNDESGESUNDHEITSAMT (BGA), COMMITTEE FOR PROPRIETARY MEDICINAL PRODUCTS (CPMP), IND, IND EXEMPTION, PHARMACOGENOMICS, HAPLOTYPE, PHASE II CLINICAL TESTS.

**Phase II Clinical Tests** The second in a series of human tests of new pharmaceuticals, mandated by the U.S. Food and Drug Administration (FDA). The primary purpose of the Phase II clinical tests is to determine the pharmaceutical's efficacy (i.e., does it work?). Successful conclusion of Phase II tests allows Phase III clinical tests to begin. See also PHASE I CLINICAL TESTING, FOOD AND DRUG ADMINISTRATION (FDA), KEFAUVER RULE, KOSEISHO, BUNDESGESUNDHEITSAMT (BGA), COMMITTEE FOR PROPRIETARY MEDICINAL PRODUCTS (CPMP), IND, IND EXEMPTION.

**Phase III Clinical Tests** The third in a series of human tests of new phamaceuticals, mandated by the U.S. Food and Drug Administration (FDA). The primary purpose of Phase III clinical tests is to verify proper dosage of a new pharmaceutical. See also PHASE I CLINICAL TESTING, PHASE II CLINICAL TESTS, FOOD AND DRUG ADMINISTRATION (FDA), KEFAUVER RULE, KOSEISHO, BUNDESGESUNDHEITSAMT (BGA), COMMITTEE FOR PROPRIETARY MEDICINAL PRODUCTS (CPMP).

**PHB** See POLYHYDROXYLBUTYLATE.

**Phenolic Hormones** A category of compounds found in the human body, that are synthesized (manufactured) by the body from certain phenolic dietary substances (phytochemicals) such as isoflavones. Research indicates that phenolic hormones act to prevent a number of cancers such as those of the prostate, breast, large bowel, etc. See also HORMONE, PHYTOCHEMICALS, ISOFLAVONES, CANCER, SELECTIVE ESTROGEN EFFECT.

**Phenomics** Utilized to refer to the relationship between genomics and phenotype/traits. See also FUNCTIONAL GENOMICS, PHENOTYPE, TRAIT, GENE FUNCTION ANALYSIS.

**Phenotype** The outward appearance (structure) or other visible characteristics of an organism (which of course, is determined by the DNA of its genotype). This also includes (and/or determines) how that organism's body responds to a given physical agent (a pharmaceutical, a toxin, sunlight, etc.). For example, genetically fair-skinned people tend to get sunburned easier/faster than other people do. See also GENOTYPE, DEOXYRIBO-NUCLEIC ACID (DNA), MORPHOLOGY, GENE, HAPLOTYPE, GENE EXPRESSION PROFILING.

**Phenylalanine (phe)** An essential amino acid. L-Phenylalanine is one of the raw materials used to manufacture NutraSweet® (NutraSweet Co.) synthetic sweetener. See also LEVOROTARY (L) ISOMER, ESSENTIAL AMINO ACIDS, STEREOISOMERS.

**Pheromones** From the Greek words *pherein*, to carry, and *hormon*, to excite, they are sex hormones emitted by insects and animals; they spread through the air by the wind and diffusion for the purposes of attracting the opposite sex. Some pheromones have been produced artificially and used in lure traps to attract and catch male insects so as to prevent their mating with females (i.e., a biological pesticide). Pheromone traps for Japanese beetles are commonplace in infested areas (e.g., when utilizing Integrated Pest Management). It is envisioned that commercial exploitation of this area of science will increase. See also HORMONE, INTEGRATED PEST MANAGEMENT (IPM).

**Philadelphia Chromosome** Refers to a particular human chromosome that is (visibly) distorted by the mutated gene that results in the disease known as chronic myelogenous leukemia (abbreviated CML, also known as chronic myeloid leukemia). That is because that gene codes for extensive production of the tyrosine kinase known as Bcr-Abl; an enzyme which causes neoplastic (aberrant) cell growth and cell division. As a result, people with CML disease tend to have 10–25 times more white blood cells than normal. The pharmaceutical known as Gleevec™ induces apoptosis — "programmed" (self-destruct) cell death — in the cells that have the Philadelphia chromosome; thus leading to cessation of CML. See also CHROMOSOMES, KARYOTYPE, KARYOTYPER, GENE, CODING SEQUENCE, MUTATION, CANCER, CELL, WHITE BLOOD CELLS, GLEEVEC™, APOPTOSIS.

**Phosphate Transporter Genes** Gene(s) within the genomes of at least some plants, which code for proteins that enable/increase the ability of those plants to extract and utilize phosphate (form of phosphorous) from the soil. Since all plants require phosphorous for proper growth and functioning, yet most plants are not inherently very adept at extracting and utilizing soil phosphate, adding (more) phosphate transporter genes to a given (crop) plant is likely to increase that plant's growth and yield (e.g., of seeds). See also GENE, GENETIC ENGINEERING.

**Phosphate-Group Energy** The decrease in free energy as one mole of a phosphorylated compound at 1.0 M concentration undergoes hydrolysis to equilibrium at pH 7.0 and 25°C (77°F). The energy that is available to do biochemical work. The energy arises from the breakage (cleavage) of a phosphate to phosphate bond. See also FREE ENERGY, HYDROLYSIS, FATS, MOLE, PHOSPHOLIPIDS.

**Phosphatidyl Choline** See LECITHIN.

**Phosphinothricin** Another name for the herbicide active ingredient glufosinate. See also GLUFOSINATE, PHOSPHINOTHRICIN ACETYLTRANS-FERASE (PAT), PAT GENE, BAR GENE.

**Phosphinothricin Acetyltransferase (PAT)** An enzyme which degrades (breaks down) phosphinothricin (also known as glufosinate), which is an active ingredient in some herbicides. PAT is naturally produced in some strains of soil bacteria (e.g., *Streptomyces viridochromogenes*). If a gene (called the "PAT gene") that codes for the production of phosphinothricin acetyltransferase is inserted via genetic engineering into a crop plant's genome, that would enable such plants to survive post-emergence applications of phosphinothricin-containing herbicides. See also ENZYME, PHOSPHINOTHRICIN, GLUFOSINATE, BACTERIA, GENE, PAT GENE, GENETIC ENGINEERING, GENOME, BAR GENE, MARKER (GENETIC MARKER).

**Phosphinotricine** See PHOSPHINOTHRICIN.

**Phosphodiesterases** A category of enzymes that inhibit apoptosis. Abbreviation for this term (category) is PDE. See also ENZYME, APOPTOSIS.

**Phospholipids** The principal class of lipids that are present in cell membranes; phospholipids are diglycerides (i.e., two fatty acids attached to a glycerol "molecular backbone") to which is also attached a phosphate group. The principal sites in plants of lipid and fatty acid biosynthesis (manufacturing) are the endoplasmic reticulum, chloroplasts, and the mitochondria. See also LIPIDS, PLASMA MEMBRANE, CELL, FATS, FATTY ACID, PHOSPHATE-GROUP ENERGY, ENDOPLASMIC RETICULUM (ER), CHLOROPLASTS, MITOCHONDRIA.

**Phosphorylation** The introduction of a phosphate group into a molecule. Formation of a phosphate derivative of a biomolecule, usually by enzymatic transfer of a phosphate group from ATP. See also ADENOSINE TRIPHOSPHATE (ATP).

**Δ Phosphorylation Potential** Abbreviated $\Delta G_p$, it is the actual free-energy change of ATP hydrolysis under a given set of conditions. See also PHOSPHORYLATION, FREE ENERGY, HYDROLYSIS, ADENOSINE TRIPHOSPHATE (ATP).

**Photon** A single unit of light energy. See also PHOTOSYNTHESIS, PHOTOSYNTHETIC PHOSPHORYLATION.

**Photoperiod** The optimum length or period of illumination required for the growth and maturation of a plant. The photoperiod is distinct from photosynthesis. See also PHYTOCHROME, CENTRAL DOGMA (NEW).

**Photophore** See BIOLUMINESCENCE.

**Photophosphorylation** See CYCLIC PHOTOPHOSPHORYLATION.

**Photorhabdus luminescens** A soil-dwelling bacterium that produces certain toxins (effective against a variety of insect pests), antibiotics, antifungal compounds, lipases, proteases, and bioluminescent (light-producing) compounds. *Photorhabdus luminescens* naturally colonizes the gut of the *Heterorhabditis* nematode which attacks certain insect pests (tobacco hornworm, mealworm, cockroaches, etc.). When that nematode enters those insects, the *Photorhabdus luminescens* is released inside the insect, which is subsequently killed via the toxins secreted by *P. luminescens*. *P. luminescens* synthesizes (manufactures) a protein that is high in content of the amino acids methionine and lysine; and that protein constitutes approximately 50% of the total protein content of *P. luminescens*. See also BACTERIA, ANTIBIOTIC, TOXIN, LIPASE, PROTEASE, BIOLUMINESCENCE, CORN, PROTEIN, METHIONINE, LYSINE.

**Photosynthesis** The synthesis (production) of bioorganic compounds (molecules) using light energy as the power source. The synthesis of carbohydrates (hexose) occurs via a complicated, multistep process involving reactions that occur both in the light (light reactions) and in the dark (dark reactions). In eucaryotic cells the photosynthetic machinery necessary to capture light energy and subsequently utilize it is contained in structures called chloroplasts, which contain the molecule that initially captures light energy, called chlorophyll. Chlorophyll appears green. Green plants synthesize carbohydrates from carbon dioxide and water, which are used as a hydrogen source. The synthesis reaction, which is light-driven, liberates oxygen in the process. Other organisms use this oxygen to sustain life. From initial carbohydrates, plants subsequently also synthesize (manufacture) other compounds (e.g., fatty acids).

Plants are not the only users of photosynthesis technology. Other organisms such as green sulfur bacteria and purple bacteria also carry out photosynthesis, but they use other compounds besides water as a hydrogen source. See also CARBOHYDRATES, CHLOROPLASTS, ORGANISM, EUCARYOTE, HEXOSE, CYCLIC PHOTOPHOSPHORYLATION, CAROTENOIDS, GOLDEN RICE, FATTY ACIDS, BACTERIA.

**Photosynthetic Phosphorylation** Also called photophosphorylation, it is the formation of ATP from the starting compounds ADP and inorganic phosphate (Pi). The formation is coupled to light-dependent electron flow in photosynthetic organisms. See also PHOTON, PHOTOSYNTHESIS, ADENOSINE TRIPHOSPHATE (ATP), ADENOSINE DIPHOSPHATE (ADP), CYCLIC PHOTOPHOSPHORYLATION.

**Phylogenetic Constraint** The limitations inherent in an organism as a result of what its ancestors were. For example, a horse will never fly and an ape will never speak, because the ancestors of neither possessed those capabilities. See also GENOTYPE, PHENOTYPE, GENOME, MORPHOLOGY.

P

**Physical Map (of genome)** A diagram showing the linear order of genes or genetic markers on the genome, with units indicating the actual distance between the genes or markers. See also GENETIC MAP, GENE, GENOME, POSITION EFFECT.

**Physiology** The branch of biology dealing with the study of the functioning of living things. The materials of physiology include all life: animals, plants, microorganisms, and viruses.

**Phytase** A digestive enzyme that is present in the digestive systems of many plant-eating animals to enable breakdown of phytate (also known as "phytic acid"). Phytase is sometimes present within the plant material consumed by animals. For example, phytase is naturally produced in the seed coat of wheat. See also ENZYME, DIGESTION (WITHIN ORGANISMS), PHYTATE, HIGH-PHYTASE CORN/SOYBEANS, LOW-PHYTATE CORN, LOW-PHYTATE SOYBEANS.

**Phytate** A chemical complex (large molecule) substance (inositol hexaphosphate) that is the dominant (i.e., 60–80%) chemical form of phosphorus present within cereal grains, oilseeds, and their byproducts. Monogastric animals (e.g., swine) cannot digest and utilize the phosphorous within phytate, because they lack the enzyme known as phytase in their digestive system so that phosphorus (phytate) is excreted into the environment. When phytase enzyme is present in the ration of a monogastric animal, at a high enough level, the monogastric animal is then able to digest the phytate (thereby "releasing" most of that phosphorus for absorption by the body of the animal). However, the (cleaved-off, "free") inositol that was "liberated" (from six phosphate atoms per molecule of phytate) can then quickly chelate ("combine" with) other minerals in the feed ration (iron, calcium, zinc, etc.). Thus, low-phytate crop varieties (i.e., containing inherently smaller amounts of inositol) are less likely to chelate important dietary minerals such as iron (which can exacerbate malnutrition in typically iron-poor diets such as in developing countries where adequate iron content/iron fortification of human diets is not common). In adult humans (e.g., those past childbearing age), the chelating ("combining"-with) property of the phytate-source inositol causes it to act as a beneficial antioxidant in the human body; which can help to protect against certain cancers (e.g., prostate cancer). See also PHYTASE, LOW-PHYTATE CORN, LOW-PHYTATE SOYBEANS, ENZYME, DIGESTION (WITHIN ORGANISMS), HIGH-PHYTASE CORN AND SOYBEANS, PROSTATE, CANCER, ANTIOXIDANTS, CHELATION.

**Phytic Acid** Also known as phytate or inositol hexaphosphate. See also PHYTATE.

**Phyto-manufacturing** Refers to the production of valuable substances (e.g., polyhydroxybutylate biodegradable plastic, industrial-process enzymes, etc.) in plants (e.g., genetically engineered plants). See also POLYHYDROXYLBUTYLATE (PHB), BIOPOLYMER, POLYHYDROXYALKANOIC ACID (PHA), EXTREMOZYMES, NUTRACEUTICALS.

**Phyto-sterols** See PHYTOSTEROLS.

**Phytoalexins** Term utilized to refer to chemical compounds (enzymes, etc.) that are produced by certain plants in response to the presence of infectious agents (e.g., fungus, bacteria) or their products. From the Greek words *phyton*, plant, and *alexein*, to defend; phytoalexins possess antimicrobial (i.e., fungus-killing, bacteria-killing) properties, so they can help plants to protect themselves against those microorganisms. See also PHYTOTOXIN, ISOFLAVONES, ALLELOPATHY, STRESS PROTEINS, PHARMACOENVIROGENETICS, ANTIBIOTIC, PHYTOCHEMICALS, ENZYME, FUNGUS, BACTERIA, ISOFLAVONES, PATHOGENIC, MICROBE, MICROBICIDE, SALICYLIC ACID (SA), PATHOGENESIS RELATED PROTEINS, SYSTEMIC ACQUIRED RESISTANCE (SAR).

**Phytochemicals** A term used to refer to certain biologically active chemical compounds that occur in fruits, vegetables, grains, herbs, flowers, bark, etc. Phytochemicals act to repel or control insects, prevent plant diseases, and control fungi and adjacent weeds. Phytochemicals also sometimes confer beneficial health effects to the animals (e.g., humans) that consume the plant (portions) containing those applicable phytochemicals. For example, vitamin C in citrus fruits, beta carotene in carrots and other orange vegetables, *d*-limonene in orange peels, tannins in green tea, capsaicin in chili peppers, n-3 (omega-3) fatty acids in soybean oil and fish oil, genistein, saponins, vitamin E, and phytosterols in soybeans, etc.

Beta carotene has been found to aid eyesight and may help prevent lung cancer. d-Limonene has been found to protect rats against breast cancer. Tannins appear to help prevent stomach cancer. Quercitin appears to help prevent prostate cancer. Capsaicin can reduce arthritis pain. N-3 (omega-3) fatty acids help to lower triglyceride levels in the blood. Genistein appears to block growth of breast cancer tumors, prostate cancer tumors, and to prevent the loss of bone density that leads to the disease osteoporosis. Tocotrienols act as antioxidants, and also inhibit synthesis of cholesterol (in humans). See also CANCER, DEXTROROTARY (D) ISOMER, FATTY ACID, LINOLENIC ACID, LINOLEIC ACID, GENISTEIN (Gen), BIOLOGICAL ACTIVITY, MOLECULAR PHARMING, FLAVONOIDS, RESVERATROL, NUTRACEUTICALS, CHOLESTEROL, N-3 FATTY ACIDS, PHYTOTOXIN, ALLELOPATHY, ANTIBIOTIC, PHYTOALEXINS, ANTIOXIDANTS, ABRIN, RICIN, *PFIESTERIA PISCICIDA*, PHYTOSTEROLS, LIGNANS, POLYPHENOLS, SAPONINS, FRUCTOSE OLIGOSACCHARIDES, LYCOPENE, LUTEIN, ANTHOCYANIN, SOYBEAN PLANT, SOYBEAN OIL, VITAMIN E, XANTHOPHYLLS, SITOSTEROLS, CAROTENOIDS, STEROLS, ALICIN, ELLAGIC ACID, PROANTHOCYANIDINS, CAFFEINE, QUERCITIN, ROSEMARINIC ACID, ZEAXANTHIN.

**Phytochrome** A protein plant pigment that serves to direct the course of plant growth and development and differentiation in a plant. The response is independent of photosynthesis, e.g., in the photoperiod (length of light period) response. See also PHOTOPERIOD, PROTEIN, PHOTOSYNTHESIS, PLANT HORMONE.

**Phytoene** See GOLDEN RICE, LYCOPENE, CAROTENOIDS.

**Phytoestrogens** Compounds possessing molecular structures somewhat similar to that of estrogen and that are naturally found in all plants on earth. As a result every vegetable, fruit, cereal and legume contains at least one type of "phytoestrogen." For example, flavones and flavonols are beneficial phytoestrogens (mostly red- and yellow-colored pigments) found in colored vegetables

and fruits (red grapes, yellow grapefruit, oranges, etc.). See also PHYTOCHEMICALS, FLAVONOIDS, FLAVONOLS, LIGNANS, SELECTIVE ESTROGEN EFFECT, ISOFLAVONES, ESTROGEN.

**Phytohormone** See PLANT HORMONE.

**Phytopharmaceuticals** See PHYTOCHEMICALS, NUTRACEUTICALS, PHYTO-MANUFACTURING.

***Phytophthora megasperma f. sp. glycinea*** A strain of *Phytophthora* fungus that can infect the soybean plant [*Glycine max* (L.) Merrill] under certain conditions, and thereby cause that soybean plant's stem and root to degrade (so-called "rot"). See also FUNGUS, PATHOGENIC, SOYBEAN PLANT, STRAIN, ISOFLAVONES.

***Phytophthora* Root Rot** A plant disease that is caused by a certain *phytophthora* fungus (*Phytophthora sojae*). Some soybean varieties are genetically resistant to as many as 21 races/strains of *phytophthora* fungi. See also FUNGUS, RPS1c GENE, RPS1k GENE, GENOTYPE, STRAIN, PATHOGENIC, SOYBEAN PLANT, RPS6 GENE, ISOFLAVONES.

***Phytophthora sojae*** See *PHYTOPHTHORA* ROOT ROT.

**Phytoplankton** Algae that float or are freely suspended in the water.

**Phytoremediation** Refers to the use of specific plants to remove contaminants or pollutants from either soils (e.g., polluted fields) or water resources (e.g., polluted lakes). For example, the Brazil water hyacinth (*Eichhornia crassipes*) naturally accumulates in its tissues toxic metals such as lead, arsenic, cadmium, mercury, nickel, copper, etc., and so has been utilized as a "biofilter" (e.g., in India). Insertion of the *Escherichia coliform* bacteria gene known as gsh 11 into the plant known as Indian mustard causes that plant to accumulate 40–90% higher amounts of cadmium (from cadmium-tainted soil) in its tissues than before; such genetically engineered plants could be utilized to extract cadmium from polluted sites. See also BIOREMEDIATION, BIORECOVERY, *ESCHERICHIA COLIFORM*, BACTERIA, GENE, GENETIC ENGINEERING.

**Phytosterols** A group of phytochemicals (i.e., solid alcohols consisting of ring-structured molecules) that are present in seeds produced by certain plants (e.g., the soybean plant *Glycine max* L.). Evidence shows that human consumption of certain phytosterols can help to prevent certain types of cancers, and can help lower total serum cholesterol and low-density lipoproteins (LDLP) levels; thereby reducing the risk of coronary heart disease (CHD). Evidence indicates that those phytosterols (e.g., campesterol, stigmasterol, beta-sitosterol) interfere with absorption of dietary cholesterol by the intestines, and decrease the body's recovery and reuse of cholesterol-containing bile salts, causing more cholesterol to be excreted from the body than previously. In 2000, the researcher Joseph Judd fed phytosterols extracted from soybeans (*Glycine max* L.) to human volunteers that were consuming a "low-fat" diet. Their total blood serum cholesterol and low-density lipoprotein (LDLP) levels decreased by more than 10% in a short time. See also PHYTOCHEMICALS, STEROLS, SITOSTANOL, SOYBEAN PLANT, LOW-DENSITY LIPOPROTEINS (LDLP), CHOLESTEROL, CAMPESTEROL, STIGMASTEROL, BETA-SITOSTEROL, SITOSTEROL, CORONARY HEART DISEASE (CHD).

**Phytotoxin** Any toxic compound produced by a plant. See also ALLELOPATHY, ANTIBIOTIC, PHYTOCHEMICALS, PHYTOALEXINS, TOXIN, ABRIN, RICIN, *PFIESTERIA PISCICIDA*, SOLANINE, GLUCOSAMINES, PSORALENE, GLUCOSINOLATES, GOSSYPOL, ALKALOIDS.

**Picogram (pg)** $10^{-12}$ gram or $3.527 \times 10^{-14}$ ounce (avoirdupoir). See also MICROGRAM.

*Picorna* A "family" of the smallest known viruses. The viruses of this family are a cause of the common cold and Hepatitis A in humans, one form of hoof and mouth disease in animals, and at least one disease in corn (maize). In 1994, Dr. Asim Dasgupta discovered a cellular molecule within ordinary baker's yeast that prevents *picorna* virus reproduction. This advance could lead to the creation of a treatment, in the future, to cure one or more of the above-mentioned diseases after infection has begun. See also VIRUS, CLADISTICS, CLADES.

**Pink Bollworm** See *PECTINOPHORA GOSSYPIELLA*.

**Pink Pigmented Facultative Methylotroph (PPFM)** A type of bacteria that is naturally present in virtually all plants. PPFM produces cytokinin, which aids the cell division (growth) process in plants. PPFM also produces a chemical substance similar to vitamin B-12. In 1996, Joe Polacco discovered that impregnation of aged seeds with PPFM improved the germination (sprouting) rate of those aged seeds. See also BACTERIA, MITOSIS, CELL DIFFERENTIATION, VITAMIN.

**Pituitary Gland** One of the endocrine glands, it lies beneath the hypothalamus (at the base of the brain). Along with the other endocrine glands, the pituitary helps control long-term bodily processes. This control is accomplished via interdependent secretion of hormones along with the other glands comprising the total endocrine system. For example, the pituitary helps control the body's growth from birth until the end of puberty by secreting growth hormone (GH). Secretion of GH by the pituitary is itself governed by the hormone known as growth hormone-releasing factor (GHRF), received by the pituitary gland from the hypothalamus.

The pituitary gland also helps control reproduction (development and growth of ovaries, timing of ovulation, maturation of oocytes, etc.) by secreting two gonadotropic (reproductive) hormones named luteinizing hormone (LH) and follicle-stimulating hormone (FSH). Secretion of LH and FSH by the pituitary is itself governed by the hormones gonadotropin-releasing hormone (GnRH, received by the pituitary from the hypothalamus) and estrogen/progesterone (received by the pituitary from the ovaries). See also ENDOCRINE GLANDS, ENDOCRINE HORMONES, HORMONE, ENDOCRINOLOGY, HYPOTHALAMUS, FOLLICLE-STIMULATING HORMONE (FSH), ESTROGEN, GROWTH HORMONE-RELEASING FACTOR (GRF or GHRF), GROWTH HORMONE (GH).

**Plant Breeder's Rights (PBR)** The intellectual property rights that are legally accorded to plant breeders by various laws, international treaties, etc. Similar to patent law for inventors. See also PLANT'S NOVEL TRAIT (PNT), PLANT VARIETY PROTECTION ACT (PVP), PLANT PROTECTION ACT, EUROPEAN PATENT CONVENTION, EUROPEAN PATENT OFFICE (EPO), U.S. PATENT AND TRADEMARK OFFICE (USPTO), UNION FOR PROTECTION OF NEW VARIETIES OF PLANTS (UPOV), COMMUNITY PLANT VARIETY OFFICE.

**Plant Hormone** An organic compound synthesized in minute quantities by certain plants. It influences and regulates plant physiological processes. Also called a phytochrome. The four general types of hormones that together influence cell division, enlargement, and differentiation are the auxins, gibberellins, kinins, and abscisic acid. See also HORMONE, GIBBERELLINS, PHYTOCHROME, GPA1, ETHYLENE, LYSOPHOSPHATIDYLETHANOLAMINE.

**Plant Protection Act** A law passed by the U.S. Congress in 1930 that enabled intellectual property protection via patents for new plants (developed by scientists) which are propagated asexually (e.g., via grafting). See also U.S. PATENT AND TRADEMARK OFFICE (USPTO), EUROPEAN PATENT CONVENTION, EUROPEAN PATENT OFFICE (EPO), PLANT'S NOVEL TRAIT (PNT), PLANT BREEDER'S RIGHTS (PBR), COMMUNITY PLANT VARIETY OFFICE, PLANT VARIETY PROTECTION ACT (PVP).

**Plant Sterols** See PHYTOSTEROLS.

**Plant Variety Protection Act (PVP)** A law passed by the U.S. Congress in 1970 that enables intellectual property protection (analogous to copyright protection) for new seed plants and seeds in America. See also U.S. PATENT AND TRADEMARK OFFICE (USPTO), EUROPEAN PATENT CONVENTION, EUROPEAN PATENT OFFICE (EPO), PLANT'S NOVEL TRAIT (PNT), PLANT BREEDER'S RIGHTS (PBR), PLANT PROTECTION ACT, UNION FOR PROTECTION OF NEW VARIETIES OF PLANTS (UPOV), COMMUNITY PLANT VARIETY OFFICE.

**Plant's Novel Trait (PNT)** The new (novel) trait added to a plant (e.g., crop plant such as cotton, corn/maize, soybean, etc.). Examples of novel traits are herbicide-tolerance (via inserted CP4 EPSPS gene, PAT gene, etc.), insect resistance (via inserted *B.t.* gene, *Photorhabdus luminescens* gene, etc.), and resistance to aluminum toxicity (via inserted CSb gene, etc.). See also TRAIT, CORN, SOYBEAN PLANT, CP4 EPSPS, GENE, PAT GENE, *B.t.*, *BACILLUS THURINGIENSIS* (*B.t.*), EVENT, CITRATE SYNTHASE (CSb) GENE, GENETIC ENGINEERING, *AGROBACTERIUM TUMEFACIENS*, *PHOTORHABDUS LUMINESCENS*.

**Plantibodies™** A trademark owned by EPIcyte Pharmaceutical, Inc. It refers to antibodies (akin to mammalian ones) produced in plants that are genetically engineered to produce those (specific) antibodies. That process (genetically engineering plants to cause them to produce plantibodies) was invented during the 1990s by Andrew Hiatt and Mich Hein. Although plants do not always glycosylate (i.e., attach oligosaccharide units to protein molecules such as these antibodies) in the same manner as animal cells, an antibody against HSV-2 pathogen expressed in genetically engineered soybean plants has proven comparable to that same antibody expressed in genetically engineered animal cells. See also ANTIBODY, GENETIC ENGINEERING, GLYCOSYLATION, OLIGOSACCHARIDES, EXPRESS, SOYBEAN PLANT, PATHOGEN, MOLECULAR PHARMING™.

**Plantigens** Antigens (e.g., of pathogenic bacteria) produced in plants which are genetically engineered to produce those (specific) antigens. That process (i.e., genetically engineering plants to cause them to produce specific antigens) can be utilized to produce edible vaccines for the pathogenic bacteria possessing those antigens. Then people could be "vaccinated" against disease merely by eating the genetically engineered plant (e.g., banana). See also ANTIGEN, PATHOGENIC, BACTERIA, VACCINE, GENETIC ENGINEERING, EDIBLE VACCINES.

**Plaque** Refers to deposits of (oxidized) cholesterol intermixed with smooth-muscle cells, lining the inside of certain blood vessels. These deposits can result in the disease atherosclerosis, and/or adversely increasing blood platelet aggregation (e.g., clotting). See also VITAMIN E, ATHEROSCLEROSIS, CHOLESTEROL, EPITHELIUM.

**Plasma** A pale, amber-colored fluid constituting the fluid portion of the blood in which are suspended the cellular elements. Plasma contains 8–9% solids. Of these, 85% are proteins consisting of three major groups, which are: fibrinogen, albumin, and globulin. The other components are the lipids, which include the neutral fats, fatty acids, lecithin, and cholesterol. Also present are sodium, chloride and bicarbonate, potassium, calcium, lycopene, and magnesium. A most essential function of plasma is the maintenance of blood pressure and the exchange (with tissues) of nutrients for waste. See also ABSORPTION, HOMEOSTASIS, LYCOPENE.

**Plasma Membrane** A thin structure that completely surrounds the cell as a "skin." It may be seen with the aid of an electron microscope. The entire membrane appears to be about 100 Angstroms (Å; 0.1 mm) thick and is composed of two dark lines, each about 30 Å thick which are, however, separated by a lighter area. This trilaminar "sandwich" structure is referred to as the unit membrane. The plasma membrane is composed of lipoidal (fat-like) material in which proteins and protein complexes and whole functional systems are embedded. In the plasma membrane are incorporated such energy-dependent transport systems as Na+ and K+ transporting ATPase and amino acid transport systems. Besides the cell, membranes surround such systems as the endoplasmic reticulum, vacuoles, lysosomes, Golgi bodies, mitochondria, chloroplasts, and the nucleus, to mention just a few. The plasma membrane and membranes in general function in part as a permeability barrier to the free movement of substances between the inside and exterior of the cell or organelles that they surround. See also CELL, PROTEIN, CECROPHINS (LYTIC PROTEINS), MAGAININS, MEMBRANES (OF A CELL), TRANSMEMBRANE PROTEINS, RECEPTORS, LIPIDS, MEMBRANE TRANSPORT.

**Plasmid** An independent, stable, self-replicating piece of DNA in bacterial cells that is not part of the normal cell genome and that never becomes integrated into the host chromosome. This is in contrast to a similar genetic element known as an episome plasmid that may exist independently of the chromosome or may become integrated into the host chromosome. Plasmids are known to confer resistance to antibiotics and may be transferred by cell-to-cell contact (by conjugation via the sex pilus) or by viral-mediated transduction. Plasmids are commonly used in recombinant DNA experiments as acceptors of foreign DNA. Known forms of plasmids include both linear and circular molecules. See also EPISOME, VECTOR, COPY NUMBER, MULTI-COPY PLASMIDS, DEOXYRIBONUCLEIC ACID (DNA), CELL, GENOME, CHROMOSOME, ANTIBIOTIC, Ti PLASMID.

**Plasmocyte** Another name for a blast cell. See also BLAST CELL.

**Plastid** An independent, stable, self-replicating piece of DNA inside a plant cell that is not part of the reproduction cell genome (i.e., in nucleus). Because there can exist up to 10,000 plastids in a given plant cell, the insertion of a gene (e.g., via genetic engineering) into plastids can result in a higher yield (of the specific protein coded for by that gene) than is achieved via insertion of the gene into the cell's nuclear DNA. See also DEOXYRIBONUCLEIC ACID (DNA), CELL, NUCLEAR DNA, COPY NUMBER, GENOME, PROMOTER, GENE, GENETIC ENGINEERING, FATS, CHLOROPLASTS.

**Platelet Activating Factor (PAF)** See CHOLINE.

**Platelet-Derived Growth Factor (PDGF)** An angiogenic growth factor produced by the blood's platelet cells which attracts the growth of capillaries into the vicinity of a fresh wound. This action releases still other growth factors, and starts the process of building a fibrin network to support the subsequent (blood) clot. PDGF is a competence factor (i.e., a growth factor that is required to make a cell able or competent to react to other growth factors). PDGF is normally contained within the platelet cells, so does not circulate in the blood in a form enabling it to be freely available to its "target cells." This "containment" of PDGF in platelets ensures site-specific delivery of the PDGF directly to a wound site so stimulus (i.e., of capillary growth) is localized to the actual wound site. After PDGF has caused the formation of the initial clot at a wound site, PDGF attracts connective tissue cells into the vicinity of the wound (to start the tissue-repair process). PDGF also acts as a mitogen (substance causing cell to divide and thus multiply) for connective tissue cells, granulocytes, and monocytes (each of which is involved in the wound's healing process). See also ANGIOGENIC GROWTH FACTORS, FIBRIN, FIBRONECTIN, PLATELETS, MITOGEN, GRANULOCYTES, MONOCYTES, CYCLOOXYGENASE.

**Platelet-Derived Wound Growth Factor (PDWGF)** See PLATELET-DERIVED GROWTH FACTOR (PDGF).

**Platelet-Derived Wound Healing Factor (PDWHF)** See PLATELET-DERIVED GROWTH FACTOR (PDGF).

P

**Platelets** Disk-shaped blood cells that stick to the (microscopically "jagged") edges of wounds. The aggregation of platelets at the wound site leads to blood clotting, forming a temporary wound covering. During this blood clotting process, the platelets release platelet-derived growth factor (PDGF) which attracts fibroblasts to the wound area (for subsequent healing process). See also FIBRIN, FIBRONECTIN, PLATELET-DERIVED GROWTH FACTOR (PDGF), FIBROBLASTS, CYCLOOXYGENASE, CHOLINE, OXIDATIVE STRESS.

**Pleiotropic** Adjective used to describe a gene that affects more than one trait (apparently unrelated) characteristic of the phenotype (appearance of an organism). For example, biologist David Ho in 1993 discovered a single gene in the barley (*Hordeum vulgare*) plant that controls the traits of the plant's height, drought resistance, strength, and time to maturity. See also GENE, GENETIC CODE, DEOXYRIBONUCLEIC ACID (DNA), INFORMATIONAL MOLECULES, PHENOTYPE.

**Pluripotent Stem Cells** Refers to those stem cells from which each of the human body's 210 different types of tissues could arise. See also STEM CELLS, STEM CELL GROWTH FACTOR (SCF), DIFFERENTIATION, HUMAN EMBRYONIC STEM CELLS.

**PNT** See PLANT'S NOVEL TRAIT (PNT).

**Point Mutation** A mutation consisting of a change of only one nucleotide in a DNA molecule. At "hot spots" (i.e., certain locations on the DNA within some organisms), numerous point mutations can occur. In the case of single-nucleotide polymorphisms (SNPs), the same point mutation occurs at the same location (on the DNA within some organisms) across a population of individuals of that organism. See also MUTATION, HEREDITY, MUTANT, MUTAGEN, DEOXYRIBONUCLEIC ACID (DNA), NUCLEOTIDE, HOT SPOTS, BASE EXCISION SEQUENCE SCANNING (BESS), ORGANISM, SITE-DIRECTED MUTAGENESIS (SDM), SINGLE-NUCLEOTIDE POLYMORPHISMS (SNPs), TRADITIONAL BREEDING METHODS.

**"Points to Consider" Document** See POINTS TO CONSIDER IN THE MANUFACTURE AND TESTING OF MONOCLONAL ANTIBODY PRODUCTS FOR HUMAN USE.

**Points to Consider in the Manufacture and Testing of Monoclonal Antibody Products for Human Use** The U.S. Food and Drug Administration's (FDA's) governing rules for IND (investigational new drug) submission for monoclonal antibody (MAb)-based pharmaceuticals. See also IND.

**Polar Group** A hydrophilic ("water loving") portion of a molecule; it may carry an electrical charge. A group that "likes" to be in the presence of water molecules or other polar compounds. See also NONPOLAR GROUP, POLARITY (CHEMICAL), POLAR MOLECULE (DIPOLE), AMPHIPATHIC MOLECULES, AMPHOTERIC COMPOUND, LIPID BILAYER.

**Polar Molecule (dipole)** A molecule in which the centers of positive and negative (electrical) charge do not coincide, so that one end of the molecule carries a positive (or partial positive) charge and the other end a negative (or partial negative) charge. See also POLARITY (CHEMICAL), POLAR GROUP, ION-EXCHANGE CHROMATOGRAPHY, NONPOLAR GROUP.

**Polar Mutation** A mutation in one gene which, because transcription occurs only in one direction, reduces the expression of subsequent genes in the same transcription unit further down the line. See also TRANSCRIPTION, TRANSLATION, EXPRESS, NUCLEIC ACIDS.

**Polarimeter** An instrument used for measuring the degree of rotation of plane-polarized light by an optically active compound/solution. See also STEREOISOMERS, OPTICAL ACTIVITY, LEVOROTARY (L) ISOMER, DEXTROROTARY (D) ISOMER.

**Polarity (chemical)** The degree to which an atom or molecule bears an electrical charge or a partial electrical charge. In general, the more polar (i.e., separation or partial separation of charge) a molecule is, the more hydrophilic ("water loving") it is. Polarity results from an uneven distribution of electrons between the atoms comprising a molecule. See also POLAR GROUP, HYDROPHILIC, POLAR MOLECULE (DIPOLE).

**Polarity (genetic)** Having to do with the one way or unidirectionality of gene transcription in an operon unit. That is, the region near the operator is always transcribed before the more distant regions. By analogy, transcription begins at the left end of an operon unit and proceeds (reads, transcribes) toward the right end of the operon unit. The distinction between the 5′ and the 3′ ends of nucleic acids. See also POLAR MUTATION, TRANSCRIPTION.

**Polyacrylamide Gel** A "sieving" gel, that is used in electrophoresis. See also POLYACRYLAMIDE GEL ELECTROPHORESIS (PAGE).

**Polyacrylamide Gel Electrophoreis (PAGE)** A form of chromatography in which molecules are separated on the basis of size and charge. The stationary phase (the polyacrylamide gel) is a polymerized version of acrylamide monomers. The gel looks and feels like Jello™. On a molecular basis it consists of an intertwined and cross-linked mesh of polyacrylamide strings. As can be imagined, there are tiny "holes" in the gel (as in a plastic mesh bag) and with enough cross-linking the size of the holes begins to approach the size of the molecules that are to be separated. Since some molecules will be larger and some smaller, some of them will be able to pass through the gel matrix more easily than others. This is part of the basis for separation. It should be noted at this point that if the gel is cross-linked enough, and because of this the holes in that gel are smaller than the molecules to be separated, then the molecules will not be able to penetrate into the gel and no separation can occur. The charge on the molecule also plays a role in the separation. Functionally, the gel serves to hold and separate the molecules. Although details are not presented here, after the gel has been prepared (poured and cross-linked), a small amount of the solution containing the molecules to be separated is placed into wells (grooves to hold the liquid) on the gel and the system is subjected to an electric current. Over the course of minutes to hours, molecules bearing different charge/mass separate. See also BIOLUMINESCENCE, CHROMATOGRAPHY, TWO-DIMENSIONAL (2D) GEL ELECTROPHORESIS, FIELD INVERSION GEL ELECTROPHORESIS (FIGE), ELECTROPHORESIS.

**Polyadenylation** The addition of a sequence of polyadenylic acid to the 3′ end of a eucaryotic mRNA after its transcription (post-transcriptional). See also MESSENGER RNA (mRNA), TRANSCRIPTION.

**Polycistronic** Coding regions representing more than one gene in mRNA (i.e., they code for two or more polypeptide chains). Many mRNA molecules in procaryotes are polycistronic. See also RIBOSOMES, PROCARYOTES.

**Polyclonal Antibodies** (used in humans) A mixture of antibody molecules (that are specific for a given antigen) that has been purified from an immunized (to that given antigen) animal's blood. Such antibodies are polyclonal in that they are the products of many different populations of antibody-producing cells (within the animal's body). Hence they differ somewhat in their precise specificity and affinity for the antigen.

Years ago, antibodies (then called antitoxin) that were purified from an immunized animal's blood (e.g., a horse) were injected into humans suffering from certain diseases (e.g., diphtheria). In these cases the pathogen. had caused disease by secreting large amounts of toxin into the victim's bloodstream. The antitoxin combined quantitatively (1:1, 2:1, 1:2, 1:3, 3:1, etc.) with, and neutralized, the toxin (for those few diseases for which it was applicable). Vaccines are now used instead, because of the adverse immune response caused by the horse's blood (antigens). See also ANTIBODY, PASSIVE IMMUNITY, MONOCLONAL ANTIBODIES (MAb), ANTIGEN, PATHOGEN, TOXIN.

**Polyclonal Response** (of immune system to a given pathogen) Because a given pathogen generally has several antigenic sites on its surface, the B lymphocytes (activated by helper T cells in response to a pathogen invading the body) synthesize several (subtly different) antibodies against that pathogen. And since the antibodies are made by different cells, the response is known as poly (many) clonal. See also PATHOGEN, ANTIGEN, ANTIBODY, HAPTEN, EPITOPE, HELPER T CELLS (T4 CELLS), LYMPHOCYTE, B LYMPHOCYTES, LYMPHOKINES.

**Polyethylene-Glycol Superoxide Dismutase (PEG-SOD)** See PEG-SOD (POLYETHYLENE GLYCOL SUPEROXIDE DISMUTASE), HUMAN SUPEROXIDE DISMUTASE (hSOD).

**Polygalacturonase (PG)** An enzyme (e.g., present in tomatoes) that starts the breakdown (softening) of the fruit tissue. Recent advances make it possible to significantly delay the softening (i.e., spoilage) process by reducing the production of polygalacturonase through genetic engineering of the plant. In 1986, William Hiatt of the American company Calgene discovered the gene for polygalacturonase. That led to the company commercializing a tomato variety that had been genetically engineered to reduce production of polygalacturonase in that variety's tomatoes (in 1994). See also EPSP SYNTHASE, GENETIC ENGINEERING, ANTISENSE (DNA SEQUENCE), ENZYME, GENE, ACC SYNTHASE.

**Polygenic** A trait or end product (e.g., in a grain-produced crop) that requires simultaneous expression of more than one gene. For example, the level of protein produced in soybeans is controlled by five genes. See also POLYHYDROXYLBUTYLATE (PHB), PROTEIN, SOYBEAN PLANT, GENE, TRAIT, SOYBEAN OIL, BCE4, ARABIDOPSIS THALIANA, PLASTID.

**Polyhydroxyalkanoates** See POLYHYDROXYALKANOIC ACID (PHA).

**Polyhydroxyalkanoic Acid (PHA)** A "family" of chemically related "energy storage" substances (i.e., polyesters) that is naturally produced by certain bacteria (90 strains known). When PHA is removed from the bacteria and purified, this substance has physical properties quite similar to thermoplastics like polystyrene. PHA can quickly be broken down by soil microorganisms, so PHA is a biodegradable plastic.

During the 1990s, Daniel Solaiman and coworkers at the U.S. Department of Agriculture developed some bacteria strains (e.g., *Bacillus thermoleovorans*) that can produce PHA utilizing vegetable oils (e.g., soybean oil) as a major part of their "diet" (energy source). The precise chemical composition (and physical characteristics) of the PHA thereby produced varies according to the particular vegetable oil that is used as the energy source for those bacteria. For example, PHA thus produced utilizing soybean oil is very amorphous (formable).

In 1994, researchers transferred genes for the production of one PHA into the weed plant *Arabidopsis thaliana* and the crop plant rapeseed (canola). In 1997, researchers transferred phaB and phaC genes into the crop plant cotton (*Gossypium hirsutum*),

which caused those transformed plants to express (produce) PHA inside the fibers (seed hair cells) in the amount of 0.34% of the fiber weight. That PHA (inside those cotton fibers) resulted in a fabric (i.e., cotton-PHA "blend") possessing better insulation properties than traditional cotton fabric. See also POLYHYDROXYLBUTYLATE (PHB), STARCH, BACTERIA, BIOPOLYMER, ARABIDOPSIS THALIANA, CANOLA, GENE, TRANSFORMATION, EXPRESS, BIODEGRADABLE, MICROORGANISM, SOYBEAN OIL.

**Polyhydroxylbutylate (PHB)** One of the PHAs, polyhydroxylbutylate is an "energy storage" substance that is naturally produced by certain bacteria, yeasts, and plants. When removed from the bacteria and purified, this substance has physical properties quite similar to thermoplastics like polystyrene. PHB can quickly be broken down by soil microorganisms, so PHB is a biodegradable plastic. Three separate enzymes are utilized by the organism in order to make the PHB molecule.

In 1994, researchers succeeded in transferring genes for PHB production into the weed plant *Arabidopsis thaliana* and the crop plant rapeseed (canola). Later (1997), researchers transferred phaB and phaC genes into the crop plant cotton (*Gossypium hirsutum*), which caused those transformed plants to express (produce) PHA inside the fibers (seed hair cells) in the amount of 0.34% of the fiber weight. That PHA (inside those cotton fibers) resulted in a fabric (i.e., cotton-PHA "blend") possessing better insulation properties than traditional cotton fabric. See also STARCH, BACTERIA, BIOPOLYMER, ENZYME, POLYGENIC, MICROORGANISM, POLYHYDROXYALKANOIC ACID (PHA), CANOLA, ARABIDOPSIS THALIANA, GENE, EXPRESS, BIODEGRADABLE.

**Polymer** A molecule possessing a regular, repeating, covalently bonded arrangement of smaller units called monomers. By analogy, a chain (polymer) that is composed of links (monomer) hooked together. See also OLIGOMER, PROTEIN, NUCLEIC ACIDS.

**Polymerase** An enzyme that catalyzes the assembly of nucleotides into RNA (RNA polymerase) and of deoxynucleotides into DNA (DNA polymerase). See also DNA POLYMERASE, RNA POLYMERASE, REVERSE TRANSCRIPTASES, DNA, RNA, TAQ.

**Polymerase Chain Reaction (PCR)** A reaction that uses the enzyme DNA polymerase to catalyze the formation of more DNA strands from an original one by the execution of repeated cycles of DNA synthesis. Functionally, this is accomplished by heating and melting double-stranded (hydrogen bonded) DNA into single-stranded (nonhydrogen bonded) DNA and producing an oligonucleotide primer complementary to each DNA strand. The primers bind to the DNA and mark it in such a way that the addition of DNA polymerase and deoxynucleoside triphosphates cause a new strand of DNA to form which is complementary to the target section of DNA. The process described previously is repeated (trait, product, etc.) again and again to produce millions of copies (amplicons) of the desired strand of DNA. PCR and its registered trademarks are the property of F. Hoffmann-La Roche & Co. AG, Basel, Switzerland. See also POLYMERASE CHAIN REACTION (PCR) TECHNIQUE, NESTED PCR, DEOXYRIBONUCLEIC ACID (DNA), DNA PROBE, PROBE, Q-BETA REPLICASE TECHNIQUE, COCLONING (OF MOLECULES), POSITIVE AND NEGATIVE SELECTION (PNS), AMPLICON, NESTED PCR, PRIMER (DNA).

**Polymerase Chain Reaction (PCR) Technique** Developed in 1984 and 1985 by Kary B. Mullis, Randall K. Saiki, Stephen J. Scharf, Fred A. Faloona, Glenn Horn, Henry A. Erlich, and Norman Arnheim, the PCR technique is an *in vitro* method that greatly amplifies (makes millions of copies of) DNA sequences that otherwise could not be detected or studied. It can be utilized to amplify a given DNA sequence that constitutes less than one part per million of initial sample (e.g., a 100-base-pair target DNA sequence within the genome of one of the higher organisms, which can contain up to 500 million base pairs). The procedure alleviates the necessity of *in vivo* replication of a target DNA sequence, or of replication of one-of-a-kind tiny DNA samples (e.g., from a crime scene). See also *IN VITRO*, *IN VIVO*, POLYMERASE CHAIN REACTION (PCR), AMPLICON, NESTED PCR, DEOXYRIBONUCLEIC ACID (DNA), BASE PAIR (bp), GENOME, SEQUENCE (OF A DNA MOLECULE), TAQ, DNA POLYMERASE, PRIMER (DNA).

**Polymorphism (chemical)** The property of a chemical substance crystallizing (or simply existing) in two or more forms having different structures. For example, diamond and graphite are two different structures (manifestations) of the element carbon. Deoxyribonucleic acid (DNA) is a polymorphic compound because the polymer can take on different forms. See also A-DNA, B-DNA, Z-DNA, DEOXYRIBONUCLEIC ACID (DNA), DNA PROFILING, POLYMORPHISM (GENETIC).

**Polymorphism (genetic)** A name applied to a condition in which a species of plant or animal is represented by several distinct, nonintegrating forms or types unrelated to age or sex. The differences are often in coloration, though any characteristic of the organism may be involved (e.g., nuclei shape for polymorphonuclear leukocytes). See also POLYMORPHONUCLEAR LEUKOCYTES (PMN), POLYMORPHONUCLEAR GRANULOCYTES, SINGLE-NUCLEOTIDE POLYMORPHISMS (SNPs), POLYMORPHISM (CHEMICAL).

**Polymorphonuclear Granulocytes** Neutrophils, eosinophils, and basophils are collectively known as polymorphonuclear granulocytes. This is due to the fact that collectively their nuclei are segmented into lobes and they have granule-like inclusions within their cytoplasm. See also GRANULOCYTES, BASOPHILS, EOSINOPHILS, NEUTROPHILS, CYTOPLASM.

**Polymorphonuclear Leukocytes (PMN)** Formerly named microphages, they are phagocytic (i.e., foreign particle-ingesting) white blood cells that have a lobed nucleus. For example, during an attack of the common cold (when virus first invades mucous membranes of the human nose), the body responds by making Interleukin-8 (IL-8); a glycoprotein that attracts large quantities of polymorphonuclear leukocytes to the mucous membranes of the nose (to try to combat the infection). Another example is when polymorphonuclear leukocytes (PMN) migrate into a female pig's uterus within 6 hours after semen is introduced via breeding. PMN remove excess sperm and bacteria, resulting in a "friendly" environment for embryos to develop in the uterus. See also CELLULAR IMMUNE RESPONSE, LEUKOCYTES,

P

POLYMORPHISM (GENETIC), VIRUS, BACTERIA, GLYCOPROTEIN, INTERLEUKIN-8 (IL-8), CELL, NUCLEUS, PLASMA MEMBRANE.

**Polypeptide (protein)** A molecular chain of amino acids linked by peptide bonds. Synonymous with protein. Via the synthesis (of this "chain") performed by ribosomes, each polypeptide (protein) in nature is the ultimate expression product of a gene. All of the amino acids commonly found in proteins have an asymmetric carbon atom, except the amino acid glycine. Thus, the polypeptide is potentially chiral in nature. See also PROTEIN, AMINO ACID, GENE, PEPTIDE, STEREOISOMERS, CHIRAL COMPOUND, EXPRESS, RIBOSOMES, POLYRIBOSOME (POLYSOME), MESSENGER RNA (mRNA).

**Polyphenols** Phytochemicals (e.g., naturally found in coffee, certain types of grapes, certain red wines, green tea, cocoa, etc.) that act as antioxidants when consumed by humans. For example, polyphenols are naturally produced within the beans of the cocoa (cacao) tree (*Theobroma cacao*), and thus are present in chocolate made from those beans. Polyphenols naturally produced in apples have been shown to inhibit certain bacteria in the human mouth from producing the particular glucans that lead to a buildup of plaque on teeth; prevention of such plaque build-up may help prevent cavities from forming in teeth. See also PHYTOCHEMICALS, FLAVONOIDS, ATHEROSCLEROSIS, ANTIOXIDANTS, OXIDATIVE STRESS, PHENOLIC HORMONES, NUTRACEUTICALS, BACTERIA, GLUCANS.

**Polyribosome (polysome)** A complex of a messenger RNA (mRNA) molecule on which ribosomes (ribosomal RNA; rRNA) are anchored. A number of ribosomes bound to only a single mRNA molecule. One mRNA molecule hence functions as a template for a number of polypeptide chains at one time. See also RIBOSOMES, rRNA (ribosomal RNA), MESSENGER RNA (mRNA).

**Polysaccharides** Linear and/or branched (structure) macromolecules (large molecules) composed of many monosaccharide units (monomers such as glucose) linked by glycosidic bonds. See also GLYCOSIDE, MONOSACCHARIDES, AMYLOSE, AMYLOPECTIN.

**Polysome** See POLYRIBOSOME.

**Polyunsaturated Fatty Acids (PUFA)** Unsaturated fatty acids, possessing more than one molecular double bond in their molecular "backbone" (i.e., they contain at least two less than the maximum possible number of hydrogen atoms). Enzymes (e.g., Δ 12 desaturase) present in some oilseed plants (soybean, canola, corn/maize, etc.) convert some monounsaturated fatty acids (e.g., oleic acid) to some polyunsaturated fatty acids (e.g., linoleic acid), within their developing seeds. For example, soybean oil contains (historical average) 60% polyunsaturated fatty acids. Extensive research shows that polyunsaturated fatty acids (PUFA) impart a variety of health benefits to humans that consume them. In general, those health benefits include anti-inflammatory, anti-hypertensive (i.e., prevention of high blood pressure), reduction in cancer risk, reduction in the blood cholesterol levels, reduction in the risk of coronary heart disease (CHD), plus aiding in the development of retina and brain tissues.

For example, the n-3 ("omega-3") PUFAs possess antithrombotic effects and also reduce blood concentrations of triglycerides. High dietary levels (in human diet) of the n-6 ("omega-6") PUFAs have been related to a decreased risk of coronary heart disease (CHD). Research indicates that some of the beneficial effects of PUFAs occur via PUFA interactions with several types of nuclear receptors (present in cells of some human tissues), which results in (PUFA) modulation of certain gene(s) expression in those cells. See also UNSATURATED FATTY ACID, ESSENTIAL FATTY ACIDS, THROMBOSIS, TRIGLYCERIDES, CORONARY HEART DISEASE (CHD), CANCER, N-3 FATTY ACIDS, SOYBEAN OIL, N-6 FATTY ACIDS, ENZYME, DOCOSAHEXANOIC ACID (DHA), HIGHLY UNSATURATED FATTY ACIDS (HUFA), EICOSAPENTANOIC ACID (EPA), CONJUGATED LINOLEIC ACID (CLA), CELL, GENE, RECEPTORS, NUCLEAR RECEPTORS, DEOXYRIBONUCLEIC ACID (DNA), EXPRESS, GENE EXPRESSION, TRANSCRIPTION, TRANSCRIPTION FACTORS, SOYBEAN PLANT, OLEIC ACID, LINOLEIC ACID, LINOLENIC ACID.

**Porcine Somatotropin (PST)** A hormone, produced in the pituitary gland of pigs, that increases a swine's muscle tissue production efficiency. Injecting this hormone causes a faster growing, leaner pig.

**Porphyrins** Complex nitrogenous compounds containing four substituted pyrroles covalently joined into a ring structure. When complexed with a central metal atom it is called a metalloporphyrin.

**Position Effect** A change in the expression of a gene brought about by its translocation to a new site in the genome. For example, a previously active gene may become inactive if placed on a new site in the genome. See also GENOME, TRANSLATION, GENETIC MAP, MAP DISTANCE, PROMOTER.

**Positional Cloning** A technique used by researchers to zero in on the gene (s) responsible for a given trait or disease. A genetic map of the organism's genome is used to make an educated guess as to the precise location of the gene of interest (e.g., near marker "x" or "y", etc.). Then those guessed genes are cloned, inserted into living organisms or cells, and tested to see if the guessed gene causes expression of the protein of interest (e.g., a protein that causes the disease that the researcher is attempting to cure). See also CLONE (A MOLECULE), GENE, GENE AMPLIFICATION, GENE DELIVERY, DNA PROBE, GENE MACHINE, GENETIC ENGINEERING, GENETIC MAP, GENETIC MARKER, GENOME, MAP DISTANCE, FUNCTIONAL GENOMICS, POSITION EFFECT, EXPRESS.

**Positive and Negative Selection (PNS)** A separation technique; a technique to speed up the task of selecting, from thousands of laboratory specimens, the few cells with precisely the desired genetic changes induced (via genetic engineering). The thousands of genetically altered cells are brought about (produced) by genetic engineering experiments. Many genetic alterations are accomplished by injecting or flooding (specimen) cells with fragments of new genetic material (genes). A few cells are produced that have precisely the desired genetic changes among a large number of cells that do not have the desired changes. Sort of like a "needle in a haystack." By analogy, the few cells possessing the desired trait represent the needles while the multitude of cells not possessing the trait represent the hay.

In order to isolate the few desired cells, the needles must be separated from the hay. PNS gets rid of the nondesired cells and leaves only the cells possessing the desired genetic change. This is accomplished in the following way. The pieces of newly injected genetic material are composed not only of the desired sequence of DNA, but also another piece of DNA (known as a marker) which renders only those cells possessing the desired (genetic) change resistant to certain antibiotic drugs (such as neomycin) and certain antiviral drugs (e.g., Ganciclovir™). When all of the engineered cells are exposed to the drug (which normally kills all of the cells) only those cells possessing the desired genetic change (and the concomitant piece of DNA providing drug resistance) survive and hence are "selected." The other cells not having the drug resistance are selected against, and die. See also GENETIC ENGINEERING, GENE, MARKER (GENETIC MARKER), Q-BETA REPLICASE TECHNIQUE, POLYMERASE CHAIN REACTION (PCR) TECHNIQUE.

**Positive Supercoiling** Occurs in double-stranded cyclic DNA molecules having no breaks at all in either strand. If the double helix (of DNA) is wound further in the same direction as the winding of the two strands of the double helix molecule, then the circular duplex itself takes on superhelical turns. By analogy, supercoiling or superhelicity may be described as follows. A piece of rope composed of two or three smaller strands of rope 197 Positive Supercoiling are wound around each other to yield the finished rope. This is equivalent to the normal double-stranded DNA. If the ends of the rope are then joined or tied together and the resultant circle of rope is again wound in the same direction as the winding that produced the rope in the first place, supercoils will be formed and the rope will become a much thicker (supercoiled), but shorter, piece of rope. See also DOUBLE HELIX.

**Post-Transcriptional Processing (Modification) of RNAs** The enzyme-catalyzed processing or structural modifications that RNAs such as mRNAs, rRNAs, and tRNAs must undergo before they are functionally finished products. For example, in eucaryotes a block of poly A containing at least 200 AMP residues is enzymatically attached to the 3′ end of mRNA in the nucleus of the cell. The mRNAs with the "tail" are then transferred to the cytoplasm and the tail enzymatically removed to form the functional mRNAs. It is believed that the poly A tail aids in the transfer of the complex and/or targets the complex to the cytoplasm. See also POST-TRANSLATIONAL MODIFICATION OF PROTEIN mRNA, rRNA, tRNA.

**Post-Translational Modification of Protein** Enzymatic processing of a polypeptide chain after its translation from its mRNA, i.e., addition of carbohydrate moieties to the protein or the removal of a portion of the polypeptide chain in order to produce a functional protein in the correct environment. See also POLYPEPTIDE (PROTEIN), MOIETY, MESSENGER RNA (mRNA), ENZYME, RIBOSOMES, CARBOHYDRATES, GLYCOPROTEIN.

**Potato Late Blight** A fungal disease of the potato plant (*Solanum tuberosum*) caused by the fungus *Phytophthora infestans*. During the 1840s, this plant disease struck the potato crops of Ireland and Europe, leading to the starvation of more than one million people (principally in Ireland, because that nation was very dependent on potatoes for food). See also FUNGUS.

**PPA** See PLANT PROTECTION ACT.

**PPFM** See PINK PIGMENTED FACULTATIVE METHYLOTROPH.

**PPO** Acronym for Protoporphyrinogen Oxidase. See also ACURON™ GENE.

**PR Proteins** See PATHOGENESIS RELATED PROTEINS.

**Prebiotics** Chemical compounds or microorganisms (e.g., yeasts) — administered alone or in combination (e.g., in the feed rations of animals) — that (generally) act to stimulate growth of beneficial types of bacteria within the digestive system of animals (e.g., livestock). Those compounds can include some organic acids (propionic acid, malic acid, etc.). For example, adding certain strains of yeast (culture) and malate (malic acid) to cattle feed rations has been shown to stimulate *Selenomonas ruminantium* bacteria (growth) in the rumen (i.e., the "first stomach" in cattle). *Selenomonas ruminantium* tend to constitute 22–51% of the total bacteria in a typical rumen, and are important for optimal digestion (e.g., of the grass eaten by that animal).

Inulin, and several fructose oligosaccharides, etc. act as prebiotics in the human digestive system (e.g., by stimulating growth of *Bifidus* species of bacteria in the digestive system). For animal feed rations, in addition to fructose oligosaccharides, transgalacto-oligosaccharides may be added, to also act as prebiotics. See also PROBIOTICS, YEAST, BACTERIA, *BIFIDUS*, INULIN, FRUCTOSE OLIGOSACCHARIDES, TRANSGALACTO-OLIGOSCCHARIDES, STRAIN.

**Pribnow Box** The consensus sequence T-A-T-A-A-T-G centered about 10 base pairs before the starting point of bacterial genes. It is a part of the promoter and is especially important in binding RNA polymerase. See also RNA POLYMERASE, TATA HOMOLOGY, HOMEOBOX, PROMOTER, BASE PAIR (bp).

**Primary Structure** Refers to the sequence of amino acids in a protein "molecular" chain, or to the linear sequence of nucleotides in a polynucleotide (RNA or DNA) molecular chain. See also POLYPEPTIDE (PROTEIN), AMINO ACID, PROTEIN, STRUCTURAL GENE, STRUCTURAL GENOMICS, NUCLEOTIDE, PROTEOMICS, DEOXYRIBONUCLEIC ACID (DNA), RIBONUCLEIC ACID (RNA).

**Primer (DNA)** A short sequence deoxyribonucleic acid (DNA) that is paired with one strand of the template DNA, in the Polymerase Chain Reaction (PCR) technique. In PCR testing (e.g., a paternity test), the primer is selected to be complementary to the analytically relevant sequence of DNA. It is the growing end of the DNA chain and it simply provides a free 3′-OH end at which the enzyme DNA polymerase adds on deoxyribonucleotide units (monomers). Which deoxyribonucleotide is added is dictated by base pairing to the template DNA chain. Without a DNA primer sequence a new DNA chain cannot form, since DNA polymerase is not able to initiate DNA chains. See also DEOXYRIBONUCLEIC ACID (DNA), SEQUENCE (OF A DNA MOLECULE), TEMPLATE, COMPLEMENTARY (MOLECULAR GENETICS), DOUBLE HELIX, POLYMERASE, POLYMERASE CHAIN REACTION (PCR),

POLYMERASE CHAIN REACTION (PCR) TECHNIQUE, NESTED PCR.

**Prion** Proteinaceous structures (molecules) found in the plasma membrane (surface) of cells, in the brains of all vertebrate animals. In 1982, Dr. Stanley Prusiner discovered that misshapen (mutated) versions can cause the neurodegenerative disease Bovine Spongiform Encephalopathy (BSE) in cattle, and the neurodegenerative diseases Creutzfeld-Jakob Disease (CJD), kuru, Gerstmann-Straussler-Scheinker Syndrome, and Fatal Familial Insomnia (FFI) in humans. Dr. Prusiner named these molecules prions for "proteinaceous infected particle," because, unlike infectious pathogenic bacteria or viruses, prions do not contain DNA. The dye named Congo Red, and IDX (a derivative of the chemotherapeutic doxorubicin) have shown some ability to slow prion-caused neurodegeneration. See also PROTEIN, CELL, PLASMA MEMBRANE, MUTANT, BACTERIA, DEOXYRIBONUCLEIC ACID (DNA), PROTEIN STRUCTURE, BSE, PROTO-ONCOGENES, STRESS PROTEINS, MONOCLONAL ANTIBODIES.

**Proanthocyanidins** The chemical components within North American cranberries (*Vaccinium macrocarpon*) and blueberries (genus *Vaccinium*) that impart health benefits to humans who consume those cranberries/blueberries. For example, when humans consume cranberries, these chemical compounds prevent *Escherichia coli* bacteria from adhering to the cells lining the human urinary tract (thereby helping to prevent some urinary tract infections). See also ANTHOCYANIDINS, PHYTOCHEMICALS, NUTRACEUTICALS, CELL, *ESCHERICHIA COLIFORM* (*E. COLI*).

**Probe** A relatively small molecule that can be used to sense the presence and condition of a specific protein, DNA fragment, RNA fragment, or nucleic acid by a unique interaction with that macromolecule. See also DNA PROBE, HYBRIDIZATION (MOLECULAR GENETICS), BACTERIAL ARTIFICIAL CHROMOSOMES (BAC), YEAST ARTIFICIAL CHROMOSOMES (YAC), HUMAN ARTIFICIAL CHROMOSOMES (HAC), MARKER ASSISTED SELECTION, SOUTHERN BLOT ANALYSIS, FLUORESCENCE *IN SITU* HYBRIDIZATION (FISH).

**Probiotics** Compounds that (generally) act to stimulate growth of beneficial types of bacteria within the digestive system of animals (e.g., livestock). For example, organic acids (propionic acid, acetic acid, lactic acid, citric acid, etc.) act to inhibit the growth/multiplication of pathogens (disease-causing microorganisms) in the digestive system of monogastric (i.e., single-stomach) animals such as poultry and swine. Those acids are able to pass through the outer cell membrane (plasma membrane) of pathogenic bacteria and fungi. Once inside those pathogens' cells, the acids dissociate, and acidify the cells' interior (which disrupts the cells' protein synthesis, growth, and replication of the pathogen). See also PREBIOTICS, *BIFIDUS*, CITRIC ACID, FRUCTOSE OLIGOSACCHARIDES, PATHOGEN, MICROORGANISM, BACTERIA, FUNGUS, CELL, ACID, PLASMA MEMBRANE.

**Procaryotes** Simple organisms that lack a distinct nuclear membrane and other organelles. Many structural systems are different between procaryotes and eucaryotes, including the DNA arrangement, composition of membranes, the respiratory chain, the photosynthetic apparatus, ribosome size, the presence or lack of cytoplasmic streaming, the cell wall, flagella, the mode of sexual reproduction, and the presence or lack of vacuoles. Some representative procaryotes are the bacteria and blue-green algae. See also EUCARYOTE.

**Process Validation** (for production of a pharmaceutical) Defined by the U.S. Food and Drug Administration (FDA) as "Establishing documented evidence which provides a high degree of assurance that a specific process will consistently produce a (pharmaceutical) product meeting pre-determined specifications and quality characteristics." See also FOOD AND DRUG ADMINISTRATION (FDA), GOOD MANUFACTURING PRACTICES (GMP), GOOD LABORATORY PRACTICES (GLP), cGMP.

**Progesterone** A female sex hormone, secreted by the ovaries, that supports pregnancy and lactation (milk production). See also HORMONE, PITUITARY GLAND, ESTROGEN.

**Programmed Cell Death** See p53 GENE, APOPTOSIS.

**Prokaryotes** See PROCARYOTES.

**Promoter** The region on DNA to which RNA polymerase binds and initiates transcription (of RNA). The promoter "promotes" the transcription (expression) of that gene, but the promoter's impact on the timing/degree of gene expression is itself regulated by the molecules that bind to the promoter. For example, the "binding" of RNA polymerase causes transcription of RNA to begin, and the "binding" to promoter of other STATs (i.e., signal transducers and activators of transcription) can regulate the degree to which a given gene is expressed. A promoter is a region of DNA (deoxyribonucleic acid) which lies "upstream" of the transcriptional initiation site of a gene. The promoter controls where (which portion of a plant, which organ within an animal, etc.) and when (which stage in the lifetime of an organism) the gene is expressed. For example, the promoter named "Bce4" is "seed-specific" [i.e., it only "promotes" the expression of a given gene's product (protein, fatty acid, amino acids, etc.) within a plant's seed]. See also POLYMERASE, GENE, EXPRESS, RNA POLYMERASE, CONTROL SEQUENCES, GENE EXPRESSION, BCE4, PLASTID, DEOXYRIBONUCLEIC ACID (DNA), POLYGENIC, TRANSCRIPTION, CAULIFLOWER MOSAIC VIRUS 35S PROMOTER, SIGNAL TRANSDUCERS AND ACTIVATORS OF TRANSCRIPTION (STATs).

**Proof-Reading** Any mechanism for correcting errors in nucleic acid synthesis that involves scrutiny of individual (chemical) units after they have been added to the (molecular) chain. This function is carried out by a 3' to 5' exonuclease, among others. Proof-reading dramatically increases the fidelity of the base pairing mechanism. See also SEQUENCING (OF DNA MOLECULES).

**Propionic Acid** See PROBIOTICS, *BIFIDUS*.

**Prostaglandin Endoperoxide Synthase** An enzyme that can exist in several different forms within the human body to catalyze the production of prostaglandins. See also ENZYME, CYCLOOXYGENASE, ARACHIDONIC ACID, ISOZYMES, PROSTAGLANDINS, HIGHLY UNSATURATED FATTY ACIDS (HUFA).

**Prostaglandins** A group of cyclic (i.e., circle-shaped) fatty acids that act as hormones in the body (promote inflammation during infections, help promote maintenance of the tissues of the stomach/kidney/intestines, etc.). Originally isolated from sheep and human prostates, prostaglandins are synthesized (manufactured) by the body via chemical reactions catalyzed by the enzymes cyclooxygenase/prostaglandin endoperoxide synthase; usually from arachidonic acid (also docosahexanoic acid). See also PROSTAGLANDIN ENDOPEROXIDE SYNTHASE, CYCLOOXYGENASE, ARACHIDONIC ACID, FATTY ACID, HORMONE, ENZYME, HIGHLY UNSAURATED FATTY ACIDS (HUFA), DOCOSAHEXANOIC ACID (DHA).

**Prostate** The gland in the body of males that produces the liquid which carries sperm into the females (during mating). In older human males, the prostate will often become enlarged (e.g., by "antagonism" when estrogen molecules circulating in the blood contact its surface). Via the selective estrogen effect, isoflavones (e.g., from soybeans) consumed by such males can displace and replace those estrogen molecules from the surface of the prostate (thereby preventing enlargement). See also ESTROGEN, ISOFLAVONES, SELECTIVE ESTROGEN EFFECT.

**Prostate-Specific Antigen (PSA)** An antigen whose concentration increases significantly 5 to 10 years prior to the (clinical) diagnosis of prostate cancer. This means that PSA level measurements can be utilized in diagnosis of prostate cancer before symptoms appear. However, a series of tests is required in order to accurately gauge the probability of cancer because PSA levels can also be elevated when a man develops a noncancerous enlarged prostate. See also ANTIGEN, TUMOR, TUMOR-ASSOCIATED ANTIGENS, PROSTATE.

**Prosthetic Group** A heat-stable metal ion or an organic group (other than an amino acid) that is covalently bonded to the apoenzyme protein. It is required for enzyme function. The term is now largely obsolete. See also ION, AMINO ACID, PROTEIN, ENZYME, APOENZYME, COENZYME.

**Protease** An enzyme that catalyzes the hydrolytic cleavage (breakdown) of proteins. By analogy, the enzyme breaks the link (peptide bond) holding a chain together. Proteases represent a whole class of protein-degrading enzymes. See also HYDROLYTIC CLEAVAGE, ENZYME, PEPTIDE BOND, TRYPSIN, CHYMOTRYPSIN, LACTOFERRIN.

**Protease Nexin I (PN-I)** A protein that acts as an inhibitor of protease. See also PROTEASE, PROTEIN, PROTEASE NEXIN II (PN-II).

**Protease Nexin II (PN-II)** A protein that is thought to regulate important activities in the body and brain by inhibiting specific enzymes and interacting with certain body cells. PN-II is formed from a precursor molecule known as beta-amyloid, via metabolic processing of the beta-amyloid. Recent research indicates that incorrect metabolic processing of beta-amyloid by the body results in amyloid plaques in the brain. The amyloid plaques are generally found in victims of Alzheimer's disease, and directly correlate (in number) with the degree of dementia. See also PROTEASE NEXIN I (PN-I), REGULATORY ENZYME, PROTEIN, ENZYME, INHIBITION, METABOLISM.

**Proteasomes** Refers to enzymatic/catalytic bodies present within all mammal cells that activate certain transcription factors, are involved in causing the cell to "present" antigens (i.e., from pathogens that invaded that cell) on the cell's surface, and perform various other cellular functions. For example, the 26S proteasome degrades (breaks down) all ubiquinated (ubiquitin-"tagged") proteins in that cell. See also ENZYME, PROTEIN, CELL, TRANSCRIPTION FACTORS, ANTIGEN, PATHOGEN, UBIQUITIN.

**Protein** Coined in 1838 by Jons Berzelius. From the Greek word *proteios*, meaning the first or the most important or of the first rank. Any of a class of high molecular weight polymer compounds composed of a variety of α-amino acids joined by peptide linkages. Via the synthesis (of this "chain") performed by ribosomes, each protein is the ultimate expression product of a gene. More than one protein can be expressed from a given gene (the particular protein expressed is determined by factors such as the cell's temperature or other environmental variable, presence of STATs — some of which themselves are proteins, presence of certain bacteria, etc.). During their synthesis (after emerging from the cell's ribosome), proteins

may also be phosphorylated (i.e., a "phosphate group" is added to the protein molecule), glycosylated (i.e., one or more oligosaccharides is added onto the protein molecule), acetylated (i.e., one or more "acetyl groups" is added to the protein molecule), farnesylated (i.e., a "farnesyl group" is added to the protein molecule), ubiquinated (i.e., a ubiquitin "tag" is added to the protein molecule), sulfated (i.e., a "sulfate group" is added to the protein molecule), or otherwise chemically modified. Proteins are the "workhorses" of living systems and include enzymes, antibodies, receptors, peptide hormones, etc. Proteins in living organisms respond to changing environmental and other conditions by changing their location within cells, by getting cut into (specific) pieces, by changing which (other) molecules they will bind (adhere) to, etc. All of the amino acids commonly found in (each and every one of the) proteins have an asymmetric carbon atom, except the amino acid glycine. Thus the protein is potentially chiral in nature. See also AMINO ACID, GENE, PEPTIDE, ABSOLUTE CONFIGURATION, STEREOISOMERS, CHIRAL COMPOUND, EXPRESS, OLIGOMER, PROTEIN FOLDING, MESSENGER RNA (mRNA), RIBOSOMES, POLYRIBOSOME (POLYSOME), ORGANISM, CELL, SIGNAL TRANSDUCERS AND ACTIVATORS OF TRANSCRIPTION (STATs), CENTRAL DOGMA (NEW), PHOSPHORYLATION, UBIQUITIN, GLYCOSYLATION (TO GLYCOSYLATE), FARNESYL TRANSFERASE.

**Protein Arrays** See PROTEIN MICROARRAYS.

**Protein Bioreceptors** See RECEPTORS.

**Protein C** An anticlotting (glyco) protein that prevents post-operative arterial clot formation when administered intravenously. May be synergistic (in its anticlotting effect) with tissue plasminogen activator (tPA). See also THROMBOMODULIN, TISSUE PLASMINOGEN ACTIVATOR (tPA), PROTEIN, GLYCOPROTEIN.

**Protein Chips** See PROTEIN MICROARRAYS.

**Protein Digestibility-Corrected Amino Acid Scoring (PDCAAS)** A method of expressing the quality of a given (food) protein source, in terms of its digestible protein (amino acid constituents') ability to support growth in young growing humans (i.e., if that protein supplies all needed essential

P

amino acids in their proportions required by humans, that protein scores 1.00). For example, the complete ('ideal') protein source soy protein (concentrate) has a PDCAAS of 0.99. PDCAAS has been recommended by the U.S. Food and Drug Administration (FDA), and the Food and Agricultural Organization of the United Nations/World Health Organization (FAO/WHO). See also PROTEIN, AMINO ACID, ESSENTIAL AMINO ACIDS, "IDEAL PROTEIN" CONCEPT, SOY PROTEIN, FOOD AND DRUG ADMINISTRATION (FDA), DIGESTION (WITHIN ORGANISMS), DEAMINATION.

**Protein Engineering** The selective, deliberate (re)designing and synthesis of proteins. This is done in order to cause the resultant proteins to carry out desired (new) functions. Protein engineering is accomplished by changing or interchanging individual amino acids in a normal protein. This may be done via chemical synthesis or recombinant DNA technology (genetic engineering). "Protein engineers" (actually genetic engineers) use recombinant DNA technology to alter a particular nucleoside or triplet (codon) in the DNA (genes) of a cell. In this way it is hoped that the resulting DNA codes for the different (new) amino acid in the desired location in the protein produced by that cell. See also PROTEIN, POLYPEPTIDE (PROTEIN), GENE, CODON, GENETIC ENGINEERING, AMINO ACID, ESSENTIAL AMINO ACIDS, SYNTHESIZING (OF PROTEINS).

**Protein Folding** The complex interactions of a polypeptide molecular chain with its environment and itself and other protein entities, which cause the polypeptide molecule to fold up into a highly organized, tightly packed, three-dimensional structure. Proven to occur spontaneously, by Christian B. Anfinsen during the 1960s, for protein molecules outside of living cells. This ability of polypeptide chains to fold into a great variety of topologies, combined with the large number of sequences (in the molecular chain) that can be derived from the 20 common amino acids in proteins, confers on protein molecules their great powers of recognition and selectivity.

How a protein folds up determines its chemical function. During the 1990s, it was discovered that inside living cells, "chaperone"

molecules are needed for proper protein folding to occur. These chaperones are protein molecules (e.g., certain heat-shock proteins) that form a loosely bound complex to suppress incorrect protein folding as the protein molecule is emerging from the cell's ribosome, so protein folding is both complete and correct as soon as the newly formed protein molecule is released from the cell's ribosome. See also AMINO ACID, PROTEIN, POLYPEPTIDE (PROTEIN), RIBOSOMES, CHAPERONES, PRION, ABSOLUTE CONFIGURATION, CONFORMATION, ENZYME, PROTEIN STRUCTURE.

**Protein Inclusion Bodies** See REFRACTILE BODIES (RB).

**Protein Interaction Analysis** Refers to a number of different analyses/technologies utilized to determine if a given (e.g., new) protein molecule interacts with a protein molecule whose function is already known (e.g., from previous research, or its use as a pharmaceutical). Through that analysis (e.g., inferring the "new" protein's function by its action vis à vis the "old"/known protein), useful information about the "new" protein can be gathered. See also PROTEOMICS, PROTEOME CHIP, BIOCHIPS, GENE EXPRESSION ANALYSIS, PROTEIN, GENOMICS, FUNCTIONAL GENOMICS, PROTEIN MICROARRAYS.

**Protein Kinases** Enzymes capable of phosphorylating (covalently bonding a phosphate group to) certain amino acid residues in specific proteins. Protein kinases play crucial roles in the regulation of signaling within and between cells. See also PHOSPHORYLATION, TYROSINE KINASE, ENZYME, AMINO ACID, PROTEIN, PROTEIN SIGNALING, CELL.

**Protein Microarrays** Refers to a piece of glass, plastic, or silicon onto which has been placed a number of proteins (or molecules of other chemical compounds that interact with proteins in a specific manner). These microarrays (sometimes called "biochips") can then be utilized to test (e.g., a single sample) for a wide variety of attributes or effects (on or by the protein molecules in the sample that is exposed to that microarray). See also PROTEIN, HIGH-THROUGHPUT SCREENING (HTS), TARGET-LIGAND INTERACTION SCREENING, RECEPTORS, PROTEIN INTERACTION ANALYSIS, PROTEIN STRUCTURE, PROTEOMICS,

PROTEOME CHIP, MICROARRAY (TESTING), BIO-CHIP, QUANTUM DOT.

**Protein Quality** See AMINO ACID PROFILE, PDCAAS.

**Protein Sequencer** See SEQUENCING (OF PROTEIN MOLECULES), GENE MACHINE, SEQUENCING (OF DNA MOLECULES).

**Protein Signaling** The "communication" by protein molecules (e.g., to cells) that governs their transport and localization (i.e., destination in the cell). Discovered and delineated by Guenter Blobel during the 1970s, protein signaling (e.g., via a short sequence of amino acids attached to end of newly synthesized protein molecules) results in proteins traveling to the appropriate cell compartments (e.g., organelles) and/or out of the cell (i.e,, secretion). See also PROTEIN, SIGNALING, SIGNALING MOLECULE, CELL, AMINO ACID, SIGNAL TRANSDUCTION, G-PROTEINS, RIBOSOMES, PROTEIN KINASES.

**Protein Structure** A polypeptide chain may take on a certain structure in and of itself because of the amino acid monomers it contains and their location within the chain. The chain may furthermore interact with other polypeptide chains to form larger proteins known as oligomeric proteins. In the following, the levels of protein structure normally encountered will be highlighted:

- Primary structure — refers to the backbone of the polypeptide chain and to the sequence of the amino acids of which it is comprised.
- Secondary structure — refers to the shape (recurring arrangement in space in one dimension) of the individual polypeptide chain. In some cases, because of its primary structure, the chain may take on an extended or longitudinally coiled conformation.
- Tertiary structure — refers to how the polypeptide chain (the primary structure) is bent and folded in three-dimensional space in order to form the normal tightly folded and compact structure.
- Quaternary structure — refers to how, in larger proteins made up of two or more individual polypeptide chains, the

individual polypeptide chains are arranged relative to each other. These large multipolypeptide proteins are called oligomeric proteins and the individual chains are called subunits. An example of such a protein is hemoglobin.

See also CONFORMATION, PROTEIN FOLDING, POLYPEPTIDE (PROTEIN), PROTEOMICS, CHAPERONES.

**Protein Tyrosine Kinase** See TYROSINE KINASE.

**Protein Tyrosine Kinase Inhibitor** Any compound (e.g., genistein, Gleevec™, etc.) that inhibits the action of the enzyme tyrosine kinase. See also ENZYME, INHIBITION, TYROSINE KINASE, GENISTEIN (Gen), GLEEVEC™.

**Protein-Protein Interactions** See PROTEIN, PROTEIN INTERACTION ANALYSIS, PROTEIN MICROARRAYS.

**Proteolytic Enzymes** Enzymes which catalyze the hydrolysis (breakdown) of proteins or peptides. Proteins (enzymes) that destroy the structure (by peptide bond cleavage) and hence the function of other proteins. These other proteins may or may not themselves be enzymes. See also PROTEASE, UBIQUITIN.

**Proteome Chip** A microarray ("biochip") developed by Michael Snyder et al., during 2001 which:

1. Has a large number of known sequence protein molecules (e.g., all proteins present in a given organism) attached to its surface at known locations (i.e., specific "addresses" on the microarray).

2. Utilizes specific bioactive agents such as certain lipids or biotinylated calmodulin (i.e., calmodulin molecules to which a molecule of biotin is "attached") in order to determine which of the protein molecules in #1 interacts with relevant bioactive agents. Because calmodulin is a well-known and very well-characterized calcium-binding protein (i.e., bioactive agent) involved in (known) cellular processes, the binding of calmodulin to specific protein molecules attached to the microarray/biochip provides critical information about the (cellular, protein-protein, etc.) functions and

P

interactions of those protein molecules in the organism.

3. Reveals a large amount of data concerning protein-protein interactions (e.g., via subsequent application to the microarray of dye-labeled streptavidin to identify the protein molecules via their addresses on the biochip) and protein-lipid interactions, all of which are needed, in order to determine the organism's proteome.

See also BIOCHIPS, PROTEIN MICROARRAY, PROTEIN INTERACTION ANALYSIS, TARGET-LIGAND INTERACTION SCREENING, MICROARRAY (TESTING), PROTEOME, BIOTIN, ORGANISM, AVIDIN.

**Proteomes** See PROTEOMICS.

**Proteomics** The scientific study of an organism's proteins and their role in an organism's structure, growth, health, disease (and/or the organism's resistance to disease, etc.). Those roles are predominantly due to each protein molecule's tertiary structure/conformation. Some methods utilized to determine which impact results from which protein, are:

• Chemical genetics, to compare two same-species organisms (one of which has protein, or a portion of protein, at least partially inactivated by a specific chemical).

• Gene expression analysis, to determine the protein(s) produced when a given gene is "switched on"; by measuring fluorescence of individual messenger RNA (mRNA) molecules (specific to which particular gene is "switched on" at the time), when that mRNA hybridizes (with DNA pieces corresponding to genes analyzed, that were attached to hybridization surface on the biochip).

• Gene expression analysis, to determine impact when a given gene is "knocked out"/"turned off."

• Protein interaction analysis, to determine if a newly discovered protein molecule interacts with a protein molecule whose function is already known (e.g., from previous research or use as a pharmaceutical). If the newly discovered

protein molecule interacts with one whose function is already known, it generally has the same or similar function (in living cells) as the previously known protein molecule. Thus, the function of a newly discovered human protein can sometimes be inferred from a protein molecule discovered earlier in a microorganism (e.g., via Expressed Sequence Tags).

• *In silico* biology (modeling), to compare computer-predicted events (e.g., the constituent peptides resulting from protein digestion) with actual or *in vitro* outcomes.

See also PROTEIN, PRIMARY STRUCTURE, CONFORMATION, NATIVE CONFORMATION, TERTIARY STRUCTURE, GENE, GENETIC MAP, GENOMICS, ELECTROPHORESIS, TWO-DIMENSIONAL (2D) GEL ELECTROPHORESIS, SEQUENCING (OF PROTEIN MOLECULES), GENETIC CODE, CELL, SEQUENCE (OF A PROTEIN MOLECULE), STRUCTURAL GENOMICS, FUNCTIONAL GENOMICS, COMBINATORIAL CHEMISTRY, BIOINFORMATICS, HIGH-THROUGHPUT SCREENING, BIOCHIPS, CHEMICAL GENETICS, GENE EXPRESSION ANALYSIS, FLUORESCENCE, MESSENGER RNA (mRNA), MICROORGANISM, HYBRIDIZATION (MOLECULAR BIOLOGY), HYBRIDIZATION SURFACES, EXPRESS, EXPRESSED SEQUENCE TAGS (EST), ORGANISM, POST-TRANSLATIONAL MODIFICATION OF PROTEIN, PROTEIN INTERACTION ANALYSIS, *IN SILICO* BIOLOGY, *IN VITRO*.

**Proto-Oncogenes** Cellular genes that can become cancer-producing. Proto-oncogenes are activated to oncogenes via different mechanisms, including point mutation, chromosome translocation, insertional mutation, and amplification. See also ONCOGENES, AMPLIFICATION, MUTATION.

**Protoplasm** Coined by J. E. Parkinje in 1840, it is a general term referring to the entire contents of a living cell; living substance. See also CELL.

**Protoplast** A structure consisting of the cell membrane and all of the intracellular components, but devoid of a cell wall. This (removal of cell's outer wall) can be done to plant cells via treatment with cell-wall-degrading enzymes or electroporation. Under specific conditions (e.g., electroporation), certain DNA sequences (genes) prepared by man can enter protoplasts. The cell then incorporates some or all of that DNA into its genetic complement (genome), and produces whatever product for which the newly introduced gene codes. In the case of plant protoplasts, whole plants can be regenerated from the (genetically engineered) protoplasts, resulting in plants that produce whatever product(s) for which the introduced gene(s) codes. See also CELL, ENZYME, ELECTROPORATION, GENE, GENETIC ENGINEERING, DEOXYRIBONUCLEIC ACID (DNA), CODING SEQUENCE, PROTEIN, SOYBEAN PLANT, CORN, CANOLA.

**Protoxin** A chemical compound that only becomes a toxin after it is altered in some way. For example, the *B.t.* protoxins (Cry9C, Cry1A (b), Cry1A (c), etc.) only become toxic after they are chemically altered by the alkaline environment inside the gut of certain insects. See also BACILLUS THURINGIENSIS (B.t.), B.t. KURSTAKI, CRY PROTEINS, CRY1A (b) PROTEIN, CRY1A (c) PROTEIN, CRY9C PROTEIN, B.t. ISRAELENSIS, B.t. TENEBRIONIS, TARGET (OF A HERBICIDE OR INSECTICIDE).

**Protozoa** A microscopic, single-celled animal form. A unicellular organism without a true cell wall, that obtains its food phagotropically. See also PHAGOCYTE.

**Provitamin A** See BETA CAROTENE, GOLDEN RICE.

**PRR** See PHYTOPHTHORA ROOT ROT.

**PSA** See PROSTATE-SPECIFIC ANTIGEN (PSA).

**Pseudogene** A segment of a DNA molecule that acts like a gene (i.e., it codes for a protein molecule product), but its protein product is generally not biologically active. See also DEOXYRIBONUCLEIC ACID (DNA), GENE, CODING SEQUENCE, PROTEIN, BIOLOGICAL ACTIVITY.

*Pseudomonas aeruginosa* See CITRATE SYNTHASE (CSb) GENE.

*Pseudomonas fluorescens* A normally harmless soil microorganism (bacteria) that colonizes the roots of certain plants. At least one company has incorporated the gene for a protein that is toxic to insects (taken from *Bacillus thuringiensis*) into a *Pseudomonas fluorescens*. This was done in order to confer insect resistance to the plants the roots of which the genetically engineered *Pseudomonas fluorescens* has colonized. See also BACILLUS THURINGIENSIS (B.t.), BACTERIA, WHEAT TAKE-ALL DISEASE, GENETIC ENGINEERING, ENDOPHYTE.

**Psoralen** See PSORALENE.

**Psoralene** A toxic chemical (e.g., to ward off insects) that is naturally produced by (wild type) plants related to the domesticated celery plant. See also TOXIN, PHYTOTOXIN, WILD TYPE, FOOD AND DRUG ADMINISTRATION (FDA), TRADITIONAL BREEDING METHODS.

**PST** See PORCINE SOMATOTROPIN.

**Psychrophile** An organism that requires 0°C (32°F) for growth. See also MESOPHILE, THERMOPHILE.

**PUFA** See POLYUNSATURATED FATTY ACIDS.

**Pure Culture** A culture containing only one species of microorganism. See also CULTURE, CULTURE MEDIUM.

**Purine** A basic nitrogenous heterocyclic compound found in nucleotides and nucleic acids; it contains fused pyrimidine and imidazole rings. Adenine and guanine are examples.

**PVP** See PLANT VARIETY PROTECTION ACT.

**PVPA** See PLANT VARIETY PROTECTION ACT.

**PVR** Plant Variety Rights. See also PLANT VARIETY PROTECTION ACT.

**PWGF** See PLATELET-DERIVED WOUND GROWTH FACTOR, GROWTH FACTOR.

**Pyralis** An insect that is also known as the European corn borer (*ostrinia nubialis*). See also EUROPEAN CORN BORER (ECB).

**Pyranose** The six-membered ring forms of sugars are called pyranoses. This is because they are derivatives of the heterocyclic compound pyran. See also SUGAR MOLECULES.

**Pyrexia** Fever; elevation of the body temperature above normal. See also PYROGEN.

**Pyrimidine** A heterocyclic organic compound containing nitrogen atoms at (ring) positions 1 and 3. Naturally occurring derivatives are components of nucleic acids and coenzymes, uracil, thymine, and cytosine.

**Pyrogen** A substance capable of producing pyrexia (i.e., fever). See also PYREXIA.

**Pyrophosphate Cleavage** The enzymatic removal of two phosphate groups (designated as PPi) from ATP in one piece leaving AMP as another product. This cleavage releases more energy, which can be used in certain reactions that require more of a "push" to get them going. See also ATP, ORTHOPHOSPHATE CLEAVAGE.

**Pyrrolizidine Alkaloids** A class of toxic chemical compounds which are produced naturally by certain plants, as a defense mechanism (against predators). One of the pyrrolizidine alkaloids, monocrotaline is consumed (preferentially) by the larvae (caterpillars) of the moth *Utetheisa ornatrix*. That moth subsequently utilizes the monocrotaline content of its body as a defense mechanism itself, against spiders that would otherwise eat that moth. See also ALKALOIDS, TOXIN.

P

# Q

**Q-beta Replicase** A viral RNA polymerase secreted by a bacteriophage that infects *Escherichia coli* bacteria. Q-beta replicase can copy a naturally occurring RNA (molecule) sequence (e.g., from bacteria, viruses, fungi, or tumor cells) at a geometric (i.e., very fast) rate. See also POLYMERASE, BACTERIOPHAGE, RIBONUCLEIC ACID (RNA), Q-BETA REPLICASE TECHNIQUE.

**Q-beta Replicase Technique** An RNA assay (test) that "amplifies RNA probes" that a researcher is seeking. For instance, by using the Q-beta replicase technique to assay for the presence of RNA specific to the AIDS virus, it is possible to detect an AIDS infection in a patient's blood sample long before that infection has progressed to the point where antibodies would appear in the blood. See also Q-BETA REPLICASE, RNA PROBES, RIBONUCLEIC ACID (RNA), POSITIVE AND NEGATIVE SELECTION (PNS), ASSAY, IMMUNOASSAY, ANTIBODY, POLYMERASE CHAIN REACTION (PCR) TECHNIQUE, COCLONING, WESTERN BLOT TEST.

**QCM** Acronym for Quartz Crystal Microbalances. See also QUARTZ CRYSTAL MICROBALANCES.

**QPCR** Acronym for Quantitative Polymerase Chain Reaction. Uses include gene expression analysis (i.e., quantitatively determine the amounts of each protein being expressed by a cell), genotyping, DNA quantification, etc. See also POLYMERASE CHAIN REACTION (PCR), CELL, GENE EXPRESSION PROFILING, PROTEIN, GENOTYPE, DEOXYRIBONUCLEIC ACID (DNA).

**QSAR** See QUANTITATIVE STRUCTURE-ACTIVITY RELATIONSHIP (QSAR).

**QSPR** See QUANTITATIVE STRUCTURE-PROPERTY RELATIONSHIP (QSPR).

**QTL** See QUANTITATIVE TRAIT LOCI (QTL).

**Quantitative Structure-Activity Relationship (QSAR)** A computer modeling technique that enables researchers (e.g., drug development chemists) to predict the likely activity (e.g., effect on tissue) of a new compound before that compound is actually created. QSAR is based on data from decades of research investigating the impact on "activity" of the chemical structures of thousands of thoroughly studied molecules. For example, the biological activity (i.e., bacteria-killing effectiveness) of most antibiotics correlates with their tendency to dimerize (i.e., link two molecules into a single molecular unit). See also BIOLOGICAL ACTIVITY, PHARMACOPHORE, ANTIBIOTIC, PHARMACOKINETICS, PHARMACOLOGY, ANALOGUE, RATIONAL DRUG DESIGN, *IN SILICO* SCREENING, POLYMER.

**Quantitative Structure-Property Relationship (QSPR)** A computer modeling technique that enables scientists to predict the likely properties of a new chemical compound before that chemical compound is actually created. See also QUANTITATIVE STRUCTURE-ACTIVITY RELATIONSHIP (QSAR), ANALOGUE, RATIONAL DRUG DESIGN.

**Quantitative Trait Loci (QTL)** Individual specific DNA sequences that are related to known traits (e.g., litter size in animals, egg production in birds, yield in crop plants.). See also MARKER (DNA SEQUENCE), TRAIT, LINKAGE, DEOXYRIBONUCLEIC ACID (DNA), LINKAGE GROUP, LINKAGE MAP, GENE, SEQUENCE (OF A DNA MOLECULE), MARKER ASSISTED SELECTION, CORN, HIGH-OIL CORN, RESTRICTION FRAGMENT LENGTH POLYMORPHISM (RFLP) TECHNIQUE, RANDOM AMPLIFIED POLYMORPHIC DNA (RAPD) TECHNIQUE, AFLP, SIMPLE SEQUENCE REPEAT (SSR), DNA MARKER TECHNIQUE.

Q

**Quantum Dot** A "molecular structure" that is between 1–100 nanometers in size, so it is midway between molecular and solid states. Quantum dots have been constructed of semiconductor materials, crystallites (grown via molecular beam epitaxy), etc. Quantum dots could conceivably be constructed to act as receptors (e.g., on "biochips") for specific ligands (e.g., a blood component that is only present in a diseased patient), in a way that would signal the presence of disease when a (blood) sample was passed over the quantum dot. That signal might be electronic, emission of specific-wavelength light, etc. See also NANOMETERS (nm), NANOTECHNOLOGY, RECEPTORS, MEMS (NANOTECHNOLOGY), BIOCHIP, BIOELECTRONICS, MICROARRAY (TESTING), LIGAND (IN BIOCHEMISTRY).

**Quantum Wire** A strip or "wire" of (electricity-) conducting material that is ten nanometers (nm) or less in its thickness or width. Indications from some research show that some forms of DNA molecules might be used as "quantum wires." See also NANOMETERS (NM), NANOTECHNOLOGY, DEOXYRIBONUCLEIC ACID (DNA), MEMS (NANOTECHNOLOGY), BIOELECTRONICS.

**Quartz Crystal Microbalances** Abbreviated QCM. Refers to biosensors consisting of small quartz crystals (to which are attached a source of appropriate electric current), with sensitive measurement devices utilized to detect when the "attachment" of specific molecules (e.g., viruses, DNA sequences, antigens) to the quartz (or to layers of certain materials previously deposited on the quartz surface) causes the specific oscillation frequency of that quartz crystal to change in a way that enables (electronic) identification of the specific molecule(s) that attached themselves to the QCM. See also BIOSENSORS (ELECTRONIC), VIRUS, SEQUENCE (OF A DNA MOLECULE), ANTIGEN.

**Quaternary Structure** The three-dimensional structure of an oligomeric protein; particularly the manner in which the subunit chains fit together. See also PROTEIN, OLIGOMER, CONFIGURATION, NATIVE CONFORMATION.

**Quencher Dye** See MOLECULAR BEACON.

**Quercetin** A phytochemical naturally produced in apples, onions, and some other plants. Research indicates that human consumption of quercitin helps prevent prostate and some other cancers. See also PHYTOCHEMICALS, NUTRACEUTICALS, CANCER, BIOLOGICAL ACTIVITY.

**Quick-Stop** The term used to describe how DNA mutants of *Escherichia coli* cease replication immediately when the temperature is increased to 42°C (108°F). See also *ESCHERICHIA COLIFORM (E. COLI)*.

Q

# R

**R Genes** Refers to genes within some plants that confer resistance (to certain plant diseases) through common signaling pathways involved in ("surveillance" and activation of) natural plant defense responses (e.g., SAR). For example, the gene that codes for (causes the "manufacture" of) harpin protein is only present in a few bacteria (e.g., *Erwinia amylovora*), but R genes (i.e., those responsible for "surveillance" and activation of plant defense responses) which respond to the presence of harpin are present within the genomes of numerous species of plants. Thus, the spraying of man-made harpin protein onto any of those numerous species of (crop) plants causes those particular plants to initiate a protective/defensive response (cascade) against pathogenic bacteria, viruses, fungi, and even some insects. See also GENE, SIGNALING, PATHWAY, PROTEIN, HARPIN, SPECIES, SYSTEMIC ACQUIRED RESISTANCE (SAR), PATHOGENIC, PATHOGENESIS RELATED PROTEINS, STRESS PROTEINS, CASCADE, BACTERIA, VIRUS, FUNGUS.

**RAC** See RECOMBINANT DNA ADVISORY COMMITTEE (RAC).

**Racemate** An equimolar (i.e., equal number of molecules) mixture of the D and L stereoisomers of an optically active compound. A solution of dextrorotary (D) isomer (enantiomer) will rotate the plane in which the light was polarized a specific number of degrees to the right (dextro) while a solution containing the same number of levorotary (L) isomer molecules will rotate the plane in which the light was polarized the same number of degrees to the left (levo). The difference between D and L enantiomers is that the rotations of the plane of plane-polarized light are equal in magnitude, but opposite in sign. Hence, a 50:50 mixture of both enantiomers (known as a racemic mixture) shows no optical activity. That is, a solution containing a 50:50 mixture of enantiomers will not rotate the plane of plane polarized light when it is passed through the solution. See also ENANTIOMERS, STEREOISOMERS, LEVOROTARY (L) ISOMER, DEXTROROTARY (D) ISOMER.

**Racemic (mixture)** See RACEMATE.

**Radioactive Isotope** An isotope with an unstable (atomic) nucleus that spontaneously emits radiation. The radiation emitted includes alpha particles, nucleons, electrons, and gamma rays. See also ISOTOPE.

**Radioimmunoassay** A very sensitive method of quantitating a specific antigen using a specific radiolabeled antibody. Functionally, the antibody is made radioactive by the covalent incorporation of radioactive iodine. The radioimmuno probe thus prepared is exposed to its antigen (which may be a protein, or a receptor, etc.) in excess (the exact amount will have to be determined). The radiolabeled probe then binds to the antigen and the unbound, free probe is washed away. The radioactivity is then determined (counted) and by comparison to a standard plot which has been constructed previously, the amount of antigen (binding) is determined. See also ANTIBODY, ASSAY, HORMONE, RADIOIMMUNOTECHNIQUE.

**Radioimmunotechnique** A method of using a radiolabeled antibody to quantitate a known antigen. See also RADIOIMMUNOASSAY, ANTIGEN, ANTIBODY.

**Radiolabeled** From the Latin *radiare*, to emit beams. See also LABEL (RADIOACTIVE).

**Random Amplified Polymorphic DNA (RAPD) Technique** A genetic mapping methodology that utilizes as its basis the fact that specific DNA sequences (polymorphic DNA) are "repeated" (i.e., appear in sequence) with the gene of interest. Thus, the polymorphic DNA sequences are linked to that specific gene. Their linked presence serves to facilitate genetic mapping (i.e., "location" of specific gene(s) on an organism's genome). See also GENETIC MAP, SEQUENCE (OF A DNA MOLECULE), RESTRICTION FRAGMENT LENGTH POLYMORPHISM (RFLP) TECHNIQUE, LINKAGE, DEOXYRIBONUCLEIC ACID (DNA), PHYSICAL MAP (OF GENOME), LINKAGE GROUP, MARKER (GENETIC MARKER), LINKAGE MAP, TRAIT, GENOME, GENE, QUANTITATIVE TRAIT LOCI (QTL).

**RAPD** See RANDOM AMPLIFIED POLYMORPHIC DNA TECHNIQUE.

**Rapid Microbial Detection (RMD)** A broad term used to describe the various testing products and technologies that can be utilized to quickly detect the presence of microorganisms (e.g., pathogenic bacteria in a food processing plant). These testing products are based on immunoassay, DNA probe, electrical conductance and/or impedance, bioluminescence, and enzyme-induced reactions (e.g., which produce fluorescence or a color change to indicate the presence of specific microorganism). See also BIOLUMINESCENCE, MICROBE, BACTERIA, PATHOGEN, IMMUNOASSAY, ENZYME, PROBE, DNA PROBE, ELECTROPHORESIS, HAZARD ANALYSIS AND CRITICAL POINTS (HACCP).

**ras Gene** Discovered in 1978 by Edward Scolnick, who named it ras for "rat sarcoma" (the particular diseased tissue in which he found it). The ras gene is also present in the human genome, and it is an oncogene that is believed to be responsible for up to 90% of all human pancreatic cancer, 50% of human colon cancers, 40% of lung cancers, and 30% of leukemias. The ras gene codes for the production (manufacture) of ras proteins, which help to signal each cell to divide and grow at appropriate time(s); e.g., when free EGF "attaches" to relevant cell receptor on the plasma membrane. When the ras gene has been damaged or mutated (e.g., via exposure to cigarette smoke or ultraviolet light, etc.), it codes for (causes to be manufactured in the cell's ribosome) a mutated version of the ras protein that can cause the cell to become cancerous (i.e., divide and

grow uncontrollably). See also GENE, ONCOGENES, p53 GENE, GENETIC CODE, MEIOSIS, DEOXYRIBONUCLEIC ACID (DNA), CARCINOGEN, RIBOSOMES, CANCER, TUMOR, ras PROTEIN, FARNESYL TRANSFERASE, PROTO-ONCOGENES, PROTEIN, EPIDERMAL GROWTH FACTOR (EGF), EGF RECEPTOR.

**ras Protein** A transmembrane (i.e., through the cell membrane) protein for which the ras gene codes. The ras protein end outside the cell membrane acts as a receptor for applicable growth factors (e.g., fibroblast growth factor), and conveys that signal (to divide/grow) into the cell when that chemical signal (i.e., the growth factor) touches the "receptor end" of the ras protein. When the ras gene has been damaged or mutated (e.g., via exposure to cigarette smoke or ultraviolet light), that gene causes excess ras proteins to be manufactured, which causes oversignaling of the cell to divide and grow (i.e., cell becomes cancerous). See also GENE, TRANSMEMBRANE PROTEINS, ras GENE, FIBROBLAST GROWTH FACTOR (FGF), ONCOGENES, GENETIC CODE, PROTEIN, p53 PROTEIN, MEIOSIS, CARCINOGEN, RIBOSOMES, DEOXYRIBONUCLEIC ACID (DNA), CANCER, TUMOR, PROTO-ONCOGENES, RECEPTORS, EGF RECEPTOR, CD4 PROTEIN, SIGNALING, SIGNAL TRANSDUCTION.

**Rational Drug Design** The 'engineering' (building) of chemically synthesized drugs based on knowledge of receptor modeling and drug/target interaction(s) with the aid of supercomputers/interactive graphics/etc.; the educated, creative design of the three-dimensional structure of a drug atom by atom, i.e., "from the ground up." This approach represents a major advance over the prior practice of first synthesizing large numbers of compounds (or finding them in nature), followed by thousands of tedious screenings to test for efficacy against a given disease (target). The approach of rational drug design has, however, not yet been perfected and optimized due, in part, to gaps in our knowledge of drug/receptor interaction and to gaps in our knowledge in general. See also RECEPTORS, RECEPTOR MAPPING (RM), ANALOGUE, MOLECULAR DIVERSITY, TARGET (OF A THERAPEUTIC AGENT), *IN SILICO* BIOLOGY, FREE ENERGY, *IN SILICO* SCREENING.

**RB** See REFRACTILE BODIES.

**RBS1 Gene** A gene that confers to any soybean plant (possessing that gene in its DNA) resistance to the adverse effects of the soil-borne fungus *Phialophora gregata*, which can cause the plant disease brown stem rot (BSR) in soybean plants. See also GENE, DEOXYRIBONUCLEIC ACID (DNA), BROWN STEM ROT (BSR), FUNGUS, PATHOGENIC, SOYBEAN PLANT.

**RBS3 Gene** A gene that confers to any soybean plant (possessing that gene in its DNA) resistance to the adverse effects of the soil-borne fungus *Phialophora gregata*, which can cause the plant disease known as brown stem rot (BSR) in soybean plants. See also GENE, DEOXYRIBONUCLEIC ACID (DNA), BROWN STEM ROT (BSR), FUNGUS, PATHOGENIC, SOYBEAN PLANT.

**rDNA** See RECOMBINANT DNA.

**Reactive Oxygen Species** See FREE RADICAL, OXIDATION, OXIDATIVE STRESS.

**Reading Frame** The particular nucleotide sequence that starts at a specific point and is then partitioned into codons. The reading frame may be shifted by removing or adding a nucleotide(s). This would cause a new sequence of codons to be read. For example, the sequence CATGGT is normally read as the two codons: CAT and GGT. If another adenosine nucleotide (A) were inserted between the initial C and A, producing the sequence CAATGGT, then the reading frame would have been shifted in such a way that the two new (different) codons would be CAA and TGG, which would code for something completely different. See also CODON, GENETIC CODE, FRAMESHIFT, DEOXYRIBONUCLEIC ACID (DNA), MUTATION.

**Reassociation (of DNA)** The pairing of complementary single strands (of the molecule) to form a double helix (structure). See also DOUBLE HELIX.

**RecA** The product of the RecA locus (in a gene of) *Escherichia coli*. It is a protein with dual activities, acting as a protease and also able to exchange single strands of DNA (deoxyribonucleic acid) molecules. The protease activity controls the SOS response. The nucleic acid handling facility (i.e., ability to exchange single strands of DNA) is involved in recombination/repair pathways. See also SOS RESPONSE, LOCUS, PROTEIN, RIBOSOMES, *ESCHERICHIA COLIFORM* (*E. COLI*).

**Receptor Fitting (RF)** A research method used to determine the macromolecular structure that a chemical compound (e.g., an inhibitor) must have in order to fit (in a lock-and-key fashion) into a receptor. For example, a pain inhibitor compound blocking a pain receptor on the surface of a cell. See also CD4 PROTEIN, T CELL RECEPTORS, RECEPTORS, RECEPTOR MAPPING (RM), INTERLEUKIN-1, RECEPTOR ANTAGONIST (IL-1ra), RATIONAL DRUG DESIGN.

**Receptor Mapping (RM)** A method used to guess at (determine) the three-dimensional structure of a receptor binding site extrapolating from the known structure of the molecule binding to it. This approach can be carried out because of the complementary shape of the receptor and the binding molecule. Functionally, the researcher projects the (guessed-at) properties of the receptor ligands into a mathematical model in which the profile of the receptor is predicted by complementarity (to known chemical molecular structures). The receptor mapping process requires repetitive refinement of the mathematical model to fit properties continually being discovered via the use/interaction of chemical reagents bearing the known molecular structures. See also CD4 PROTEIN, T CELL RECEPTORS, RECEPTORS, RECEPTOR FITTING (RF).

**Receptor-Mediated Endocytosis** See ENDOCYTOSIS.

**Receptors** Functional proteinaceous structures typically found in the plasma membrane (surface) of cells that tightly bind specific molecules (organic, proteins or viruses). Some (relatively rare) receptors are located inside the cell's membrane (e.g., free-floating receptor for Retin-A). Both (membrane, internal) types of receptors are a functional part of information transmission to the cell.

A general overview is that once bound, both the receptor and its "bound entity" as a complex are internalized by the cell via a process called endocytosis, in which the cell membrane in the vicinity of the bound complex invaginates. This process forms a membrane "bubble" on the inside of the cell, which then pinches off to form an endocytic

**R**

vesicle. The receptor then is released from its bound entity by cleavage in the cell's lysosomes. It is recycled (returned) to the surface of the cell (e.g., low-density lipoprotein receptors). In some cases the receptor, along with its bound molecule, may be degraded by the powerful hydrolytic enzymes found in the cell's lysosomes (e.g., insulin receptors, epidermal growth factor receptors, and nerve growth factor receptors).

Endocytosis (internalization of receptors and bound ligand such as a hormone) removes hormones from the circulation and makes the cell temporarily less responsive to them because of the decrease in the number of receptors on the surface of the cell. Hence the cell is able to respond (to a new signal). A receptor may be thought of as a butler who allows guests (in this case molecules that bind specifically to the receptor) to enter the house (cell) and who accompanies them as they enter.

Another mode of "reception" occurs when, following binding, a transmembrane protein (e.g., one of the G proteins) activates the portion of the transmembrane (i.e., through the cell membrane) protein lying inside the cell. That "activation" causes an effector inside the cell to produce a "signal" chemical inside the cell which causes the cell to react to the original external chemical signal (that bound itself to the receptor portion of the transmembrane protein). See also CD4 PROTEIN, T CELL RECEPTORS, RECEPTOR FITTING (RF), RECEPTOR MAPPING (RM), LYSOSOMES, INTERLEUKIN-1 RECEPTOR ANTAGONIST (IL-1ra), CD95 PROTEIN, TRANSFERRIN, VAGINOSIS, SIGNAL TRANSDUCTION, ENDOCYTOSIS, G PROTEINS, CELL, SIGNALING, PROTEIN, NUCLEAR RECEPTORS, HUMAN IMMUNODEFICIENCY VIRUS TYPE 1 (HIV-1), HUMAN IMMUNODEFICIENCY VIRUS TYPE 2 (HIV-2).

**Recessive (gene)** See RECESSIVE ALLELE.

**Recessive Allele** Discovered by Gregor Mendel in the 1860s, this refers to an allelic gene whose existence is obscured in the phenotype of a heterozygote by the dominant allele. In a heterozygote, the recessive allele does not produce a polypeptide; it is "switched off." In this case, the dominant allele is the one producing the polypeptide chain (via cell's ribosome). See also GENETICS, ALLELE, DOMINANT ALLELE, HOMOZYGOUS, HETEROZYGOTE, POLYPEPTIDE (protein), CELL, RIBOSOMES.

**Recombinant DNA (rDNA)** DNA formed by the joining of genes (genetic material) into a new combination. See also RECOMBINATION, GENETIC ENGINEERING.

**Recombinant DNA Advisory Committee (RAC)** The former standing U.S. national committee set up in 1974 by the U.S. National Institutes of Health (NIH) to advise the NIH director on matters regarding policy and safety issues of recombinant DNA research and development. Over time, it had evolved to become part of the American government's regulatory process for recombinant DNA research and product approval. The RAC was terminated by the director of the NIH in 1996 because the "human health and environmental safety concerns expressed at the inception (of genetic engineering/biotechnology) had not materialized." See also INTERIM OFFICE OF THE GENE TECHNOLOGY REGULATOR (IOGTR), GENE TECHNOLOGY OFFICE, GENETIC ENGINEERING, ZKBS (CENTRAL COMMITTEE ON BIOLOGICAL SAFETY), NATIONAL INSTITUTES OF HEALTH (NIH), RECOMBINANT DNA (rDNA), BIOTECHNOLOGY, RECOMBINATION, INDIAN DEPARTMENT OF BIOTECHNOLOGY, COMMISSION OF BIOMOLECULAR ENGINEERING, GENE TECHNOLOGY REGULATOR (GTR), GENETIC MANIPULATION ADVISORY COMMITTEE (GMAC).

**Recombinase** An enzyme that acts to "cut open" the strand of DNA within a cell (e.g., to "splice-out" or "splice in") a given gene. During 2000, Nam-Hai Chua and and Jianru Zuo showed that activation of the gene for recombinase (via β estradiol transcription factor) could be done to cause expression of recombinase in a manner that "spliced out" (removed) antibiotic-resistance "marker genes" from genetically engineered plants. See also ENZYME, DEOXYRIBONUCLEIC ACID (DNA), GENE, CELL, GENE SPLICING, GENETIC ENGINEERING, TRANSCRIPTION FACTORS, ANTIBIOTIC RESISTANCE, MARKER GENES (GENETIC MARKER).

**Recombination** The joining of genes, sets of genes, or parts of genes, into new combinations, either biologically or through laboratory manipulation (e.g., genetic engineering). See also GENETIC ENGINEERING, GENE, RECOMBINANT DNA (rDNA).

**Red Blood Cells** See ERYTHROCYTES.

**Redement Napole (RN) Gene** A swine gene that causes animals (possessing at least one negative allele of this gene) to produce meat which is more acidic than average, and thus that meat has a lower "water-holding" capacity. The RN gene was first identified in the Hampshire breed of swine in France. Since the 1960s, the Hampshire breed has been known to produce meat that is more acidic than average. See also GENE, ALLELE, ACID.

**Reduction (biological)** The decomposition of complex compounds and cellular structures by heterotrophic organisms. In a given ecological system, this heterotrophic decomposition serves the valuable function of recycling organic materials. This occurs because the heterotrophs absorb some of the decomposition products (for nourishment) and leave the balance of the (decomposed) substances for consumption (recycling) by other organisms. For example, bacteria break down fallen leaves on the floor of a forest, thus releasing some nutrients to be utilized by plants. See also HETEROTROPH.

**Reduction (in a chemical reaction)** The gain of (negatively charged) electrons by a chemical substance. When one substance is reduced by another, the other compound is oxidized (loses electrons) and is called the reducing agent. See also OXIDATION-REDUCTION REACTION, OXIDIZING AGENT.

**Redundancy** A term used to describe the fact that some amino acids have more than one codon (that codes for production of that amino acid). There are approximately 64 possible codons available to code for 20 amino acids. Therefore, some amino acids will be specified by more than one codon. These (extra) codons are redundant. See also CODON, GENETIC CODE, RIBOSOMES.

**Refractile Bodies (RB)** Dense, insoluble (not easily dissolved) protein bodies (i.e., clumps) produced within the cells of certain microorganisms. The refractile bodies function as a sort of natural storage device for the microorganism. They are called refractile bodies because their greater density (than the rest of the microorganism's body mass)

causes light to be refracted (bent) when it is passed through them. This bending of light causes the appearance of very bright and dark areas around the refractile body and makes them visible under a microscope.

Relatively rare in natural occurrence, refractile bodies can be induced (caused to occur) in procaryotes (e.g., bacteria) when the procaryotes are genetically engineered to produce eucaryotic (e.g., mammal) proteins. The proteins are stored in refractile bodies. For example, the *Escherichia coli* bacterium can be genetically engineered to produce bovine somatotropin (BST, a cow hormone), which is stored within refractile bodies in the bacterium. After some time of growth when a significant amount of BST has been synthesized, the *Escherichia coli* cells are disrupted (broken open), and the refractile bodies are removed by centrifugation and washed. They are then dissolved in appropriate solutions to release the protein molecules. This step denatures (unfolds, inactivates) the BST molecules and they are refolded to their native conformation (i.e., restored to the natural conformation found within the cow) in order to regain their natural activity. The protein is then formulated in such a way as to be commercially viable as a biopharmaceutical.

Refractile bodies are also known as inclusion bodies, protein inclusion bodies, and refractile inclusions.

One point of interest is that the prerequisite for the generation of a mammalian protein by (in) a living foreign system such as *E. coli* is that the system used to generate the protein (1) must not have an immune system capable of destroying the foreign protein it is making, or (2) the foreign protein made must be camouflaged or protected from any defense mechanisms possessed by the synthesizing organism. See also PROTEIN, GENETIC ENGINEERING, GENETIC CODE, PROCARYOTES, EUCARYOTE, *ESCHERICHIA COLIFORM* (*E. COLI*), BOVINE SOMATOTROPIN (BST), ULTRACENTRIFUGE, CONFORMATION, NATIVE CONFORMATION, PROTEIN FOLDING.

**Regulatory Enzyme** A highly specialized enzyme having a regulatory (controlling) function through its capacity to undergo a change in its catalytic activity. There exist two major types of regulatory enzymes: (1) covalently modulated enzymes, and (2) allosteric enzymes.

Covalently modulated enzymes can be interconverted between active and inactive (or less active) forms by the covalent attachment (or removal) of a modulating metabolite by other enzymes. Hence the activity of one enzyme can, under certain conditions, be regulated by other enzymes. Glycogen phosphorylase, an oligomeric protein with four major subunits (tetramer), is a classic example of a covalently modulated enzyme. The enzyme occurs in two forms: (1) phosphorylase a, the more active form, and (2) phosphorylase b, the less active form. In order for the enzyme to possess maximal catalytic activity (i.e., be phosphorylase a) certain serine residue on all four subunits must have a phosphate covalently attached. If, due to other regulatory signals it has received, the enzyme phosphorylase phosphatase hydrolytically cleaves and removes the phosphate group from the four subunits, the tetramer dissociates into the inactive (or much less active) dimer, phosphorylase b. Another enzyme, phosphorylase kinase, is able to rephosphorylate the four specific serine residues of the four subunits at the expense of ATP and regenerate the active phosphorylase a tetramer.

Allosteric enzymes are enzymes that possess a special site on their surfaces that is distinct from the enzyme's catalytic site and to which specific metabolites (called effectors or modulators) are reversibly and noncovalently bound. The allosteric binding site is as specific for a particular metabolite as is the catalytic site, but it cannot catalyze a reaction, only bind the effector. The binding of the effector causes a conformation change in the enzyme such that its catalytic activity is impaired or stopped. Allosteric enzymes are normally the first enzymes in, or are near the beginning of, a multienzyme system. The very last product produced by the multienzyme system (the end product) may act as a specific inhibitor of the allosteric enzyme by binding to that enzyme's allosteric site. The binding consequently causes a conformation change to occur in the enzyme, which inactivates it. A classic example of an allosteric enzyme in a multienzyme sequence is the enzyme L-threonine dehydratase, which is the initial enzyme in the enzyme sequence that catalyzes the conversion of L-threonine to L-isoleucine. This reaction occurs in five enzyme-catalyzed steps. The end product, L-isoleucine, strongly inhibits L-threonine dehydratase, the first enzyme in the five-enzyme sequence. No other intermediate in the sequence is able to inhibit the enzyme. This kind of repression is called feedback or end-product inhibition.

It should be noted that allosteric control may be negative (as in the example above) or positive. In positive control the effector binds to an allosteric site and stimulates the activity of the enzyme. Furthermore, some allosteric enzymes respond to two or more specific modulators with each modulator having its own specific binding site on the enzyme. An allosteric enzyme that has only one specific modulator is called monovalent, whereas an enzyme responding to two or more specific modulators is called polyvalent. Combinations of the above possibilities could lead to very fine tuning of the enzymes involved in the synthesis and/or degradation of metabolites.

Note that in the two examples above, the common denominator is the structural change that occurs upon execution of the mechanism. See also METABOLITE, REPRESSIBLE ENZYME.

**Regulatory Genes** Genes whose primary function is to control the state of synthesis of the products of other genes.

**Regulatory Sequence** A DNA sequence involved in regulating the expression of a gene, e.g., a promoter or operator region (in the DNA molecule). See also OPERATOR, PROMOTER, DOWN PROMOTER MUTATIONS, DOWN REGULATING, TRANSCRIPTION FACTORS.

**Remediation** The cleanup or containment (if chemicals are moving) of a hazardous waste disposal site to the satisfaction of the applicable regulatory agency [e.g., the Environmental Protection Agency (EPA)]. Such cleanup can sometimes be accomplished via use of microorganisms that have been adapted (naturally or via genetic engineering) to consume those chemical wastes present in the disposal site. See also ACCLIMATIZATION.

**Renaturation** The return to the natural structure of a protein or nucleic acid from a denatured (more random coil) state. For example, a protein may be denatured [lose its native (natural) structure] by exposure to surfactants such as SDS or to changes in the pH of the medium. If the surfactant is slowly removed, or the pH is slowly readjusted to the optimum for the protein, it will refold (snap) back into its original (native) form. See also NATIVE CONFIGURATION, DENATURATION, SDS.

**Renin** A proteolytic enzyme secreted by the juxtaglomerular cells of the kidney. Its release is stimulated by decreased arterial pressure and renal blood flow resulting from decreased extracellular fluid volume. It catalyzes the formation of angiotensin I from hypertensinogen. Angiotensin I is converted to angiotensin II by another enzyme located in the endothelial cells of the lungs. Angiotensin II then causes the increase in the force of the heartbeat and constricts the arterioles. This scenario causes a rise in the blood pressure and is thus a cause of hypertension (high blood pressure). See also HOMEOSTASIS, RENIN INHIBITORS, ATRIAL PEPTIDES.

**Renin Inhibitors** Those chemicals that act to block the hypertensive (i.e., high blood pressure-inducing) effect of the enzyme, renin. See also HOMEOSTASIS, RENIN INHIBITORS, ATRIAL PEPTIDES.

**Rennin** See CHYMOSIN.

**Reovirus** A virus containing double-stranded RNA. It is isolated from the respiratory and intestinal tracts of humans and other mammals. The prefix "reo-" is an acronym for respiratory enteric orphan. See also RETROVIRUSES.

**Reperfusion** The restoration of blood flow to an occluded (blocked) blood vessel. May be done biochemically (e.g., via tissue plasminogen activator) or via surgery. See also HUMAN SUPEROXIDE DISMUTASE (hSOD), LAZAROIDS.

**Replication (of DNA)** Reproduction of a DNA molecule (inside a cell). This process can be viewed as occurring in stages, in which the first stage consists of an enzyme

"unwinding" the double helix of the DNA molecule at a replication origin, forming a replication fork. At the replication fork, the two separated (DNA) strands serve as templates for new DNA synthesis. That new DNA synthesis is accomplished on each strand via enzymes known as DNA polymerase, which travel along each (single) strand making a second complementary strand by catalyzing the addition of DNA bases (to the new, growing strands). The end result is two new double helices (DNA molecules), each of which has one chain from the original DNA molecule and one chain that was newly synthesized by the DNA polymerase enzymes. See also DEOXYRIBONUCLEIC ACID (DNA), DNA POLYMERASE, ENZYME, REPLICATION FORK, DUPLEX, DOUBLE HELIX, BASE PAIR (bp).

**Replication (of virus)** Reproduction of the original virus. This process can be viewed as occurring in stages, in which the first stage consists of the adsorption of the virus to the host cell; penetration of the virus (or its nucleic acid) into the cell, the taking over of the cell's biomachinery and harnessing of it to replicate viral nucleic acid along with the synthesis of other virus constituents; the correct assembly of the nucleic acids and other constituents into a functional virus; followed finally by release of the virus from the confines of the cell. See also VIRUS, CELL, NUCLEIC ACIDS.

**Replication Fork** The point at which strands of parental duplex DNA are separated in a Y shape. This region represents a growing point in DNA replication. See also REPLICATION (OF DNA), DEOXYRIBONUCLEIC ACID (DNA), DUPLEX.

**Reporter Gene** A specific gene inserted into the DNA of a cell so that cell will "report" (to researchers) when signal transduction has occurred in that cell, or when a (linked) gene was successfully expressed. The gene that codes for production of the enzyme luciferase [which catalyzes bioluminescence (light production)] is one of the most commonly used reporter genes.

For example, when researchers are testing numerous candidate drugs for their ability to stop cells from (over-) producing a hormone or growth factor, the researchers need to quickly know when one of the candidate drugs has had the desired effect on the cell of interest. By prior insertion into that cell of a gene (e.g., which causes bioluminescence or a certain chemical to be produced by the cell when signal transduction has taken place), that cell "reports" (when a candidate drug has had the desired effect on the cell) by producing the bioluminescence or chemical (coded for by the reporter gene) which can be rapidly detected by the researcher (e.g., via light sensors or biosensors placed adjacent to the cell). See also GENE, GENETIC ENGINEERING, GENETIC CODE, CODING SEQUENCE, CELL, BIOLUMINESCENCE, CELL CULTURE, SIGNAL TRANSDUCTION, HORMONE, GROWTH FACTOR, BIOSENSORS (ELECTRONIC).

**Repressible Enzyme** An enzyme whose synthesis (rate of production) is inhibited (repressed) when the product that it (or the enzyme within a multienzyme sequence) synthesizes is present in high concentrations. It is a way of shutting down the synthesis of an enzyme whose product is not required because so much of it is readily available to the cell. When that enzyme product is no longer available (e.g., because the cell has consumed that product), more of the enzyme is synthesized (to catalyze production of more product). See also REPRESSION (OF AN ENZYME), REGULATORY ENZYME, ENZYME.

**Repression (of an enzyme)** The prevention of synthesis of certain enzymes when their reaction products are present. See also REPRESSIBLE ENZYME.

**Repression (of gene transcription/translation)** The inhibition of transcription (or translation) by the binding of a repressor protein to a specific site on the DNA (or RNA) molecule. The repressor molecule is the product of a repressor gene. See also REPRESSOR (PROTEIN), TRANSCRIPTION, TRANSLATION, DEOXYRIBONUCLEIC ACID (DNA).

**Repressor (protein)** The product of a regulatory gene, it is a protein that combines both with an inducer (or corepressor) and with an operator region (e.g., of DNA). See also INDUCERS, COREPRESSOR, OPERATOR, REPRESSION (OF GENE TRANSCRIPTION/TRANSLATION).

**Research Foundation for Microbiological Diseases** (includes Institute of Physical and Chemical Research) Also known as Riken. A Japanese institution that performs research on infectious diseases, among other research. See also NATIONAL INSTITUTE OF ALLERGY AND INFECTIOUS DISEASES (NIAID), KOSEISHO.

**Residue (of chemical within a foodstuff)** See MAXIMUM RESIDUE LEVEL (MRL).

**Residue (portion of a protein molecule)** See MINIMIZED PROTEINS.

**Respiration** Oxidative process in living cells in which oxygen or an inorganic compound serves as the terminal (final, ultimate) electron acceptor. Aerobic organisms obtain most of their energy from the oxidation of organic fuels. This process is known as respiration. See also OXIDATION-REDUCTION REACTION, REDUCTION (IN A CHEMICAL REACTION), OXIDATION, OXIDIZING AGENT.

**Restriction Endoglycosidases** A class of enzymes, each of which cleaves (cuts) oligosaccharides (e.g., the side chains on glycoprotein molecules) at a specific location within the chain. They are important tools in carbohydrate engineering, enabling the carbohydrate engineer to sequence (i.e., determine the structure of) existing oligosaccharides, to create different oligosaccharides, and to create different glycoproteins via removal/addition/change of the oligosaccharide chains on glycoprotein molecules. See also OLIGOSACCHARIDES, GLYCOPROTEIN, CARBOHYDRATE ENGINEERING, GLYCOSIDASES, ENDOGLYCOSIDASE, EXOGLYCOSIDASE, GLYCOFORM, GLYCOBIOLOGY, GLYCOSYLATION.

**Restriction Endonucleases** A class of enzymes that cleave (cut) DNA at a specific and unique internal location along its length. These enzymes are naturally produced by bacteria that use them as a defense mechanism against viral infection. The enzymes chop up the viral nucleic acids and hence their function is destroyed. Discovered in the late 1970s by Werner Arber, Hamilton Smith, and Daniel Nathans, restriction endonucleases are important tools in genetic engineering, enabling the biotechnologist to splice new genes into the location(s) of a molecule of DNA where a restriction endonuclease has created a gap (via cleavage of the DNA). See also VECTOR, ENZYME, POLYMERASE, GENE, GENETIC ENGINEERING, GENE SPLICING, ELECTROPHORESIS.

**Restriction Enzymes** See RESTRICTION ENDONUCLEASES.

**Restriction Fragment Length Polymorphism (RFLP) Technique** A "genetic mapping" technique that analyzes the specific sequence of bases (i.e., nucleotides) in a piece of DNA (from an organism). Since the specific sequence of bases in DNA molecules is different for each species, strain, variety, and individual (due to DNA polymorphism), RFLP can be utilized to "map" those DNA molecules (for plant breeding purposes, for criminal investigation purposes, etc.). See also GENETIC MAP, SEQUENCE (OF A DNA MOLECULE), RANDOM AMPLIFIED POLYMORPHIC DNA (RAPD) TECHNIQUE, DEOXYRIBONUCLEIC ACID (DNA), GENOME, PHYSICAL MAP (OF GENOME), LINKAGE, LINKAGE GROUP, MARKER (GENETIC MARKER), LINKAGE MAP, TRAIT, BASE PAIR (bp), DNA PROFILING, POLYMORPHISM (CHEMICAL), NUCLEIC ACIDS, GENETIC CODE, INFORMATIONAL MOLECULES.

**Restriction Map** A pictorial representation of the specific restriction sites (i.e., nucleotide sequences that are cleaved by given restriction endonucleases) in a DNA molecule (e.g., plasmid or chromosome). See also RESTRICTION SITE, RESTRICTION ENDONUCLEASES, DNA.

**Restriction Site** A nucleotide sequence (of base pairs) in a DNA molecule that is "recognized," and cleaved by a given restriction endonuclease. See also NUCLEOTIDE, SEQUENCE (OF A DNA MOLECULE), BASE PAIR (bp), DNA, RESTRICTION ENDONUCLEASES, RESTRICTION MAP.

**Resveratrol** Also known as 3,5,4 trihydroxy stilbene, it is a naturally occurring (in grapes) anti-fungal agent (e.g., against grape fungus). Resveratrol is thought to be responsible for the fact that consumption of red wine by humans helps those humans' blood fat (triglycerides) levels and blood cholesterol levels to be lowered; thereby reducing risk of cardiovascular disease. Resveratrol is a phytochemical produced by certain plants in response to "wounding" (e.g., by fungal growth on plant) or other stress. Plants that produce resveratrol include red grapes, mulberries, soybeans, and peanuts. Resveratrol inhibits cell mutations, stimulates at least one enzyme that can inactivate certain carcinogens, and (when consumed by humans) lowers blood cholesterol and blood fat levels. See also PHYTOCHEMICALS, SOYBEAN PLANT, FUNGUS, CARCINOGEN, CELL, MUTATION, TRIGLYCERIDES, CHOLESTEROL, ENZYME, ATHEROSCLEROSIS, CORONARY HEART DISEASE (CHD).

**Retinoids** A group of biologically active compounds that are chemical derivatives of vitamin A. Among other effects on living cells, some of the retinoid compounds act to deprive cancerous cells of their ability to proliferate endlessly, so these (formerly cancerous) cells then progress to a natural death (after exposure to an applicable retinoid). See also CELL, APOPTOSIS, VITAMIN, BIOLOGICAL ACTIVITY, CANCER, NEOPLASTIC GROWTH.

**Retroelements** See TRANSPOSON.

**Retroviral Vectors** Certain retroviruses used by genetic engineers to carry new genes into cells. These molecules become part of that cell's protoplasm. See also RETROVIRUSES, GENETIC ENGINEERING, VECTOR, GENE, PROTOPLASM.

**Retroviruses** (From the Latin word *retrovir*, which means backward man) Oncogenic (i.e., cancer-producing), single-stranded, diploid RNA (ribonucleic acid) viruses that contain (+) RNA in their virions and propagate through a double-helical DNA intermediate. They are known as retroviruses because their genetic information flows from RNA to DNA (reverse of normal). That is, the viruses contain an enzyme that allows the production of DNA using RNA as a template. Retroviruses can only infect cells in which DNA is replicating, such as tumor cells (since they are constantly replicating) or cells comprising the lining of the stomach (since that lining must replace itself every few days). See also ONCOGENES, DIPLOID, RIBONUCLEIC ACID (RNA), REVERSE TRANSCRIPTASES, CENTRAL DOGMA.

**Reverse Micelle (RM)** Also known as reversed micelle or inverted micelle. A spheroidal structure formed by the association of a number of amphipathic (i.e., bearing both polar and nonpolar domains) surfactant molecules dissolved in organic, nonpolar solvents such

**R**

as benzene, hexane, isooctane, and oils such as corn and sesame. The structure of an RM is the reverse of that of a micelle. Reverse micelles may be characterized by a structure in which the polar groups of the surfactant and any water present are centrally located with the surfactant hydrocarbon chains pointing outward into the surrounding hydrocarbon medium. Reverse micelles may be used to solubilize polar molecules (i.e., water, enzymes) in organic nonpolar solvents and oils. See also AMPHIPATHIC MOLECULES, MICELLE, SURFACTANT.

**Reverse Phase Chromatography (RPC)** A method of separating a mixture of proteins or nucleic acids or other molecules by specific interactions of the molecules with a hydrophobic ("water hating") immobilized phase (i.e., stationary substrate) which interacts with hydrophobic regions of the protein (or nucleic acid) molecules to achieve (preferential) separation of the mixture. See also CHROMATOGRAPHY.

**Reverse Transcriptases** Also known as RNA-directed DNA polymerases. A class of enzymes first discovered to be present in RNA tumor-virus, which allows the synthesis of DNA (complementary to the RNA) using the RNA present in the virus as a template. This is the reverse of what normally happens and hence the name. Reverse transcriptases closely resemble the DNA-directed DNA polymerases in that they require the same materials and conditions as the DNA polymerases (e.g., for RT-PCR). See also ENZYME, VIRUS, RIBONUCLEIC ACID (RNA), CENTRAL DOGMA (NEW), POLYMERASE, RT-PCR.

**Reversed Micelle** See REVERSE MICELLE (RM).

**RFLP (restriction fragment length polymorphism)** Restriction fragment length polymorphism. See also POLYMORPHISM (CHEMICAL), RESTRICTION ENDONUCLEASES, RESTRICTION FRAGMENT LENGTH POLYMORPHISM (RFLP) TECHNIQUE.

**rh** Used to denote compounds (human molecules) made through the use of recombinant DNA technology. Recombinant (r) human (h). See also rhTNF, RECOMBINANT DNA (rDNA), RECOMBINATION, GENETIC ENGINEERING.

*Rhizobium* **(bacteria)** Refers to several strains of bacteria that live in the soil and colonize the roots of certain plants (i.e., legumes) symbiotically to thereby "fix" nitrogen from the air (i.e., change gaseous nitrogen into the chemical form that can be used by plants). For the legume known as the soybean plant (*Glycine max* L.), the relevant strain of the bacteria is *Rhizobium japonicum*. For the legume known as the alfalfa plant, the relevant strain of the bacteria is *Sinorhizobium meliloti*. See also BACTERIA, NITROGEN FIXATON, NODULATION, SOYBEAN PLANT, SYMBIOTIC, PHARMACOENVIROGENETICS.

**Rhizoremediation** See PHYTOREMEDIATION, *RHIZOBIUM* (BACTERIA).

**Rho Factor** A protein involved in (chemically) assisting *Escherichia coli* RNA polymerase in the termination of transcription at certain (rho-dependent) sites on the DNA molecule. See also TRANSCRIPTION, POLYMERASE, *ESCHERICHIA COLIFORM* (*E. COLI*).

**rhTNF** Recombinant human TNF. See also TUMOR NECROSIS FACTOR (TNF).

**RIA** See RADIOIMMUNOASSAY.

**Ribonucleic Acid (RNA)** A long-chain, usually single-stranded nucleic acid consisting of repeating nucleotide units containing four kinds of heterocyclic, organic bases: adenine, cytosine, guanine, and uracil. These bases are conjugated to the pentose sugar ribose and held in sequence by phosphodiester (chemical) bonds.

The primary function of RNA is protein synthesis within a cell. However, RNA is involved in various ways in the processes of expression and repression of hereditary information. The three main functionally distinct varieties of RNA molecules are: (1) messenger RNA (mRNA), which is involved in the transmission of DNA information, (2) ribosomal RNA (rRNA), which makes up the physical machinery of the synthetic process, and (3) transfer RNA (tRNA), which also constitutes another functional part of the machinery of protein synthesis. See also HEREDITY, GENETIC CODE, RIBOSOMES, INFORMATIONAL MOLECULES, NANOTECHNOLOGY.

**Ribose** D-Ribose, a five-carbon-atom monosaccharide (i.e., a sugar). It is important to life because it and the closely allied compound deoxyribose form a part of the molecules that constitute the backbone of nucleic acids. See also NUCLEIC ACIDS, MONOSACCHARIDES.

**Ribosomal RNA** See rRNA (RIBOSOMAL RNA).

**Ribosomes** The molecular "machines" within cells that coordinate the interplay of tRNAs, mRNAs, and proteins in the complex process of protein synthesis (manufacture). RNA constitutes nearly two-thirds of the mass of these large (mega-Dalton) molecular assemblies, which are technically ribozymes (i.e., an enzyme in which the catalysis is performed by RNA).

The formation of a ribosome (in the endoplasmic reticulum of a cell) from individual RNA and protein molecules is largely a self-assembly process, because all of the information needed for the correct assembly of this structure is contained in the primary structure of its (molecular) components. The assembly process is ordered and proceeds in stages. Many ribosomes (in a given cell) can simultaneously translate an mRNA molecule. The structure, consisting of a group of ribosomes bound to an mRNA molecule actively synthesizing protein, is called a polyribosome or a polysome. The ribosomes in this (polysome) unit operate independently of each other, each synthesizing a complete polypeptide (protein) "molecular chain." See also PROTEIN, POLYPEPTIDE (PROTEIN), PROTEIN SIGNALING, PROTEIN FOLDING, POLYCISTRONIC, PROTEIN STRUCTURE, PRIMARY STRUCTURE, TRANSCRIPTION, TRANSCRIPTION UNIT, MESSENGER RNA (mRNA), CELL, ENDOPLASMIC RETICULUM (ER), TRANSFER RNA (tRNA), rRNA (RIBOSOMAL rRNA), DALTON, SELF-ASSEMBLY (OF A LARGE MOLECULAR STRUCTURE), RIBOZYMES, RIBONUCLEIC ACID (RNA).

**Ribozymes** Discovered by Thomas Cech and Sidney Altman, they are RNA molecules that act as enzymes; that is, possess catalytic activity and can specifically cleave (cut) other RNA molecules. The ribozyme (RNA) molecule and the other RNA molecule come together, whereupon the ribozyme molecule cuts the other RNA molecule at a specific defined (three-base) site. Because the ribozyme molecule acts as an enzyme in this reaction, the ribozyme molecule is not consumed or destroyed, but goes on to similarly "cut" other RNA molecules. During 2000,

Thomas Steitz and Peter Moore, et al., proved that ribosomes (i.e., the cell's internal protein-synthesis "machinery") are functionally ribozymes. See also RIBONUCLEIC ACID (RNA), CATALYTIC RNA, BASE (NUCLEOTIDE), ENZYME, CELL, RIBOSOMES, CATALYST.

**Ricin** A lethal-to-cells protein naturally produced in castor beans. In 1994, Robert J. Ferl and Paul C. Sehnke genetically engineered a tobacco plant to produce ricin. Attached to a pharmaceutical "guided missile" or "magic bullet" such as a monoclonal antibody or the CD4 protein, ricin is potentially useful for treatment against some tumors and has been investigated as a possible treatment against acquired immune deficiency syndrome (AIDS). See also IMMUNOTOXIN, MONOCLONAL ANTIBODIES (MAb), CELL, CD4 PROTEIN, GENETIC ENGINEERING, FUSION PROTEIN, FUSION TOXIN, SOLUBLE CD4, PHYTOCHEMICALS, "MAGIC BULLET".

**Riken** Japan's Institute for Physical and Chemical Research. See also RESEARCH FOUNDATION FOR MICROBIOLOGICAL DISEASES.

**RMD** See RAPID MICROBIAL DETECTION.

**RN Gene** See REDEMENT NAPOLE (RN) GENE.

**RNA** See RIBONUCLEIC ACID (RNA).

**RNA Polymerase** An enzyme that catalyzes the synthesis of a complementary mRNA (messenger RNA) molecule from a DNA (deoxyribonucleic acid) template in the presence of a mixture of the four ribonucleotides (ATP, UTP, GTP, and CTP). Also called transcriptase. See also CENTRAL DOGMA, POLYMERASE, DNA POLYMERASE, PROMOTER.

**RNA Probes** See DNA PROBE.

**RNA Transcriptase** See RNA POLYMERASE.

**RNA Vectors** An RNA (ribonucleic acid) vehicle for transferring genetic information from one cell to another. See also VECTOR, RETROVIRAL VECTORS.

**Rootworm** See CORN ROOTWORM.

**ROS** Acronym for Reactive Oxygen Species. See also FREE RADICAL.

**Rosemarinic Acid** A phenolic compound (naturally found in some plants) that acts as an antioxidant in the body's tissues when consumed by humans. For example, rosemarinic acid is naturally produced in the edible herbs *Origanum vulgare* and *Salvia officinalis*. See also PHYTOCHEMICALS, ANTIOXIDANTS, OXIDATIVE STRESS, NUTRACEUTICALS.

**Roving Gene** See JUMPING GENES, TRANSPOSI-
TION, TRANSPOSASE, GENE, GENOME, DEOXYRIBO-
NUCLEIC ACID (DNA).

**Rps1c Gene** A gene that confers to any soybean plant (possessing that gene in its DNA)
resistance to several strains/races of *phytophthora* root rot (PRR) disease. See also
GENE, DEOXYRIBONUCLEIC ACID (DNA), SOYBEAN
PLANT, *PHYTOPHTHORA* ROOT ROT.

**Rps1k Gene** A gene that confers to any soybean plant (possessing that gene in its DNA)
resistance to as many as 21 strains/races of
*phytophthora* root rot (PRR) disease. See
also GENE, DEOXYRIBONUCLEIC ACID (DNA), PHY-
TOPHTHORA ROOT ROT, SOYBEAN PLANT.

**Rps6 Gene** A gene that confers to any soybean
plant (possessing that gene in its DNA) resistance to some strains/races of *phytophthora*
root rot (PRR) disease. See also GENE, DEOXY-
RIBONUCLEIC ACID (DNA), *PHYTOPHTHORA* ROOT
ROT, SOYBEAN PLANT.

**rRNA (ribosomal RNA)** The nucleic acid
component of ribosomes, making up approximately two-thirds of the mass of the bacteria
*Escherichia coli* ribosome, and approximately one-half of the mass of mammalian
ribosomes. Ribosomal RNA accounts for
nearly 80% of the RNA content of the bacterial cell. See also NUCLEIC ACIDS, RIBOSOMES,
*ESCHERICHIA COLIFORM* (*E. COLI*), RIBONUCLEIC
ACID (RNA).

**RT-PCR** Acronym for Reverse Transcriptase
Polymerase Chain Reaction technique. See
also REVERSE TRANSCRIPTASES, DNA POLY-
MERASE, POLYMERASE CHAIN REACTION (PCR),
POLYMERASE CHAIN REACTION (PCR) TECHNIQUE.

**Rubitecan** A pharmaceutical that either shrinks
or halts the growth of pancreatic cancer
tumors in humans. The pharmacophore (i.e.,
active portion of molecule) in rubitecan was
derived from a Chinese flowering tree (*Camp-
totheca acuminata*); thus that "family" of
drugs is known as camptothecins. Camptothecins inhibit a critical enzyme required for
cell division to occur (thus it inhibits rapidly
growing tumors). See also CANCER, PANCREAS,
TUMOR, PHARMACOPHORE, ENZYME.

**Rumen (of cattle)** See PREBIOTICS.

**Rusts** Various fungal diseases (*puccinia* spp.)
that attack small grain plants such as wheat,
corn/maize, sorghum, oats, barley and rye.
Its visual appearance is like that of rust on
the surfaces of those plants. See also FUNGUS,
WHEAT, CORN.

R

# S

**S1 Nuclease** An enzyme that specifically degrades (destroys) single-stranded sequences of DNA. See also RESTRICTION ENDONUCLEASES, ENZYME, DEOXYRIBONUCLEIC ACID (DNA).

**SAAND** Acronym for Selective Apoptotic Anti-Neoplastic Drug. See also SELECTIVE APOPTOTIC ANTI-NEOPLASTIC DRUG (SAAND).

**SAGB** Senior Advisory Group on Biotechnology. See also SENIOR ADVISORY GROUP ON BIOTECHNOLOGY (SAGB).

**Salicylic Acid (SA)** SA is a signaling molecule in Systemic Acquired Resistance (SAR) when SAR is triggered in plants (e.g., via spray application of COBRA R herbicide to soybean plants, via spray application of harpin protein to various plants, via chewing by insects on the leaves of tomato plants, and/or the entry-into-plant of certain pathogenic bacteria/fungi, etc.). See also SYSTEMIC ACQUIRED RESISTANCE (SAR), SIGNALING MOLECULE, SOYBEAN PLANT, HARPIN, FUNGUS, PATHOGEN, PROTEIN, PATHOGENESIS RELATED PROTEINS, JASMONIC ACID.

**Salinity Tolerance** See SALT TOLERANCE.

**Salmonella** A genus of bacteria, consisting of more than 2,400 serovars (strains/types) classified within two species (*Salmonella enterica* and *Salmonella bongori*). All of these serovars are potentially pathogenic (disease-causing) to humans. For example, some variants of *Salmonella typhimurium* can cause typhoid fever. The nontyphoid strains of *Salmonella* generally cause enterocolitis; although that enterocolitis can lead to more serious systemic infections. *Salmonella enteritidis* and *Salmonella typhimurium* are increasingly causing outbreaks of foodborne illnesses (e.g., when foods are not washed or cooked thoroughly enough prior to consumption by humans). See also BACTERIA, PATHOGEN, PATHOGENIC, STRAIN, COMMENSAL.

***Salmonella enteritidis* (Se)** A pathogenic strain of Salmonella bacteria that can cause fatal infections in poultry and humans (e.g., when undercooked eggs are eaten by humans). See also BACTERIA, PATHOGEN, PATHOGENIC, STRAIN, *SALMONELLA*.

***Salmonella typhimurium*** A pathogenic strain of *Salmonella* bacteria, which can cause disease in humans (e.g., when contaminated food is not washed and cooked enough prior to consumption). See also BACTERIA, PATHOGEN, PATHOGENIC, STRAIN, COMMENSAL.

**Salt Tolerance** Refers to the trait (of a plant) that enables a plant to grow/survive in soil that contains a high level of salt. For example, during 2001, Eduardo Blumwald and Hong-Xia Zhang inserted a salt-tolerance gene from *Arabidopsis thaliana* into a tomato plant (*Lycopersicon esculentum*) and thereby made that tomato plant resistant to salt concentrations up to 200 mM (far higher than it could previously survive). That (*Arabidopsis* origin) gene enables the tomato plant to extract salt from the soil, and then sequester and store the salt in vacuoles (i.e., small compartments) within its leaf cells. See also *ARABIDOPSIS THALIANA*, VACUOLES.

**Salting Out** A technique used for forcing (dissolved) proteins out of a solution by increasing the concentration of salt in the solution. The $Na^+$ and $Cl^-$ ions derived from the salt compete for and "tie up" water molecules that are solubilizing the protein molecules, thereby rendering them insoluble or more insoluble.

**SAM** See SAM-K GENE.

**Sam-K Gene** A gene naturally present within the *E. coli* bacteriophage T3. If the sam-k gene is inserted via genetic engineering into a (fruit crop) plant's genome, that causes greatly reduced production of the chemical compound S-adenosylmethionine (SAM) in that plant's fruit. Because the SAM is normally converted (chemically) into 1-aminocyclopropane-1-carboxylic acid (ACC) in the fruits of traditional vaieties of (fruit crop) plants, such sam-k gene-containing plants produce fruits which ripen/soften far slower than fruit from traditional varieties of those plants; which can reduce spoilage/loss in the harvest and transport of such fruit. That is because ACC is required for fruits to produce ethylene, the plant hormone which triggers (over-) ripening/softening of fruit. See also GENE, BACTERIOPHAGE, *ESCHERICHIA COLIFORM* (*E. COLI*), GENETIC ENGINEERING, GENOME, ACC, ACC SYNTHASE.

**Sanitary and Phytosanitary (SPS) Agreement** The agreement to GATT/WTO via which WTO member nations agreed to base their technical barriers (regarding some imports, designed for the protection of human health or the control of animal and plant pests/diseases) only on an assessment of actual risks posed by the particular import in question; and to utilize only scientific methods in assessing those risks. See also SANITARY AND PHYTOSANITARY (SPS) MEASURES, WORLD TRADE ORGANIZATION (WTO), SPS.

**Sanitary and Phytosanitary (SPS) Measures** Technical barriers (i.e., against some imports) that are designed for the protection of human health or the control of animal and plant pests/diseases. In the Sanitary and Phytosanitary (SPS) Agreement to GATT/WTO, the WTO member nations agreed to base their SPS measures only on an assessment of actual risks posed by the particular import in question, and to utilize only scientific methods in assessing those risks. See also SANITARY AND PHYTOSANITARY (SPS) AGREEMENT, WORLD TRADE ORGANIZATION (WTO), SPS.

**Saponification** Alkaline hydrolysis of triacyl glycerols to yield fatty acid salts. The molecules thus produced are known as surfactants (surface active agents), commonly called soap. The process of soapmaking. See also HYDROLYSIS.

**Saponins** A group of phytochemicals (i.e., sugars linked to a triterpene or a steroid molecular subunit) produced by certain plants (the soybean plant, spinach plant, tomatoes, potatoes, ginseng plant, etc.). Evidence suggests that human consumption of saponins (e.g., produced in soybeans) can help to lower a person's blood content of low-density lipoproteins (LDLP) and can help prevent certain types of cancer. See also PHYTOCHEMICALS, SUGAR MOLECULES, SOYBEAN PLANT, LOW-DENSITY LIPOPROTEINS (LDLP), CANCER, STEROID.

**SAR** Acronym for Systemic Acquired Resistance. See also SYSTEMIC ACQUIRED RESISTANCE (SAR).

**Satellite DNA** Many tandem repeats (identical or related) of a short basic repeating unit (in the DNA molecule). See also DEOXYRIBONUCLEIC ACID (DNA).

**Saturated Fatty Acids (SAFA)** Fatty acids containing fully saturated alkyl chains (on their molecules). This means that the carbon atoms comprising the chains are held together by one carbon-to-carbon bond and not two or three. High levels of dietary SAFA have been related to increased blood cholesterol levels, which tends to lead to coronary heart disease (CHD) in humans. The sole exception is stearic acid (also known as stearate), which research has shown has no impact on the blood cholesterol levels of humans that consume it. Beef fat typically contains approximately 54% saturated fatty acids; sheep fat typically contains approximately 58% saturated fatty acids; pork fat typically contains approximately 45% saturated fatty acids; chicken fat typically contains approximately 32% saturated fatty acids.

In general, fats possessing the highest levels of saturated fatty acids tend to be solid at room temperature; and those fats possessing the highest levels of unsaturated fatty acids tend to be liquid at room temperature. That rule of thumb was the original "dividing line" between the terms "fats" and "oils," respectively. See also FATTY ACID, DEHYDROGENATION, CHOLESTEROL, MONOUNSATURATED FATS, SAPONIFICATION, LPAAT PROTEIN, UNSATURATED FATTY ACID, POLYUNSATURATED FATTY

ACIDS (PUFA), CORONARY HEART DISEASE (CHD), PALMITIC ACID, STEARATE (STEARIC ACID), HIGH-STEARATE SOYBEANS, HIGH-STEARATE CANOLA.

**Saxitoxins** Paralytic poisons produced by certain shellfish. See also RICIN.

**SBO** Soybean oil.

**Scab** See FUSARIUM.

**Scale-Up** The transition step in moving a (chemical) process from experimental (e.g., "test tube," small, bench) scale to a larger scale producing more or much more product that the bench scale (e.g., production of tons/year in a chemical plant). A process may require a number of scale-ups, with each scale-up producing more product than the last one.

**Scanning Tunneling Electron Microscopy** See ELECTRON MICROSCOPY (EM).

**SCP** See SINGLE-CELL PROTEIN (SCP).

**SDM** Site-directed mutagenesis. See also SITE-DIRECTED MUTAGENESIS (SDM).

**SDS** Sodium dodecyl sulfate. Also known as sodium lauryl sulfate (SLS). A surfactant commonly used in biochemical and biotechnological applications for the solubilization of membrane components and hard-to-solubilize (dissolve) molecules. For example, it is often utilized at high concentration in water solution (e.g., along with potassium acetate) to dissolve plant DNA samples (e.g., when a scientist wants to sequence that sample of plant DNA). The SDS/PA in water solution helps the scientist to separate out contaminants commonly present in samples from plant tissues (polysaccharides, proteins, etc.) because DNA molecules are much more soluble in SDS/PA solution than are those contaminant molecules. Above a critical concentration (CMC), SDS forms micelles in water which are thought to be responsible for its solubilizing action. SDS is also used in such items as shampoo. See also CRITICAL MICELLE CONCENTRATION, MICELLE, REVERSE MICELLE (RM), PROTEIN, MEMBRANE (OF A CELL), SURFACTANT, DEOXYRIBONUCLEIC ACID (DNA), POLYSACCHARIDES, SEQUENCING (OF DNA MOLECULES), HEXADECYLTRIMETHYLAMMONIUM BROMIDE (CTAB).

**Seed-Specific Promoter** See PROMOTER.

**"Seedless" Fruits** See TRIPLOID.

**Selectable Marker Genes** See MARKER (GENETIC MARKER).

**Selectins** Also called LEC-CAMs (leukocyte-cell adhesion molecules). A class of molecular structurally related lectins that mediate (control, cause, etc.) the contacts between a variety of cells (e.g., leukocytes and endothelial cells), and function as cellular adhesion receptors. See also RECEPTORS, LECTINS, ADHESION MOLECULE, LEUKOCYTES, ENDOTHELIAL CELLS, ENDOTHELIUM, SIGNAL TRANSDUCTION.

**Selective Apoptotic Anti-Neoplastic Drug (SAAND)** A category of pharmaceuticals that acts to prevent neoplastic growth (i.e., cancer) by allowing normal cell apoptosis to occur again (e.g., by blocking an enzyme that is hindering normal apoptosis) in abnormal precancerous cells and cancerous cells. Examples of SAANDs include sulindac, which blocks phosphodiesterases (enzymes). See also NEOPLASTIC GROWTH, CANCER, TUMOR, APOPTOSIS, CELL, ENZYME, PHOSPHODIESTERASES.

**Selective Estrogen Effect** A term used to describe how certain phytochemicals (flavones, flavonols, isoflavones, etc.) and pharmaceuticals (Evista/raloxifene, tamoxifen, etc.) possessing molecular structures that are similar to estrogen (a hormone) impart some beneficial effect on the human body when consumed by humans, without any of the adverse impacts of estrogen (e.g., promotion of the growth of certain tumors by estrogen). See also PHYTOCHEMICALS, FLAVONOLS, ISOFLAVONES, FLAVONOIDS, ESTROGEN, PHYTOESTROGENS, PROSTATE, GENISTEIN (Gen).

**Selective Estrogen Receptor Modulators** Abbreviated SERM. This term refers to chemical compounds (isoflavones, the pharmaceuticals Evista/raloxifene and tamoxifen, etc.) which impart some beneficial effect on the human body when consumed by humans, without any of the adverse impacts of estrogen (e.g., promotion of the growth of certain tumors by estrogen). See also SELECTIVE ESTROGEN EFFECT, ESTROGEN, ISOFLAVONES, PHYTOCHEMICALS.

**Self-Assembly (of a large molecular structure)** The essentially automatic ordering and assembly of certain molecules into a large structure. Examples of such large molecular structures (often called supramolecular structures or supramolecular assemblies) include micelles, reverse micelles, ribosomes, peptide nanotubes, and Tobacco Mosaic Virus (TMV).

The first discovery of a self-assembling active biological structure occurred in 1955, when Heinz Frankel-Conrat and Robley Williams showed that TMV will reassemble into functioning, infectious virus particles (after TMV has been dissociated into its components via immersion in concentrated acetic acid). In the future, it is hoped that man will be able to "direct" the self-assembly of molecular structures which will:

- Serve as "cages" to carefully protect and deliver sensitive/unstable pharmaceuticals to targeted tissues within the body.
- Serve as "crucibles" (i.e., reaction vessels) for small-scale chemical reactions to occur within.
- Serve as computer logic or memory devices (i.e., bioelectronics).
- Serve as antibiotics.

For example, during the 1990s, M. Reza Ghadiri created "peptide nanotubes" made via self-assembly of certain peptides into tubes (cylinders) of nanometer dimensions. These peptide nanotubes are "membrane active" (i.e., insert one end of themselves into the outer membrane of a cell), and cause the cell (e.g., pathogenic bacteria) contents to "leak out," which kills the bacteria. See also MICELLE, REVERSE MICELLE (RM), RIBOSOMES, TOBACCO MOSAIC VIRUS (TMV), NANOCRYSTAL MOLECULES, NANOSCIENCE, NANOTECHNOLOGY, NANOMETERS (NM), BIOELECTRONICS, ANTIBIOTIC, PATHOGEN, BACTERIA.

**Semisynthetic Catalytic Antibody** An antibody produced (e.g., via monoclonal antibody techniques) in response to a carefully selected antigen (i.e., one of the molecules involved in the chemical reaction that you are trying to catalyze). Such an antibody is then made to be catalytic by "attaching" a (molecular) group that is known to catalyze the desired chemical reaction. This attaching is done either via chemical modification of the antibody, or via genetic engineering of the cell (DNA) that produces that antibody.

See also CATALYST, ANTIBODY, CATALYTIC ANTIBODY, SITE-DIRECTED MUTAGENESIS (SDM), MONOCLONAL ANTIBODIES (MAb), ANTIGEN, GENETIC ENGINEERING, ABZYMES.

**Senior Advisory Group on Biotechnology (SAGB)** An association of approximately 35 of the largest European companies that are engaged in at least some form of genetic engineering research or production. Similar to America's Biotechnology Industry Organization (BIO), the SAGB works with governments and the public to promote safe and rational advancement of genetic engineering and biotechnology. It was formed in 1989 and is based in Brussels, Belgium. See also BIOTECHNOLOGY, GENETIC ENGINEERING, RECOMBINANT DNA (rDNA), JAPAN BIO-INDUSTRY ASSOCIATION, INTERNATIONAL FOOD BIOTECHNOLOGY COUNCIL (IFBC), BIOTECHNOLOGY INDUSTRY ORGANIZATION (BIO).

**Sense** Normal (forward) orientation of DNA sequence (gene) in genome. See also GENE SILENCING, ANTISENSE (DNA SEQUENCE).

**Sepsis** Also known as systemic inflammatory response syndrome, this life-threatening condition ("septic shock") occurs when the body's immune system over-responds to infection (e.g., by gram-negative bacteria) in which release of bacterial endotoxin (lipopolysaccharide, or LPS) occurs. Those immune system cells (e.g., macrophages, etc.) overproduce numerous inflammatory agents (e.g., cytokines), which induce fever, shock, and sometimes organ failure. See also GRAM-NEGATIVE (G-), BACTERIA, CYTOKINES, ENDOTOXIN, MACROPHAGE.

**Septic Shock** See SEPSIS.

**Sequence (of a DNA molecule)** The specific nucleic acids that comprise a given segment of a DNA molecule. See also DEOXYRIBONUCLEIC ACID (DNA), GENETIC CODE, GENE, CHROMOSOMES, NUCLEIC ACIDS, CONTROL SEQUENCES, SEQUENCING (OF DNA MOLECULES), STRUCTURAL GENOMICS, COMPLEMENTARY (MOLECULAR GENETICS).

**Sequence (of a protein molecule)** The specific amino acids (and the order in which they are coupled together) that comprise a given segment of a protein molecule. See also PROTEIN, AMINO ACID, STRUCTURAL GENE, GENOMICS,

STRUCTURAL GENOMICS, SEQUENCING (OF PROTEIN MOLECULES).

**Sequence Map** A pictorial representation of the sequence of amino acids in a protein molecule, the sequence of nucleic acids in a DNA molecule, or the sequence of oligosaccharide components in a glycoprotein/carbohydrate molecule. See also SEQUENCING (OF DNA MOLECULES), SEQUENCING (OF PROTEIN MOLECULES), SEQUENCING (OF OLIGOSACCHARIDES), SEQUENCE (OF A DNA MOLECULE), SEQUENCE (OF A PROTEIN MOLECULE), RESTRICTION MAP.

**Sequencing (of DNA molecules)** The process used to obtain the sequential arrangement of nucleotides in the DNA backbone. The cleavage into fragments (followed by separation of those fragments, which can then be sequenced individually) of DNA molecules by one of several methods: (1) a chemical cleavage method followed by polyacrylamide gel electrophoresis (PAGE) or capillary electrophoresis, (2) a method consisting of controlled interruption of enzymatic replication methods followed by PAGE, (3) a didexyl method utilizing fluorescent "tag" atoms attached to the DNA fragments, followed by use of spectrophotometry to identify the respective DNA fragments by their differing "tags" (which fluoresce at different wavelengths). This (fluorescent tag) variant of the dideoxy method can be automated to "decipher" large DNA molecules (i.e., genomes). Such automated machines are sometimes called "gene machines." Sequencing of DNA was first done in the mid-1970s by Frederick Sanger. See also POLYACRYLAMIDE GEL ELECTROPHORESIS (PAGE), GENE MACHINE, DEOXYRIBONUCLEIC ACID (DNA), SEQUENCE (OF A DNA MOLECULE), BASE EXCISION SEQUENCE SCANNING (BESS), SHOTGUN SEQUENCING, NANOPORE, NEAR-INFRARED SPECTROSCOPY (NIR), COMPARATIVE SEQUENCING, BIOCHIPS.

**Sequencing (of oligosaccharides)** See RESTRICTION ENDOGLYCOSIDASES, SEQUENCE MAP.

**Sequencing (of protein molecules)** The process used to obtain the sequential arrangement of amino acids in a protein molecule. See also PROTEIN, AMINO ACID, SEQUENCE (OF A PROTEIN MOLECULE).

**Sequon** A (potential) site on a protein molecule's "backbone" where a sugar molecule (or a chain of sugar molecules, i.e., an oligosaccharide) may be attached. See also PROTEIN, SUGAR MOLECULES, GLYCOPROTEIN, GLYCOGEN, GLYCOSYLATION, PROTEIN ENGINEERING, OLIGOSACCHARIDES.

**Serine (ser)** A nonessential amino acid; a biosynthetic precursor of several metabolites, including cysteine, glycine, and choline. In 1999, Solomon H. Snyder, Herman Wolosker, and Seth Blackshaw conducted research that showed that some mammals synthesize (manufacture) D-serine within their brains, and it functions as a neurotransmitter there. See also ESSENTIAL AMINO ACIDS, METABOLITE, CYSTEINE (cys), GLYCINE (Gly), CHOLINE, NEUROTRANSMITTER.

**Seroconversion** The development of antibodies (specific to that disease-causing microorganism) in response to vaccination or natural exposure to a disease-causing microorganism. See also SEROLOGY, ANTIBODY, IMMUNOGLOBULIN, HUMORAL IMMUNITY, PATHOGEN, POLYCLONAL ANTIBODIES, PASSIVE IMMUNITY.

**Serologist** See SEROLOGY.

**Serology** A subdiscipline of immunology, concerned with the properties and reactions of blood sera. It includes the diverse techniques used for the "test tube" measurement of antibody-antigen reactions, including blood typing (e.g., for transfusions). See also MAJOR HISTOCOMPATIBILITY COMPLEX (MHC), OLIGOSACCHARIDES, SERUM LIFETIME.

**Seronegative** Refers to negative results of a serology test. See also SEROLOGY, HUMORAL IMMUNITY, ANTIBODY.

**Serotonin** An important neurochemical whose effects upon the human brain include mood elevation. Production of serotonin in the brain is increased by ingestion of the amino acid tryptophan (a chemical precursor to serotonin). Elevation of brain levels of serotonin can also be caused by consumption of the herb known as Saint John's Wort (*Hypericum perforatum*), or by consumption of certain pharmaceuticals such as the antidepressants Prozac™ (trademarked product of Eli Lilly & Company), Zoloft™ (trademarked product of Pfizer, Inc.), or Paxil™ (trademarked product of Smithkline Beecham PLC). In 1997, Marianne Regard and Theodor Landis discovered that humans

**S**

afflicted with hemorrhagic lesions in the brain (cause of abnormal serotonin activation/production) often became "passionate culinary afficionados." See also TRYPTOPHAN (trp), ESSENTIAL AMINO ACIDS, BLOOD-BRAIN BARRIER (BBB), NEUROTRANSMITTER.

**Serotypes** A variety (sub-strain) of a microorganism that is distinguished from others (in the strain) via its serological effects (within immune system of the host organism it inhabits). See also BACTERIA, STRAIN, *E. COLI* 0157:H7, SEROLOGY, HUMAN IMMUNODEFICIENCY VIRUS TYPE 1 (HIV-1), HUMAN IMMUNODEFICIENCY VIRUS TYPE 2 (HIV-2).

**Serum** Blood plasma that has had its clotting factor removed. See also FACTOR VIII, FACTOR IX, PLASMA.

**Serum Half Life** See SERUM LIFETIME.

**Serum Immune Response** See HUMORAL IMMUNITY.

**Serum Lifetime** The average length of time that a molecule circulates in an organism's bloodstream before it is cleared from the bloodstream. See also IMMUNE RESPONSE, ANTIGEN.

**Sessile** (Micro)organisms that are attached to a (support) substrate directly by their base; not attached via an intervening peduncle (i.e., stalk). Can also refer to fruit or leaves that are attached directly to the main stem or branch of a plant. See also VAGILE.

**Sex Chromosomes** Those chromosomes whose content is different in the two sexes of a given species. They are usually labeled X and Y (or W and Z); one sex has XX (or WW), the other sex has XY (or WZ). XX (WW) is female and XY (WZ) is male.

**Sexual Conjugation** An infrequent occurrence in which two adjacent bacteria stretch out portions of their (cell) membranes to touch one another, fuse, and then pass transposons, jumping genes, or plasmids to each other. See also ASEXUAL, BACTERIA, CELL, CONJUGATION, PLASMID, TRANSPOSON, JUMPING GENES.

**Shotgun Cloning Method** A technique for obtaining the desired gene that involves "chopping up" the entire genetic complement of a cell using restriction enzymes, then attaching each (resultant) DNA fragment to a vector and transferring it into a bacterium, and finally screening those (engineered) bacteria to locate the bacteria that are producing the desired product (e.g., a protein). See also GENETIC ENGINEERING, GENOME, RESTRICTION ENDONUCLEASES, VECTOR.

**Shotgun Sequencing** Sometimes called Whole-genome Shotgun Sequencing. A technology for rapid sequencing of DNA, in which an organism's genome (DNA) is first fragmented ("broken up"), and then randomly selected pieces of the DNA are individually sequenced. Those individual pieces' sequences must subsequently be "bridged" (i.e., "assembled" in an overlapping end-by-end pattern) in order to assemble a complete map (e.g., of an organism's chromosome or genome). See also SEQUENCING (OF DNA MOLECULES), DEOXYRIBONUCLEIC ACID (DNA), SEQUENCE (OF A DNA MOLECULE), GENOME, DNA "BRIDGES", CHROMOSOME, GENETIC MAP.

**Shuttle Vector** A vector capable of replicating in two unrelated species. See also VECTOR, REPLICATION (OF VIRUS).

**Signal Transducers and Activators of Transcription (STATs)** Molecules that cause signal transduction to occur (i.e., when a hormone or other chemical "binds" to it), or molecules that cause transcription to occur (i.e., when transcription factor(s) "bind" to it). STATs can be attached to solid surfaces (e.g., in a bioassay or biosensor) for use in such research applications as high-throughput screening. See also SIGNAL TRANSDUCTION, HORMONE, TRANSCRIPTION FACTORS, BIOCHIPS, BIOSENSORS (ELECTRONIC), BIOASSAY, HIGH-THROUGHPUT SCREENING (HTS), MICROARRAY (TESTING), TARGET (OF A HERBICIDE OR INSECTICIDE), CASCADE.

**Signal Transduction** The "reception" and "conversion" of a "chemical message" (e.g., hormone) by a cell. For example, G-proteins (which are embedded in the surface membrane of certain cells, but extend through to outside and inside of the membrane) accomplish signal transduction. When a hormone, drug, neurotransmitter, or other signal chemical binds (i.e., "docks") to the receptor (on the exterior of the cell's plasma membrane), the receptor activates the G-protein, causing an effector inside cell to produce a "signal" chemical inside the cell, which then reacts to the original external chemical signal received. See also CELL, PLASMA MEMBRANE, TRANSMEMBRANE PROTEINS, RECEPTORS, EGF RECEPTOR, RAS GENE, NUCLEAR RECEPTORS, SIGNALING, G-PROTEINS, MAST CELLS, CD95 PROTEIN, HORMONE, SUBSTANCE P, LECITHIN, CASCADE.

**Signaling** The "communication" that occurs between and within cells of an organism, e.g., via hormones, nitric acid, etc. Such signaling "tells" certain cells to grow, change, or produce specific proteins at specific times. See also RECEPTORS, PROTEIN, NUCLEAR RECEPTORS, G-PROTEINS, SIGNAL TRANSDUCTION (SIGNAL), TRANSDUCTION (GENE), CD95 PROTEIN, HORMONE, PARKINSON'S DISEASE, HARPIN, SUBSTANCE P, LECITHIN, NITRIC OXIDE, SIGNAL TRANSDUCERS AND ACTIVATORS OF TRANSCRIPTION (STATs), PROTEIN SIGNALING, CASCADE, CHOLINE.

**Signaling Molecule** A molecule utilized to "signal" (communicate) with cells, or to deliver a signal to other organisms (e.g., a signal by the soybean plant to attract beneficial *Rhizobium* bacteria to colonize the roots of that soybean plant).

For example, the young offspring of fleas can remain immature (larvae) for up to 2 years in the absence of a food source, until carbon dioxide molecules and heat from a nearby mammal (potential host/food source) signal them to mature into adults in order to prey on the mammal. Another example: the larvae of North American tree frogs are signaled by chemicals released into a pond's water when the first such frog larva is killed by a (predatory) dragonfly nymph (i.e., when those dragonflies first arrive each year at a given pond, to prey on the frog larvae). That chemical "signal" causes all of the North American tree frog larvae in that pond to subsequently grow tails that are twice as large as were grown by them prior to that chemical signal, to facilitate their escape from the dragonfly nymphs. See also SIGNALING, NITRIC OXIDE, G-PROTEINS, HORMONE, SUBSTANCE P, LEUKOTRIENES, ISOFLAVONES, SOYBEAN PLANT, *RHIZOBIUM* (BACTERIA), HARPIN, OCTADECANOID/JASMONATE SIGNAL COMPLEX, SALICYLIC ACID (SA).

**Signaling Protein** See SIGNALING MOLECULE.

**Silencing** See GENE SILENCING.

**Silent Mutation** A mutation in a gene that causes no detectable change in the biological characteristics of that gene's product (e.g., a protein). See also EXPRESS, GENE, PROTEIN.

**Silk** A natural, protein polymer with a predominance of alanine and glycine amino acids. Silk is produced by silkworms that have fed on mulberry tree leaves. The body of a silkworm can retain proteins (i.e., raw material for silk) amounting to as much as 20% of its body weight. It is thought that silk may be altered, via genetic engineering of silkworms, to produce fibers of very high strength. See also GENETIC ENGINEERING, PROTEIN ENGINEERING, AMINO ACID.

**Simple Protein** A protein that yields only amino acids on hydrolysis (i.e., cleavage of the protein molecule into fragments), and does not have other molecular constituents such as lipids or polysaccharide attachments. See also PROTEIN, AMINO ACID, GLYCOPROTEIN, LIPIDS, POLYSACCHARIDES.

**Simple Sequence Repeat (SSR) DNA Marker Technique** A "genetic mapping" technique which utilizes the fact that microsatellite sequences "repeat" (appear repeatedly in sequence within the DNA molecule) in a manner enabling them to be used as "markers." See also GENETIC MAP, SEQUENCE (OF A DNA MOLECULE), RANDOM AMPLIFIED POLYMORPHIC DNA (RAPD) TECHNIQUE, RESTRICTION FRAGMENT LENGTH POLYMORPHISM (RFLP) TECHNIQUE, DEOXYRIBONUCLEIC ACID (DNA), PHYSICAL MAP (OF GENOME), LINKAGE, LINKAGE GROUP, MARKER (GENETIC MARKER), LINKAGE MAP, TRAIT, MICROSATELLITE DNA, QUANTITATIVE TRAIT LOCI (QTL).

**Single-Cell Protein (SCP)** Protein derived from single-celled organisms with a high protein content. Yeast is an example. Generally used in regard to those organisms that are edible by domesticated animals or humans. Single-Domain Antibodies (dAbs) VH "heavy chains" (portion of antibody molecules) produced by genetically engineered *Escherichia coli* cells that act to bind antigens in a manner similar to antibodies or monoclonal antibodies (MAbs). Similar to MAbs, dAbs can be produced in large quantities, to be used as human or animal therapeutics (e.g., to combat diseases). See also

**S**

ANTIBODY, MONOCLONAL ANTIBODIES (MAb), ANTIGEN, ESCHERICHIA COLI.

**Single-nucleotide Polymorphisms (SNPs)** Variations (in individual nucleotides) that occur within DNA at the rate of approximately one in every 1,300 base pairs in most organisms (approximately one in every 100 base pairs in humans' DNA). SNPs usually occur in the same genomic location (e.g., on the organism's DNA) in different individuals. These variations account for:

- Diversity within a given species (e.g., black cattle and white cattle, different strains/serotypes within a given bacteria species, etc.)
- Some genetic diseases [e.g., the disease cystic fibrosis is due to one SNP, the disease known as familial dysautonomia is due to one SNP, the disease known as (Duchenne) muscular dystrophy is due to one SNP, etc.]
- The body's response to certain pharmaceuticals and food ingredients (e.g., the diuretic drug thiazide works to control hypertension in 60% of U.S. African Americans, but only 8% of U.S. Caucasian people, due to one SNP)

Certain pharmaceuticals do not have the desired effect in some groups of humans possessing certain specific "grouped SNPs" known as haplotypes. Because those "groupings of SNPs" are linked (i.e., tend to "travel together" as a group within the genetics of a given population), they can collectively confer a given "multiple-SNP-trait" to an identifiable subpopulation of individuals. For example, the pharmaceuticals acetaminophen, aspirin, and Valium remain in the bodies of women (who constitute a haplotype for that pharmacogenomic trait) longer than in men.

Methods utilized to identify SNPs include examination of the DNA of populations of individuals with and without a given (genetically related) disease and with and without a given trait. "SNP mapping" is a "genetic mapping" technique that utilizes the fact that individual nucleotides (within a DNA molecule) can exist in different forms (for a particular "site"/location on that DNA molecule), which enables such SNPs to be utilized as "markers." One example would be to track a given SNP vs. occurrence of genetically related disease in a given human population. See also POINT MUTATION, DEOXYRIBONUCLEIC ACID (DNA), SEQUENCE (OF A DNA MOLECULE), NUCLEOTIDE, POLYMORPHISM (GENETIC), GENETICS, GENETIC MAP, PHYSICAL MAP (OF GENOME), GENOME, TRAIT, MARKER (GENETIC MARKER), QUANTITATIVE TRAIT LOCI (QTL), DIVERSITY (WITHIN A SPECIES), BASE PAIR (bp), TRANSVERSION, CYSTIC FIBROSIS TRANSMEMBRANE REGULATOR PROTEIN (CFTR), MUSCULAR DYSTROPHY (MD), PHARMACOGENETICS, PHARMACOGENOMICS, HAPLOTYPE, SNP MAP, TOXICOGENETICS, ORGANISM.

**Site-Directed Mutagenesis (SDM)** A technique that can be used to make a protein that differs slightly in its structure from the protein normally produced (by an organism or cell). A single mutation (in the cell's DNA) is caused by hybridizing the region in a codon to be mutated with a short, synthetic oligonucleotide. This causes the codon to code for a different specific amino acid in the protein gene product. Site-directed mutagenesis holds the potential to enable man to create modified (engineered) proteins that have desirable properties not currently available in the proteins produced by existing organisms. See also MUTANT, MUTATION, POINT MUTATION, PROTEIN, GENE, INFORMATIONAL MOLECULES, HEREDITY, GENETIC CODE, GENETIC MAP, AMINO ACID, DEOXYRIBONUCLEIC ACID (DNA), CODON, OLIGONUCLEOTIDE, PROTEIN ENGINEERING.

**Sitostanol** A chemical (ester) derived from sitosterol (a sterol present in pine trees), and fibers (e.g., the hull or seed coat) of corn/maize (*Zea mays*) or soybeans (*Glycine max* L.). When sitostanol is consumed by humans in sufficient quantities, it causes their total serum cholesterol and their low-density lipoprotein (LDLP) levels to be lowered by approximately 10%, via inhibition (i.e., the sitostanol is preferentially absorbed by the gastrointestinal system instead of cholesterol). During 2000, the U.S. Food and Drug Administration approved a (label) health claim that associates consumption of

S

sitostanols with reduced blood cholesterol content and with reduced coronary heart disease (CHD). See also ABSORPTION, DIGESTION (WITHIN ORGANISMS), SOYBEAN PLANT, LOW-DENSITY LIPOPROTEINS (LDLP), SERUM LIFETIME, CHOLESTEROL, STEROLS, PHYTOSTEROLS, SITO-STEROL, CORONARY HEART DISEASE (CHD).

**Sitosterol** A phytosterol that is naturally produced in fibers within soybean (*Glycine max* L.) hulls, pumpkin seeds, pine trees, fibers of corn/maize (*Zea mays*) seed coats, etc. Sitosterol can exist in several different molecular forms (known as alpha a, beta b, etc.). A human diet containing large amounts of sitosterol and/or certain other phytosterols (campesterol, stigmasterol, etc.) has been shown to lower total serum (blood) cholesterol and low-density lipoprotein (LDLP) levels; and thereby lower the risk of coronary heart disease (CHD).

Evidence indicates that certain phytosterols (including sitosterol) interfere with absorption of cholesterol by the intestines, and decrease the body's recovery and reuse of cholesterol-containing bile salts, which causes more cholesterol to be excreted from the body than previously. During 2000, the U.S. Food and Drug Administration approved a (label) health claim that associates consumption of sitosterols with reduced blood cholesterol content and with reduced coronary heart disease (CHD). See also PHY-TOSTEROLS, SOYBEAN PLANT, CORN, STEROLS, SITOSTANOL, CAMPESTEROL, STIGMASTEROL, COR-ONARY HEART DISEASE (CHD), BETA-SITOSTEROL, CHOLESTEROL.

**SK** See SUBSTANCE K.

**Slime** An extracellular (i.e., outside of the cell) material produced by some (micro)organisms and characterized by a slimy consistency. The slime is of varied chemical composition. However, usual components are polysaccharides (polysugars) and specific protein molecules.

**Smut** See TELETHIA CONTROVERSIA KOON SMUT.

**SNP** See SINGLE-NUCLEOTIDE POLYMORPHISMS (SNPs).

**SNP MAP** A group of known/detailed SNPs (single-nucleotide polymorphisms), superimposed onto the genome map of an organism (e.g., to facilitate genetic/population studies, such as of genetically related disease susceptibility). See also SINGLE-NUCLEOTIDE POLY-MORPHISMS (SNPs), ORGANISM, GENOME, GENOMIC SCIENCES, MAPPING (OF GENOME), MAP DISTANCE, MARKER (DNA SEQUENCE).

**SNP MARKERS** See SINGLE-NUCLEOTIDE POLYMORPHISMS (SNPs).

**Sodium Dodecyl Sulfate** See SDS.

**Sodium Lauryl Sulfate** See SDS.

**Solanine** A glycoside neurotoxin naturally present at low levels within potatoes. As a result, solanine is present at detectable levels in the bloodstream of humans who consume potatoes. The U.S. Food and Drug Administration (FDA) prohibits the sale in the U.S. of potatoes which contain more (than a very low level of solanine); e.g., the naturally present level in potatoes can unfortunately increase in potatoes that are exposed to direct sunlight. See also TOXIN, PHYTOTOXIN, CHACONINE, GLYCOSIDE, WILD TYPE, FOOD AND DRUG ADMINISTRATION (FDA), TRADITIONAL BREEDING METHODS.

**Solid-Phase Synthesis** See SYNTHESIZING (OF PROTEINS), SYNTHESIZING (OF DNA MOLECULES).

**Soluble CD4** A synthetic version of the CD4 protein that may interfere with the ability of HIV (i.e., AIDS) viruses to infect human immune system cells with the acquired immune deficiency syndrome (AIDS) virus. See also CD4 PROTEIN, ADHESION MOLECULE, SELECTINS, LECTINS, PROTEIN.

**Soluble Fiber** See WATER SOLUBLE FIBER.

**Somaclonal Variation** The genetic variation (i.e., new traits) that results from the growing of entire new plants from plant cells or tissues (e.g., maintained in culture). Frequently encountered when plants are regenerated (grown) from plant cells that have been altered via genetic engineering. However, somaclonal variation (i.e., new genetic traits) can occur even when plants are regenerated from cells that were part of the same original plant. See also CELL CULTURE, SOMATIC VARI-ANTS, CLONE (AN ORGANISM), *AGROBACTERIUM TUMEFACIENS*, BIOLISTIC® GENE GUN, "EXPLOSION" METHOD, SHOTGUN METHOD.

**Somatacrin** See also GROWTH HORMONE-RELEASING FACTOR (GRF or GHRF).

S

**Somatic Cells** All eucaryote body cells except the gametes and the cells from which they develop. See also GAMETE, OOCYTES.

**Somatic Variants** Regenerated plants (i.e., clones) derived (produced) from cells that originally came from the same plant, but are not genetically identical. Such plants (clones) are called "sports" or somatic variants because they vary (genetically) from the "parent" plant. Sometimes such somatic variants are developed by man to become a new plant variety (e.g., the nectarine is an example of this). See also SOMACLONAL VARIATION, CELL CULTURE, CLONE (AN ORGANISM), GENOTYPE.

**Somatomedins** A family of peptides that mediates the action of growth hormone on skeletal tissue, and stimulates bone formation. See also HUMAN GROWTH HORMONE (HGH), PEPTIDE, BONE MORPHOGENETIC PROTEINS (BMP).

**Somatostatin** A 14 amino acid peptide that inhibits the release of growth hormone. See also HUMAN GROWTH HORMONE (HGH), GROWTH HORMONE-RELEASING FACTOR (GRF or GHRF), PEPTIDE.

**Somatotropin** Category of hormone that is produced naturally in the bodies of all mammals, including man. See also HORMONE, GROWTH HORMONE, BOVINE SOMATOTROPIN (BST), PORCINE SOMATOTROPIN (PST).

**SOS Protein** See SOS RESPONSE (IN *ESCHERICHIA COLI* BACTERIA).

**SOS Response (in *Escherichia coli* bacteria)** The "switching on" of genetic repair machinery in this bacteria when its DNA has been damaged (e.g., by radiation). See also *ESCHERICHIA COLIFORM* (*E. COLI*).

**Southern Blot Analysis** A test that is performed on biological samples such as plant DNA (e.g., to ascertain if "inserted" DNA is present in particular plant cells). Gel electrophoresis is used to separate the DNA fragments according to size, and then those fragments are transferred to a filter (blot). Radiolabeled DNA probes or RNA probes are added, and the ones which are complementary to each of the (separated, on blot) fragments will hybridize to those respective DNA fragments. The location (on the blot) and "radioactive label" of those hybridized probes can then be utilized to determine the nature of the DNA that was in those plant cells. See also DEOXYRIBONUCLEIC ACID (DNA), RIBONUCLEIC ACID (RNA), GENETIC ENGINEERING, ELECTROPHORESIS, TWO-DIMENSIONAL (2D) GEL ELECTROPHORESIS, POLYACRYAMIDE GEL ELECTROPHORESIS (PAGE), RADIOLABELED, DNA PROBE, COMPLEMENTARY (MOLECULAR GENETICS), HYBRIDIZATION (MOLECULAR GENETICS), RADIOIMMUNOASSAY.

**Southern Corn Rootworm** Latin name *Diabrotica undecimpunctata hawardii*. See also CORN ROOTWORM.

**Soy Protein** An edible protein (after heat processing) produced within its beans (seeds) by the soybean plant (botanical name *Glycine max* (L.) Merrill). When removed from soybeans via crushing, extrusion, or other process(es) involving adequate heat treatment, soy protein is (historical average) composed of 2.5% cysteine, 3.4% histidine, 5.2% isoleucine, 8.2% leucine, 6.8% lysine, 1.1% methionine, 5.6% phenylalanine, 4.2% threonine, 1.3% tryptophan, 4.2% tyrosine, 5.4% valine, 4% alanine, 7.7% arginine, 6.9% aspartic acid, 19% glutamic acid, 3.7% glycine, 0.1% 4-hydroxyproline, 5.3% proline, and 5.4% serine. Soy protein (concentrate) is a complete ("ideal") protein (i.e., it provides all essential amino acids) for humans. It is a good dietary source of calcium, with an absorption rate equivalent to milk.

In its initial form (i.e., following crushing/extrusion from soybeans as described above), the soy protein is known as soybean meal, and contains a bit less than half protein by weight. If the soy is washed with water (following crushing/extrusion) to remove soluble polysaccharides (e.g., the carbohydrates known as stachyose, raffinose, etc.), the resultant soy protein is known as soy protein concentrate and contains approximately 60% protein by weight. If the soy is washed with water-and-alkali solution, followed by isoelectric precipitation of the soluble protein, the result is "isolated soy protein" (ISP), often known as soy protein isolate or soy isolate. In 1999, the U.S. FDA approved a (label) health claim that associates consumption of soy protein with reduced blood cholesterol content and with reduced coronary heart disease (CHD) in humans. See also SOYBEAN PLANT, TRYPSIN INHIBITORS, PROTEIN, CHOLESTEROL, CORONARY HEART DISEASE (CHD), AMINO ACID, ESSENTIAL AMINO ACIDS, "IDEAL PROTEIN" CONCEPT, PROTEIN DIGESTIBILITY-CORRECTED AMINO ACID SCORING (PDCAAS), STACHYOSE.

**Soybean Aphid** An aphid (*Aphis glycines*) native to China, but accidentally introduced into the U.S. during the 1990s (apparently via aphid eggs adhering to an ornamental plant). It feeds on the sap of the soybean plant (*Glycine max* L.). See also SOYBEAN PLANT.

**Soybean Cyst Nematodes (SCN)** Microscopic roundworms (*Heterodera glycines*) living in the soil, which feed parasitically on roots of the soybean plant. The nematodes use a spear-like mouthpart, called a stylet, to puncture the plant's root cells so the nematodes can eat their cell contents. That root damage causes the soybean's growth to be stunted, and the plants turn yellow because of a reduction in nodule formation by the nitrogen-fixing *Rhizobium* bacteria (which normally colonize roots of soybean plants). SCN can combine with a fungus (*Fusarium solani*) to cause a soybean plant disease known as "sudden death syndrome."

As part of Integrated Pest Management (IPM), farmers can utilize naturally resistant soybean varieties (e.g., CystX) and/or the parasitic *Pasteuria* bacteria to help control the soybean cyst nematodes. The *Pasteuria* bacteria must attach their spores (for reproduction) to juvenile nematodes, so that the *Pasteuria* offspring can consume the SCN when the spores later germinate. See also SOYBEAN PLANT, NITROGEN FIXATION, BACTERIA, *RHIZOBIUM* (BACTERIA), FUNGUS, SUDDEN DEATH SYNDROME, ALLELOPATHY, ISOFLAVONES, NEMATODES, CystX.

**Soybean Meal** See SOYBEAN PLANT, SOY PROTEIN.

**Soybean Oil** An edible oil that is produced within its beans (seeds) by the soybean plant (botanical name *Glycine max* (L.) Merrill). When removed from soybeans via crushing and refining processes, soybean oil is (historical average) composed of 60.8% polyunsaturated fatty acids (PUFA), 24.5% monounsaturated fatty acids, and 15.1% saturated fatty acids. However, soybean vari-eties have recently been created that possess as little as 7% saturated fatty acids. See also POLYUNSATURATED FATTY ACIDS (PUFA), FATTY ACID, ESSENTIAL FATTY ACIDS, LECITHIN, HYDRO-GENATION, SOYBEAN PLANT, HIGH-OLEIC OIL SOY-BEANS, LOW-LINOLENIC OIL SOYBEANS, LINOLENIC ACID, OLEIC ACID, LINOLEIC ACID, MONOUNSATURATED FATTY ACIDS, SATURATED FATTY ACIDS, CONJUGATED LINOLEIC ACID.

**Soybean Plant** Botanical name *Glycine max* (L.) Merrill. A green, bushy legume that is the world's single largest provider of protein and edible oil for mankind's use. This summer annual plant varies in height from less than a foot (0.3 meter) to more than three feet (one meter) tall. The seeds (soybeans) are borne in pods, and historically have contained 13–26% oil and 38–45% protein (on a moisture-free basis). Its leaves contain some carotenoids. The soybean plant has approximately 80,000 genes. It is a self-pollinating plant (i.e., there are male and female reproductive structures on the same plant — so it is monoecious). Soybean oil contains a total of 327 mg/100 g of the plant sterols (phytosterols) campesterol, stigmasterol, and beta-sitosterol (β-sitosterol). Soybeans contain the highest amount of isoflavones of any plant (seeds) i.e., up to 0.3% of each soybean's dry weight. The traditional soybean, possessing (average) 20% oil content, contains an average of 3% stachyose within its meal (i.e., the solids remaining after the soybean oil is removed). See also FATTY ACID, PROTEIN, SOY PROTEIN, LECITHIN, NITROGEN FIX-ATION, NODULATION, SOYBEAN OIL, SOYBEAN CYST NEMATODES (SCN), RESVERATROL, BROWN STEM ROT, *PHYTOPHTHORA* ROOT ROT, PEROXI-DASE, ISOFLAVONES, LOW-STACHYOSE SOYBEANS, GENISTEIN (Gen), LIPOXYGENASE (LOX), SAPONINS, CANOLA, CHLOROPLAST TRANSIT PEPTIDE (CTP), HERBICIDE-TOLERANT CROP, LOX NULL, PHY-TOSTEROLS, PHYTOCHEMICALS, CORN ROOTWORM, NITRIC OXIDE (NO), MONOECIOUS, ALLELOPATHY, LOW-LINOLENIC OIL SOYBEANS, HIGH-OLEIC OIL SOYBEANS, HIGH-PHYTASE (SOYBEANS), HIGH-ISOFLAVONE SOYBEANS, HIGH-STEARATE SOY-BEANS, HIGH-SUCROSE SOYBEANS, WATER SOLU-BLE FIBER, LOW-PHYTATE SOYBEANS, *PHYTOPHTHORA MEGASPERMA* f. sp. *GLYCINEA*, STACHYOSE, PHYTOSTEROLS, CAMPESTEROL,

S

STIGMASTEROL, SITOSTEROL, BETA-SITOSTEROL (β-SITOSTEROL), SITOSTANOL, CORONARY HEART DISEASE (CHD), SOYBEAN APHID, CAROTENOIDS, TOCOPHEROLS, *RHIZOBIUM* (BACTERIA), PHARMACOENVIROGENETICS.

**SP** See SUBSTANCE P.

**Species** A single type (taxonomic group) or organism as determined by the distinguishing characteristics used for the particular group of life forms (e.g., the horse is one species among the mammals). While the horse is easily distinguished from other, obviously non-similar mammals, such as humans (e.g., due to the horse's four legs vs. the human's two legs and two arms), it is less easy to distinguish a horse from a more closely related animal such as a donkey or a zebra. The so-called "boundary between different species" is determined by human assessment/categorization (e.g., whether systematics or cladistics are utilized by those doing the species categorizations and definitions), and sometimes changes when more information becomes known at a later date (e.g., if new 2D electrophoresis tests reveal certain types to be genetically related or not). See also STRAIN, SYSTEMATICS, CLADISTICS, CONSERVED, DIVERSITY (within a species), ELECTROPHORESIS, TWO-DIMENSIONAL (2D) GEL ELECTROPHORESIS.

**Species Specific** Refers to a compound (e.g., a protein) or a disease (e.g., a viral infection) or some other effect that only acts in/on one specific species of organism. For example, the antibiotic penicillin kills bacteria by blocking an enzyme that is critical for growth and repair of the bacterial cell wall (i.e., peptidoglycan layer), but penicillin does not harm other species (e.g., man).

Bovine somatotropin is a protein hormone that increases the growth rate of young cattle and also increases the efficiency of mature cows in converting their feed into milk. Bovine somatotropin has no effect on humans, and (if eaten) is simply digested like any other food protein. It appears that most growth hormones are species specific. See also SPECIES, HORMONE, PENICILLIN G (BENZYLPENICILLIN).

**Specific Activity** An enzyme unit defined as the number of moles of substrate converted to product by an enzyme preparation per unit time under specified conditions of pH, substrate concentration, temperature, etc. Specific enzyme activity units may be expressed as: moles of product produced/minute/mg of protein used (or mole of enzyme used if the preparation is pure). See also MOLE, ENZYME, SUBSTRATE (CHEMICAL).

**Spectrophotometer** An instrument that measures the concentration of a compound that has been dissolved in a solvent (water, alcohol, etc.). The instrument shines a light through the solution, measures the fraction of the light that is absorbed by the solution, and calculates the concentration from that absorbance value. See also OPTICAL DENSITY (OD), ABSORBANCE (A).

**Splice Variants** Refers to all possible gene transcripts (e.g., arising from alternative splicing). See also TRANSCRIPTOME.

**Splicing** The removal of introns and joining of exons in RNA (e.g., genes). Thus, introns are spliced out, while exons are spliced together. See also EXON, INTRON, GENETIC ENGINEERING, RIBONUCLEIC ACID (RNA), CENTRAL DOGMA (NEW).

**Splicing Junctions** The sequences (in RNA molecules) of nucleotides immediately surrounding the exon-intron boundaries. See also EXON, INTRON, SPLICING, NUCLEOTIDE.

**Spontaneous Assembly** See SELF-ASSEMBLY.

**SPS** Acronym for the Sanitary and Phytosanitary Standards Agreement of the World Trade Organization (WTO), a multinational trading agreement that "sets the rules" governing international trade. Sanitary (i.e., human and animal) and phytosanitary (i.e., plant) standards are important in preventing the transfer of diseases from one nation to another via international trade. SPS standards are designed to protect animal, plant, and human life/health (within WTO member countries) from:

- Entry of pests (insects, weeds, etc.)
- Entry of disease-carrying organisms (e.g., European Corn Borer)
- Entry of disease-causing organisms (e.g., *Aspergillus flavus*)
- Toxins, contaminants, or disease-causing organisms in foods, beverages, or feedstuffs

WTO member nations are required to base their SPS standards as much as possible on existing (e.g., Codex Alimentarius, IPPC, and OIE) international sanitary/phytosanitary standards and practices. See also SANITARY AND PHYTOSANITARY (SPS) AGREEMENT, SANITARY AND PHYTOSANITARY (SPS) AGREEMENT, INTERNATIONAL PLANT PROTECTION CONVENTION (IPPC), INTERNATIONAL OFFICE OF EPIZOOTICS (OIE), CODEX ALIMENTARIUS COMMISSION, MAXIMUM RESIDUE LEVEL (MRL), WORLD TRADE ORGANIZATION (WTO), EUROPEAN CORN BORER (ECB), *ASPERGILLUS FLAVUS*.

**Squalamine** A potent antimicrobial agent (steroid, antibiotic) discovered in the tissues of the dogfish shark in 1992. It has been found to be active against a broad spectrum of bacteria, protozoa, and fungi. Squalamine was chemically synthesized by man in 1993. See also MAGAININS, STEROID, FUNGUS, BACTERIA, BACTERIOCINS, PROTOZOA, ANTIBIOTIC.

**Squalene** A sterol that is produced in some plants. See also STEROLS.

**SRB (sulfate reducing bacterium)** Any organism that metabolically reduces sulfate to $H_2S$ (hydrogen sulfide). This includes a variety of microorganisms. See also REDUCTION (IN A CHEMICAL REACTION), METABOLISM, MICROORGANISM, FERROBACTERIA.

**SSR** See SIMPLE SEQUENCE REPEAT (SSR) DNA MARKER TECHNIQUE.

**Stacchyose** See STACHYOSE.

**Stachyose** A carbohydrate (oligosaccharide) naturally produced in soybeans (and some other plants). Stachyose is relatively insoluble in water, and much less available for digestion by monogastric animals (e.g., swine, poultry) than the other carbohydrate components within soybeans. See also CARBOHYDRATES (SACCHARIDES), LOW STACHYOSE SOYBEANS, OLIGOSACCHARIDES, SOYBEAN PLANT.

**"Stacked" Genes** Refers to the insertion of two or more (synthetic) genes into the genome of an organism. One example would be of a plant into which a gene from *Bacillus thuringiensis* (*B.t.*) and a gene for resistance to a specific herbicide have been inserted. See also GENE, BIOTECHNOLOGY, GENETIC ENGINEERING, *BACILLUS THURINGIENSIS* (*B.t.*), *B.t. KURSTAKI*, GENETICALLY ENGINEERED MICROBIAL PESTICIDES (GEMP), EPSP SYNTHASE, PAT GENE, BAR GENE.

**Staggered Cuts** Scissions (cuts) made in duplex DNA when the two strands of DNA that make up the duplex DNA are cleaved at different points near each other by restriction endonucleases. What is produced is a single-stranded structure (in which the single strands are a number of nucleotide bases long) with a double-stranded core section. This core section is much longer than the single-stranded region. See also DEOXYRIBONUCLEIC ACID (DNA), RESTRICTION ENDONUCLEASES, STICKY ENDS.

**Stanol Ester** See SITOSTANOL.

**Stanol Fatty Acid Esters** See SITOSTANOL, FATTY ACID.

**Starch** A polymer of glucose molecules (i.e., a polysaccharide) used by plants to store energy. Plants produce starch in two different molecular forms, amylopectin and amylose. For example, the starch content in traditional corn (maize) kernels averages 72–76% amylopectin and 24–28% amylose. Starch is broken down by enzymes (amylases) to yield glucose, which can be used as an energy source. The analogous polymer used by mammalian systems is called glycogen or, in old usage, "animal starch." See also GLUCOSE (GLc), ENZYME, AMYLASE, CORN, AMYLOSE, AMYLOPECTIN.

**Startpoint** Refers to the position on a DNA molecule corresponding to the first base incorporated into mRNA. See also DEOXYRIBONUCLEIC ACID (DNA), MESSENGER RNA (mRNA), EXON, RIBONUCLEIC ACID (RNA).

**Stearate (stearic acid)** A saturated fatty acid, containing 18 carbon atoms in its molecular "backbone," that is essentially neutral in effect on coronary heart disease in humans (i.e., doesn't appreciably increase low-density lipoproteins in the bloodstream). Because of the heart disease neutrality, stearate-containing oils (e.g., high-stearate soybean oil) are an acceptable cooking oil choice, with the resistance to oxidation/breakdown of a saturated fatty acid, but no bloodstream-cholesterol increasing effect.

In the mid-1990s, the American Cocoa Research Institute/Chocolate Manufacturers Association filed a petition with the U.S. Food and Drug Administration (FDA) to

**S**

differentiate stearate (on food product labels) from the other saturated long-chain fatty acids used as food ingredients.

In order to make milk, dairy cows require more stearic acid than a conventional digestive system alone could provide from the cow's (mainly carbohydrate) diet. Therefore, cows utilize microorganisms living in their rumen (a special sort of pre-stomach) to convert carbohydrate (grass) to stearic acid. Thus, high-performance dairy cows might benefit from a diet that contains high-stearate soybeans, if their milk output is limited by dietary stearate availability. See also FATTY ACID, LOW-DENSITY LIPOPROTEINS (LDLP), SATURATED FATTY ACIDS, FOOD AND DRUG ADMINISTRATION (FDA), HIGH-STEARATE SOYBEANS, FATS, ENOYL-ACYL PROTEIN REDUCTASE, HIGH-STEARATE CANOLA.

**Stearic Acid** See STEARATE.

**Stearoyl-ACP Desaturase** A "family" of enzymes that is naturally produced in oilseed plants. They play the central role in determining the ratio of saturated to unsaturated fatty acids (in the vegetable oils produced from such plants). See also FATS, FATTY ACID, ENZYME, GENETIC ENGINEERING, GENETIC CODE, LAURATE, HIGH-STEARATE SOYBEANS, HIGH-STEARATE CANOLA.

**Stem Cell Growth Factor (SCF)** A growth factor (glycoprotein hormone) that acts upon stem cells in a wide variety of ways to increase growth, proliferation, and maturity (into red blood cells or white blood cells). See also STEM CELLS, GROWTH FACTOR, HORMONE, GLYCOPROTEIN, DIFFERENTIATION, TOTIPOTENT STEM CELLS, COLONY STIMULATING FACTORS (CSFs).

**Stem Cell One** The single stem cell in the bone marrow of a fetus from which every immune system cell in the adult is subsequently derived. The primordial stem cell is stimulated to develop into the mature immune system's differentiated, specialized cells by interleukin-7. See also STEM CELLS, TOTIPOTENT STEM CELLS, INTERLEUKIN-7 (IL-7), EMBRYONIC STEM CELLS, DIFFERENTIATION.

**Stem Cells** Certain cells — present in the bodies of mammals even prior to birth, although also present in adult mammals — that can grow/differentiate into different cells/tissues of the (adult organism) body. For example, bone marrow (stem) cells, some of which eventually mature into red blood cells or white blood cells. The stem cells that remain in the bone marrow maintain their own numbers by self-renewal divisions, yielding more cells to start the maturation process. This maturation process is stimulated and controlled by stem cell growth factor (SCF), granulocyte colony stimulating factor (G-CSF), and by granulocyte-macrophage colony stimulating factor (GM-CSF).

During 2000, research by Richard Childs showed that stem cells (i.e., collected from a sibling's bloodsteam) transplanted into a patient suffering from kidney cancer could induce generation of a "new" immune system which could help stop/reverse the kidney cancer. See also CELL, MULTIPOTENT ADULT STEM CELLS, ECTODERMAL ADULT STEM CELLS, ENDODERMAL ADULT STEM CELLS, MESODERMAL ADULT STEM CELLS, HEMATOPOIETIC STEM CELLS, RED BLOOD CELLS, WHITE BLOOD CELLS, BASOPHILS, STEM CELL ONE, STEM CELL GROWTH FACTOR (SCF), TOTIPOTENT STEM CELLS, TOTIPOTENCY, EMBRYONIC STEM CELLS, DIFFERENTIATION, IMMUNE RESPONSE, CANCER.

**Stereoisomers** Molecules that have the same structural formula but different spatial arrangements of dissimilar groups (of atoms) bonded to a common atom (in the molecule). Many of the physical and chemical properties of stereoisomers are the same, but there are differences in the crystal structures, in the direction in which they rotate polarized light (which has been passed through a solution of the stereoisomer), and in their use in an enzyme-catalyzed (biological) reaction. See also RACEMATE, POLARIMETER, DEXTROROTARY (D) ISOMER, EPIMERS, ISOMER, LEVOROTARY (L) ISOMER, ISOMERASE, DIASTEREOISOMERS.

**Steric Hindrance** Refers to the compression that a group (chemical entity) suffers by being too close to its nonbonded neighbors. If an enzyme and a substrate try to come together in order to react, but the substrate has on it a bulky group that disallows close contact between the two (because the group bumps into the enzyme), then the reaction will not occur because of steric hindrance.

Seen another way, two chemical groups bump into each other and cannot get by each other because they are held in place by the bonds binding them to other atoms. Hindrance of movement or activity occurs because chemical groups bump into each other and cannot occupy the same space. See also REPRESSION (OF AN ENZYME), INHIBITION, COREPRESSOR.

**Sterile (environment)** One that is free of any living organisms or spores. For example, a hypodermic needle that has been sterilized (e.g., by heating it) and is free of living microorganisms is said to be sterile.

**Sterile (organism)** One that is unable to reproduce. For example, a bull that has been castrated is rendered sterile. See also TRIPLOID, BARNASE.

**Sterilization** See STERILE (ENVIRONMENT), STERILE (ORGANISM).

**Steroid** A chemical compound composed of a series of four carbon rings joined together to form a (molecular) structural unit called cyclopentanoperhydrophenanthrene. Any of a group of naturally occurring, fat-soluble substances essential to life, usually classed as lipids.

Steroids of importance to the body are the sterols, which are bile acids (produced by the liver, characterized by the presence of a carboxyl group in the molecule's side chain), and the hormones of the sex glands and the adrenal cortex. In addition, the plant kingdom possesses a wide variety of steroid glycosides. See also GLYCOSIDE, LIPIDS, HORMONE, CHOLESTEROL, STEROLS, SAPONINS.

**Sterols** Solid alcohols consisting of ring-structured molecules (i.e., a 'ring' made of atoms). Evidence suggests that human consumption of certain phytosterols (i.e., sterols produced in plant seeds) can help to prevent certain types of cancers, and can help lower levels of total blood serum cholesterol and low-density lipoproteins (LDLP); thereby reducing risk of coronary heart disease (CHD). Evidence indicates that those phytosterols interfere with absorption of cholesterol by the intestines, and they decrease the body's recovery and reuse of cholesterol-containing bile salts, which causes more cholesterol to be excreted from the body.

During 2000, researcher Joseph Judd fed phytosterols extracted from soybeans (*Glycine max* L.) to human volunteers who were already consuming a "low fat" diet. Their total blood serum cholesterol and low-density lipoprotein (LDLP) levels decreased by more than 10%, in a short time. During 2001, the U.S. FDA approved a (label) health claim that associates the consumption of plant sterols with reduced blood cholesterol content, and with reduced coronary heart disease (CHD). Some of the sterols known to impart health benefits when consumed by humans include β-sitosterol (beta-sitosterol) and squalene. See also PHYTOSTEROLS, STEROID, CHOLESTEROL, BILE, SITOSTANOL, SOYBEAN PLANT, CAMPESTEROL, STIGMASTEROL, BETA-SITOSTEROL, CORONARY HEART DISEASE (CHD), LOW-DENSITY LIPOPROTEINS (LDLP), FOOD AND DRUG ADMINISTRATION (FDA).

**Sticky Ends** Complementary single strands of DNA (deoxyribonucleic acid) that protrude from opposite ends of a DNA duplex or from ends of different DNA duplex molecules. They can be generated by staggered cuts in DNA. They are called "sticky" because the exposed single strands can bind (stick) to complementary single strands on another DNA molecule. A hybrid piece of DNA is hence produced (by that binding). See also STAGGERED CUTS, HYBRIDIZATION (MOLECULAR GENETICS), DUPLEX, ANNEAL, DEOXYRIBONUCLEIC ACID (DNA), BLUNT-END LIGATION, RESTRICTION ENDONUCLEASES.

**Stigmasterol** A phytosterol produced within the seeds of the soybean plant (*Glycine max* L.), among others. Evidence indicates that human consumption of stigmasterol helps reduce levels of total serum cholesterol and low-density lipoproteins (LDLP); thereby lowering risk of coronary heart disease (CHD). Evidence indicates that certain phytosterols (including stigmasterol) interfere with absorption of cholesterol by the intestines, and decrease the body's recovery and reuse of cholesterol-containing bile salts; which causes more cholesterol to be excreted from the body. See also PHYTOSTEROLS, PHYTOCHEMICALS, STEROLS, SOYBEAN PLANT, CHOLESTEROL, CAMPESTEROL, BETA-SITOSTEROL, CORONARY HEART DISEASE (CHD).

S

**Stomatal Pores** See GPA1, ABSCISIC ACID.

**Strain** A group or organisms of the same species that possess(es) distinctive genetic characteristics that set it apart from others within the same species, but whose differences are not "severe" enough for it to be considered a different breed or variety (of that species). The basic taxonomic unit of microbiology. Can also be used to designate a population of cells derived from a single cell. See also SPECIES, CELL, CLONE (AN ORGANISM).

*Streptococcus* Refers to bacteria of the genus *Streptococcus*. See also BACTERIA, GENUS, STREPTOCOCCUS MUTANS.

*Streptococcus mutans* The strain of *Streptococcus* bacteria that grows on the surface of teeth and can contribute to causing tooth "decay." See also STRAIN, BACTERIA, STREPTOCOCCUS.

**Stress Proteins** Discovered by Italian biologist Ferruchio Ritossa in the 1960s, these molecules are also called heat-shock proteins. Proteins made by many organisms' (plant, bacteria, and mammal) cells when those cells are stressed by environmental conditions such as certain chemicals, pathogens, or heat.

When corn/maize (*Zea mays* L.) is stressed during its growing season by high nighttime temperatures, that plant switches from its normal production of ("immune system" defense) chitinase to production of heat-shock (i.e., stress) proteins, instead.

Stress proteins are also produced by tuberculosis and leprosy bacteria after these bacteria have invaded (infected) cells in the human body, in an attempt by those bacteria to mimic the stress proteins that (mammal) cells would normally manufacture to repair damage done to the (mammal) cells. This mimicry makes it more difficult for the immune system to recognize and attack those pathogenic bacteria (and/or repair misshaped protein molecules in the body's cells). Similarly, production of stress proteins helps some types of cancer cells to avoid being attacked by the immune system. Because consumption of genistein by humans causes a reduction in the production of stress proteins, genistein may thereby help the human immune system destroy cancerous cells. In 1996, Richard I. Morimoto discovered that two stress proteins known as HSP 90 and HSP 70 help ensure that certain crucial proteins in cells are folded into the configuration/conformation needed by that cell. See also ANTIGEN, IMMUNE RESPONSE, PATHOGEN, PROTEIN, PROTEIN FOLDING, CONFORMATION, CHAPERONES, PROTEIN STRUCTURE, ABSOLUTE CONFIGURATION, PRION, CHITINASE, AFLATOXIN, GENISTEIN, CANCER, LIPOXYGENASE (LOX), PHYTOALEXINS.

**Stromelysin (MMP-3)** A collagenase (enzyme) that "clears a path" through living tissue, ahead of tumor cells, thereby enabling a cancer to spread within the body. See also COLLAGENASE, ENZYME, CANCER, TUMOR.

**Structural Biology** See STRUCTURAL GENE.

**Structural Gene** A gene that codes for any RNA (ribonucleic acid) or protein product other than a regulator molecule. It determines the primary sequences (i.e., the amino acid sequences) of a polypeptide (protein). See also GENE, EXPRESS, POLYPEPTIDE (PROTEIN), AMINO ACID, PRIMARY STRUCTURE, RIBONUCLEIC ACID (RNA).

**Structural Genomics** Study of, or discovery of, where (gene) sequences are located within the genome, and what (DNA) subunits comprise those sequences. See also GENE, SEQUENCE (OF A DNA MOLECULE), DEOXYRIBONUCLEIC ACID (DNA), SEQUENCING (OF DNA MOLECULES), GENOME, GENOMICS, PRIMARY STRUCTURE.

**STS Sulfonylurea (Herbicide)-Tolerant Soybeans** These are soybeans that have been bred (via insertion of ALS gene by traditional breeding methods) to resist the (weed killing) effects of sulfonylurea-based herbicides. The ALS gene was discovered by Scott Sebastian in 1986. See also GENE, GENETIC ENGINEERING, HTC, ALS, ALS GENE, BAR GENE, PAT GENE, EPSP SYNTHASE, GLYPHOSATE OXIDASE, HERBICIDE-TOLERANT CROP.

**Stx** Shiga-like toxins. See also TOXIN, TOXIGENIC *E. COLI*, ENTEROHEMORRHAGIC *E. COLI*, *ESCHERICHIA COLIFORM* 0157:H7 (*E. COLI* 0157:H7).

**Substance K** See TACHYKININS.

**Substance P** A neuropeptide (i.e., peptide produced by cells of the nervous system) which is involved in activation of the immune system, pain sensation, and (when in excess) some psychiatric disorders. In the case of chronic, intractable pain (hypersensitivity), approximately 1% of the nerve cells in the human spine processes substance P (thereby "transmitting" its pain message via signal transduction). In 1997, Patrick Mantyh showed that killing those (1%) cells relieved chronic pain hypersensitivity without impairing sense of touch or normal (beneficial) pain sensation, in humans. See also TACHYKININS, PROTEIN, POLYPEPTIDE (PROTEIN), SIGNAL TRANSDUCTION, SIGNALING, PEPTIDE, NEUROTRANSMITTER.

**Substantial Equivalence** See CANOLA, ORGANIZATION FOR ECONOMIC COOPERATION AND DEVELOPMENT (OECD).

**Substantially Equivalent** See SUBSTANTIAL EQUIVALENCE.

**Substrate (chemical)** The substance acted upon by an enzyme. For example, the enzyme amylase catalyzes the breakdown of starch molecules into glucose polysaccharide molecules; starch is the substrate (of the enzyme amylase). See also ENZYME, AMYLASE, CATALYST, SUBSTRATE (STRUCTURAL).

**Substrate (in chromatography)** The (usually solid or gel) substance that attracts and non-covalently binds (interacts) with one or more of the molecules in a solution that is passed over that substrate (e.g., in a chromatography column). This preferential binding (interaction with the substrate) enables one or more of the solution's molecular ingredients to be separated from the other(s). See also CHROMATOGRAPHY.

**Substrate (structural)** The substance (support) to which the agent of interest (a molecule) is attached. For example, some catalyst molecules are chemically attached to nonreactive solids to preserve the catalyst from being flushed away when the chemical substrate (the molecule to be converted by the catalyst) is washed by the catalyst immobilized on the structural substrate. See also SUBSTRATE (CHEMICAL), CATALYST, HYBRIDIZATION SURFACES.

**Sudden Death Syndrome** A plant disease caused by the *Fusarium solani* fungus, that sometimes afflicts soybean plants. See also SOYBEAN PLANT, SOYBEAN CYST NEMATODES (SCN).

**Sugar Molecules** See OLIGOSACCHARIDES, POLYSACCHARIDES, MONOSACCHARIDES, CARBOHYDRATES, ALDOSE, GLYCOBIOLOGY, PYRANOSE, GLUCOSE (GLc), FURANOSE, GLYCOPROTEIN.

**Suicide Genes** See GENE, p53 GENE, APOPTOSIS.

**Sulfate Reducing Bacterium** See SRB (SULFATE REDUCING BACTERIUM).

**Sulforaphane** A compound naturally produced within cruciferous plants such as broccoli and cabbage. Research indicates that human consumption of significant amounts of sulforaphane helps lower the risk of several cancers. See also NUTRACEUTICALS, PHYTOCHEMICALS, CANCER.

**Sulfosate** An active ingredient in some herbicides, it kills plants (e.g., weeds) by inhibiting the crucial plant enzyme EPSP Synthase. Chemically, sulfosate is a trimethylsulfonium salt of the same organic acid as glyphosate, so sulfoste can be applied over crops (e.g., soybeans) that have been genetically engineered to be tolerant to glyphosate-based herbicides. See also ENZYME, EPSP SYNTHASE, CP4 EPSPS, GLYPHOSATE, ACID, SOYBEAN PLANT, HERBICIDE-TOLERANT CROP, GENETIC ENGINEERING.

**Superantigens** Certain types of antigens that activate a large proportion of an organism's immune system T cells. These superantigens, which thus overactivate the organism's immune system, are thought to be responsible for some autoimmune diseases (in which T cells attack and destroy the organism's own, healthy tissues). See also ANTIGEN, T CELLS, AUTOIMMUNE DISEASE.

**Supercoiling** Also known as superhelicity. The coiling of a closed duplex DNA (deoxyribonucleic acid molecule) in space so that it crosses over its own axis. See also DEOXYRIBONUCLEIC ACID (DNA), HELIX, DUPLEX, DOUBLE HELIX, POSITIVE SUPERCOILING.

**Supercritical Carbon Dioxide** A solvent that, when combined with water and an appropriate surfactant (e.g., fluoroethers), forms a solvent system that can effectively dissolve large biological molecules without causing those molecules to lose biological activity. Carbon dioxide is a gas at normal (atmospheric) pressure and ambient temperature, but in its supercritical state — temperature above 31.3°C (88°F) and pressure greater than 72.9 atmospheres — carbon dioxide becomes a dense (sort of) liquid. Some coffee processors have used supercritical carbon dioxide as a solvent to remove caffeine from coffee.

In 1995, Keith Johnston added the surfactant ammonium carboxylate perfluoropolyether to a supercritical carbon dioxide system containing water and proved that the large biological molecule bovine serum albumin dissolved inside the micelles that form via water droplet surrounded by fluoroether molecules. Subsequent to that, Eric Beckman proved that the protease subtilisin Carlsberg can be extracted from crude (impure) cell broth because that protease preferentially dissolves in a supercritical carbon dioxide/water system containing fluoroether amphiphiles as surfactants. See also BIOLOGICAL ACTIVITY, SURFACTANT, MICELLE, REVERSE MICELLE (RM), BROTH, PROTEASE, SUPERCRITICAL FLUID, ALBUMIN.

**Supercritical Fluid** Refers to a material that has been heated to a temperature above its (normal atmospheric pressure) boiling point, but which is kept in a state that resembles a liquid via the application of high pressure. Less commonly, refers to a liquid that has been cooled to a temperature below its normal freezing point, but which is kept in a liquid state by various means. For example, water will remain "liquid" up to a temperature of 375°C (617°F) if it is placed under enough pressure. Ammonia will remain "liquid" up to a temperature of 133°C (271°F) if it is placed under enough pressure, despite the fact that ammonia normally becomes a gas (at standard atmospheric pressure) whenever the temperature is higher than −33.35°C (−30°F).

One predatory mite (*Alaskozetes antarcticus*) living in Antarctica is able to survive subfreezing temperatures by preventing ice crystals from forming (i.e., supercritical water) inside its body, even when the environmental temperature is below the freezing point (i.e., supercritical). Most supercritical fluids have unique physical properties (e.g., they are often better solvents than their true liquid forms). Some supercritical fluids (e.g., supercritical carbon dioxide) can be used to

extract biological molecules (e.g., chlorophyll) from mixtures (e.g., ground-up plant leaves). After the biological molecule has dissolved out of the mixture, the biological molecule is recovered by releasing pressure so the carbon dioxide returns to gaseous form, and drifts away. See also SUPERCRITICAL CARBON DIOXIDE.

**Superoxide Dismutase (SOD)** See HUMAN SUPEROXIDE DISMUTASE (hSOD).

**Suppressor Gene** A gene that can reverse the effect of a specific type of mutation in other genes, such as a premature termination sequence. See also GENE, TRANSWITCH®.

**Suppressor Mutation** A mutation that totally or partially restores a function lost by a primary mutation. It is located at a site in the gene different from the site of the primary mutation. See also GENE.

**Suppressor T Cells** Those T cells (thymus-derived lymphocytes) that are triggered (after other types of T cells and other immune system cells have successfully fought off an infection) to slow down gradually and halt the body's immune response (to the now-conquered pathogen). Discovered by Tomio Tada in 1971, suppressor T cells inhibit B cell activity. Failure to halt the immune response in time could lead to harm to the body by its own immune system. The B and T lymphocytes are indistinguishable in size and general morphology. Only the existence or nonexistence of certain proteins on their cell surfaces distinguishes the two classes of lymphocytes. See also CELLULAR IMMUNE RESPONSE, PATHOGEN, B LYMPHOCYTES, T CELLS, AUTOIMMUNE DISEASE.

**Supramolecular Assembly** Refers to a very large molecular structure. See also SELF-ASSEMBLY (OF A LARGE MOLECULAR STRUCTURE).

**Surfactant** Acronym for surface active agent. Amphipathic molecules (i.e., molecules that contain both a polar and nonpolar domain) which, due to their unique properties, position themselves at interfacial regions (surfaces) such as an oil/water interface. When surfactants are dissolved above a certain critical concentration in either water or nonpolar solvents, they may form micelles or reverse micelles, respectively. Surfactants are commonly used to solubilize cell membrane components and other hard-to-solubilize molecules. See also AMPHIPATHIC MOLECULES, AMPHIPHILIC MOLECULES, MICELLE, REVERSE MICELLE (RM), SDS, ADJUVANT (TO A HERBICIDE).

**Sustainable Development** Defined in the 1987 United Nations report "our common future" to be development (e.g., economic development) that meets the needs of the present without compromising the ability of future generations to meet their own needs. See also CONSERVATION TILLAGE, GLOMALIN, NO-TILLAGE CROP PRODUCTION, LOW-TILLAGE CROP PRODUCTION, EARTHWORMS.

**Switch Proteins** Refers to certain protein molecules that signal a plant when environmental conditions are so dry (or cold, etc.) that the plant needs to protect itself (via extreme measures) to survive. See also TREHALOSE, PROTEIN, SIGNALING, TRANSCRIPTION FACTORS, CBF1, SEQUENCE (OF A DNA MOLECULE), REGULATORY SEQUENCE.

**Switching (e.g., on/off) of Genes** See GENE, GENETIC CODE, CODING SEQUENCE, DEOXYRIBONUCLEIC ACID (DNA), SEQUENCE (OF A DNA MOLECULE), REGULATORY SEQUENCE, TRANSCRIPTION FACTORS, CBF1, COLD HARDENING, CESSATION CASSETTE.

**Syk Protein** See MAST CELLS.

**Symbiotic** Refers to the mutually beneficial living together of organisms, in an intimate association or union. For example, lichen are a life form consisting of algae and a fungus growing together as a unit on a solid surface (e.g., a tree trunk or a rock). Each helps the other to survive and grow. See also ALGAE, FUNGUS, *RHIZOBIUM* (BACTERIA), PHARMACOENVIROGENETICS, ANTIBIOSIS.

**Synthase** See ACC SYNTHASE, EPSP SYNTHASE, ENZYME, CP4 EPSPS, CITRATE SYNTHASE (CSb) GENE, GLUTAMINE SYNTHETASE, ALS GENE.

**Synthesizing (of DNA molecules)** The building (i.e., polymerization manufacture) of a known sequence of nucleotides into a chain called an oligonucleotide (of which genes are made) or DNA (deoxyribonucleic acid). Invented by Har Goribind Khorana and his colleagues at the University of Wisconsin, Madison, in 1968, this process enables scientists to create genes or gene fragments for use in research. In 1973, Robert Bruce

Merrifield developed a means to partially automate the oligonucleotide assembly process. This led to automated machines that can now rapidly manufacture a gene fragment, gene, or DNA probe. See also GENE MACHINE, NUCLEOTIDE, OLIGOMER, OLIGONUCLEOTIDE, SYNTHESIZING (OF PROTEINS), DEOXYRIBONUCLEIC ACID (DNA), DNA PROBE, SYNTHESIZING (OF OLIGOSACCHARIDES).

**Synthesizing (of oligosaccharides)** Chemical synthesis (manufacture) of a known oligosaccharide (structure). For example, a synthesis of a defined-sequence oligosaccharide (molecular) "branch" at a specific site on a glycoprotein in order to "cover up" an antigenic site on that glycoprotein molecule (e.g., so the glycoprotein can be used as a pharmaceutical). See also OLIGOSACCHARIDES, GLYCOPROTEIN, ANTIGEN, ANTIGENIC DETERMINANT, RESTRICTION ENDOGLYCOSIDASES.

**Synthesizing (of proteins)** Chemical synthesis (manufacture) of a known protein molecule. Devised based upon the solid phase synthesis methodology developed by Robert Bruce Merrifield in 1963, the desired proteins are assembled by repetitive coupling of the constituent amino acids to a growing polypeptide backbone, which itself is attached to a polymeric support (substrate). This procedure has been automated, so it is now possible to make proteins via automated synthesizers. See also PROTEIN, POLYPEPTIDE (PROTEIN), AMINO ACID, SUBSTRATE (STRUCTURAL), COMBINATORIAL CHEMISTRY, SYNTHESIZING (OF DNA MOLECULES).

**Synthetase** See SYNTHASE.

**Systematic Activated Resistance** See SYSTEMIC ACQUIRED RESISTANCE (SAR).

**Systematics** An extension of taxonomy, it is the scientific classification of living organisms.

**Systemic Acquired Resistance (SAR)** Discovered in 1992 (applicable to harpininduced SAR) and in 1996 by J.A. Ryals, U.H. Neuenschwander, M.G. Willits, A. Molina, H.-Y. Steiner, and M.D. Hunt, SAR is a sort of "immune (cascade) response" by a plant, against infection (by bacteria, fungus, etc.). One example of this is the production of stress proteins or pathogenesis-related proteins when certain plants are attacked by certain pathogens. Via such SAR

**S**

response triggered by low-level fungal or viral infection, many plants successfully resist fungal/bacterial/viral attacks.

In 1998, the U.S. Environmental Protection Agency (EPA) approved one herbicide (COBRAR owned by Valent Corp.), whose active ingredient is the chemical lactofin, to be applied to soybean plants "at or near bloom stage" in order to trigger SAR against white mold disease. In 2000, the U.S. EPA approved harpin protein to be applied to some crops in order to trigger SAR against certain plant diseases. See also PATHOGENESIS RELATED PROTEINS, PHYTOALEXINS, R GENES, ISOFLAVONES, SOYBEAN PLANT, FUNGUS, IMMUNE RESPONSE, VIRUS, PATHOGEN, STRESS PROTEINS, SALICYLIC ACID (SA), JASMONIC ACID, HARPIN, CASCADE, WHITE MOLD DISEASE.

**Systemic Inflammatory Response Syndrome**
See SEPSIS.

S

# T

**T Cell Growth Factor (TCGF)** Also known as Interleukin-2. See also INTERLEUKIN-2 (IL-2).

**T Cell Modulating Peptide (TCMP)** A short protein chain that is thought to restrain certain types of T cells from attacking an (arthritis) afflicted patient's tissues (mainly cartilage). Arthritis is caused by the sufferer's own immune system attacking the body's cartilage tissues. See also CYTOTOXIC T CELLS, HELPER T CELLS (T4 CELLS), LYMPHOCYTE, SUPPRESSOR T CELLS, T CELL RECEPTORS, AUTOIMMUNE DISEASE, TUMOR NECROSIS FACTOR (TNF).

**T Cell Receptors** Antibody-like transmembrane (i.e., across the cell's surface membrane) proteins located on the surface of T cells. These trigger the (cellular) immune response that is mounted by T cells when these receptors bind to antigens (foreign pieces of antigenic protein) that have been "presented" to these receptors by an MHC protein which itself is located on the surface of phagocytic (i.e., scavenging, pathogen-ingesting) B lymphocyte. Antibodies in the blood recognize native antigen macromolecules (large molecules), whereas T cell receptors recognize fragments derived from those antigen macromolecules (upon presentation at the surface of B lymphocytes following ingestion and digestion by the B lymphocytes). See also ANTIBODY, ANTIGEN, MAJOR HISTOCOMPATIBILITY COMPLEX (MHC), PROTEIN, T CELLS, CELLULAR IMMUNE RESPONSE, PHAGOCYTE, B LYMPHOCYTES, CYTOTOXIC T CELLS, HELPER T CELLS, SUPPRESSOR T CELLS.

**T Cells** A class of (thymus-derived) lymphocytes that include helper T cells (also known as T helper cells or $T_H$ cells), suppressor T cells, and cytotoxic T cells (also known as killer cells or CTL for cytotoxic T lymphocyte). These cells mediate (i.e., control/direct) the cellular response of the human immune system in very complex ways. T cells are involved in the activation of B cells. See also CELLULAR IMMUNE RESPONSE, CYTOTOXIC T CELLS, HELPER T CELLS (T4 CELLS), LYMPHOCYTE, SUPPRESSOR T CELLS, T CELL RECEPTORS, T CELL MODULATING PEPTIDE (TCMP), ALLERGIES (FOODBORNE), DENDRITIC CELLS, LEUKOTRIENES.

**T Lymphocytes** See T CELLS, LYMPHOCYTE, LYMPHOKINES, THYMUS.

**T-DNA** See Ti PLASMID.

**t-IND** Treatment Investigational New Drug Application to the U.S. Food and Drug Administration (FDA). See also "TREATMENT" IND REGULATIONS.

**t-IND Treatment** Investigational New Drug Application to the U.S. Food and Drug Administration (FDA). See also "TREATMENT" IND REGULATIONS.

**T3** See SAM-K GENE.

**T4 Cells** See HELPER T CELLS (T4 CELLS).

**Tachykinins** A class of neuropeptides (i.e., peptides produced by cells of the nervous system; neurons) that includes neurokinin A, neurokinin B, eledoisin, physalaemin, kassinin, substance P, and substance K. Some of these neuropeptides (e.g., Substance P) are picked up by mast cells, lymphocytes, and/or monocytes; and cause those three types of immune system cells to release certain lymphokines (e.g., tumor necrosis factor, interleukin-1 etc.), thus activating the immune system. See also MAST CELLS, LYMPHOCYTE, MONOCYTES, TUMOR NECROSIS FACTOR (TNF), INTERLEUKIN-1 (IL-1).

**TAG** See TRIACYLGLYCEROLS.

***Taq* DNA Polymerase** A 94 kilodalton DNA polymerase, which was originally isolated from the thermophilic bacteria *Thermus aquaticus*. Commonly utilized to catalyze PCR reactions due to its heat resistance (needed for thermal cycles utilized in the PCR technique). See also DNA POLYMERASE, POLYMERASE, KILODALTON (Kd), DEOXYRIBO-NUCLEIC ACID (DNA), BACTERIA, THERMOPHILIC BACTERIA, PCR, POLYMERASE CHAIN REACTION (PCR) TECHNIQUE.

**Target (of a herbicide or insecticide)** The molecule (receptor, enzyme, etc.) within a weed plant or within a pest insect that a given herbicide or insecticide is "aimed" at (e.g., when scientists are conducting research aimed at creating that herbicide or insecticide). For example, glyphosate-containing herbicides act on the (target) crucial plant enzyme EPSP synthase. For example, insect-resistant transgenic plants containing "*B.t.* gene(s)" act on (target) receptors inside the digestive system of specific insect species via the *B.t.* protoxin. See also RECEPTORS, ENZYME, GLYPHOSATE, EPSP SYNTHASE, TRANSGENIC (ORGANISM), PROTOXIN, HERBICIDE-TOLERANT CROP, PAT GENE, GLUTAMINE, GLUTAMINE SYN-THETASE, CORN, BIOLOGICAL ACTIVITY, TARGET-LIGAND INTERACTION SCREENING.

**Target (of a therapeutic agent)** The mole-cule (receptor) or moiety that a given drug or therapeutic regimen (e.g., gene delivery) is "aimed" at (i.e., when scientists are working to create/discover that drug or regimen). Tar-gets can be normally occurring constituents of the body (receptors, enzymes, factors, hor-mones, ion channels, nuclear receptors, DNA, etc.), nonnormal constituents of the body (tumors, antigens on tumor surfaces, etc.), or (external, invading) pathogenic agents (microorganisms, viruses, parasites, etc.). See also ENZYME, FACTOR, HORMONE, ION CHANNELS, NUCLEAR RECEPTORS, DEOXYRIBONUCLEIC ACID (DNA), TUMOR, MICROORGANISM, BIOLOGICAL ACTIVITY, PATHOGEN, PATHOGENIC, VIRUS, PHAR-MACOPHORE, GENE DELIVERY, RECEPTORS, MOIETY, COMBINATORIAL CHEMISTRY, COMBINATORIAL BIOLOGY, SIGNALING, SIGNAL TRANSDUCTION, G-PROTEINS, TUMOR NECROSIS FACTOR (TNF), HIGH-THROUGHPUT SCREENING (HTS), TARGET-LIGAND INTERACTION SCREENING, BIOCHIPS.

**Target-Ligand Interaction Screening** A meth-odology of high-throughput screening (HTS)

that is utilized to screen a large number of candidates (e.g., compounds) based upon their interaction (e.g., chemical "binding") to a preselected "target" (e.g., molecule present within a cell membrane, molecule placed on a biochip or other bioassay to facilitate HTS, molecule present on the sur-face of a nematode utilized in HTS, etc.). See also HIGH-THROUGHPUT SCREENING (HTS), TARGET (OF A THERAPEUTIC AGENT), TARGET (OF A HERBICIDE OR INSECTICIDE), COMBINATORIAL CHEMISTRY, COMBINATORIAL BIOLOGY, LIGAND (IN BIOCHEMISTRY), RECEPTORS, SIGNAL TRANS-DUCTION, NUCLEAR RECEPTORS, BIOCHIP, SIGNAL TRANSDUCERS AND ACTIVATORS OF TRANSCRIP-TION (STATs), *CAENORHABDITIS ELEGANS* (*C. ELEGANS*).

**TAT** The name of a protein that helps the HIV ("AIDS virus") to cross the human cell plasma membrane, thereby enabling infec-tion of those cells by HIV (human immun-odeficiency virus). TAT is the main activator of HIV gene expression in cells; it is a pro-tein which complexes with TAR (a 60-nucle-otide sequence found in all viral messenger ribonucleic acid) to mediate synthesis of proteins (in an infected cell) necessary for HIV to reproduce. See also TATA HOMOLOGY, HUMAN IMMUNODEFICIENCY VIRUS TYPE 1 (HIV-1), HUMAN IMMUNODEFICIENCY VIRUS TYPE 2 (HIV-2), GENE, EXPRESS, NUCLEOTIDE, MESSEN-GER RNA (mRNA), VIRUS, PROTEIN, PLASMA MEM-BRANE, CELL.

**TATA Homology** An adenine-thymidine-rich (gene) sequence present 20–30 nucleotides "upstream" of the transcription start site on most eucaryotic protein coding genes; it is required for correct expression. Recent research indicates that blocking this portion of the (gene) sequence may inhibit ability of the AIDS virus to reproduce. See also GENE, GENETIC CODE, NUCLEOTIDE, ADENINE, SEQUENCE (OF A DNA MOLECULE), TAT, TRANSCRIPTION, STARTPOINT, EUCARYOTE, CODING SEQUENCE, HOMOLOGY, PRIBNOW BOX, PROMOTER, SEQUENCE (OF A PROTEIN MOLECULE).

**Taxol** A phytochemical that is naturally pro-duced in some plants and functions to protect those plants from the plant pathogen known as water mold. Coined during the 1960s by Monroe E. Wall when it was originally isolated from the Pacific yew tree (genus *Taxus*), Taxol™ is now a trademark of the Bristol-Myers Squibb Co. refering to the antitumor pharmaceutical sold by the company. The active compound from Pacific yew tree is now known as paclitaxel. Both Taxol™ and paclitaxel act by binding and stabilizing microtubules in cells (thereby halting/preventing the uncontrolled cell growth/proliferation that is cancer). See also CHEMOTHERAPY, PACLITAXEL, CANCER, CELL, MICROTUBULES, TUBULIN.

**TBT** Acronym for the Technical Barriers to Trade (TBT) Agreement to WTO. See also TECHNICAL BARRIERS TO TRADE (TBT) AGREEMENT, WORLD TRADE ORGANIZATION (WTO).

**TCGF** See T CELL GROWTH FACTOR (TCGF).

**TCK Smut** See TELETHIA CONTROVERSIA KOON SMUT.

**Technical Barriers To Trade (TBT) Agreement** The agreement to GATT/WTO, via which WTO member nations agreed to base their import (restrictive) regulations and standards (e.g., mandatory packaging, package marking, testing, certification, labeling requirements, etc.) — known as TBT measures — only on scientific assessments of actual risks (i.e., for those TBT measures intended to protect human health, animal and plant health, or the environment) and to require only those TBT measures that do not create unnecessary obstacles to international trade. See also WORLD TRADE ORGANIZATION (WTO), SPS, SANITARY AND PHYTOSANITARY (SPS) AGREEMENT, SANITARY AND PHYTOSANITARY (SPS) MEASURES, TECHNICAL BARRIERS TO TRADE (TBT MEASURES).

**Technical Barriers To Trade (TBT) Measures** These are (restrictive) import regulations and standards (e.g., mandatory packaging, package marking, testing, certification and labeling requirements, etc.). Some of them are designed to protect human health, animal and plant health, and/or the environment. In the Technical Barriers to Trade (TBT) Agreement to GATT/WTO, the WTO member nations agreed to base their TBT measures only on requirements that do not create unnecessary obstacles to international trade. See also TECHNICAL BARRIERS TO TRADE (TBT) AGREEMENT, SPS, WORLD TRADE ORGANIZATION (WTO).

**Technology Protection System** See CESSATION CASSETTE.

***Telethia Controversia Koon* Smut** A fungal disease that sometimes afflicts wheat (*Triticum aestivum*) plants. See also FUNGUS, WHEAT.

**Telomerase** An enzyme that enables the "repair" of telomeres (thereby stabilizing their length, and preventing "shortening" of the telomeres). The telomerase enzyme is only present in cancerous cells (thereby enabling the "immortality" of cancerous cells). Human telomerase contains an RNA component and a catalytic-protein component (i.e., a member of the reverse transcriptase "family" of enzymes). See also REVERSE TRANSCRIPTASES, CANCER, NEOPLASTIC GROWTH, ZYGOTE, TELOMERES, ENZYME, ONCOGENES, HYBRIDOMA, MONOCLONAL ANTIBODIES (MAb), AGING.

**Telomeres** DNA sequences, that do not code for proteins, which are located at the (end) tips of chromosomes. Telomeres consist of the sequence GGGGTT repeated many times. With the exception of certain types of cells (e.g., zygotes, cancerous cells, "immortal" hybridoma cells), portions of each telomere "break off " each time the cell containing that chromosome divides. This "shortening" process serves to limit the lifetime (i.e., number of replications) of those (noncancerous, nonzygote, nonhybridoma, etc.) cells. See also DEOXYRIBONUCLEIC ACID (DNA), CODING SEQUENCE, PROTEIN, CHROMOSOMES, SEQUENCE (OF A DNA MOLECULE), TELOMERASE, MITOSIS, MITOGEN, CANCER, GAMETE, AGING, RETINOIDS, HYBRIDOMA, ZYGOTE.

**Template** In general terms, it is a mold or pattern that can be copied or its shape reproduced. When used with reference to molecular dimensions, it is a macromolecular mold or pattern for the synthesis of another macromolecule. See also DEOXYRIBONUCLEIC ACID (DNA), STRUCTURAL GENE, INFORMATIONAL MOLECULES, HEREDITY, GENE, GENETIC CODE, GENETIC MAP, BIOSENSORS (CHEMICAL), GENOSENSORS, RIBONUCLEIC ACID (RNA), GENE REPAIR (DONE BY MAN), CODON, EXON, CHIMERAPLASTY, NANOTECHNOLOGY, PRIMER (DNA).

**T**

**Teosinte** A wild plant (*Zea diploperennis*), native to the country of Mexico, which is related to (domesticated) corn/maize (*Zea mays* L.). See also CORN, WILD TYPE.

**Termination Codon** Also known as terminator sequence. One of three triplet sequences (U-A-G, U-A-A, or U-G-A) found in DNA molecules (genes) that cause termination of protein synthesis; they are also called nonsense codons. The sequences cause the termination of the peptide chain and its release in free form. See also CODING SEQUENCE, CODON, DEOXYRIBONUCLEIC ACID (DNA), GENETIC CODE, NONSENSE CODON, SEQUENCING (OF DNA MOLECULES), CONTROL SEQUENCES.

**Terminator** See TERMINATION CODON.

**Terminator Cassette** See CESSATION CASSETTE.

**Terminator Sequence** See TERMINATION CODON.

**Tertiary Structure** The three-dimensional folding of the polypeptide (i.e., protein) molecular chains that characterizes a protein molecule in its native state. See also PROTEIN STRUCTURE, PROTEIN, POLYPEPTIDE (PROTEIN), CONFORMATION, PROTEIN FOLDING, NATIVE CONFORMATION, PROTEOMICS, TRANSCRIPTOME.

**Testosterone** An androgen (steroid hormone) that is biochemically synthesized (made) from androstenedione, which is itself synthesized from progesterone. Testosterone is responsible for the development of male secondary sex characteristics in humans such as greater strength, larger body size, facial hair, a deeper voice, etc. See also STEROID, ESTROGEN.

**Tetrahydrofolic Acid** The reduced, active coenzyme form of the vitamin folic acid; involved in $C_1$ transfers. Tetrahydrofolate (also known as $FH_4$) serves as an intermediate carrier (molecule) of methyl, hydroxymethyl, or formyl groups (all containing one carbon atom) in a relatively large number of enzymatic reactions in which such one-carbon groups are transferred from one metabolite to another.

**TG** See TRIGLYCERIDES.

**TGA** The government regulatory agency charged with approving all pharmaceutical products sold within Australia. See also FOOD AND DRUG ADMINISTRATION (FDA), KOSEISHO, COMMITTEE FOR PROPRIETARY MEDICINAL PRODUCTS (CPMP), EUROPEAN MEDICINES EVALUATION AGENCY (EMEA), MEDICINES CONTROL AGENCY (MCA), COMMITTEE ON SAFETY IN MEDICINES, BUNDESGESUNDHEITSAMT (BGA), GENE TECHNOLOGY OFFICE.

**TGF** See TRANSFORMING GROWTH FACTOR-ALPHA (TGF-ALPHA), TRANSFORMING GROWTH FACTOR-BETA (TGF-BETA).

**Thale Cress** Common name for *Arabidopsis thaliana*. See also *ARABIDOPSIS THALIANA*.

**Thermoduric** An organism that can survive high temperatures but does not necessarily grow at such temperatures. See also THERMOPHILE, MESOPHILE, EXTREMOPHILIC BACTERIA, PSYCHROPHILE.

**Thermophile** An organism whose optimum temperature for growth is close to, or exceeds, the boiling point of water (100°C, 212°F). See also EXTREMOPHILIC BACTERIA, THERMOPHILIC BACTERIA, THERMODURIC, MESOPHILE, PSYCHROPHILE, EUCARYOTE.

**Thermophilic Bacteria** Literally "heat loving" bacteria. They are a category of thermophiles generally found near geothermal vents beneath bodies of water. See also THERMOPHILE, THERMODURIC, EXTREMOPHILIC BACTERIA, MESOPHILE, PSYCHROPHILE.

**Thioesterase** A "family" of enzymes naturally produced within some plants, such as the California bay tree (*Umbellularia californica*). Thioesterase catalyzes those plants' production of the fatty acid laurate. See also FATS, FATTY ACID, LAUROYL-ACP THIOESTERASE, ENZYME, LAURATE, CANOLA, HIGH-LAURATE CANOLA.

**Thiol Group** Refers to a specific chemical entity (on a molecule). See also CYSTEINE (CYS), CYSTINE.

**Thioredoxin** See ALLERGIES (FOODBORNE).

**Threonine (thr)** A crystalline, α-amino acid considered essential for normal growth of animals. It is biosynthesized (made) from aspartic acid and is a precursor of isoleucine in microorganisms. See also ESSENTIAL AMINO ACIDS.

**Thrombin** The key to thrombus (blood clot) formation. Thrombin is a proteolytic enzyme that cleaves fibrinogen into (molecular) pieces, which then spontaneously assemble themselves into fibrin, which forms a clot. See also THROMBUS, THROMBOSIS, THROMBOMODULIN, THROMBOLYTIC AGENTS, FIBRIN, FIBRINOLYTIC AGENTS, CASCADE.

**Thrombolytic Agents** Bloodborne compounds (such as tissue plasminogen activator) that work to disintegrate (break up or lyse) blood clots. See also FIBRIN, FIBRINOLYTIC AGENTS, TISSUE PLASMINOGEN ACTIVATOR (tPA).

**Thrombomodulin** A cell surface protein found on endothelial cells that plays a key role in modulating the final step in the coagulation process. After thrombin binds to thrombomodulin, thrombin loses its ability to cleave fibrinogen to form fibrin. In addition, once thrombin binds to thrombomodulin, thrombin's activation of protein C is increased 200-fold and this activated protein C then degrades factors Va and VIIIa which are both required for the production of thrombin from prothrombin. Hence, thrombomodulin modulates the activity of the enzyme thrombin causing a cessation of full-blown clotting activity. See also THROMBIN, PROTEIN, PROTEIN C, THROMBOSIS, PATHWAY, PATHWAY FEEDBACK MECHANISMS.

**Thrombosis** The intravascular (i.e., inside of blood vessel) formation of a blood clot. See also THROMBIN, THROMBUS, THROMBOLYTIC AGENTS, TRIGLYCERIDES, FIBRIN, FIBRINOLYTIC AGENTS, TISSUE PLASMINOGEN ACTIVATOR (tPA).

**Thrombus** The blood clot itself. The mass of blood coagulated *in situ* in the heart or other blood vessel. For example, such a clot causes a heart attack when the coagulation occurs in the vessels feeding the heart. See also THROMBIN, THROMBOSIS, THROMBOLYTIC AGENTS, FIBRIN, TRIGLYCERIDES, FIBRINOLYTIC AGENTS.

**Thymine (thy)** A pyrimidine component of nucleic acid first isolated from the thymus. Its hydrogen-bonding counterpart in RNA is uracil. See also NUCLEIC ACIDS, PYRIMIDINE, BASE (NUCLEOTIDE), THYMUS, RIBONUCLEIC ACID (RNA).

**Thymoleptics** A class of drugs that primarily exerts its effect on the brain influencing "feeling" and behavior.

**Thymus** A gland that enables cells of the immune system of mammals to mature. In humans, it lies behind the breast bone and extends upward as far as the thyroid gland. The thymus is the place in the body where T lymphocytes are "taught" to distinguish foreign (e.g., pathogen) antigens from "self" cell antigens, to avoid immune responses in which the body's immune system attacks organs and other cells within the body (resulting in autoimmune diesease). Any T lymphocytes that remain "autoreactive" (i.e., would tend to attack "self" cells, such as organs in the body) are destroyed by the thymus via a cytotoxic mechanism.

An example of an autoimmune disease is multiple sclerosis (MS), where the body's acetylcholine receptors are attacked by the body's immune system. Since acetylcholine is crucial in the transmission of nerve impulses to the body's muscles, such destruction of acetylcholine receptors results in loss of control of the body's muscles. See also T LYMPHOCYTES, CYTOTOXIC, RECEPTORS, T CELLS, IMMUNE RESPONSE, PATHOGEN, ANTIGEN, NEUROTRANSMITTER, ACETYLCHOLINE, AUTOIMMUNE DISEASE.

**Thyroid Gland** A gland that is found on both sides of the trachea ("windpipe") in humans. This gland secretes the hormone thyroxine, which increases the rate of metabolism. See also THYROID STIMULATING HORMONE (TSH), GRAVE'S DISEASE.

**Thyroid Stimulating Hormone (TSH)** A hormone that causes the thyroid gland to secrete additional amounts of thyroxine. See also THYROID GLAND, GRAVE'S DISEASE.

**Ti Plasmid** Abbreviation for tumor-inducing plasmid or tumor induction plasmid. It is the plasmid of *Agrobacterium tumefaciens* bacteria that naturally has part of its DNA transferred to a plant when *Agrobacterium tumefaciens* infects that plant (e.g., via a wound in the plant). After it has been transferred into the plant, that Ti plasmid DNA segment (now known as T-DNA or transferred DNA) inserts itself into the plant's DNA, where it causes cells to grow into tumor-like structures known as galls. The Ti plasmid can be modified so that it can be utilized (by genetic engineers) to insert genes from other organisms into plants. See also PLASMID, BACTERIA, *AGROBACTERIUM TUMEFACIENS*, CELL, DEOXYRIBONUCLEIC ACID (DNA), GENE, GENETIC ENGINEERING.

**Tissue Culture** The growth and maintenance (by researchers) of cells from higher organisms *in vitro*, i.e., in a sterile test tube or petri dish environment which contains the nutrients necessary for cell growth. One use of tissue culture is to produce disease free offspring from certain (valuable, high quality) crop plants. Another use of tissue culture methods is for "embryo rescue" to enable "wide crosses" between two different species of plants. In that procedure, pollen from one plant species (e.g., a wild plant possessing disease resistance) is induced to fertilize a plant from another species (e.g., a domesticated crop). The resultant fertilized plant embryo, which would not grow on its own, is "rescued" via tissue culture methods. Following maturation, that wide cross (i.e., a hybrid plant from two species that normally would not cross) produces fertile seeds on its own without any need for further intervention by man. See also CELL, ORGANISM, CULTURE MEDIUM, SPECIES, HYBRIDIZATION (PLANT GENETICS).

**Tissue Plasminogen Activator (tPA)** A glycoprotein that possesses thrombolytic (i.e., blood clot-dissolving) activity. It is used as a drug to dissolve clots and acts by first binding to fibrin (clots). It then activates (i.e., proteolytically cleaves) plasminogen (molecules) to yield plasmin, a bloodborne enzyme that itself cleaves molecular bonds in the fibrin clot. The plasmin molecules diffuse through the fibrin clot and cause the clot to dissolve rapidly. With the dissolution of the clot, blood flow to the formerly blocked blood vessel (e.g., the heart) is restored. See also THROMBUS, THROMBIN, THROMBOLYTIC AGENTS, GLYCOPROTEIN, FIBRIN, FIBRINOLYTIC AGENTS.

**TKI** See TYROSINE KINASE INHIBITORS.

**TME (N)** Abbreviation for "true metabolizable energy (corrected for nitrogen)"; a measure of the amount of energy that a given animal (e.g., chicken) can extract from a given feed ration. See also METABOLISM, CHEMOMETRICS, CALORIE.

**TMEn** See TME (N).

**Tobacco Budworm** See *HELIOTHIS VIRESCENS* (*H. VIRESCENS*).

**Tobacco Hornworm** Caterpillars (pupae) of the Lepidopteran insect *Manduca sexta*. Tobacco Hornworm is susceptible to Cry1A(b) protein (e.g., they are killed if they eat plants genetically engineered to contain Cry1A(b) protein). See also CRY1A(b) PROTEIN.

**Tobacco Mosaic Virus (TMV)** One the of smallest viruses, consisting of some 2,200 chains of identical polypeptides and a molecule of RNA. All of the genetic/heredity information of the Tobacco Mosaic Virus is contained in its RNA. The first discovery of a self-assembling, active biological structure occurred in 1955, when Heinz Frankel-Conrat and Robley Williams showed that TMV will reassemble into functioning, infectious virus particles (after the TMV has been dissociated into its components via immersion in concentrated acetic acid). The TMV virus infects the leaves of tomato and tobacco plants, causing disease. Tobacco plants can be genetically engineered to resist TMV infection. A tomato plant, genetically engineered to resist TMV infection, has been commercially available since 1992. See also GENETIC ENGINEERING, CAPSID, VIRUS, RNA, POLYPEPTIDE (PROTEIN), GENE, INFORMATIONAL MOLECULES, HEREDITY, SELF-ASSEMBLY (OF A LARGE MOLECULAR STRUCTURE).

**Tocopherols** A "family" of different molecular forms of vitamin E; each of which has a saturated phytyl "tail" attached (to the "backbone" of the molecule). Commercial tocopherols are extracted from soybeans, although some are also naturally present in canola and sunflower. See also VITAMIN, SOYBEAN PLANT, VITAMIN E.

**Tocotrienols** A "family" of different molecular forms of vitamin E; each of which has an unsaturated isoprenoid "side chain" attached (to the "backbone" of the molecule). Tocotrienols are naturally present in cereal grains (e.g., oats, barley, rye, and rice bran). See also VITAMIN, ISOPRENE, VITAMIN E.

**Tomato** A green bushy plant, botanical name *Lycopersicon esculentum*. The wild type is native to South America, but the (domesticated) tomato is grown worldwide today. Its fruit, known as tomatoes, are a natural source of the antioxidant carotenoid lycopene, a phytochemical whose consumption has been linked to a reduction in coronary heart disease and some cancers (e.g., prostate cancer). See also LYCOPENE, PHYTOCHEMICALS,

ANTIOXIDANTS, CANCER, CAROTENOIDS, CORONARY HEART DISEASE (CHD), WILD TYPE.

**Tomato Fruitworm** See the link. See also HELICOVERPA ZEA (H. ZEA).

**Topotaxis** See TROPISM.

**TOS** See TRANSGALACTO-OLIGOSACCHARIDES.

**Totipotency** The ability to grow/differentiate into all of the types of cells/tissues constituting an (adult) organism's body. See also STEM CELL ONE, CELL, ZYGOTE, CELL-DIFFERENTIATION, CELL-DIFFERENTIATION PROTEINS, TOTIPOTENT STEM CELLS.

**Totipotent Stem Cells** Bone marrow cells that (when signaled) mature into both red blood cells and white blood cells. Receptors on the surface of totipotent stem cells "grasp" passing blood cell growth factors (e.g., Interleukin-7, Stem Cell Growth Factor, etc.), bringing them inside these stem cells and thus causing the maturation and differentiation into red and white blood cells. These receptors are called FLK-Z receptors. See also STEM CELL ONE, STEM CELLS, WHITE BLOOD CELLS, GROWTH FACTOR, RECEPTORS, CELL-DIFFERENTIATION PROTEINS, CELL DIFFERENTIATION, CELL.

**Toxic Substances Control Act (TSCA)** A 1976 American federal law under which the U.S. Environmental Protection Agency (EPA) has regulated the release of genetically engineered organisms (e.g., bacteria or plants) that produce natural insecticides. This is based on legal analogy to synthetic chemical insecticides, which are clearly regulated under TSCA. See also OAB (OFFICE OF AGRICULTURAL BIOTECHNOLOGY), FEDERAL INSECTICIDE FUNGICIDE AND RODENTICIDE ACT (FIFRA), GENETICALLY ENGINEERED MICROBIAL PESTICIDES (GEMP), WHEAT TAKE-ALL DISEASE, BACILLUS THURINGIENSIS (B.t.).

**Toxicogenomics** A branch of toxicology that deals with the reactions between toxins and the specific differences in response of different organisms due to their different genomes/DNA (of the different individuals that consume the same toxin). For example, some rare humans can tolerate eating certain poisonous mushrooms (which sicken or kill all other humans that consume those particular mushroom species).

During 2001, Fred Gould, David Heckel, and Linda Gahan showed that a rare, recessive gene (allele) known as BtR-4 could confer (to tobacco budworms possessing two copies of that particular gene) resistance to at least some of the "cry" proteins (which kill all other tobacco budworms that consume those "cry proteins"). The subgroup of all those individuals whose DNA (genome) causes their bodies to resist the effects of a given toxin, is known as a haplotype. A haplotype could (theoretically) be as small as one individual, because the particular resistance-to-toxin could result from one single-nucleotide polymorphism (SNP). See also GENE, GENOMICS, PHARMACOGENOMICS, TOXIN, GENOME, DEOXYRIBONUCLEIC ACID (DNA), HAPLOTYPE, SINGLE-NUCLEOTIDE POLYMORPHISMS (SNPs), RECESSIVE ALLELE, CRY PROTEINS, TOBACCO BUDWORM.

**Toxigenic E. coli** See ENTEROHEMORRHAGIC E. COLI, ESCHERICHIA COLIFORM 0157:H7 (E. COLI 0157:H7).

**Toxin** A substance (e.g., produced in some cases by fungi, weeds, ants, or disease-causing microorganisms) which is poisonous to certain other living organisms. See also ANTITOXIN, ABRIN, RICIN, COLICINS, BACTERIOCINS, ESCHERICHIA COLIFORM 0157:H7 (E. COLI 0157:H7), ENTEROHEMORRHAGIC E. COLI, PFIESTERIA PISCICIDA, PHYTOTOXIN, PHOTORHABDUS LUMINESCENS, ENTEROTOXIN, GLUCOSINOLATES, ALKALOIDS, AFLATOXIN, MYCOTOXINS, FUNGUS,.

**TPS** See TECHNOLOGY PROTECTION SYSTEM.

**Tracer** (radioactive isotopic method) A metabolite that is labeled by incorporation of an isotopic atom into its structure. The metabolic fate of the labeled metabolite can then be traced in intact organisms. That is, one is able to ascertain where (in what kind of structure) the metabolite ends up as well as the transformation products (intermediate molecules) that were involved in its formation.

Certain atoms of a given metabolite are labeled. This is done by substituting radioactive isotopes for the atom in question. Because an atom is replaced by an isotope, the metabolite as a whole is chemically and biologically indistinguishable from its normal analog. The presence of the isotope allows the metabolite and its transformation products to be detected and measured. Without this technique, many aspects of metabolism could not have been studied. These include: the process of photosynthesis, metabolic turnover rates, and the biosynthesis of proteins and nucleic acids. See also REASSOCIATION (OF DNA), RADIO-ACTIVE ISOTOPE, RADIOIMMUNOASSAY.

**Traditional Breeding Methods** A phrase utilized by some people to refer to some or most techniques/technologies utilzed by crop plant breeders prior to some arbitrarily chosen date (after which some people feel that "genetic engineering" arrived abruptly). For example, in 1992 Tim Croughan discovered a single rice (*Oryza sativa*) plant that had survived (what should have been a lethal dose of) an imidazolinone-based herbicide, due to a (mutated) gene in its DNA that made it resistant to imidazolinones. That plant was then propagated via straightforward breeding to yield seeds still sown today. Many years ago, some other crops similarly were given new traits (e.g., herbicide tolerance, compositional improvements, etc.) via mutation breeding (i.e., soaking seeds or pollen in mutation-causing chemicals, or bombarding seeds with ionizing radiation to cause random genetic mutations, followed by grow-out and selection of the particular mutation desired such as herbicide tolerance, as described above).

Other crops were given new traits via crossing them with related wild plants, which occasionally resulted in extremely high levels of natural toxicants in those plants/seeds (solanine, psoralene, etc.). Still others were given new traits via wide-crossing them with other domesticated species (e.g., the tangelo is a hybrid of the grapefruit and the tangerine). The U.S. Food and Drug Administration (FDA) regulates all new crop plants similarly (e.g., also requires testing of plants produced via "traditional breeding methods" for the potential presence of introduced or increased natural toxicants). See also GENETIC ENGINEERING, HERBICIDE-TOLERANT CROP, GENETICS, MUTATION, MUTATION BREEDING, TRAIT, CANOLA, SOYBEAN PLANT, CORN, SOLANINE, PSORALENE, FOOD AND DRUG ADMINISTRATION (FDA), BARLEY, HYBRIDIZATION (PLANT GENETICS), MARKER (DNA SEQUENCE), MARKER ASSISTED SELECTION, POINT MUTATION, SOMACLONAL VARIATION, SOMATIC VARIANTS, WIDE CROSS, EMBRYO RESCUE, TISSUE CULTURE.

**Traditional Breeding Techniques** See TRADITIONAL BREEDING METHODS.

**Trait** A characteristic of an organism, which manifests itself in the phenotype (physically). Many traits are the result of the expression of a single gene, but some are polygenic (result from simultaneous expression of more than one gene). For example, the level of protein content in soybeans is controlled by five genes. See also PHENOTYPE, GENOTYPE, EXPRESS, GENE, POLYGENIC, PROTEIN, CALLIPYGE.

***trans* Fatty Acids** One of the two isomeric forms that fatty acids can exist in. *Trans* fatty acids are naturally present in some meat and dairy products (which constitute approximately 5% of the average American diet). See also FATTY ACID, ISOMER, STEREOISOMERS, HYDROGENATION.

***trans*-Acting Protein** A *trans*-acting protein has the exceptional property of acting (having an effect) only on the molecule of DNA (deoxyribonucleic acid) from which it was expressed. See also EXPRESS, *cis*-ACTING PROTEIN.

**Transactivating Protein** See VIRAL TRANSACTIVATING PROTEIN.

**Transaminase** A large group of enzymes that catalyze the transfer of the amino group from any one of at least 12 amino acids to a keto acid to form another amino acid. Also known as aminotransferases. See also ENZYME, AMINO ACID.

**Transamination** The reaction of the enzymatic removal and transfer of an amino group from one specific compound to another. See also TRANSAMINASE, AMINO ACID.

**Transcript** Term used to refer to the various segment(s) of messenger RNA (mRNA) that result from transcription of a gene. See also GENE, TRANSCRIPTION, MESSENGER RNA (mRNA), TRANSCRIPTOME, CENTRAL DOGMA (NEW).

**Transcription** The enzyme-catalyzed process whereby the genetic information contained in one strand of DNA (deoxyribonucleic acid) is used as a template to specify and produce a complementary mRNA strand. Transcription may be thought of as a rewriting of the information contained in DNA into RNA. The language, however, is the same — both are nucleic acid-based. This is in contrast to translation, in which the information is translated from one language (RNA, nucleic acid-based) into another language (protein, amino acid-based). See also GENE EXPRESSION, TRANSLATION, MESSENGER RNA (mRNA), GENETIC CODE, DEOXYRIBONUCLEIC ACID (DNA), TRANSCRIPTION FACTORS, TRANSCRIPTION UNIT, ANTICODING STRAND.

**Transcription Factors** Proteins and/or other chemical compounds that interact with each other, and with regulatory sequences within DNA (when immediately adjacent to the DNA in a cell), to either facilitate ("turn on") or inhibit ("turn off ") the activity (i.e., coding for proteins) of that DNA's genes. Transcription factors hold potential to:

- Cure diseases (e.g., by blocking the deleterious effects of certain disease-causing genes).
- To assist farmers in crop protection (e.g., by switching on the genes that cause crop plants to initiate "cold hardening," or certain types of insect resistance mechanisms).
- To improve human health (e.g., PUFA modulation of genes, modulation of genes by some vitamins, etc.).

Some transcription factors are an integral component in certain gene expression cascades. For example, a gene expression cascade is initiated by the first gene causing expression of a transcription factor, which then itself interacts with the cell's DNA to either cause or speed-up yet another gene expression. The protein resulting from that second gene expression is yet another transcription factor which triggers another (i.e., third) gene expression, and so on. See also PROTEIN, GENETIC CODE, CODING SEQUENCE, DEOXYRIBONUCLEIC ACID (DNA), CELL, INHIBITION, GENE, p53 GENE, TRANSCRIPTION, p53 PROTEIN, CBF1, COLD HARDENING, REGULATORY SEQUENCE, EXPRESS, GENE EXPRESSION, GENE EXPRESSION CASCADE, DOWN REGULATING, VITAMIN, POLYUNSATURATED FATTY ACIDS (PUFA), RECOMBINASE.

**Transcription Unit** A group of genes that code for functionally related RNA molecules or protein molecules. This group of genes is expressed (transcribed) together (as a unit, thus the name). See also EXPRESS, GENE, TRANSCRIPTION, TRANSLATION, GENETIC CODE, CODING SEQUENCE, DEOXYRIBONUCLEIC ACID (DNA), RIBONUCLEIC ACID (RNA), RIBOSOMES.

**Transcriptome** Refers to the entire (complete, possible) set of all gene transcripts (i.e., mRNA segments resulting from gene transcription process) in a given organism. Also to knowledge of their roles in that organism's structure, growth, health, disease (and/or that organism's resistance to disease), etc. Those roles are predominantly due to the impact of each protein molecule (i.e., resulting from the mRNA segments being translated in cells' ribosomes); which is itself due to the protein molecule's composition and its tertiary conformation (which determines the protein's impact in the organism's tissues, metabolism, etc.).

More than one protein can result from each gene in an organism's genome, due to:

- Interactions between genes.
- Interactions between genes and their (protein) products.
- Interactions between genes and some environmental factors.

Mechanistically, this results in different proteins being produced (during translation process) via:

- Alternative splicing of the mRNA transcript. For example, a single intronic base substitution that is present within the IKAP gene (i.e., the allele responsible) for the disease known as familial dysautonomia affects the splicing of the IKAP transcript (i.e., the mRNA segment that determines which specific protein is subsequently "manufactured" by the ribosomes).
- Varying translation start or stop site (on the gene).
- Frameshifting (i.e., different set of triplet codons in the mRNA/transcript is translated by the ribosome).

T

See also GENE, TRANSCRIPT, MESSENGER RNA (mRNA), CODING SEQUENCE, TRANSLATION, CODON, PROTEIN, GENOME, GENETIC CODE, CENTRAL DOGMA (NEW), ORGANISM, CONFORMATION, METABOLISM, TERTIARY STRUCTURE, INTRON, BASE.

**Transduction (gene)** The transfer of bacterial genes (DNA) from one bacterium to another by means of a (temperature or defective) bacterial virus (bacteriophage). There exist two kinds of transduction: specialized and general. In the case of specialized transduction, a restricted group of host genes becomes integrated into the virus genome. These "guest" genes usually replace some of the virus genes and are subsequently transferred to a second bacterium. In the case of generalized transduction, host genes become part of the mature virus particle in place of, or in addition to, the virus DNA. However, in this case the genes can come from virtually any portion of the host genome and this material does not become directly integrated into the virus genome. In the case of plants, the vector can be *Agrobacterium tumefaciens*. See also BACTERIOPHAGE, VECTOR, GENETIC CODE, *AGROBACTERIUM TUMEFACIENS*, RETROVIRAL VECTORS, GENE DELIVERY, TRANSFECTION.

**Transduction (signal)** See SIGNAL TRANSDUCTION.

**Transfection** This term has several different meanings, depending on the context in which it is used: A word utilized generally to refer to insertion of DNA segments (genes) into cells (via electroporation, endocytosis, etc.); a special case of transformation in which an appropriate recipient strain of bacteria is exposed to (free) DNA isolated from a transducing phage with the "take up" of that DNA by some of the bacteria and consequent production and release of complete virus particles. The process involves the direct transfer of genetic material from donor to recipient. See also MARKER (GENETIC MARKER), TRANSFORMATION, ELECTROPORATION, GENE, VIRUS, CELL, BACTERIA, DEOXYRIBONUCLEIC ACID (DNA), TRANSDUCTION (gene).

**Transfer RNA (tRNA)** A class of relatively small RNA (ribonucleic acid) molecules of molecular weight 23,000 to about 30,000. tRNA molecules act as carriers of specific amino acids during the process of protein synthesis. Each of the 20 amino acids found in proteins has at least one specific corresponding tRNA. The tRNA binds covalently with its specific amino acid and "leads" it to the ribosome for incorporation into the growing peptide chain. See also RIBONUCLEIC ACID (RNA), MOLECULAR WEIGHT, AMINO ACID, MESSENGER RNA (mRNA).

**Transferases** Enzymes that catalyze the transfer of functional groups to molecules (from other molecules). See also TRANSAMINASE, ENZYME, HEDGEHOG PROTEINS, GLYCOSYLTRANSFERASES.

**Transferred DNA** See Ti PLASMID.

**Transferrin** The protein molecule responsible for transporting iron (molecules) to tissues throughout the body, via the circulatory system. See also PROTEIN, TRANSFERRIN RECEPTOR, HEME, BLOOD-BRAIN BARRIER (BBB).

**Transferrin Receptor** The receptor molecule (located on the surface of cells throughout the body) responsible for binding to transferrin molecules, then bringing those iron-rich transferrin molecules into the cell where the iron is released to be used by the cell. See also TRANSFERRIN, RECEPTORS, HEME, BLOOD-BRAIN BARRIER (BBB).

**Transformation** The process in which free DNA is transferred directly into a competent recipient cell. The direct transfer of genetic material from donor to recipient. The acquisition (e.g., by bacteria cells) of new genetic markers (new traits coded for by the new DNA) via the process of transformation. See also DEOXYRIBONUCLEIC ACID (DNA), TRANSFECTION, MARKER (GENETIC MARKER).

**Transforming Growth Factor-Alpha (TGF-alpha)** An angiogenic growth factor produced by tumor cells. It is able to induce specific malignant characteristics in normal cells (such as fibroblasts), thereby "transforming" those cells. TGF-alpha appears to possess a variety of potentially useful pharmaceutical properties, such as powerful stimulation of scar tissue formation following wounding of a tissue, as indicated by preliminary research. See also TRANSFORMING GROWTH FACTOR-BETA (TGF-BETA), GROWTH FACTOR, NERVE GROWTH FACTOR (NGF), TUMOR, FIBROBLASTS, ANGIOGENIC GROWTH FACTORS.

**Transforming Growth Factor-Beta (TGF-beta)** An angiogenic growth factor produced by tumor cells, it is able to induce specific malignant characteristics in normal cells (such as fibroblasts), thereby "transforming" those cells. TGF-beta stimulates blood vessel growth, even though it inhibits the division of endothelial cells. TGF-beta is a strong "attracting agent" for macrophages (i.e., TGF-beta is chemotactic), and appears to be responsible for the high concentrations of macrophages often found in tumors. TGF-beta has shown immunosuppressive activity (i.e., it suppresses the immune system). For example, transforming growth factor-beta works together with osteoinductive factor (OIF) to promote bone-formation by first causing connective tissue cells to grow together to form a matrix of cartilage (e.g., across a bone break); bone cells slowly replace that cartilage. See also TRANSFORMING GROWTH FACTOR-ALPHA (TGF-ALPHA), GROWTH FACTOR, OSTEOINDUCTIVE FACTOR (OIF), IMMUN-OSUPPRESSIVE, NERVE GROWTH FACTOR (NGF), TUMOR, FIBROBLASTS, ANGIOGENIC GROWTH FACTORS, MITOGEN, ENDOTHELIAL CELLS, CHEMOTAXIS, MACROPHAGE.

**Transgalacto-oligosaccharides** A "family" of oligosaccharides (produced via enzymatic conversion of lactose, using β-glucosidase enzyme); some of which help to foster the growth of beneficial bifidobacteria in the lower colon of monogastric animals (humans, swine, etc.). See also OLIGOSACCHARIDES, PREBIOTICS, BACTERIA, BIFIDOBACTERIA, *BIFIDUS*, ENZYME.

**Transgene** A "package" of genetic material (i.e., DNA) that is inserted into the genome of a cell via gene splicing techniques. May include promoter(s), leader sequence, termination codon, etc. See also DEOXYRIBO-NUCLEIC ACID (DNA), GENE SPLICING, GENOME, LEADER SEQUENCE, PROMOTER, GENETIC CODE, TERMINATION CODON (SEQUENCE), GENETIC ENGINEERING, CASSETTE.

**Transgenic** An organism whose gamete cells (sperm/egg) contain genetic material originally derived from an organism other than the parents, or in addition to the parental genetic material. See also GENETIC ENGINEERING, GAMETE, NUCLEAR TRANSFER.

**Transgressive Segregation** A plant breeding (propagation) technique, in which genetically very different members of the same species are mated with each other. The offspring of that mating can be more healthy, productive (e.g., fast growing), and uniform than their parents, a phenomenon known as "hybrid vigor." See also GENETICS, SPECIES, F1 HYBRIDS, HYBRIDIZATION (PLANT GENETICS).

**Transit Peptide** A peptide that, when fused to a protein, acts to transport that protein between compartments within eucaryotic cells. Once inside the "destination compartment," the transit peptide is cleaved off the protein and that protein is then free (to do its designed task). See also PEPTIDE, PROTEIN, EUCARYOTE, CELL, FUSION PROTEIN, GATED TRANSPORT, VESICULAR TRANSPORT, CHLOROPLAST TRANSIT PEPTIDE (CTP).

**Transition** Refers to the replacement (i.e., in DNA or RNA molecule) of one purine by another purine; or one pyrimidine by another pyrimidine. See also PURINE, PYRIMIDINE, DEOXYRIBONUNCLEIC ACID (DNA), RIBONUCLEIC ACID (RNA), BASE SUBSTITUTION.

**Transition State** (in a chemical reaction) That point in the chemical reaction at which the reactants (i.e., chemical entities about to react with each other) have been "brought to the brink." It is a point in the chemical reaction process in which an "activated condition" is reached. From this point the probability of the reaction going to completion and producing a product is very high. The transition state separates (energetically) products from reactants. It is viewed as being at the top of the energy barrier separating reactants and products. The reacting species in the transition state can, because of their location at the "top" of the energy barrier, "fall" to either products or reactants. See also CATALYST, ENDERGONIC REACTION, ACTIVATION ENERGY, FREE ENERGY, CATALYTIC ANTIBODY, SEMISYNTHETIC CATALYTIC ANTIBODY, EXERGONIC REACTION.

**Translation** The process whereby the genetic information present in an mRNA molecule directs the order of incorporation of specific amino acids, and hence the growth of the polypeptide chain during protein synthesis. One can think of translation as the process of translating one language into another. In this particular case the nucleic acid-based language represented by mRNA is translated into the amino acid-based language of proteins. See also CODING SEQUENCE, CODON, RIBOSOMES, MESSENGER RNA (mRNA), PROTEIN, GENE, GENETIC CODE.

**Translocation** Genetic mutation in which a section of a chromosome "breaks off" and moves to a new (abnormal) position in that (or a different) chromosome. See also GENE, CHROMOSOMES, GENETIC CODE, CODING SEQUENCE, TRANSPOSITION, DEOXYRIBONUCLEIC ACID (DNA), MUTATION, INTROGRESSION, JUMPING GENES, HOT SPOTS.

**Transmembrane Proteins** Refers to those protein molecules that extend from one side of a cell membrane to the other side of that membrane. For example, G-proteins are transmembrane proteins that act to accomplish signal transduction (i.e., convey "signal" from outside the cell to one or more internal cell parts). EGF receptors bind to EGF molecules (e.g., passing-by in the blood), then both enter the cell (through the cell membrane) together, where the EGF stimulates growth/division of that cell. See also PROTEIN, CELL, PLASMA MEMBRANE, RECEPTORS, MEMBRANE (OF A CELL), MEMBRANE TRANSPORT, ABC TRANSPORTERS, EGF RECEPTOR, G-PROTEINS, CECROPHINS (LYTIC PROTEINS), MAGAININS, SIGNAL TRANSDUCTION, SIGNALING, EPIDERMAL GROWTH FACTOR (EGF).

**Transposable Element** See TRANSPOSON.

**Transposase** An enzyme required for transposition to occur. It is coded for by the transposon known as the P element. See also TRANSPOSITION, TRANSPOSON, ENZYME, GENETIC CODE, CODING SEQUENCE.

**Transposition** Movement of a gene or set of genes from one site in the genome to another without a reciprocal exchange (of DNA). See also GENE, JUMPING GENES, GENOME, TRANSPOSON, TRANSPOSASE, HOT SPOTS, DEOXYRIBONUCLEIC ACID (DNA).

**Transposon** A DNA (deoxyribonucleic acid) sequence (segment of molecule) able to replicate and insert one copy (of itself ) at a new location in the genome (i.e., a transposition of location). Discovered in 1950 by geneticist Barbara McClintock in corn (maize)

plants (*Zea mays* L.); and in bacteria a decade later by Joshua Lederberg.

Transposons can either carry genes along one organism's genome, or even into another organism's genome (e.g., via sexual conjugation, in bacteria). By such sexual conjugation, transposons can carry genes that confer new phenotypic properties (e.g., resistance to certain antibiotics, for a given bacterial cell). See also DEOXYRIBONUCLEIC ACID (DNA), REPLICATION (OF VIRUS), GENOME, TRANSPOSITION, TRANSPOSASE, SEQUENCE (OF A DNA MOLECULE), CORN, JUMPING GENES, GENE, SEXUAL CONJUGATION, PHENOTYPE, CONJUGATION.

**Transversion** The substitution of a purine for a pyramidine, or of a pyramidine for a purine (at a specific site, within a given nucleotide in a molecule of DNA). See also NUCLEOTIDE, DEOXYRIBONUCLEIC ACID (DNA), SINGLE-NUCLEOTIDE POLYMORPHISMS (SNPs), MUTATION, BASE SUBSTITUTION.

**TRANSWITCH®** A "sense" technology used to "turn off " (suppress) a gene (e.g., the one that causes tomato to ripen) that causes an unwanted effect (e.g., premature softening of tomato). TRANSWITCH® and its registered trademark are owned by DNA Plant Technology Corp. See also GENE SILENCING, SUPPRESSOR GENE, SENSE.

**"Treatment" IND Regulations** Food and Drug Administration (FDA) regulations promulgated in 1987, to provide a more rapid formal pharmaceutical approval mechanism than the usual IND (Investigational New Drug) regulatory approval process. Its purpose is to enable drug developers to provide promising experimental drugs to patients suffering from immediately life-threatening diseases or certain serious conditions (e.g., acquired immune deficiency syndrome, or AIDS) before complete data on that drug's efficacy or toxicity are available. See also IND, FOOD AND DRUG ADMINISTRATION (FDA), DELANEY CLAUSE, KOSEISHO, COMMITTEE FOR PROPRIETARY MEDICINAL PRODUCTS (CPMP).

**Treatment Investigational New Drug** See "TREATMENT" IND REGULATIONS.

**Trehalose** A disaccharide (simple sugar) that is naturally synthesized (manufactured) by many plants and animals in response to the stresses of freezing, heating, or drying. That is because trehalose protects certain proteins (needed for life) and prevents loss of crucial volatile (i.e., easily evaporated) compounds from organisms during those stressful (dry, frozen, or hot) conditions. Trehalose also provides a source of quick energy after the stressful conditions have passed. That is why dried baker's yeast (which contains up to 20% trehalose by weight) can be stored in its dry state for many years, yet quickly leavens bread dough within minutes of being rehydrated (i.e., rewetted).

Trehalose accomplishes this protection by forming a nonhygroscopic "glass" on the surfaces of cells and large molecules. It immobilizes and stabilizes large molecules (e.g., proteins), but still allows water to diffuse out so complete drying can occur. Thus, trehalose holds potential as a food additive to keep proteins (e.g., eggs) fresh in the dried form. In 1991, the U.K. approved trehalose for use in food, and the U.S. approved its use in 2001. Trehalose hydrolyzes (e.g., during digestion) into two molecules of glucose. See also DISACCHARIDES, PROTEIN, GLUCOSE (GLc), HYDROLYSIS, CONFORMATION, "SWITCH" PROTEINS, TERTIARY STRUCTURE, PROTEIN FOLDING.

**Tremorgenic Indole Alkaloids** A "family" of toxic alkaloids (chemical compounds) that are naturally produced (within some plants) by certain fungi (which sometimes grow in those plants). For example, the alkaloid known as Penitrem D is produced by certain fungi which grow in some grass species. It causes tremors, weakness, lack of coordination, and convulsions in animals that consume those fungus-infested grasses. See also ALKALOIDS, TOXIN, FUNGUS, ENDOPHYTE.

**Triacylglycerols** See TRIGLYCERIDES.

***Trichoderma harzianum*** A microorganism that possesses (natural) fungicide activity. See also *BACILLUS THURINGIENSIS* (*B.t.*), WHEAT TAKE-ALL DISEASE, FUNGUS, FUNGICIDE.

**Trichosanthin** An enzyme extracted from a specific Chinese plant. It has been discovered to "cut apart" the ribosomes in some cells infected with the HIV (i.e., AIDS) virus, thus potentially stopping the virus and preventing infection of additional cells. See also RIBOSOMES, ACQUIRED IMMUNE DEFICIENCY SYNDROME (AIDS), ENZYME, PROTEIN, HUMAN IMMUNODEFICIENCY VIRUS TYPE 1 (HIV-1), HUMAN IMMUNODEFICIENCY VIRUS TYPE 2 (HIV-2).

**Triglycerides** The primary constituent of fats or oils; triglycerides are molecules that consist of three fatty acids attached to a glycerol "molecular backbone." More accurately called triacylglycerols, although long-term historical usage of "triglycerides" has made the latter term more common. Similarly, the term "diglyceride" is often used to refer to those molecules which consist of two fatty acids attached to a glycerol "molecular backbone." "Diglycerides" (more accurately called diacylglycerols) can result from the splitting-off (i.e., hydrolysis) of one fatty acid from a triacylglycerol ("triglyceride") molecule (e.g., during fat breakdown/oxidation); or from the combination of two fatty acids with glycerol (e.g., during synthesis of fats). The "triglyceride level" in human bloodstream refers to the blood's content of noncholesterol total fats. Research during the 1990s provided evidence that high blood levels of triglycerides in humans (e.g., immediately after meals) contribute to thrombosis. See also FATS, THROMBOSIS, FATTY ACID, SATURATED FATTY ACIDS (SAFA), LPAAT PROTEIN, UNSATURATED FATTY ACID, HYDROLYSIS, OXIDATION (of fats/oils/lipids), ADIPOCYTES, FRUCTOSE OLIGOSACCHARIDES, *BIFIDUS*, POLYUNSATURATED FATTY ACIDS (PUFA), DIACYLGLYCEROLS.

**Triploid** Refers to organisms that possess three sets of chromosomes, instead of the normal two sets. Conversion of a diploid (i.e., two sets of chromosomes) organism to triploid can be done by man (certain fish, "seedless" grapes, etc.). For example, fish are ordinarily diploid. By exposing fish eggs to certain specific combinations of temperature and pressure, immediately after fertilization of those eggs, scientists can cause the resultant fish to become triploid. Triploid fish are unable to reproduce. This sterility is desired by man, in order to prevent certain fish (e.g., those that have been genetically engineered) from mating with wild fish. Such induced (triploid) sterility also prevents the (genetically engineered) fish from wasting energy on the act of reproduction, so they grow faster and larger. That transfer (of energy use from reproduction to growth) also holds true for "seedless" grapes, watermelons, etc. See also DIPLOID, CHROMOSOMES, WHEAT.

**tRNA** See TRANSFER RNA (tRNA).

**Tropism** Orientation movement of a sessile organism in response to a stimulus. Movement of curvature due to an external stimulus that determines the direction of movement. Also known as topotaxis. See also SESSILE, CHEMOTAXIS.

**Trypsin** A proteolytic (protein molecular chain-cutting) enzyme produced by the pancreas, to facilitate digestion within certain animals. Trypsin cleaves polypeptide (protein) molecular chains on the carboxyl (group) side of arginine and lysine units (residues); and it is often utilized by man to break apart protein molecules (e.g., to enable scientists to study that protein's constituent peptides). See also ARGININE (arg), LYSINE (lys), PROTEIN, PEPTIDE, POLYPEPTIDE (PROTEIN), PROTEOLYTIC ENZYMES, PROTEASES, CHYMOTRYPSIN, TRYPSIN INHIBITORS, DIGESTION (WITHIN ORGANISMS), COWPEA TRYPSIN INHIBITOR (CpTI).

**Trypsin Inhibitors** Compounds present in certain legumes (soybeans, etc.) that inhibit the activity (i.e., protein-cleavage, which aids digestion) of proteases (i.e., protein-cleaving enzymes such as trypsin or chymotrypsin) in the digestive systems of monogastric (single-stomach) animals (which include swine, poultry, and humans). Trypsin inhibitors present in traditional varieties of soybeans (botanical name *Glycine max* (L.) Merrill) include:

- the Kunitz trypsin inhibitor (TI), which was first isolated and crystallized by M. Kunitz in 1945. It combines tightly with molecules of trypsin on a 1:1 basis, and thereby reduces the rate of protein-cleavage effected by the trypsin enzyme; which inhibits the animal's digestion of protein(s).
- the Bowman-Birk trypsin inhibitor (BB T.I.), first described by D. E. Bowman in 1944. It combines with molecules of trypsin and chymotripsin, and thereby reduces the rate of protein-cleavage effected by the trypsin and

chymotrypsin enzymes, which inhibits the animal's digestion of protein(s).

NOTE: During 2000, research by Frank Meyskins and William Armstrong indicated that consumption of BB T.I. in a manner that 'bathes' mouth tissues in it (for extended period of time) inhibits the development of the precancerous mouth lesions that can become oral cancer.

- certain free fatty acids and their acyl CoA esters, which reduce the rate of protein-cleavage effected by the trypsin enzyme inhibiting the animal's digestion of protein(s).

Heating of soybeans to a temperature of 212° (100°C) for 15 min causes these trypsin inhibitors to be rendered inactive in soybeans, so the animal's digestion is unimpeded when it is fed soy that has been thus heated. See also TRYPSIN, CHYMOTRYPSIN, SOYBEAN PLANT, PROTEIN, PROTEASES, ENZYME, PROTEOLYTIC ENZYMES, DIGESTION (WITHIN ORGANISMS), POLYPEPTIDE (PROTEIN), BIOLOGICAL ACTIVITY, ACYL CoA, COWPEA TRYPSIN INHIBITOR (CpTI), ORAL CANCER.

**Tryptophan (trp)** An essential amino acid, it is a precursor of the important biochemical molecules indoleacetic acid, serotonin, and nicotinic acid. L-Tryptophan is used as a common feed additive for livestock to ensure that their diet includes an adequate amount of this essential amino acid. See also ESSENTIAL AMINO ACIDS, STEREOISOMERS, SEROTONIN, AMINO ACID.

**TSH** See THYROID STIMULATING HORMONE (TSH).

**Tuberculosis** See MYCOBACTERIUM TUBERCULOSIS.

**Tubulin** A cell protein required for cell mitosis (i.e., the cell-reproduction process in which a cell divides into two identical cells). When the drugs paclitaxel or Taxol™ are administered to the body (e.g., in chemotherapy), they bind tubulin; which halts cell division and causes apoptosis in the affected cells (e.g., tumor cells) by binding Bc1-2 (a protein that prevents apoptosis in cells). See also CELL, PROTEIN, MITOSIS, PACLITAXEL, TAXOL, CANCER, CHEMOTHERAPY, APOPTOSIS.

**Tumor** A mass of abnormal tissue that resembles normal tissues in structure, but which fulfills no useful function (to the organism) and grows at the expense of the body. Tumors may be malignant or benign. Malignant tumors (which infiltrate adjacent healthy tissues) can result from oncogenes and/or carcinogens. They can eventually kill their host if unchecked. Epidermal growth factor encourages rapid cell growth in more than 50% of human tumors. See also CANCER, ANGIOGENESIS, ONCOGENES, PROTO-ONCOGENES, CELL, CARCINOGEN, TYROSINE KINASE, TYROSINE KINASE INHIBITORS (TKI), ATP SYNTHASE, EPIDERMAL GROWTH FACTOR (EGF).

**Tumor Necrosis Factor (TNF)** Literally, tumor death factor. A cytokine (protein that helps regulate the immune system) that has shown potential to combat (kill) malignant (cancer) tumors. Tumor necrosis factor was discovered to be 10,000 times more toxic in humans than in rodents, where it had been tested for toxicity prior to human clinical tests. This example illustrates one potential pitfall of nontarget animal testing in that sometimes animal testing does not accurately reflect or foretell what will happen in humans. Another drawback to using TNF as a drug to combat human tumors is the fact that it is one of the substances released (in the disease rheumatoid arthritis) that destroys tissue in the joints. When released as part of the AIDS (disease), TNF causes cachexia, which is a "wasting away" of the body due to the body's inability to process nutrients received via digestion. See also CYTOKINES, LYMPHOKINES, NECROSIS, TUMOR, TUMOR-INFILTRATING LYMPHOCYTES (TIL CELLS), PROTEIN, AUTOIMMUNE DISEASE, T CELL MODULATING PEPTIDE (TCMP), DIGESTION (WITHIN ORGANISMS).

**Tumor-Associated Antigens** Discovered by Thierry Boon in 1991, these are distinctive protein molecules that are produced in the surface membrane of tumor cells. These protein molecules are used by the body's cytotoxic T cells to recognize (and destroy) tumor cells, so such proteins hold promise for use in vaccines. See also MAJOR HISTOCOMPATIBILITY COMPLEX (MHC), MACROPHAGE, TUMOR, T CELL RECEPTORS, ANTIGEN, T CELLS, PROTEIN, CELL, CYTOTOXIC T CELLS, HUMAN LEUKOCYTE ANTIGENS (HLA).

**Tumor-Infiltrating Lymphocytes (TIL cells)** The white blood cells of a cancer patient which have been:

1. Taken from that patient's tumor (where those white blood cells had been attempting to combat the cancer, albeit unsuccessfully).
2. Stimulated with doses of interleukin-2 (to make the lymphocytes more effective against the cancer).
3. Multiplied *in vitro* (i.e., outside of the patient's body) to make them more numerous (and thus more likely to successfully combat the cancer).

When these "souped up" lymphocytes (white blood cells) are reintroduced into that same patient's body, the lymphocytes (now called TIL cells because they have been "souped up") attack the cancer tumor (malignant growth) more vigorously than before. See also TUMOR, WHITE BLOOD CELLS, LYMPHOCYTE, LYMPHOKINES, T CELLS, CYTOTOXIC T CELLS.

**Tumor-Suppressor Genes** Also called anticancer genes. Genes within a cell's DNA that code for (cause to be manufactured in cell's ribosomes) proteins that hold the cell's growth in check. If these genes are damaged (e.g., by radiation, by a carcinogen, or by chance accident in normal cell division), they no longer hold cell growth in check — and the cell becomes malignant (if the cell's DNA also contains a gene called an oncogene). Oncogenes must be present for the cell to become malignant, but oncogenes cannot cause a cell to become malignant until a tumor-suppressor gene is damaged. As with all genes, tumor-suppressor genes are inherited in two copies (alleles, one from each parent) and either copy can code for the proteins necessary for cell growth control. However, an organism born with one defective copy of a tumor-suppressor gene (or in whom one copy is damaged early in life) is especially prone to cancer (malignancy). See also GENE, p53 GENE, GENETIC CODE, MEIOSIS, DEOXYRIBONUCLEIC ACID (DNA), CARCINOGEN, RIBOSOMES, ONCOGENES, CANCER, TUMOR, PROTO-ONCOGENES, PROTEIN.

**T**

**Tumor-Suppressor Proteins** Proteins that are coded for (caused to be manufactured in the cell's ribosomes) by tumor-suppressor genes (e.g., the p53 gene). Such proteins (e.g., the p53 protein) then act upon the cell's DNA in order to prevent uncontrolled cell growth and division (i.e., cancer). See also TUMOR-SUPPRESSOR GENES, GENE, p53 GENE, PROTEIN, GENETIC CODE, MEIOSIS, DEOXYRIBONUCLEIC ACID (DNA), RIBOSOMES, ONCOGENES, CANCER, TUMOR, CELL, PROTO-ONCOGENES.

**Turnover Number** The number of molecules of a product produced per minute by a single-enzyme molecule when that enzyme is working at its maximum rate. That is, the number of substrate molecules converted into a product by one enzyme molecule per minute when that enzyme is "going (catalyzing) as fast as it can." See also ENZYME, TRANSFERASES, PROTEASE, PROTEIN KINASES, PROTEOLYTIC ENZYMES, TRANSAMINASE.

**Two-Dimensional (2D) Gel Electrophoresis** A technology/methodology developed during the 1970s to separate the various proteins within a given biological sample, prior to their analysis. The proteins are moved by applying an electrical field. The sample is moved through two different gels (i.e., two different dimensions). The initial gel has a pH gradient that separates the different proteins based on their respective isoelectric points. The second gel (dimension) the sample is moved through is a gel that separates the protein molecules based on their individual molecular weights. That gel acts as a "molecular sieve" (i.e., smaller proteins move faster — and farther — than larger proteins do through this gel; in a fixed amount of time).

A fixed-time gel run (i.e., with appropriate gel and the appropriate electrical field applied to the gel) leaves a scientist with approximately 1,000 "spots" (of protein molecules) on the gel. Each "spot" is a collection of the molecules of one protein (or of several proteins with similar molecular weights) from the original sample (mixture). To identify the protein(s) in the "spots," the scientist stains them, then assesses the entire gel with an electronic image scanner (or he assesses it visually). From the pattern (coupled with intensity) of the "spots," two such gels could be utilized to confirm if two organisms were the same species/strain/variety, or to determine the differences (in gene expression) between samples of diseased vs. healthy tissues. See also PROTEIN, GEL, ELECTROPHORESIS, AGAROSE, POLYACRYLAMIDE GEL ELECTROPHORESIS (PAGE), PAGE, ISOELECTRIC POINT, GENE EXPRESSION ANALYSIS, PROTEOMICS, MOLECULAR WEIGHT, SPECIES.

**Type I Diabetes** The form of diabetes disease that usually strikes young people (thus, it was formerly known as juvenile or insulin-dependent diabetes). This disease is characterized by the body's immune system destroying the insulin-producing cells (Beta cells) of the pancreas. If not treated in time (i.e., via insulin injections), the person can die suddenly. Even when treated, the person is at increased risk of blindness, atherosclerosis, coronary heart disease, heart attack, stroke, and kidney disease. See also DIABETES, BETA CELLS, PANCREAS, INSULIN, INSULIN-DEPENDENT DIABETES MELLITIS (IDDM), CALPAIN-10, ATHEROSCLEROSIS, CORONARY HEART DISEASE (CHD), TYPE II DIABETES.

**Type II Diabetes** The form of diabetes disease that usually strikes people who are more than 40 years old. Also known as adult-onset diabetes, this disease is characterized by the body's tissues becoming insensitive to insulin. Effects on the body include increased likelihood of blindness, atherosclerosis, coronary heart disease, heart attack, stroke, and kidney disease. See also DIABETES, INSULIN-DEPENDENT DIABETES MELLITIS (IDDM), INSULIN, CALPAIN-10, ATHEROSCLEROSIS, CORONARY HEART DISEASE (CHD), TYPE I DIABETES, INOSITOL.

**Type Specimen** The actual physical specimen (e.g., a stuffed lizard or a dried insect) that a scientist (who describes and names a previously unknown species) must place in a museum (or other recognized repository) in order to have the right to name that newly discovered species. This "officially deposited specimen" is required for three purposes:

1. So that comparisons can later be made if there is ever a doubt whether another "new" species is simply a member of

this same species (and thus already named)

2. So that taxonomists (who determine and keep the official scientific names by which scientists must refer to each of the world's organisms) can name each of the newly discovered species in accordance with the complex rules of the International Codes for Nomenclature. Examples of such names in this glossary are *Arabidopsis thaliana*, *Escherichia coli*, and *Agrobacterium tumefaciens*.

3. So that patent claims for genetically engineered organisms can later be enforced.

See also SPECIES, STRAIN, CLADISTICS, CHAKRA-BARTY DECISION, AMERICAN TYPE CULTURE COLLECTION (ATCC), CONSULTATIVE GROUP ON INTERNATIONAL AGRICULTURAL RESEARCH (CGIAR).

**Tyrosine (tyr)** A phenolic α-amino acid. It is a precursor of the hormones epinephrine, norepinephrine, thyroxine, and triiodothyronine. It is also a precursor of the molecule known as melanin (which is the pigment of a suntan). See also AMINO ACID, HORMONE.

**Tyrosine Kinase Inhibitors (TKI)** Refers to various chemical compounds that inhibit the activity of tyrosine kinase enzyme (inside the body). One example of TKI is genistein. Because the activity of tyrosine kinase helps cancerous (tumor) cells to metastasize (spread/grow), consumption by humans of relevant TKI acts to help prevent (spreading of) certain cancers. See also ENZYME, TYROSINE KINASE, PROTEIN TYROSINE KINASE INHIBITOR, BIOLOGICAL ACTIVITY, CANCER, CELL, TUMOR, GENISTEIN, ISOFLAVONES.

T

# U

**U.S. Patent and Trademark Office (USPTO)** The Washington, D.C.-based American Government agency responsible for common patent protection matters for all of America's 50 states and its territorial possessions. The USPTO allows the patenting of new and unique microbes, plants, and animals, as well as the new and unique methods to produce such biotechnology advances. See also EUROPEAN PATENT OFFICE (EPO), CHAKRABARTY DECISION, MICROBE, GENETIC ENGINEERING, PLANT'S NOVEL TRAIT (PNT), PLANT BREEDER'S RIGHTS (PBR), BIOTECHNOLOGY, AMERICAN TYPE CULTURE COLLECTION (ATCC).

**Ubiquitin** A small protein present in all eucaryotic cells (ubiquitous) that plays an important role in "tagging" other proteins destined (marked) for destruction (via proteolytic cleavage). Such proteins are then broken down and removed because they are damaged or no longer needed by the body. Such "tagged" protein molecules are said to have been ubiquitinated. See also EUCARYOTE, PROTEIN, PROTEOLYTIC ENZYMES, PROTEASOMES, DENATURATION.

**Ubiquitinated** See UBIQUITIN.

**Ultracentrifuge** A high-speed centrifuge that can attain revolving speeds up to 85,000 rpm and centrifugal fields up to 500,000 times gravity. The machine is used to sediment (i.e., cause to settle out) and hence separate macromolecules (large molecules) and macromolecular structures in a mixture/solution. In general, a centrifuge is a machine that whirls test tubes around rapidly, like a merry-go-round, to force the heavier suspended materials (in the solutions in the test tubes) to the bottoms of those test tubes before the lighter material.

**Ultrafiltration** A (mixture) separation methodology that uses the ability of synthetic semipermeable membranes (possessing appropriate physical and chemical natures) to discriminate between molecules in the mixture, primarily on the basis of the molecules' size and shape. Invented and developed by Dr. Roy J. Taylor in the 1950s and 1960s, ultrafiltration is typically utilized for the separation of relatively high-molecular-weight solutes (e.g., proteins, gums, polymers, and other complex organic molecules) and colloidally dispersed substances (e.g., minerals, microorganisms, etc.) from their solvents (e.g., water). See also DIALYSIS, MEMBRANE TRANSPORT, MICROORGANISM, MOLECULAR WEIGHT, PROTEIN, POLYMER, HOLLOW FIBER SEPARATION.

**Union for Protection of New Varieties of Plants (UPOV)** A group of the world's countries that have jointly agreed to mutually protect the intellectual property (of owners, breeders) that is inherent in new plant varieties developed by man. These intellectual property protections are often collectively referred to as "Breeder's Rights." Established in 1961, the secretariat for this union (UPOV) is in Geneva, Switzerland. See also PLANT VARIETY PROTECTION ACT (PVP), U.S. PATENT AND TRADEMARK OFFICE (USPTO), PLANT'S NOVEL TRAIT (PNT), PLANT BREEDER'S RIGHTS (PBR), EUROPEAN PATENT CONVENTION, EUROPEAN PATENT OFFICE (EPO), MUTUAL RECOGNITION AGREEMENTS (MRAs), COMMUNITY PLANT VARIETY OFFICE.

**Units (U)** A measure (quantitation) of biological activity of a substance, as defined by various standardized assays (tests). See also ASSAY, BIOASSAY.

**Unsaturated Fatty Acid** A fatty acid containing one or more double bonds (between individual atoms of the molecule). See also

FATTY ACID, DESATURASE, MONOUNSATURATED FATS, POLYUNSATURATED FATTY ACIDS (PUFA).

**UPOV** See UNION FOR PROTECTION OF NEW VARIETIES OF PLANTS (UPOV).

**Uracil** A pyrimidine base, important as a component of ribonucleic acid (RNA). Its hydrogen-bonding counterpart in DNA is thymine. See also PYRIMIDINE, RIBONUCLEIC ACID (RNA), BASE (NUCLEOTIDE), DEOXYRIBONUCLEIC ACID (DNA).

**Urokinase** A thrombolytic (i.e., clot-dissolving) enzyme used as a bio-pharmaceutical. See also THROMBOLYTIC AGENTS, TISSUE PLASMINOGEN ACTIVATOR (tPA), FIBRINOLYTIC AGENTS.

**USPTO** See U.S. PATENT AND TRADEMARK OFFICE (USPTO).

# V

**Vaccine** Any substance, bearing antigens on its surface, that causes activation of an animal's immune system without causing actual disease. The animal's immune system components (e.g., antibodies) are then prepared to quickly vanquish those particular pathogens when they later enter the body. See also DNA VACCINES, "NAKED" GENE, "EDIBLE VACCINES", ANTIGEN, CELLULAR IMMUNE RESPONSE, HUMORAL IMMUNITY.

*Vaccinia* A nonpathogenic virus believed to be a (modified) form of the virus that causes cowpox. *Vaccinia* readily accepts genes (inserted into its genome via genetic engineering) from pathogenic viruses, so it can be used to make vaccines that do not possess the risk inherent in attenuated-virus vaccines (i.e., that the attenuated virus "revives" and causes disease). Such genetically engineered *vaccinia* codes for (presents) the proteins of the pathogenic virus on its surface, which activates the immune system (e.g., of vaccinated animal) to produce antibodies against that pathogenic virus. See also VACCINE, PATHOGENIC, VIRUS, GENE, GENE DELIVERY, GENETIC ENGINEERING, ATTENUATED (PATHOGENS), ANTIBODY, MACROPHAGE, COMPLEMENT CASCADE, CELLULAR IMMUNE RESPONSE, PHAGOCYTE.

**Vacuoles** A membrane-bound sac within a cell, within which water, food, waste, or salt, etc. are temporarily stored. Also pigments, in certain plant cells. See also PLASMA MEMBRANE, CELL, ANTHOCYANIDINS.

**VAD** Acronym for vitamin A deficiency. See also GOLDEN RICE, VITAMIN, BETA CAROTENE, CAROTENOIDS.

**Vagile** Wandering or roaming (e.g., a microorganism that is not attached to a solid support tends to "wander" through its environment as it gets pushed about by currents of air or liquid). See also SESSILE, VAGILITY.

**Vagility** The ability of organisms to disseminate (e.g., spread throughout a given habitat). See also VAGILE.

**Vaginosis** The process whereby a cell internalizes an entity (such as a virus or a protein) that has bound to the cell's outer membrane. Once that "bound entity" is inside the cell, the cell membrane fuses together again. See also NUCLEAR RECEPTORS, RECEPTORS, ENDOCYTOSIS, TRANSFERRIN, VIRUS, BLOOD-BRAIN BARRIER (BBB).

**Validation** See PROCESS VALIDATION.

**Valine (val)** An amino acid considered essential for normal growth of animals. It is biosynthesized (made) from pyruvic acid. See also AMINO ACID, ESSENTIAL AMINO ACIDS, ALS GENE.

**Value-Added Grains** See VALUE-ENHANCED GRAINS.

**Value-Enhanced Grains** Those grains that possess novel traits that are economically valuable (e.g., higher-than-normal protein content, better quality protein, higher-than-normal oil content, etc.). For example, high-oil corn (maize) possesses a kernel oil content of 5.8% or greater, vs. oil content of 3.5% or less for traditional No. 2 yellow corn. Glutamate dehydrogenase (GDH) corn (maize) possesses a kernel protein content that tends to be approximately 10% greater than the protein content of traditional corn (maize) varieties. High-amylose corn possesses a kernel amylose content of 50% or more of the total kernel starch, etc. See also HIGH-OIL CORN, PROTEIN, AMYLOSE, HIGH-AMYLOSE CORN, OPAGUE-2, FLOURY-2, GENETIC ENGINEERING, LOW-PHYTATE CORN, LOW-PHYTATE SOYBEANS, TRAIT, HIGH-LYSINE CORN, HIGH-METHIONINE CORN, HIGH-PHYTASE CORN AND

SOYBEANS, HIGH-OLEIC OIL SOYBEANS, HIGH-STEARATE SOYBEANS, GLUTAMATE DEHYDROGE-NASE, HIGH-SUCROSE SOYBEANS, HIGH-LAURATE CANOLA, HIGH-LACTOFERRIN RICE.

**Van der Waals Forces** The relatively weak forces of attraction between molecules that contribute to intermolecular bonding (i.e., binding together two or more adjacent molecules). Historically, it was thought that van der Waals forces were always weaker than the hydrogen bond forces responsible for intramolecular bonding. However, in 1995, Dr. Alfred French discovered that van der Waals forces are primarily responsible for holding together a mass of cellulose molecules, with hydrogen bonding playing a lesser role.

During 2000, Kellar Autumn discovered that van der Waals forces (acting between foot skin hairs and the surface climbed) are responsible for enabling the Tokay gecko (*Gecko gecko*) to climb vertical surfaces and also to hang upside down. These forces work (to "adhere" a gecko's foot) even underwater or in a vacuum. See also CELLULOSE, CELLULASE, MOLECULAR WEIGHT, WEAK INTERACTIONS.

**Vascular Endothelial Growth Factor (VEGF)** A human growth factor (GF) that causes growth/proliferation of blood vessels/endothelium and endothelial cells. See also GROWTH FACTOR, ENDOTHELIUM, ENDOTHE-LIAL CELLS.

**Vector** The agent used (by researchers) to carry new genes into cells. Plasmids currently are the biological vectors of choice; though viruses and other biological vectors such as *Agrobacterium tumefaciens* bacteria or BACs are increasingly being used for this purpose. Nonbiological vectors include the metal microparticles (coated with genes) which are "shot" into cells by the Biolistic® gene gun. See also PLASMID, GENE, CELL, RET-ROVIRAL VECTORS, PROTOPLASM, *AGROBACTERIUM TUMEFACIENS*, BACTERIA, BIOLISTIC® GENE GUN, MICROPARTICLES, BACULOVIRUS EXPRESSION VECTORS (BEVs), BAC.

**Vertical Gene Transfer** See OUTCROSSING.

**Very Low-Density Lipoproteins (VLDL)** VLDLs and LDLPs are the specific lipoproteins that are most likely to deposit cholesterol on artery walls inside the human body,

which increases risk of coronary heart disease (CHD). See also LOW-DENSITY LIPOPRO-TEINS (LDLP), LIPOPROTEIN, APOLIPOPROTEINS, CHOLESTEROL.

**Vesicle** A small vacuole. See also VESICULAR TRANSPORT, VACUOLES.

**Vesicular Transport** (of a protein) One of three means for a protein molecule to pass between compartments within eucaryotic cells. The compartment "wall" (membrane) possesses a "sensor" (receptor) that detects the presence of correct protein (e.g., after that protein has been synthesized in the cell's ribosomes), then bulges outward along with that protein molecule. The membrane bulge containing protein then "breaks off" and carries (transports) the protein to its destination in another compartment in the cell. See also PROTEIN, EUCARYOTE, CELL, RIBOSOMES, MICRO-TUBULES, SIGNALING, VAGINOSIS, ENDOCYTOSIS, GATED TRANSPORT.

**Viral Transactivating Protein** The specific protein used by a lytic virus to "switch on" the cascade of gene regulation by which that virus "takes over" a healthy cell and subverts its molecular processes (machinery) to produce virus components. This (transactivating) protein is key to the whole lytic cycle of the virus and therefore a potential target for therapeutic intervention. See also LYTIC INFECTION, VIRUS, PROTEIN, CELL, GENE CASCADE.

**Virtual HTS** See *IN SILICO* SCREENING, HIGH-THROUGHPUT SCREENING (HTS).

**Virus** A simple, noncellular particle (entity) that can reproduce only inside living cells (of other organisms), which was first proved to exist in 1892 by Dimitry Ivanovsky. The simple structure of viruses is their most important characteristic. Most viruses consist only of a genetic material — either DNA (deoxyribonucleic acid) or RNA (ribonucleic acid) — and a protein coating. This (combination) material is categorized as a nucleoprotein. Some viruses also have membranous envelopes (coatings).

Viruses are "alive" in that they can reproduce themselves — although only by taking over a cell's "synthetic genetic machinery" — but they have none of the other characteristics of living organisms. Viruses cause a large variety of significant diseases in plants and animals, including humans. They present a philosophical problem to those who would speak of living and nonliving systems because in and of itself a virus is not "alive" as we know life, but rather represents "life potential" or "symbiotic life." See also VACCINIA, NUCLEOPROTEINS, RETROVIRUSES, TOBACCO MOSAIC VIRUS (TMV), VIRAL TRANSACTIVATING PROTEIN, GENE DELIVERY, ADENOVIRUS.

**Viscosity** A measure of a liquid's resistance to flow, as expressed in units called poise (P; grams per cm per sec). The degree of "thickness" or "syrupiness" of a liquid.

**Vitafoods** See NUTRACEUTICALS.

**Vitamers** See VITAMIN.

**Vitamin** The modern term descended from the original phrase "vital amine" (or "vitamine"), which was coined by Casimir Funk in the early 1900s. Most vitamins are actually "families" of chemically related isomers (i.e., vitamers) which cause same or similar metabolic impact (benefit) in most animals (including humans) that consume those vitamins. Some compounds are vitamins for certain species of animals, but are not for certain other species. In general, a vitamin is an organic compound required in tiny amounts (for the optimal growth, proper biological functioning, and maintenance of health of an organism).

Vitamins are commonly classified into two categories, the fat soluble and the water soluble. Vitamins A, D, E, and K are fat soluble whereas vitamin C (ascorbic acid) and members of the vitamin B complex group are water soluble. In general, the vitamins play catalytic and regulatory roles in the body's metabolism. Among the water-soluble vitamins, the B vitamins apparently function as coenzymes (nonprotein parts of enzymes). Vitamin C's coenzyme role, if any, has not been established. Part of the importance of vitamin C to the body may arise from its strong antioxidant action. The functions of the fat-soluble vitamins are less well understood. Some of them, too, may contribute to enzyme activity; and others are essential to the functioning of cellular membranes (on surface of cells).

Some vitamins act as transcription factors. Vitamin A is able to regulate the expression of certain genes in the embryos of mammals, via one of its metabolites; retinoic acid. Those embryo cells contain nuclear receptors (which bring the retinoic acid "signal" from outside into the cell's nucleus) on their cell membrane surface. The retinoic acid then (via the nuclear receptors) regulates the expression of the genes that cause embryonic cell differentiation into complex body structures, such as legs and arms, of the growing embryo. See also ENZYME, CATALYST, COENZYME, METABOLISM, METABOLITE, GENE, EXPRESS, BETA CAROTENE, EMBRYOLOGY, RETINOIDS, PROTEIN, CELL, RECEPTORS, SIGNALING, CHOLINE, SIGNALING MOLECULES, SIGNAL TRANSDUCTION, NUCLEAR RECEPTORS, LYCOPENE, LUTEIN, FATS, TRANSCRIPTION FACTORS, SPECIES, AVIDIN, VITAMIN E, BIOTIN, TOCOPHEROLS, TOCOTRIENOLS, ANTIOXIDANTS, INOSITOL.

**Vitamin E** Refers to a group of related, naturally occurring compounds consisting of tocopherol and tocotrienol "families." It is a fat-soluble vitamin with antioxidant properties (i.e., helps prevent lipids in the body from breaking down). Vitamin E is especially effective at preventing oxidation of low-density lipoproteins (so-called "bad cholesterol"), whose oxidation products (e.g., beta hydroxycholesterol) can be deposited onto the interior walls of blood vessels (e.g., arteries) in the form of plaque — which can result in the disease atherosclerosis — and/or adversely increasing blood platelet aggregation (e.g., clotting). Vitamin E occurs naturally in soybeans, cereal grains, etc., so it can be considered a phytochemical. In 2000, the Institute of Medicine of the U.S. National Academy of Sciences issued a report that called for an increase in the amount of vitamin E consumed each day, to improve citizens' health. See also VITAMIN, OXIDATIVE STRESS, ANTIOXIDANTS, PHYTOCHEMICALS, OXIDATION, LIPIDS, CHOLESTEROL, LOW-DENSITY LIPOPROTEINS (LDLP), ATHEROSCLEROSIS, PLAQUE, PLATELETS, PHYTOCHEMICALS, NATIONAL ACADEMY OF SCIENCES (NAS), TOCOPHEROLS, TOCOTRIENOLS, SOYBEAN PLANT.

**Volicitin** A chemical compound produced by Beet Armyworm caterpillars (*Spodoptera* *exigua*) after they have consumed some linoleic acid (in plants they chew on, such as corn/maize). The body cells of Beet Armyworm caterpillars conjugate (i.e., chemically join together) the linoleic acid molecules onto glutamine molecules. The conjugated molecule, consisting of one linoleic acid (molecule) joined to one glutamine (molecule), is known as volicitin. When Beet Armyworm caterpillars subsequently chew on corn/maize plants, some volicitin is inadvertently inserted by those caterpillars into the tissue of the corn (maize). That volicitin causes the corn (maize) plant to emit certain volatile compounds that attract type(s) of wasps which are natural enemies of the Beet Armyworm; leading them to attack those Beet Armyworm caterpillars (which are feeding on the maize/corn). See also LINOLEIC ACID, CORN, GLUTAMINE, CELL, OCTADECANOID/JASMONATE SIGNAL COMPLEX.

**Vomitoxin** See *FUSARIUM*, MYCOTOXINS.

# W

**Water Activity ($A_w$)** A measure of the "free" unbound water (e.g., in a processed food product) available to sustain the growth of microorganisms (spoilage) and/or to sustain undesired chemical reactions (e.g., "staling" of baked food products). Most bacteria are unable to grow in foods possessing a water activity below 0.90. Most yeasts and molds that cause spoilage cannot grow in foods possessing a water activity below 0.80. Sugars can be added to certain foods in order to increase $A_w$, as they "bind up" the (formerly) free water present. See also MICROORGANISM, HYDROPHILIC, BACTERIA, YEAST, *PENICILLIUM*.

**Water Soluble Fiber** Food fiber (e.g., oat fiber, barley fiber, soybean fiber) that dissolves in water. It apparently absorbs low-density lipoproteins (LDLP) in the intestine, before the fiber passes from the body; plus it inhibits absorption of LDLP by the body's intestinal walls due to increasing the viscosity of the intestine's contents. Those two effects thus lower the amount of "bad" cholesterol (i.e., LDLP can lead to hardening/blockage of arteries) in the body and thereby coronary heart disease (CHD). Additional to those two effects, water soluble fiber also absorbs/binds bile acid and causes it to be excreted along with that water soluble fiber. That helps to lower cholesterol levels in the body (bloodstream), because the liver synthesizes (manufactures) more bile acids (to replace those absorbed and removed by the fiber) from cholesterol. Water soluble fiber from oat bran is a polysaccharide known as beta-glucan; composed entirely of glucose (molecular units). U.S. FDA regulations also include gums, pectins, mucilages, and certain hemicelluloses in the category of water soluble fiber.

Soybean flour/meal is also a source of water soluble fiber.

In 1997, the U.S. FDA approved a (label) health claim that associates consumption of oat fiber with reduced blood cholesterol content and with reduced coronary heart disease (CHD). In 1998, the U.S. FDA approved a (label) health claim that associates soluble fiber from psyllium husks with reduced risk of coronary heart disease (CHD). See also HIGH-DENSITY LIPOPROTEINS (HDLPs), LOW-DENSITY LIPOPROTEINS (LDLP), POLYSACCHARIDES, GLUCOSE (GLc), FOOD AND DRUG ADMINISTRATION (FDA), ATHEROSCLEROSIS, CORONARY HEART DISEASE (CHD), SOYBEAN MEAL, SOYBEAN PLANT, CHOLESTEROL, PLAQUE.

**Waxy Corn** Refers to corn (maize) hybrids that produce kernels in which the starch contained within those kernels is at least 99% amylopectin, versus the average of 72–76% amylopectin in traditional corn starch. See also CORN, STARCH, AMYLOPECTIN.

**Waxy Wheat** Refers to varieties of wheat (*Triticum aestivum*) that produce a higher amylopectin content, and thus a lower amylose content in the starch within their seeds than traditional varieties of wheat. For example, bread flour made from waxy wheat would contain 0–3% amylose, vs. 24–27% amylose in bread flour made from traditional varieties of wheat. Because bread made from such waxy (i.e., lower amylose) wheat becomes firm at a much slower rate than bread made from traditional wheat varieties, bread made from waxy wheats would probably require less shortening (added to the flour) to keep that bread soft. See also WHEAT, STARCH, AMYLOSE, AMYLOPECTIN.

**Weak Interactions** The forces between atoms that are less strong than the forces involved

W

in a covalent (chemical) bond (between two atoms). Weak interactions include ionic (chemical) bonds, hydrogen bonds, and van der Waals forces. See also VAN DER WAALS FORCES.

**Weevils** A term describing a number of insects that consume grains (i.e., grown and used by man). Many of the weevils consume (and proliferate in) stored grains, and stored grain products (e.g., flour). One example of a weevil is the insect known as the pea weevil, which lays its eggs on pea pods or dried peas. When the larvae hatch, they burrow into the pod and eat the peas inside. The insect *Theocolax elegans* attacks the larvae of maize weevils (*Sitophilus granarius*, *Triboleum castaneum*), rice weevils (*Sitophilus oryzae*), and the lesser grain borer. Thus it could potentially be added to grain storage bins (silos) as part of an Integrated Pest Management (IPM) program. See also INTEGRATED PEST MANAGEMENT (IPM), BIOTIN, AVIDIN, ALPHA AMYLASE INHIBITOR-1.

**Western Blot Test** A test performed on biological samples such as blood (after centrifugation to remove red blood cells from the blood) to detect AIDS antibodies individually. Gel electrophoresis is used to separate the AIDS antigen proteins of killed (known) AIDS viruses. Next the protein bands (resulting from the gel electrophoresis) are exposed to the blood being tested and (AIDS) antibodies stick to specific individual antigens (bands) which are then identified (as being present in the tested blood) via dyes. See also ACQUIRED IMMUNE DEFICIENCY SYNDROME (AIDS), ANTIBODY, ANTIGEN, ELECTROPHORESIS, POLYACRYLAMIDE GEL ELECTROPHORESIS (PAGE), BASOPHILIC, BUFFY COAT (CELLS).

**Western Corn Rootworm** Latin name *Diabrotica virgifera virgifera* LeConte. See also CORN ROOTWORM.

**Wheat** Refers to a family of related small grains descended from the natural crossing of three Middle East grasses (*Triticum monococcum*, *Aegilops speltoids*, and *Triticum tauscii*) centuries ago. As a result, wheat's genome is triploid (i.e., it incorporates three complete sets of deoxyribonucleic acid (DNA)), and contains approximately 17 billion base pairs (bp). Wheat is historically an annual plant that can attain a height of four feet (1.2 meters), although variations (e.g., shorter) have been bred. The Latin name for traditional (bread) wheat is *Triticum aestivum*, and for durum (pasta) wheat is *Triticum durum* desf. Historically, wheat kernels have contained 15% or less protein. Most of the rest of the kernel is composed of starch (amylose and amylopectin). See also GENOME, DEOXYRIBONUCLEIC ACID (DNA), BASE PAIR (bp), HYBRIDIZATION (PLANT GENETICS), TRIPLOID, WHEAT TAKE-ALL DISEASE, WHEAT SCAB, KARNAL BUNT, WHEAT HEAD BLIGHT, GLUTEN, GLUTENIN, PROTEIN, STARCH, AMYLOSE, AMYLOPECTIN. *TELETHIA CONTROVERSIA KOON* SMUT.

**Wheat Head Blight** See *FUSARIUM*.

**Wheat Scab** See *FUSARIUM*.

**Wheat Take-All Disease** A fungal disease that attacks wheat (*Triticum aestivum*) plant roots, and causes dry rot and premature death of the plant. Certain strains of *Brassica* plants and *Pseudomonas* bacteria produce compounds that can act as natural antifungal agents against the wheat take-all fungus. See also FUNGUS, BACTERIA, GENETICALLY ENGINEERED MICROBIAL PESTICIDES (GEMP), *BRASSICA*, ALLELOPATHY.

**Whiskers™** A trademarked method for inserting DNA (genes) into plants cells, so that those plant cells will then incorporate that new DNA and express the protein(s) coded for by that DNA. Developed by ICI Seeds Inc. (Garst Seed Company) in 1993, Whiskers™ is an alternative to other methods of inserting DNA into plant cells (e.g., the Biolistic® Gene Gun, *Agrobacterium tumefaciens*, the "Shotgun" Method, etc.); it consists of needle-like crystals ("whiskers") of silicon carbide. The crystals are placed into a container along with the plant cells, then mixed at high speed, which causes the crystals to pierce the plant cell walls with microscopic "holes" (passages). Then the new DNA (gene) is added, which causes the DNA to flow into the plant cells. The plant cells then incorporate the new gene(s); and thus they have been genetically engineered. See also BIOLISTIC® GENE GUN, *AGROBACTERIUM TUMEFACIENS*, "SHOTGUN" METHOD, GENETIC ENGINEERING, GENE, BIOSEEDS, CODING

SEQUENCE, PROTEIN, CELL, DEOXYRIBONUCLEIC ACID (DNA).

**White Blood Cells** See LEUKOCYTES (white blood cells).

**White Corpuscles** See LEUKOCYTES.

**White Mold Disease** The common name that refers to a plant disease caused under certain conditions (e.g., moist, humid, etc.) by the *Sclerotinia sclerotiorum* fungus. In 1998, the U.S. Environmental Protection Agency (EPA) approved one herbicide (COBRAR, owned by Valent Corporation), whose active ingredient is the chemical lactofin, to be applied to soybean plants "at or near bloom stage" in order to trigger systemic acquired resistance (SAR, a sort of "immune response") in those soybean plants against white mold disease. Use of No-tillage Crop Production (methodology) for some crops helps to reduce the incidence of white mold disease. See also FUNGUS, SYSTEMIC ACQUIRED RESISTANCE (SAR), NO-TILLAGE CROP PRODUCTION, SOYBEAN PLANT.

**Whole-Genome Shotgun Sequencing** See SHOTGUN SEQUENCING.

**Wide Cross** Refers to the plant breeding technologies/techniques utilized to cross two plant species that would not normally cross in nature. See also TRADITIONAL BREEDING METHODS, TISSUE CULTURE, SPECIES.

**Wide Spectrum** See GRAM STAIN.

**Wild Type** The traditional/historical form of an organism as it is ordinarily encountered in nature, in contrast to domesticated strains, natural mutant, or laboratory mutant individuals (organisms). One example of a measurable difference between the two types is that wild strains of animals respond to the presence of EMF fields (e.g., weak magnetic fields such as those generated near power transmission cables), but laboratory strains of the same animals do not. See also STRAIN, MUTANT, PHENOTYPE, GENOTYPE, PSORALENE, SOLANINE.

**Wobble** The ability of the third base in a tRNA (transfer RNA) anticodon to hydrogen bond with any of two or three bases at the 3′ end of a codon. This wobble (nonspecificity) allows a single tRNA species to recognize several different codons. See also TRANSFER RNA (tRNA), CODON, BASE PAIR (bp), REDUNDANCY.

**World Trade Organization (WTO)** The international organization composed of the more than 100 nations that signed the General Agreement on Tariffs and Trade (GATT), which contained 38 Articles that lay out the rules and procedures which signatory countries must observe in their conduct of international trade and trade policy. GATT was WTO's predecessor body. The WTO permits signatory countries to ban specific imports from other countries in order to protect the health of humans, animals, or plants. Such import bans are allowed based on the (GATT/WTO) Agreement on Sanitary and Phytosanitary Measures, or the Agreement on Technical Barriers to Trade; which were approved in 1994 by GATT.

WTO was established on January 1, 1995. The WTO's Agreement on Sanitary and Phytosanitary (SPS) Measures requires that such import bans must be based on sound internationally agreed science. WTO recognizes only the following three international science organizations in order to resolve SPS disputes between member nations:

1. Codex Alimentarius Commission — for foods and food ingredients.
2. International Plant Protection Convention (IPPC) — for plants.
3. International Office of Epizootics (OIE) — for animal diseases.

See also SPS, CODEX ALIMENTARIUS COMMISSION, INTERNATIONAL PLANT PROTECTION CONVENTION (IPPC), INTERNATIONAL OFFICE OF EPIZOOTICS (OIE).

**WP 900** See Z-DNA.

**WTO** See WORLD TRADE ORGANIZATION (WTO).

# X

**X Chromosome** A sex chromosome that usually occurs paired in each female cell, and single (i.e., unpaired) in each male cell in those species in which the male typically has two unlike sex chromosomes (e.g., humans). See also CHROMOSOMES, IMPRINTING.

**X-ray Crystallography** The use of diffraction patterns produced by X-ray scattering from crystals (of a given material's molecules) to determine the three-dimensional structure of the molecules. See also CONFIGURATION, CONFORMATION, TERTIARY STRUCTURE, PROTEIN FOLDING.

**Xanthine Oxidase** An enzyme responsible for production of free radicals in the body. See also HUMAN SUPEROXIDE DISMUTASE (hSOD).

**Xanthophylls** A "family" of carotenoids (i.e., plant-produced pigments that act as protective antioxidants in photosynthetic plants, and in the bodies of animals that consume those cartenoids). Among other plants, xanthophylls are produced by yellow carrots. Consumption of xanthophylls by humans and animals assists development of healthy eye tissue. Research indicates that consumption of xanthophylls by humans helps prevent lung cancer and some other cancers. See also CAROTENOIDS, ANTIOXIDANTS, OXIDATIVE STRESS, CANCER.

**Xenobiotic Compounds** Those compounds (e.g., veterinary drugs, agrochemical herbicides, etc.) designed for use in an ecosystem comprised of more than one species. For example, herbicides intended to kill weeds but leave commercial crops undamaged or veterinary drugs that are intended to kill parasitic worms but leave the host livestock unharmed.

**Xenogeneic Organs** From the Greek word *xenos*, stranger. Xenogeneic literally means "strange genes." Refers to genetically engineered (e.g., "humanized") organs that have been grown within an animal of another species. For example, several companies are working to engineer and grow — inside swine — a number of organs to be transplanted into humans that need those organs (e.g., due to loss of their own organs via disease or accident). If successful, this would free human organ transplant recipients from having to use immunosuppressive drugs continually in order to keep their body from "rejecting" the new organ. See also IMMUNOSUPPRESSIVE, GRAFT-VERSUS-HOST DISEASE (GVHD), CYCLOSPORIN, MAJOR HISTOCOMPATIBILITY COMPLEX (MHC), GENETIC ENGINEERING.

**Xenogenesis** The (theoretical) production of offspring that are genetically different from, and genotypically unrelated to, either of the parents. See also GENOTYPE, TRANSGENIC, HEREDITY, GENETICS, MEIOSIS, GENETIC CODE.

**Xenogenetic Organs** See XENOGENEIC ORGANS.

**Xenogenic Organs** See XENOGENEIC ORGANS.

**Xenograft** See XENOTRANSPLANT.

**Xenotransplant** From the Greek word *xenos*, stranger. Xenotransplant is the implantation of an organ or limb from one species to another organism in a different species. When performed in animals, "rejection" of the transplant by the recipient's immune system is a common response. See also GRAFT-VERSUS-HOST DISEASE (GVHD), XENOGENEIC ORGANS.

X

# Y

**Y Chromosome** A sex chromosome that is characteristic of male zygotes (and cells) in species in which the male typically has two unlike sex chromosomes. See also CHROMOSOMES.

**YAC** See YEAST ARTIFICIAL CHROMOSOMES (YAC).

**Yeast** A fungus of the family *Saccharomycetaceae* that is used by man especially in the making of alcoholic liquors and as a leavening agent in bread making. Some strains of yeast cells are also commonly used in bioprocesses, because they are relatively simple to genetically engineer (via recombinant DNA) and relatively easy to propagate (via fermentation) to yield desired products (e.g., proteins). See also FUNGUS, STRAIN, PREBIOTICS, FERMENTATION, GENETIC ENGINEERING, YEAST ARTIFICIAL CHROMOSOMES (YAC), RECOMBINANT DNA (rDNA).

**Yeast Artificial Chromosomes (YAC)** Pieces of DNA (usually human DNA) that have been cloned (made) inside living yeast cells. While most bacterial vectors cannot carry DNA pieces that are larger than 50 base pairs, YACs can typically carry DNA pieces that are as large as several hundred base pairs. See also YEAST, CHROMOSOMES, HUMAN ARTIFICIAL CHROMOSOMES (HAC), BACTERIAL ARTIFICIAL CHROMOSOMES (BAC), *ARABIDOPSIS THALIANA*, DEOXYRIBONUCLEIC ACID (DNA), CLONE (A MOLECULE), VECTOR, BASE PAIR (bp), MEGA-YEAST ARTIFICIAL CHROMOSOMES (mega YAC).

**Yeast Episomal Plasmid (YEP)** A cloning vehicle used for introduction of constructions (i.e., genes and pieces of genetic material) into certain yeast strains at high copy number. YEP can replicate in both *Escherichia coli* and certain yeast strains. See also PLASMID, CASSETTE, CLONE (AN ORGANISM), GENE, GENETIC ENGINEERING, *ESCHERICHIA COLIFORM (E. COLI)*, COPY NUMBER.

# Z

**Z-DNA** A left-handed helix (molecular structure) of DNA, in contrast to A-DNA and B-DNA which are right-handed helix structures. The difference is in the direction of the double-helix twist. Z-DNA has the most base pairs per turn (in the helix), and so has the least twisted structure; it is very "skinny" and its name is taken from the zigzag path that the sugar-phosphate "backbone" follows along the helix. This is quite different from the smoothly curving path of the backbone of B-DNA. The Z-form of DNA has been found in polymers that have an alternating purine-pyrimidine sequence.

One possible biological importance of Z-DNA is that it is much more stable at lower salt concentrations, and there is a possibility that the Z-DNA form (of DNA within cells) is the cause of certain diseases (e.g., certain cancers). During 2000, Jonathan Chaires, Waldemar Priebe, and John Trent showed that WP 900 (i.e., the enantiomer of daunorubicin, a natural chemical compound which inhibits cancer) binds tightly (and selectively) to a Z-DNA polymer. See also CELL, DEOXYRIBONUCLEIC ACID (DNA), B-DNA, HELIX, DOUBLE HELIX, A-DNA, PURINE, BASE PAIR (bp), PYRIMIDINE, ENANTIOMERS, CANCER.

**Zearalenone** One of the mycotoxins (i.e., toxins produced by a fungus), it causes reproductive difficulties in swine (e.g., reduced sperm production, halting of estrus, etc.) when consumed by animals (e.g., in contaminated grain such as corn/maize). Zearalenone is produced by certain strains of *Fusarium* fungi when climate (moisture and temperature) conditions during the grain growing season, combined with entry points (e.g., holes chewed into the grain plants by insects) facilitate growth of those *Fusarium* strains in grain. See also TOXIN, MYCOTOXINS, FUNGUS, STRAIN, *FUSARIUM*, LACTONASE.

**Zeaxanthin** A carotenoid (i.e., "light harvesting" compound utilized in photosynthesis) that is naturally produced in Brussels sprouts, summer squash, maize, avocado, green beans, and dark green leafy vegetables. Zeaxanthin is a phytochemical/nutraceutical whose consumption by humans has been shown to reduce risk of the disease age-related macular degeneration, a leading cause of blindness in elderly people. See also CAROTENOIDS, PHOTOSYNTHESIS, PHYTOCHEMICALS, NUTRACEUTICALS.

**Zinc Finger Proteins** Protein molecules bearing at least one "finger shaped" molecular appurtenance which acts to either repress or activate transcription (i.e., of the gene the "finger" touches within a DNA molecule). Thus, they could potentially be utilized in functional genomics (i.e., to study the specific function of a given gene). See also FUNCTIONAL GENOMICS, PROTEIN, GENE, TRANSCRIPTION, REPRESSION (of gene transcription), PROMOTER, DEOXYRIBONUCLEIC ACID (DNA).

**ZKBS (Central Committee on Biological Safety)** The advisory body on safety in gene-splicing labs and plants for the German Government's Ministry of Health. It is the German counterpart of the American government's Recombinant DNA Advisory Committee (RAC), Australia's Genetic Manipulation Advisory Committee (GMAC), Brazil's National Biosafety Commission (CTNBio), and the Kenya Biosafety Council. The ZKBS is composed of 10 experts from the biology and ecology sectors, trade union representatives, and representatives from the industrial sector and environmental pressure groups. The ZKBS advises the Ministry of Health and the individual

German States (Länder), that regulate all recombinant DNA (i.e., gene-splicing) activities in Germany. See also GENETIC MANIPULATION ADVISORY COMMITTEE (GMAC), CTNBio, KENYA BIOSAFETY COUNCIL, GENE TECHNOLOGY OFFICE, RECOMBINANT DNA ADVISORY COMMITTEE (RAC), GENETIC ENGINEERING, RECOMBINANT DNA (rDNA), RECOMBINATION, BIOTECHNOLOGY, INDIAN DEPARTMENT OF BIOTECHNOLOGY, COMMISSION OF BIOMOLECULAR ENGINEERING.

**Zoonoses** Diseases that are communicable from animals to humans.

**Zoonotic** See ZOONOSES.

**Zygote** A fertilized egg formed as a result of the union of the male (sperm) and female (egg) sex cells. The zygote gives rise to the placenta (lining of the uterus) in addition to growing into the adult (organism) body. See also X CHROMOSOME, Y CHROMOSOME, TELOMERES, GAMETE, ORGANISM, CELL, CELL DIFFERENTIATION.

**Zyme Systems** Chemical reactions characterized by the presence of an inactive precursor of an enzyme. The enzyme is activated via another enzyme that normally removes an extra piece of peptide chain at a physiologically appropriate time and place. See also ZYMOGENS, FIBRIN, DIGESTION (WITHIN ORGANISMS).

**Zymogens** The enzymatically inactive precursors of certain proteolytic enzymes. The enzymes are inactive because they contain an extra piece of peptide chain. When this peptide is hydrolyzed (clipped away) by another proteolytic enzyme, the zymogen is converted into the normal, active enzyme. The reason for the existence of zymogens may be to protect the cell, its machinery, and/or the place of manufacture within the cell from the potentially harmful or lethal effects of an active, proteolytic enzyme. In other words, the strategy is to activate the enzyme only when, and especially where, it is needed. See also PROTEOLYTIC ENZYMES, FIBRIN, ZYME SYSTEMS, LIPOPROTEIN-ASSOCIATED COAGULATION (CLOT) INHIBITOR (LACI).

Z